PROGRESS IN BRAIN RESEARCH

VOLUME 167

STRESS HORMONES AND POST TRAUMATIC STRESS DISORDER

BASIC STUDIES AND CLINICAL PERSPECTIVES

Other volumes in PROGRESS IN BRAIN RESEARCH

Volume 129: Nervous System Plasticity and Chronic Pain, by J. Sandkühler, B. Bromm and G.F. Gebhart (Eds.) – 2000, ISBN 0-444-50509-1.
Volume 130: Advances in Neural Population Coding, by M.A.L. Nicolelis (Ed.) – 2001, ISBN 0-444-50110-X.
Volume 131: Concepts and Challenges in Retinal Biology, by H. Kolb, H. Ripps and S. Wu (Eds.) – 2001, ISBN 0-444-50677-2.
Volume 132: Glial Cell Function, by B. Castellano López and M. Nieto-Sampedro (Eds.) – 2001, ISBN 0-444-50508-3.
Volume 133: The Maternal Brain. Neurobiological and Neuroendocrine Adaptation and Disorders in Pregnancy and Post Partum, by J.A. Russell, A.J. Douglas, R.J. Windle and C.D. Ingram (Eds.) – 2001, ISBN 0-444-50548-2.
Volume 134: Vision: From Neurons to Cognition, by C. Casanova and M. Ptito (Eds.) – 2001, ISBN 0-444-50586-5.
Volume 135: Do Seizures Damage the Brain, by A. Pitkänen and T. Sutula (Eds.) – 2002, ISBN 0-444-50814-7.
Volume 136: Changing Views of Cajal's Neuron, by E.C. Azmitia, J. DeFelipe, E.G. Jones, P. Rakic and C.E. Ribak (Eds.) – 2002, ISBN 0-444-50815-5.
Volume 137: Spinal Cord Trauma: Regeneration, Neural Repair and Functional Recovery, by L. McKerracher, G. Doucet and S. Rossignol (Eds.) – 2002, ISBN 0-444-50817-1.
Volume 138: Plasticity in the Adult Brain: From Genes to Neurotherapy, by M.A. Hofman, G.J. Boer, A.J.G.D. Holtmaat, E.J.W. Van Someren, J. Verhaagen and D.F. Swaab (Eds.) – 2002, ISBN 0-444-50981-X.
Volume 139: Vasopressin and Oxytocin: From Genes to Clinical Applications, by D. Poulain, S. Oliet and D. Theodosis (Eds.) – 2002, ISBN 0-444-50982-8.
Volume 140: The Brain's Eye, by J. Hyönä, D.P. Munoz, W. Heide and R. Radach (Eds.) – 2002, ISBN 0-444-51097-4.
Volume 141: Gonadotropin-Releasing Hormone: Molecules and Receptors, by I.S. Parhar (Ed.) – 2002, ISBN 0-444-50979-8.
Volume 142: Neural Control of Space Coding, and Action Production, by C. Prablanc, D. Pélisson and Y. Rossetti (Eds.) – 2003, ISBN 0-444-509771.
Volume 143: Brain Mechanisms for the Integration of Posture and Movement, by S. Mori, D.G. Stuart and M. Wiesendanger (Eds.) – 2004, ISBN 0-444-513892.
Volume 144: The Roots of Visual Awareness, by C.A. Heywood, A.D. Milner and C. Blakemore (Eds.) – 2004, ISBN 0-444-50978-X.
Volume 145: Acetylcholine in the Cerebral Cortex, by L. Descarries, K. Krnjević and M. Steriade (Eds.) – 2004, ISBN 0-444-51125-3.
Volume 146: NGF and Related Molecules in Health and Disease, by L. Aloe and L. Calzà (Eds.) – 2004, ISBN 0-444-51472-4.
Volume 147: Development, Dynamics and Pathology of Neuronal Networks: From Molecules to Functional Circuits, by J. Van Pelt, M. Kamermans, C.N. Levelt, A. Van Ooyen, G.J.A. Ramakers and P.R. Roelfsema (Eds.) – 2005, ISBN 0-444-51663-8.
Volume 148: Creating Coordination in the Cerebellum, by C.I. De Zeeuw and F. Cicirata (Eds.) – 2005, ISBN 0-444-51754-5.
Volume 149: Cortical Function: A View from the Thalamus, by V.A. Casagrande, R.W. Guillery and S.M. Sherman (Eds.) – 2005, ISBN 0-444-51679-4.
Volume 150: The Boundaries of Consciousness: Neurobiology and Neuropathology, by Steven Laureys (Ed.) – 2005, ISBN 0-444-51851-7.
Volume 151: Neuroanatomy of the Oculomotor System, by J.A. Büttner-Ennever (Ed.) – 2006, ISBN 0-444-51696-4.
Volume 152: Autonomic Dysfunction after Spinal Cord Injury, by L.C. Weaver and C. Polosa (Eds.) – 2006, ISBN 0-444-51925-4.
Volume 153: Hypothalamic Integration of Energy Metabolism, by A. Kalsbeek, E. Fliers, M.A. Hofman, D.F. Swaab, E.J.W. Van Someren and R. M. Buijs (Eds.) – 2006, ISBN 978-0-444-52261-0.
Volume 154: Visual Perception, Part 1, Fundamentals of Vision: Low and Mid-Level Processes in Perception, by S. Martinez-Conde, S.L. Macknik, L.M. Martinez, J.M. Alonso and P.U. Tse (Eds.) – 2006, ISBN 978-0-444-52966-4.
Volume 155: Visual Perception, Part 2, Fundamentals of Awareness, Multi-Sensory Integration and High-Order Perception, by S. Martinez-Conde, S.L. Macknik, L.M. Martinez, J.M. Alonso and P.U. Tse (Eds.) – 2006, ISBN 978-0-444-51927-6.
Volume 156: Understanding Emotions, by S. Anders, G. Ende, M. Junghofer, J. Kissler and D. Wildgruber (Eds.) – 2006, ISBN 978-0-444-52182-8.
Volume 157: Reprogramming of the Brain, by A.R. Møller (Ed.) – 2006, ISBN 978-0-444-51602-2.
Volume 158: Functional Genomics and Proteomics in the Clinical Neurosciences, by S.E. Hemby and S. Bahn (Eds.) – 2006, ISBN 978-0-444-51853-8.
Volume 159: Event-Related Dynamics of Brain Oscillations, by C. Neuper and W. Klimesch (Eds.) – 2006, ISBN 978-0-444-52183-5.
Volume 160: GABA and the Basal Ganglia: From Molecules to Systems, by J.M. Tepper, E.D. Abercrombie and J.P. Bolam (Eds.) – 2007, ISBN 978-0-444-52184-2.
Volume 161: Neurotrauma: New Insights into Pathology and Treatment, by J.T. Weber and A.I.R. Maas (Eds.) – 2007, ISBN 978-0-444-53017-2.
Volume 162: Neurobiology of Hyperthermia, by H.S. Sharma (Ed.) – 2007, ISBN 978-0-444-51926-9.
Volume 163: The Dentate Gyrus: A Comprehensive Guide to Structure, Function, and Clinical Implications, by H. Scharfman (Ed.) – 2007, ISBN 978-0-444-53015-8.
Volume 164: From Action to Cognition, by C. von Hofsten and K. Rosander (Eds.) – 2007, ISBN 978-0-444-53016-5.
Volume 165: Computational Neuroscience: Theoretical Insights into Brain Function, by P. Cisek, T. Drew and J.F. Kalaska (Eds.) – 2007, ISBN 978-0-444-52823-0.
Volume 166: Tinnitus: Pathophysiology and Treatment, by B. Langguth, G. Hajak, T. Kleinjung, A. Cacace and A.R. Møller (Eds.) – 2007, ISBN 978-0-444-53167-4.

PROGRESS IN BRAIN RESEARCH

VOLUME 167

STRESS HORMONES AND POST TRAUMATIC STRESS DISORDER

BASIC STUDIES AND CLINICAL PERSPECTIVES

EDITED BY

E. RONALD DE KLOET and MELLY S. OITZL
Department of Medical Pharmacology, LACDR/LUMC, Leiden University, 2300 RA Leiden, The Netherlands

ERIC VERMETTEN
Military Mental Health Research Center/Rudolf Magnus Institute of Neuroscience, Department of Psychiatry, UMCU, 3509 AA Utrecht, The Netherlands

ELSEVIER

AMSTERDAM – BOSTON – HEIDELBERG – LONDON – NEW YORK – OXFORD
PARIS – SAN DIEGO – SAN FRANCISCO – SINGAPORE – SYDNEY – TOKYO

Elsevier
Radarweg 29, PO Box 211, 1000 AE Amsterdam, The Netherlands
Linacre House, Jordan Hill, Oxford OX2 8DP, UK

First edition 2008

Library of Congress Cataloging-in-Publication Data
A catalog record for this book is available from the Library of Congress

British Library Cataloguing in Publication Data
A catalogue record for this book is available from the British Library

ISBN: 978-0-444-53140-7 (this volume)
ISSN: 0079-6123 (Series)

For information on all Elsevier publications
visit our website at books.elsevier.com

Transferred to Digital Print 2008
Printed and bound by CPI Antony Rowe, Eastbourne

List of contributors

I. Akirav, Department of Psychology and the Brain and Behavior Research Center, University of Haifa, Haifa 31905, Israel

A. Avital, Department of Psychology and the Brain and Behavior Research Center, University of Haifa, Haifa 31905, Israel

A. Barsegyan, Center for the Neurobiology of Learning and Memory, Department of Neurobiology and Behavior, University of California, Irvine CA 92697-3800, USA

D.M. Benedek, Department of Psychiatry and Center for the Study of Traumatic Stress, Uniformed Services University, Bethesda, MD 20814, USA

I.A.E. Bicanic, University Medical Center Utrecht, Center for the Treatment of Psychotrauma in Children and Adolescents, P.O. Box 85090, 3508 AB Utrecht, The Netherlands

L.M. Bierer, Department of Psychiatry, Mount Sinai School of Medicine and Bronx Veterans Affairs, James J. Peters VAMC, 116-A, OOMH-PTSD, 130 West Kingsbridge Road, New York, NY 10468, USA

M.H. Braakman, Institute of Mental Health Care, De Gelderse Roos and Department of Psychiatry, Radboud University, Nijmegen Medical Center, Wolfheze 2, 6874 BE Wolfheze, The Netherlands

J.D. Bremner, Departments of Psychiatry and Behavioral Sciences and Radiology, Emory University, 1256 Briarcliff Road, Room 308e, Mailstop 1256/001/AT, Atlanta, GA 30306, USA

C.R. Brewin, Subdepartment of Clinical Health Psychology, University College London, Gower Street, London WC1E 6BT, United Kingdom

V. Brinks, Division of Medical Pharmacology, LACDR/LUMC, University of Leiden, P.O. Box 9502, 2300 RA Leiden, The Netherlands

S. Chourbaji, Central Institute of Mental Health (ZI), J 5, D-68159 Mannheim, Germany

C.S. de Kloet, Altrecht Institute for Mental Health Care, Oude Arnhemseweg 260, 3705 BK Zeist, The Netherlands

E.R. de Kloet, Division of Medical Pharmacology, LACDR/LUMC, University of Leiden, P.O. Box 9502, 2300 RA Leiden, The Netherlands

D.J.-F. de Quervain, Division of Psychiatry Research, University of Zurich, Lenggstr. 31, 8032 Zurich, Switzerland

R.H. de Rijk, Department of Medical Pharmacology, LACDR/LUMC, P.O. Box 9502, 2300 RA Leiden, The Netherlands

T. Ehling, Neuroimaging Center (NIC/BCN), University of Groningen, Groningen, and Mental Health Care Drenthe, Department of Outpatient Services, Altingerweg 1, 9411 PA Beilen, The Netherlands

B. Elzinga, Section of Clinical and Health Psychology, University of Leiden, Wassenaarseweg 52, 2333 AK Leiden, The Netherlands

W. Everaerd, Department of Clinical Psychology and Cognitive Sciences Center, University of Amsterdam, Roetersstraat 15, 1018 WB Amsterdam, The Netherlands

C.S. Fullerton, Department of Psychiatry and Center for the Study of Traumatic Stress, Uniformed Services University, Bethesda, MD 20814, USA

P. Gass, Central Institute of Mental Health (ZI), J 5, D-68159 Mannheim, Germany

E. Geuze, Military Mental Health Research Centre, Ministry of Defense, P.O. Box 90.000, 3509 AA Utrecht, The Netherlands

T.A. Grieger, Department of Psychiatry and Center for the Study of Traumatic Stress, Uniformed Services University, Bethesda, MD 20814, USA
M.R. Gunnar, Institute of Child Development, University of Minnesota, 51 E. River Road, Minneapolis, MN 55455, USA
C.J. Heijnen, Laboratory of Psychoneuroimmunology, Division of Perinatology and Gynaecology, University Medical Center Utrecht, P.O. Box 85090, 3508 AB Utrecht, The Netherlands
H.C. Holloway, Department of Psychiatry and Center for the Study of Traumatic Stress, Uniformed Services University, Bethesda, MD 20814, USA
C.J. Hough, Department of Psychiatry and Center for the Study of Traumatic Stress, Uniformed Services University, Bethesda, MD 20814, USA
M. Jöels, SILS-CNS, University of Amsterdam, Kruislaan 320, 1098 SM Amsterdam, The Netherlands
H. Karst, SILS-CNS, University of Amsterdam, Kruislaan 320, 1098 SM Amsterdam, The Netherlands
N. Kato, Department of Neuropsychiatry, Graduate School of Medicine, The University of Tokyo, 7-3-1 Hongo, Bunkyo-ku, Tokyo 113-8655, Japan, *Present address*: Department of Psychiatry, Showa University School of Medicine, Karasuyama Mental Hospital, Setagaya-Ku, Tokyo 157-8577, Japan
A. Kavelaars, Laboratory of Psychoneuroimmunology, Division of Perinatology and Gynaecology, University Medical Center Utrecht, P.O. Box 85090, 3508 AB Utrecht, The Netherlands
A. Kavushansky, Department of Psychology and the Brain and Behavior Research Center, University of Haifa, Haifa 31905, Israel
E.R. Klaassens, Department of Psychiatry, Leiden University Medical Center (LUMC), Albinusdreef 2, 2333 ZA Leiden, The Netherlands
F.A.M. Kortmann, Department of Psychiatry, Radboud University, Nijmegen Medical Center, P.O. Box 9101, NL-6500 HB Nijmegen, The Netherlands
A.P. Krikke, Wilhelmina Hospital Assen, Department of Radiology, Europaweg-Zuid 1, 9401 RK Assen, The Netherlands
H. Krugers, SILS-CNS, University of Amsterdam, Kruislaan 320, 1098 SM Amsterdam, The Netherlands
S. Lee, Center for the Neurobiology of Learning and Memory, Department of Neurobiology and Behavior, University of California, Irvine CA 92697-3800, USA
E.G.W.M. Lentjes, Laboratory of Endocrinology, University Medical Center Utrecht, Utrecht, The Netherlands
H. Li, Department of Psychiatry and Center for the Study of Traumatic Stress, Uniformed Services University, Bethesda, MD 20814, USA
I. Liberzon, Department of Psychiatry, University of Michigan, 1500 E. Medical Center Dr., MCHC, F6135, Ann Arbor, MI 48109, USA
M. Meijer, University Medical Center Utrecht, Department of Pediatrics, P.O. Box 85090, 3508 AB Utrecht, The Netherlands
P.M.C. Mommersteeg, Department of Medical Psychology, Tilburg University, Faculty of Social and Behavioral Sciences, P.O. Box 90153, 5000 LE Tilburg, The Netherlands
E.R.S. Nijenhuis, Mental Health Care Drenthe, Top Referent Trauma Center, Beilerstraat 197, 9401 PJ Assen, The Netherlands
M.S. Oitzl, Division of Medical Pharmacology, LACDR-LUMC, University of Leiden, P.O. Box 9502, 2300 RA Leiden, The Netherlands
M. Olff, Psychotrauma Center, MFO Psychiatrie AMC/De Meren, Tafelbergweg 25, 1105 BC Amsterdam, The Netherlands
R.K. Pitman, Massachusetts General Hospital-East, Room 2616 Building 149, 13th Street, Charleston, MA 02129, USA
K.N. Quevedo, Institute of Child Development, University of Minnesota, 51 E. River Road, Minneapolis, MN 55455, USA

G. Richter-Levin, Department of Psychology and the Brain and Behavior Research Center, University of Haifa, Haifa 31905, Israel

S.A.R.B. Rombouts, Department of Radiology, Leiden University Medical Center, and Leiden Institute for Brain and Cognition (LIBC) Department of Psychology, Leiden, The Netherlands

B. Roozendaal, Center for the Neurobiology of Learning and Memory, Department of Neurobiology and Behavior, University of California, Irvine CA 92697-3800, USA

G. Schelling, Klinikum Grosshadern, Department of Anaesthesiology, Ludwig-Maximilians University, 81377 Munich, Germany

C. Schmahl, Department of Psychosomatic Medicine and Psychotherapy, Central Institute of Mental Health, Mannheim, Germany

J.R. Seckl, Endocrinology Unit, Centre for Cardiovascular Science, Queen's Medical Research Institute, 47 Little France Crescent, Edinburgh EH16 4TJ, UK

R.H. Segman, Department of Psychiatry, Hadassah University Hospital, P.O. Box 12000, Ein Kerem, Jerusalem 90815, Israel

A.Y. Shalev, Department of Psychiatry, Hadassah University Hospital, P.O. Box 12000, Ein Kerem, Jerusalem 90815, Israel

G. Sinnema, Universtiy Medical Center Utrecht, Department of Pediatric Psychology, P.O. Box 85090, 3508 AB Utrecht, The Netherlands

C.S. Sripada, Department of Psychiatry, University of Michigan, 1500 E. Medical Center Dr., MCHC, F6135, Ann Arbor, MI 48109, USA

G.K. Stalla, Max Planck Institute of Psychiatry, Kraepelinstrasse 10, 80804 Munich, Germany

N.U. Takemura, Department of Neuropsychiatry, Graduate School of Medicine, The University of Tokyo, 7-3-1 Hongo, Bunkyo-ku, Tokyo 113-8655, Japan

M.M. Tsoory, Department of Psychology and the Brain and Behavior Research Center, Univeristy of Haifa, Haifa 31905, Israel

R.J. Ursano, Department of Psychiatry and Center for the Study of Traumatic Stress, Uniformed Services University, Bethesda, MD 20814, USA

E.M. van de Putte, Department of Pediatrics, University Medical Center Utrecht, P.O. Box 85090, 3508 AB Utrecht, The Netherlands

W. van den Brink, Department of Psychiatry, Academic Medical Center, University of Amsterdam, P.O. Box 75867, 1070 AW Amsterdam, The Netherlands

L.J.P. van Doornen, Department of Clinical and Health Psychology, Utrecht University, P.O. Box 80140, 3508 TC Utrecht, The Netherlands

A.H. van Stegeren, Department of Clinical Psychology and Cognitive Sciences Center, University of Amsterdam, Roetersstraat 15, 1018 WB Amsterdam, The Netherlands

T. van Veen, Department of Psychiatry, Leiden University Medical Center (LUMC), Albinusdreef 2, 2333 ZA Leiden, The Netherlands

I.M. van Vliet, Department of Psychiatry, Leiden University Medical Center, P.O. Box 9600, 2300 RC Leiden, The Netherlands

G. Veen, Department of Psychiatry, Leiden University Medical Center, P.O. Box 9600, 2300 RC Leiden, The Netherlands

R.J. Verkes, Department of Psychiatry, Radboud University Nijmegen Medical Center, P.O. Box 9101, NL-6500 HB Nijmegen, The Netherlands

E. Vermetten, Military Mental Health Research Center, Central Military Hospital-Q3 and Rudolf Magnus Institute of Neuroscience, Department of Psychiatry, University Medical Center, Mailbox B01206, Heidelberglaan 100, 3584 CX Utrecht, The Netherlands

M.A. Vogt, Central Institute of Mental Health (ZI), J 5, D-68159 Mannheim, Germany

R.M. Vouimba, Department of Psychology and the Brain and Behavior Research Center, University of Haifa, Haifa 31905, Israel

H.G.M. Westenberg, Rudolf Magnus Institute of Neurosciences, Department of Psychiatry, University Medical Center, Postbus 85500, 3508 GA Utrecht, The Netherlands

O. Wiegert, Swammerdam Institute for Life Sciences, SILS-CNS, Universtiy of Amsterdam, Kruislaan 320, 1098 SM Amsterdam, The Netherlands

O.T. Wolf, Department of Psychology, T5-221, Universitatsstrasse. 25, 33615 Bielefeld, Germany

R. Yehuda, Department of Psychiatry, Mount Sinai School of Medicine and Bronx Veterans Affairs, James J. Peters VAMC, 116-A, OOMH-PTSD, 130 West Kingsbridge Road, New York, NY 10468, USA

L. Zhang, Department of Psychiatry and Center for the Study of Traumatic Stress, Uniformed Services University, Bethesda, MD 20814, USA

F.G. Zitman, Department of Psychiatry, Leiden University Medical Center, P.O. Box 9600, 2300 RC Leiden, The Netherlands

Preface

It is a strange almost eerie coincidence that this Volume, dedicated to 'Stress', went to press at the time that Seymour (Gig) Levine passed away at the age of 82 early November 2007. Gig was a living legend in stress research who in the mid-1950s already published seminal papers on the long-term outcome of experiences in early life. For half a century Gig was pioneering the stress field, writing influential papers and living for new data. We are fortunate to have had Gig in this meeting. And we know he was present from his articulate comments. For instance at the end of the General Discussion Section III, where Gig concludes a debate with: 'One other thing: I just want to add that the world is complex'. Typically Gig. A monument has fallen, but he died with his boots on.

This volume of Progress in Brain Research is based on the Colloquium on 'Stress Hormones and Post Traumatic Stress Disorder (PTSD): Basic Studies and Clinical Perspectives' that was organized in Amsterdam under auspices of the Royal Netherlands Academy of Arts and Sciences. There are at least three reasons we feel this book on PTSD is urgent right now.

First, the concept that containment of fear-motivated stress reactions is inadequate in patients suffering from PTSD. This concept is particularly inspiring for translational research using imaging technology and genetics. The current results using these techniques allow synthesis and new hypotheses on the pathophysiology of PTSD.

Second, the evidence that genetic predisposition and priming early life events generate phenotypes with inadequate containment of the stress response. This has led to novel animal models that allow us to study the mechanism underlying pathogenesis of PTSD.

Third, the notion that an aberrant stress system may provide a rational therapeutic strategy to limit vulnerability and/or to promote resilience in the compromised brain. New findings have revealed the importance of dosage, timing and context in which drugs based on stress mediator pharmacology become promising in treatment of PTSD, particularly if aberrant stress regulation is restored.

As a result the participants of this Colloquium have created a remarkable book addressing fundamental questions from different angles, such as: Why do some individuals develop PTSD, but others actually gain strength from identical stressful situations? Are these manifestations of vulnerability and resilience triggered by identical mechanisms?

The chapters and commentaries written by the expert Colloquium speakers are highlighted by transcripts of intense debates and sharp opinions, which we have printed in this book to the benefit of the reader. Then, the final section in the book shows how the creative atmosphere of the accompanying Masterclass has led young scientists to present their latest exciting results and plans.

We sincerely hope that this Progress in Brain Research volume will be an inspiring resource for basic and clinical scientists interested in PTSD and stress research.

E. Ronald de Kloet
Melly S. Oitzl
Eric Vermetten

Participants Masterclass on 'Stress Hormones and PTSD', Amsterdam, August 30, 2006

From left to right:

Standing: Mario Braakman, Carien de Kloet, Elbert Geuze, Thomas Ehling, Ron de Kloet, Ellen Klaassens, Gig Levine, Gerthe Veen, Melly Oitzl, Olof Wiegert, Eric Vermetten, Gal Richter-Levin

Sitting: Iva Bicanic, Vera Brinks, Paula Mommersteeg.

Participants KNAW Colloquium 'Stress Hormones and PTSD' Amsterdam, August 28–29, 2006

Left to right on stairs: Anda van Stegeren, Gal Richter-Levin, Nobumasa Kato, Israel Liberzon, Frans Zitman, Marian Joëls, Ron dc Kloet and Roel de Rijk.

Left to right in front: Harm Krugers, Carmen Sandi, Bob Ursano, Mechiel Korte, Eric Vermetten, Bauke Buwalda, Thomas Rinne, Dominique de Quervain, Mathias Schmidt, Onno van der Hart, Karin Roelofs, Ingrid Philippens, Bernet Elzinga, Marcel van den Hout, Rolf Kleber, Onno Meijer, Gig Levine, Gustav Schelling, Melly Oitzl, Peter Gass, Rianne Stam, Megan Gunnar, Benno Roozendaal, Chris Brewin, Roger Pitman, Rachel Yehuda, Doug Bremner, Jonathan Seckl, Arieh Shalev

Acknowledgements

The Colloquium and Masterclass on ‘Stress Hormones and Post Traumatic Stress Disorder: Basic Studies and Clinical Perspectives’ was organized under the auspices of the Royal Netherlands Academy of Arts and Sciences, in cooperation with the Ministry of Defence of The Netherlands. Financial support was also obtained from H. Lundbeck A/S, Copenhagen, Denmark.

Contents

V. Short Communications

SECTION I

The Vulnerable Phenotype: Molecules and Cells

E.R. de Kloet, M.S. Oitzl & E. Vermetten (Eds.)
Progress in Brain Research, Vol. 167
ISSN 0079-6123

CHAPTER 1

Stress-induced changes in hippocampal function

Marian Joëls*, Harm Krugers and Henk Karst

SILS-CNS, University of Amsterdam, Amsterdam, The Netherlands

Abstract: Exposure of an organism to stress leads to activation of the sympatho-adrenomedullary system and the hypothalamo-pituitary-adrenal axis. Consequently, levels of noradrenaline, peptides like vasopressin and CRH, and corticosteroid hormones in the brain rise. These hormones affect brain function at those sites where receptors are enriched, like the hippocampus, lateral septum, amygdala nuclei, and prefrontal cortex. During the initial phase of the stress response, when hormone levels are high, these compounds mostly enhance excitability and promote long-term potentiation. Later on, when hormone levels have subsided but gene-mediated effects of corticosteroids start to appear, the excitability is normalized to the pre-stress level, in the CA1 hippocampal area, but possibly less so in the dentate gyrus and amygdala. A disturbed balance between these early and late phases of the stress response as well as a shift toward the relative contribution of the dentate/amygdala pathways may explain why the normal restorative capacity fails in vulnerable people experiencing a life-threatening situation, which could contribute to the development of PTSD.

Keywords: mineralocorticoid; glucocorticoid; CA1 area; dentate gyrus; electrophysiology; corticosterone

Introduction

Exposure of an organism to a stressful situation, which is perceived through the brain, leads to activation of two systems: the sympatho-adrenomedullary system and the hypothalamo-pituitary-adrenal (HPA) system (for review, see de Kloet et al., 2005; Fig. 1). Activation of the former results in enhanced circulating levels of adrenaline. Via the HPA axis, levels of peptides — like corticotrophin releasing hormone (CRH) and adrenocorticotrophin — as well as steroid hormones that are released from the adrenal cortex will rise. The two systems and their prime actors serve to optimally face the stressful situation to which the organism is exposed by changing its bodily functions. Due to the negative feedback action of corticosteroids at the level of the pituitary gland and the hypothalamus (Fig. 1), these functions will be restored again to their normal level of activation over the course of some hours, an essential aspect of an optimal stress response.

During the stress response, not only peripheral organs are changed in their function, but also the brain. Adrenaline can, via vagal afferents, activate noradrenergic neurons in the nucleus tractus solitarii and locus coeruleus (Roosevelt et al., 2006). This will lead to enhanced release of noradrenaline from synaptic terminals, and subsequent activation of specific adrenoceptors. Peptides like vasopressin and CRH are released in the brain from hypothalamic as well as extrahypothalamic sources. Moreover, corticosteroid hormones can easily enter the brain due to their lipophylic properties.

*Corresponding author. Tel.: +31-20-5257626;
Fax: +31-20-5257709; E-mail: joels@science.uva.nl

DOI: 10.1016/S0079-6123(07)67001-0

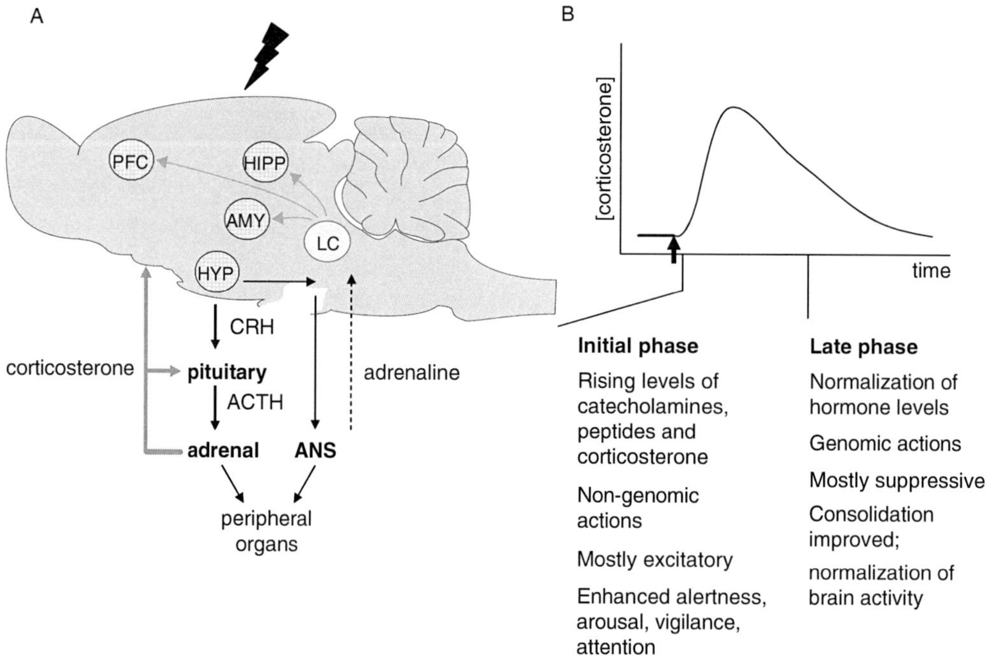

Fig. 1. (A) Exposure of a rat to stress may activate many brain regions (depending on the type of stressor), including the amygdala (AMY), hippocampus (HIPP), and prefrontal cortex (PFC). These areas project to the hypothalamus (HYP). Stimulation of cells in the hypothalamus leads to the activation of the fast-acting sympatho-adrenomedullar system (lower right) and the slower-acting hypothalamo-pituitary-adrenal axis (lower left). Both systems not only affect the function of peripheral organs but also feed back to the brain, via adrenaline and corticosterone respectively. Adrenaline can, via intermediate steps involving the nucleus tractus solitarius, give rise to central release of noradrenaline (NA) from the locus coeruleus (LC), which then exerts widespread influence on other areas such as the amygdala, prefrontal cortex, and hippocampus. Corticosterone is distributed throughout the brain but acts only at those sites where receptors are enriched. ANS, autonomic nervous system; ACTH, adrenocorticotrophin hormone; CRH, corticotrophin releasing hormone. (B) Shortly after stress exposure (arrow), levels of corticosterone start to rise; they peak after 30–45 min and are normalized ~2 h after the stress exposure started. During the initial phase (i.e., when hormone levels are high), non-genomic actions can alter brain function. In general these are predicted to result in enhanced excitability. This may contribute to the enhanced arousal, vigilance, alertness, and attention during this phase. After ~1–2 h, when hormone levels start to normalize again, genomic effects particularly by corticosteroids start to develop. These effects can last for several hours. They help to consolidate and preserve the information about the stressful event and normalize the brain activity to pre-stress levels.

Catecholamines and peptides act through G-protein coupled receptors. As a result their actions are accomplished within seconds and terminate when the ligand is no longer bound to its receptor, although secondary delayed actions may occur, e.g., via cAMP response element-binding (CREB) protein. Corticosteroid receptors, by contrast, generally act as transcriptional regulators (Pascual-Le Tallec and Lombes, 2005; Zhou and Cidlowski, 2005). Receptor dimers can bind directly to hormone response elements in the promoter region of responsive genes. In addition, monomers can bind to other transcription factors and in this way interfere with (generally transrepress) the activity of the latter.

Within the brain corticosteroid hormones bind to two types of receptors: the mineralocorticoid receptor (MR) and the glucocorticoid receptor (GR) (for review, see de Kloet, 1991). The MR has a very high affinity for the endogenous hormone corticosterone (in most rodents; cortisol in humans). Consequently, these receptors are to a large extent already occupied when the circadian release of corticosteroids is at its trough, i.e., just before the onset of the inactive period. MRs have a restricted distribution in the brain and are mainly confined to limbic regions such as the lateral septum and all subareas of the hippocampus. By comparison, expression in most other parts of the brain is rather low. GRs, on the other hand, are ubiquitously distributed, both in neurons and glia, although particularly high levels of expression are found in several parts of the brain including the paraventricular nucleus of the hypothalamus and the dentate gyrus as well as the cornus ammoni 1 (CA1) region of the hippocampus. This receptor has a lower affinity for corticosterone and cortisol. At the trough of the circadian release pattern GRs will only be activated to a limited extent. These receptors fill up toward the circadian peak or after exposure to stress. The differential degree of occupation of MRs and GRs is particularly relevant for those neurons that co-express the two receptor-types, e.g., pyramidal cells in the hippocampal CA1 area. Via both receptors corticosterone will slowly change brain function (usually taking >30–60 min) and give rise to long-lasting effects. Recently, though, it has become clear that corticosteroids can also act through membrane receptors in a non-genomic fashion (Di et al., 2003; Karst et al., 2005). These receptors could differ from the nuclear receptors, but in at least one case it was shown that the presence of the "classical" MR gene is necessary to see rapid non-genomic hormone effects (Karst et al., 2005).

The joint actions of catecholamines, specific peptides, and corticosteroids will thus affect the functioning of the brain, both in a rapid and delayed fashion. Collectively this contributes to the cognitive aspects of the stress response. It helps the organism to be alert, focus its attention, compare the current situation with experiences in the past, and determine the appropriate behavioral strategy. Moreover, it is important to store information about the event for future use. Importantly, for these central aspects of the stress response too it is essential to normalize activity eventually, so that the state of behavioral "red alert" is not maintained when no longer of use. This chapter will highlight the neurobiological mechanisms by which catecholamines, peptides, and corticosteroids in concert can achieve these various aspects of the central stress response.

The initial phase of the stress response

As pointed out above, levels of specific catecholamines, peptides, and corticosteroids in the brain rise, e.g., in the central and medial amygdala, the medial prefrontal cortex, and lateral septum (Morilak et al., 2005). These compounds have mixed effects on cellular excitability depending on the receptor mediating their actions but mostly exert an excitatory influence on the cells they reach. They not only increase the firing activity of cells, but also facilitate long-term potentiation (LTP), i.e., the long-lasting strengthening of synaptic contacts that is thought to contribute to learning and memory processes (Lynch, 2004).

Thus, noradrenaline has been reported to decrease excitatory transmission in various brain areas, including the amygdala and hippocampal formation, through activation of α-adrenoceptors (Croce et al., 2003; DeBock et al., 2003) but enhances excitatory transmission via β-receptors (Huang et al., 1996; Ferry et al., 1997; Croce et al., 2003); the inhibitory effects via α-adrenoceptors in the basolateral amygdala are suppressed by exposure to a physical stressor (Braga et al., 2004). In addition, LTP is clearly enhanced by β-agonists as well as noradrenaline itself (Hopkins and Johnston, 1988; Katsuki et al., 1997; Izumi and Zorumski, 1999; Walling et al., 2004).

Also for CRH the influence on excitability strongly depends on the receptor involved. In the central amygdala CRH acting via CRH-R1 depresses but urocortin via CRH-R2 enhances excitatory transmission (Rainnie et al., 1992; Liu et al., 2004); a similar effect was seen in the dentate gyrus (Wang et al., 2000) but the reverse was seen in the lateral septum (Liu et al., 2004). Moreover, facilitatory effects of CRH on LTP were reported in the hippocampus (Blank et al., 2002). Interestingly, it has been known for a long time already that vasopressin also enhances glutamatergic transmission and LTP in the lateral septum (Joëls and Urban, 1982; Van den Hooff et al., 1989; Van den Hooff and Urban, 1990) and hippocampus (Urban and Killian, 1990; Rong et al., 1993).

Very recently it was found that corticosterone, in addition to its well-documented slow and gene-mediated effects, exerts rapid non-genomic actions in the hippocampus (Karst et al., 2005). Within minutes corticosterone enhances the frequency but not the amplitude nor the kinetic properties of miniature excitatory postsynaptic potentials (mEPSCs) in CA1 pyramidal neurons. As paired-pulse facilitation was decreased, it was concluded that corticosterone promotes the release probability of glutamate-containing vesicles in afferent projections to the CA1 area. This is in agreement with a microdialysis study showing that corticosterone rapidly leads to more release of glutamate but not of γ-aminobutyric acid (GABA) in the CA1 region (Venero and Borrell, 1999). Unexpectedly, the MR agonist aldosterone was very potent in mimicking the effect of corticosterone, while the MR antagonist spironolactone fully blocked the effect of corticosterone on mEPSC frequency. GR agonists and antagonists were ineffective. In accordance, no enhancement in mEPSC frequency by corticosterone was observed in CA1 cells from forebrain-specific MR knockout mice, while the effect was still observed in forebrain-specific GR knockouts.

LTP is also affected in a rapid manner by corticosterone (Wiegert et al., 2006). Brief administration of corticosterone around the time of LTP induction in the CA1 area was reported to facilitate synaptic plasticity, an effect that is particularly visible during the early stages of potentiation. There is a critical time-window for this facilitation, as brief application of corticosterone 30 min before LTP induction was ineffective. Interestingly, these facilitatory actions by corticosterone could not be blocked by antagonists of the classical steroid receptors, i.e., neither by an MR antagonist, nor by a GR antagonist. It is therefore not clear at present whether the MR-dependent increase in glutamate release probability contributes to the corticosterone-induced enhancement of LTP. A role of the MR in these rapid facilitatory effects on LTP is suggested by studies in the dentate gyrus, where swim stress was found to maintain LTP, via a rapid mitogen-activated protein kinase (MAPK)-dependent process (Ahmed et al., 2006) involving the MR (Korz and Frey, 2003).

It should be realized, though, that during the initial phase of the central stress response, levels of all of these catecholamines and hormones are elevated more or less at the same time. They do not act independently but in concert and most likely affect each other's efficacy. This is exemplified by the fact that stress leads to local release of CRH in the locus coeruleus (Valentino et al., 1991), where the hormone activates noradrenergic cells (Jedema and Grace, 2004). Not only do these actors influence each other's availability, they also converge on the same effector mechanism, in this case glutamatergic synapses involved in LTP. Several examples are known now where LTP is modulated in a relatively rapid fashion both by noradrenaline and corticosterone; this has been studied particularly in the dentate gyrus. The data so far seem to indicate that corticosterone facilitates the effect of noradrenaline, provided that the two compounds are locally present around the same time (Akirav and Richter-Levin, 2002; Korz and Frey, 2005; Pu et al., 2007). This congrues with an extensive behavioral line showing that corticosterone facilitates adrenergic effects on inhibitory avoidance behavior provided adrenergic stimulation and glucocorticoids are administered within a time-window of no more than 30 min (for review, see Roozendaal, 2003).

All in all, the initial phase of the central stress response is characterized by rapid rises in levels of specific catecholamines, peptides, and steroids in brain. These compounds have complex effects on cellular activity but generally seem to excite neurons and promote LTP, a process that is thought to be critically involved in the encoding of information. They also appear to facilitate each other's efficacy. Behavioral studies suggest that noradrenaline is the essential component in this phase while other actors like corticosterone play a more modulatory role (Roozendaal, 2003). Their loci of action will be strongly determined by the sites where high densities of receptors are encountered. For some of the peptides, like CRH and vasopressin, enrichment of receptors can be found in parts of the limbic system, such as the lateral septum, amygdala nuclei, hippocampus, and prefrontal cortex. Interestingly, for at least one case of rapid effects by corticosterone, it was shown that classical MRs are essential. These receptors too are highly enriched in the very same limbic regions. It is therefore tempting to assume that catecholamines, peptides, and corticosteroids in concert facilitate the activity and encoding of information in limbic regions that play an important role in focused attention, determination of behavioral strategy, and consolidation of emotional and spatial information.

The late phase of the stress response

As soon as corticosteroid levels rise following stress exposure, a gene-mediated cascade of events is started of which the functional consequences are not immediately evident since they need at least 30–60 min to develop. Recent large scale transcript analysis in hippocampal tissue has shown that already 1 h after a pulse of corticosterone abundant transrepression can be discerned (Morsink et al., 2006). Although at present little is known about the delay introduced by the translational step, these data with microarrays indicate that changes in hippocampal protein levels can occur 1–2 h after stress exposure. This delay is highly relevant, because it effectively means that gene-mediated corticosteroid effects develop by the time the actual concentration of catecholamines, peptides, and corticosteroid hormones is almost back to the prestress level.

The genomic effects of corticosterone that develop after stress exposure are primarily mediated by GRs, as these receptors (in contrast to MRs) are still abundantly available when hormone levels rise. Over the past decade numerous GR-dependent effects have been described, although this by no means indicates that GR effects are non-selective. Most of the studies were performed in the CA1 hippocampal area (Joëls, 2001).

Cells that are recorded under "resting" conditions are not visibly affected by stress or changes in corticosterone level (Joëls and de Kloet, 1989; Kerr et al., 1989). Neither the resting membrane potential nor the input resistance was reported to be changed. However, when CA1 cells are moved away from their resting state, e.g., by current injection or through the actions of neurotransmitters, corticosteroid effects become apparent (Joëls and de Kloet, 1994). Thus, it was observed that voltage dependent calcium (Ca) currents are very sensitive to circulating corticosterone levels. In the absence of corticosterone Ca-current amplitude is large (Karst et al., 1994). Under conditions of predominant MR activation Ca-currents are small, while they increase when GRs are activated (in addition to MRs; Kerr et al., 1992; Karst et al., 1994; Joëls et al., 2003). Consequently, a U-shaped dose dependency is seen for corticosteroid effects on Ca-current amplitude (Joëls and de Kloet, 1994; Joëls, 2006). The GR-dependent increase in Ca-currents was seen >2 h after exposing a rat to a stressor in vivo (Joëls et al., 2003). The increase in current amplitude appears to be caused by a doubling of L-type channels in the membrane that can be activated upon depolarization (Chameau et al., 2007). Although the increase critically depends on binding of GR homodimers to the DNA (Karst et al., 2000), GRs most likely do not target the gene encoding for the α1-subunits that form the pore of the L-type channels. Rather, GRs seem to activate a more indirect pathway that may involve altered trafficking of Ca-channels from the intracellular compartment to the membrane (Chameau et al., 2007). Other voltage dependent ion conductances, e.g., for potassium or sodium do show some sensitivity to corticosteroids, but these effects are relatively minor (for review, see Joëls, 2001).

The altered Ca-influx has functional consequences for CA1 neurons. When combined with other challenges to the system (e.g., caused by epileptic seizures), it may form a risk factor for pathology (Karst et al., 1999). However, even under "normal" physiological conditions, the enhanced Ca-influx changes the network function. A well-documented example concerns the increase in Ca-dependent K-conductances. The latter underlie firing frequency accommodation during periods of depolarization and a lingering afterhyperpolarization of the membrane once the depolarization period has come to an end (Faber and Sah, 2003). This would mean that a GR-induced enhancement of the Ca-influx increases Ca-dependent K-conductances and hence attenuates the transmission of steady excitatory signals through the CA1 area. While this has indeed been observed (Joëls and de Kloet, 1989; Kerr et al., 1989), it is still not fully resolved whether the attenuation of firing frequency is exclusively due to an enhanced Ca-influx or also depends on direct modulation of the Ca-dependent K-conductance or even on impaired Ca-extrusion following GR activation (Sidiropoulou et al., 2007).

Similar to what was reported for the Ca-current amplitude, hyperpolarizing responses to serotonin (5-HT) also display a U-shaped dose dependency on the corticosterone concentration. Responses to 5-HT are large in the absence of corticosterone, small with predominant MR activation, and become large again when GRs are activated, e.g., some hours after stress (Joëls et al., 1991; Birnstiel and Beck, 1995; Hesen and Joëls, 1996b; Hesen et al., 1996). The increased 5-HT dependent hyperpolarization after stress drives cells away from their firing threshold and hence potentially attenuates excitability. A similar attenuation of excitability also follows from GR effects on noradrenergic activation. Noradrenaline, via β-adrenoceptors, reduces Ca-dependent K-conductances in CA1 cells, which results in enhanced firing frequency during periods of depolarization. The β-adrenoceptor dependent increased excitability of CA1 cells was found to be impaired by delayed effects through the GR (Joëls and de Kloet, 1989). An exception to the generally reduced excitability seen some hours after GR activation is formed by the small but significant increase by corticosterone of responses to muscarinic agonists (Hesen and Joëls, 1996a). As the latter slightly depolarize CA1 cells, GR-dependent augmentation of muscarinic actions would lead to an increase in excitability.

The effect of corticosteroids on ionotropic receptors is somewhat confusing at present. On the one hand there are several studies which report a GR-dependent decrease in glutamatergic responses (e.g., Vidal et al., 1986; Rey et al., 1987). In many cases, though, these effects were only seen with extremely high doses of the hormone or appeared to depend on the metabolic status of the tissue (Joëls and de Kloet, 1993). By contrast, a recent study reported that within a restricted time-window of 2–4 h after GR activation, the amplitude of α-amino-3-hydroxy-5-methyl-4-isoxazolepropionic acid (AMPA) receptor-mediated responses is enhanced, while *N*-methyl-D-aspartate (NMDA) receptor mediated events are unchanged (Karst and Joëls, 2005). This was also reported in a preliminary study on dopaminergic neurons in the ventral tegmental area (Daftary and Saal, 2006).

Many studies have looked at the effect of GRs or stress on LTP. The consensus is that in the CA1 area GR activation impairs NMDA-receptor dependent forms of LTP, with a delay of some hours (Diamond et al., 1992; Pavlides et al., 1996; Mesches et al., 1999; Alfarez et al., 2002; Kim and Diamond, 2002; Wiegert et al., 2005). Such impairment may result from the GR-dependent increase of Ca-dependent K-conductances (Sah and Bekkers, 1996), a Ca-dependent attenuation of NMDA-receptor function (Rosenmund et al., 1995), but could also be caused by GR-dependent changes in endogenous AMPA receptor function that preclude subsequent exogenously induced LTP. Interestingly, another type of LTP, which depends on voltage-dependent Ca-currents, was found to be increased by GR activation (Krugers et al., 2005). Effectively this means that the balance between NMDA-receptor and voltage dependent Ca current types of LTP is shifted in the direction of the latter by GR activation.

Taken together, it seems likely that the late phase of the stress response is mostly governed by corticosterone, the main actor of the HPA-axis. Studies on CA1 hippocampal cells indicate that 1–2 h after stress exposure, gene-mediated GR effects mostly lead to reduced excitability of hippocampal cells and impaired capacity to induce (exogenous) LTP. While noradrenergic actions were facilitated during the initial phase of the stress response, they are impaired during the late phase. With the development of gene-mediated corticosteroid actions, the initial phase of arousal is terminated and hippocampal activity restored to pre-stress levels (Fig. 2). Presumably, events that happen some time after the initial stress exposure and cause patterned high-frequency input to the very same areas that had been exposed earlier to catecholamines, peptides, and corticosterone (i.e., during the initial phase of the stress response) need to be very salient in order to evoke prolonged strengthening of synaptic contacts and overwrite the message conveyed by the initial stressor (Diamond et al., 2005; Joëls et al., 2006). In this

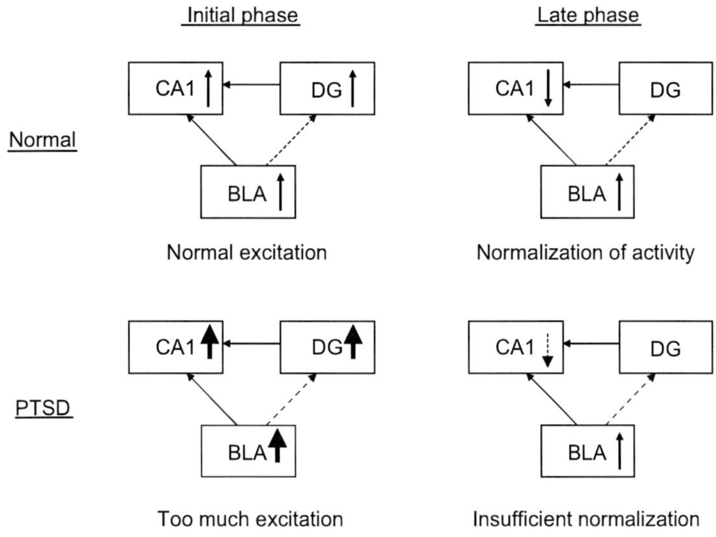

Fig. 2. Hypothetic scheme of the main changes in activity in the CA1 hippocampal area, dentate gyrus (DG) and basolateral amygdala (BLA) during the initial and late phase of the stress response. These three areas are interconnected as indicated by the arrows. The stippled arrow between the BLA and DG indicates an indirect projection. Under normal conditions, monoamines, peptides, and corticosterone primarily activate neurons in the CA1 area, DG, and BLA during the initial phase of the stress response, as indicated by the upward arrows. Later on, genomic effects cause normalization of activity in the CA1 area. In the DG relatively few genomic effects are seen after stress. In the BLA, gene-mediated corticosteroid effects may even result in excitatory actions. The net effect on the circuit will depend on the nature of the stressor. In stress situations that involve a strong emotional component normalization of brain activity during the late phase may be less prominent so that the event is better remembered. In the situation of PTSD, overall corticosteroid release is reduced while the sympathetic system is strongly activated. During the initial phase of the stress response this is hypothesized to result in strong activation of limbic areas, in particular of the BLA. During the late phase normalization especially in the CA1 area is insufficient, due to lower levels of corticosteroids.

way information that was acquired during the early phase of the stress response is preserved and can be consolidated.

Some care, however, with these generalizations is at place. Nearly all of the above insight is based on studies performed in the CA1 hippocampal area. It is by no means clear at present whether it also holds true for other parts of the brain that play a role in central aspects of the stress response, such as the amygdala, prefrontal cortex, or the dentate gyrus. With respect to the latter, it was found that GR activation evokes far less profound effects than seen in the CA1 region (Joëls, 2006); the explanation for this is at present unclear, as both areas abundantly express GRs (in addition to MRs). Opposite effects to the ones here described for the CA1 cells were in some cases seen in the basolateral amygdala, e.g., with respect to LTP (Vouimba et al., 2004; Duvarci and Pare, 2007), showing delayed excitatory effects of stress. Exposure to stressors with a very strong emotional component, which emphatically involve the amygdalar complex (Sigurdsson et al., 2007), may then lack the normalization seen in the CA1 area. Clearly, an overall understanding of the mechanistic underpinning regarding the central aspects of the stress response can only be acquired if more information is available about the cellular effects of catecholamines, peptides, and corticosteroids in each of the areas involved in processing of stressful situations.

Relevance for PTSD

The above-mentioned studies were generally performed in the standard male young-adult laboratory rat, which is group-housed and grows up in an environment with few challenges. However, there are many examples emphasizing that genetic background and life history are important factors in determining the cellular response to an acute stressor in adulthood. For instance, the suppressive effects via the CRH-R2 in the lateral septum can be switched from an inhibitory to an excitatory effect when animals have been chronically exposed to cocaine (Liu et al., 2005), so that processes that are normally adaptive fail (Orozco-Cabal et al., 2006). Likewise, the apparent lack of genomic effect by GR activation in dentate granule cells is turned into a delayed GR-dependent excitatory effect when rats have experienced a 21 days period of unpredictable stress prior to recording (Karst and Joëls, 2003). These and similar changes may form the mechanistic underpinning why individuals with a particular life history, especially during the vulnerable postnatal period when brain circuits are still in full development (Ladd et al., 2000), respond differently and/or more strongly to a given stressful event than others, as also seems to be the case in post traumatic stress disorder (PTSD).

Characteristic for PTSD is that individuals are exposed to an extremely strong emotional stressor that is (directly or indirectly) life-threatening. This will cause very profound activation of amygdalar cells. In this respect it is relevant that several studies have documented opposite (gene-mediated) effects of corticosteroids in the CA hippocampal complex versus the amygdala/dentate complex (e.g., Kavushansky et al., 2006). Moreover, it has been proposed that vulnerable phenotypes display a hypofunction of the pituitary-adrenal axis, combined with a sympathetic hyperdrive (Yehuda, 2006). In the hippocampus this is predicted (Fig. 2) to result in a shift from a situation with an appropriate balance between the early and late — i.e., normalizing — phase of the stress response, to a situation where the early excitatory phase (which is mainly governed by the sympathetic drive) is no longer restrained by the late phase (which heavily depends on HPA function). At the same time the contribution of the amygdala/dentate areas, where GR activation does not reduce but rather seems to enhance activity, in the central processing of the stressful information is largely increased. It is this combination of life history, vulnerable phenotype, and emotional stressor that appears to affect brain circuits such that acutely threatening information is engrained and not restrained by the normal adaptive actions exerted by corticosteroid hormones. At present, much of these considerations are still theoretical. They will need to be tested in much more detail by specific designs addressing the here-mentioned assumptions.

Abbreviations

AMPA	α-amino-3-hydroxy-5-methyl-4-isoxazolepropionic acid
CA1	cornus ammoni 1
CREB	cAMP response element-binding
CRH	corticotrophin releasing hormone
GABA	γ-aminobutyric acid
GR	glucocorticoid receptor
HPA	hypothalamo-pituitary-adrenal
5-HT	5-hydroxytryptamine (serotonin)
LTP	long-term potentiation
MAPK	mitogen-activated protein kinase
mEPSC	miniature excitatory postsynaptic current
MR	mineralocorticoid receptor
NMDA	*N*-methyl-D-aspartate

References

Ahmed, T., Frey, J.U. and Korz, V. (2006) Long-term effects of brief acute stress on cellular signaling and hippocampal LTP. J. Neurosci., 26: 3951–3958.

Akirav, I. and Richter-Levin, G. (2002) Mechanisms of amygdala modulation of hippocampal plasticity. J. Neurosci., 22: 9912–9921.

Alfarez, D.N., Wiegert, O., Joëls, M. and Krugers, H.J. (2002) Corticosterone and stress reduce synaptic potentiation in mouse hippocampal slices with mild stimulation. Neuroscience, 115: 1119–1126.

Birnstiel, S. and Beck, S.G. (1995) Modulation of the 5-hydroxytryptamine4 receptor-mediated response by short-term and long-term administration of corticosterone in rat CA1 hippocampal pyramidal neurons. J. Pharmacol. Exp. Ther., 273: 1132–1138.

Blank, T., Nijholt, I., Eckart, K. and Spiess, J. (2002) Priming of long-term potentiation in mouse hippocampus by corticotropin-releasing factor and acute stress: implications for hippocampus-dependent learning. J. Neurosci., 22: 3788–3794.

Braga, M.F., Aroniadou-Anderjaska, V., Manion, S.T., Hough, C.J. and Li, H. (2004) Stress impairs alpha(1A) adrenoceptor-mediated noradrenergic facilitation of GABAergic transmission in the basolateral amygdala. Neuropsychopharmacology, 29: 45–58.

Chameau, P., Qin, Y., Spijker, S., Smit, G. and Joëls, M. (2007) Glucocorticoids specifically enhance L-type calcium current amplitude and affect calcium channel subunit expression in the mouse hippocampus. J. Neurophysiol., 97: 5–14.

Croce, A., Astier, H., Recasens, M. and Vignes, M. (2003) Opposite effects of alpha 1- and beta-adrenoceptor stimulation on both glutamate- and gamma-aminobutyric acid-mediated spontaneous transmission in cultured rat hippocampal neurons. J. Neurosci. Res., 71: 516–525.

Daftary, S.S. and Saal, D.B. (2006) Glucocorticoid-dependent potentiation of synaptic strength in midbrain dopamine neurons. SfN Meeting, Atlanta, Abstract no. 58.12.

DeBock, F., Kurz, J., Azad, S.C., Parsons, C.G., Hapfelmeier, G., Zieglgansberger, W. and Rammes, G. (2003) Alpha2-adrenoreceptor activation inhibits LTP and LTD in the basolateral amygdala: involvement of Gi/o-protein-mediated modulation of Ca^{2+}-channels and inwardly rectifying K^{+}-channels in LTD. Eur. J. Neurosci., 17: 1411–1424.

Di, S., Malcher-Lopes, R., Halmos, K.C. and Tasker, J.G. (2003) Nongenomic glucocorticoid inhibition via endocannabinoid release in the hypothalamus: a fast feedback mechanism. J. Neurosci., 23: 4850–4857.

Diamond, D.M., Bennett, M.C., Fleshner, M. and Rose, G.M. (1992) Inverted-U relationship between the level of peripheral corticosterone and the magnitude of hippocampal primed burst potentiation. Hippocampus, 2: 421–430.

Diamond, D.M., Park, C.R., Campbell, A.M. and Woodson, J.C. (2005) Competitive interactions between endogenous LTD and LTP in the hippocampus underlie the storage of emotional memories and stress-induced amnesia. Hippocampus, 15: 1006–1025.

Duvarci, S. and Pare, D. (2007) Glucocorticoids enhance the excitability of principal basolateral amygdala neurons. J. Neurosci., 27: 4482–4491.

Faber, E.S. and Sah, P. (2003) Calcium-activated potassium channels: multiple contributions to neuronal function. Neuroscientist, 9: 181–194.

Ferry, B., Magistretti, P.J. and Pralong, E. (1997) Noradrenaline modulates glutamate-mediated neurotransmission in the rat basolateral amygdala in vitro. Eur. J. Neurosci., 9: 1356–1364.

Hesen, W. and Joëls, M. (1996a) Cholinergic responsiveness of rat CA1 hippocampal neurons in vitro: modulation by corticosterone and stress. Stress, 1: 65–72.

Hesen, W. and Joëls, M. (1996b) Modulation of 5HT1A responsiveness in CA1 pyramidal neurons by in vivo activation of corticosteroid receptors. J. Neuroendocrinol., 8: 433–438.

Hesen, W., Karst, H., Meijer, O., Cole, T.J., Schmid, W., de Kloet, E.R., Schütz, G. and Joëls, M. (1996) Hippocampal cell responses in mice with a targeted glucocorticoid receptor gene disruption. J. Neurosci., 16: 6766–6774.

Hopkins, W.F. and Johnston, D. (1988) Noradrenergic enhancement of long-term potentiation at mossy fiber synapses in the hippocampus. J. Neurophysiol., 59: 667–687.

Huang, C.C., Hsu, K.S. and Gean, P.W. (1996) Isoproterenol potentiates synaptic transmission primarily by enhancing presynaptic calcium influx via P- and/or Q-type calcium channels in the rat amygdala. J. Neurosci., 16: 1026–1033.

Izumi, Y. and Zorumski, C.F. (1999) Norepinephrine promotes long-term potentiation in the adult rat hippocampus in vitro. Synapse, 31: 196–202.

Jedema, H.P. and Grace, A.A. (2004) Corticotropin-releasing hormone directly activates noradrenergic neurons of the locus ceruleus recorded in vitro. J. Neurosci., 24: 9703–9713.

Joëls, M. (2001) Corticosteroid actions in the hippocampus. J. Neuroendocrinol., 13: 657–669.

Joëls, M. (2006) Corticosteroid effects in the brain: U-shape it. Trends Pharmacol. Sci., 27: 244–250.

Joëls, M., Hesen, W. and de Kloet, E.R. (1991) Mineralocorticoid hormones suppress serotonin-induced hyperpolarization of rat hippocampal CA1 neurons. J. Neurosci., 11: 2288–2294.

Joëls, M. and de Kloet, E.R. (1989) Effects of glucocorticoids and norepinephrine on the excitability in the hippocampus. Science, 245: 1502–1505.

Joëls, M. and de Kloet, E.R. (1993) Corticosteroid actions on amino acid-mediated transmission in rat CA1 hippocampal cells. J. Neurosci., 13: 4082–4090.

Joëls, M. and de Kloet, E.R. (1994) Mineralocorticoid and glucocorticoid receptors in the brain: implications for ion permeability and transmitter systems. Prog. Neurobiol., 43: 1–36.

Joëls, M. and Urban, I.J. (1982) Arginine-vasopressin enhances the responses of lateral septal neurons in the rat to excitatory amino acids and fimbria-fornix stimuli. Brain Res., 311: 201–209.

Joëls, M., Velzing, E., Nair, S., Verkuyl, J.M. and Karst, H. (2003) Acute stress increases calcium current amplitude in rat hippocampus: temporal changes in physiology and gene expression. Eur. J. Neurosci., 18: 1315–1324.

Joëls, M., Wiegert, O., Oitzl, M.S. and Krugers, H.J. (2006) Learning under stress: how does it work? Trends Cogn. Sci., 10: 152–158.

Karst, H., Berger, S., Turiault, M., Tronche, F., Schütz, G. and Joëls, M. (2005) Mineralocorticoid receptors are indispensable for nongenomic modulation of hippocampal glutamate transmission by corticosterone. Proc. Natl. Acad. Sci. U.S.A., 102: 19204–19207.

Karst, H. and Joëls, M. (2003) Effect of chronic stress on synaptic currents in rat hippocampal dentate gyrus neurons. J. Neurophysiol., 89: 625–633.

Karst, H. and Joëls, M. (2005) Corticosterone slowly enhances miniature excitatory postsynaptic current amplitude in mice CA1 hippocampal cells. J. Neurophysiol., 94: 3479–3486.

Karst, H., Karten, Y.J., Reichardt, H.M., de Kloet, E.R., Schütz, G. and Joëls, M. (2000) Corticosteroid actions in hippocampus require DNA binding of glucocorticoid receptor homodimers. Nat. Neurosci., 3: 977–978.

Karst, H., de Kloet, E.R. and Joëls, M. (1999) Episodic corticosterone treatment accelerates kindling epileptogenesis and triggers long-term changes in hippocampal CA1 cells, in the fully kindled state. Eur. J. Neurosci., 11: 889–898.

Karst, H., Wadman, W.J. and Joëls, M. (1994) Corticosteroid receptor-dependent modulation of calcium currents in rat hippocampal CA1 neurons. Brain Res., 649: 234–242.

Katsuki, H., Izumi, Y. and Zorumski, C.F. (1997) Noradrenergic regulation of synaptic plasticity in the hippocampal CA1 region. J. Neurophysiol., 77: 3013–3020.

Kavushansky, A., Vouimba, R.M., Cohen, H. and Richter-Levin, G. (2006) Activity and plasticity in the CA1, the dentate gyrus, and the amygdala following controllable vs. uncontrollable water stress. Hippocampus, 16: 35–42.

Kerr, D.S., Campbell, L.W., Hao, S.Y. and Landfield, P.W. (1989) Corticosteroid modulation of hippocampal potentials: increased effect with aging. Science, 245: 1505–1509.

Kerr, D.S., Campbell, L.W., Thibault, O. and Landfield, P.W. (1992) Hippocampal glucocorticoid receptor activation enhances voltage-dependent Ca^{2+} conductances: relevance to brain aging. Proc. Natl. Acad. Sci. U.S.A., 89: 8527–8531.

Kim, J.J. and Diamond, D.M. (2002) The stressed hippocampus, synaptic plasticity and lost memories. Nat. Rev. Neurosci., 3: 453–462.

de Kloet, E.R. (1991) Brain corticosteroid receptor balance and homeostatic control. Front. Neuroendocrinol., 12: 95–164.

de Kloet, E.R., Joëls, M. and Holsboer, F. (2005) Stress and the brain: from adaptation to disease. Nat. Rev. Neurosci., 6: 463–475.

Korz, V. and Frey, J.U. (2003) Stress-related modulation of hippocampal long-term potentiation in rats: involvement of adrenal steroid receptors. J. Neurosci., 23: 7281–7287.

Korz, V. and Frey, J.U. (2005) Bidirectional modulation of hippocampal long-term potentiation under stress and no-stress conditions in basolateral amygdala-lesioned and intact rats. J. Neurosci., 25: 7393–7400.

Krugers, H.J., Alfarez, D.N., Karst, H., Paraskouhi, K., van Gemert, N. and Joëls, M. (2005) Corticosterone shifts different forms of synaptic potentiation in opposite directions. Hippocampus, 15: 697–703.

Ladd, C.O., Huot, R.L., Thrivikraman, K.V., Nemeroff, C.B., Meaney, M.J. and Plostky, P.M. (2000) Long-term behavioral and neuroendocrine adaptations to adverse early experience. Prog. Brain Res., 122: 81–103.

Liu, J., Neugebauer, V., Grigoriadis, D.E., Rivier, J., Vale, W.W., Shinnick-Gallagher, P. and Gallagher, J.P. (2004) Corticotropin-releasing factor and urocortin I modulate excitatory glutamatergic synaptic transmission. J. Neurosci., 24: 4020–4029.

Liu, J., Yu, B., Orozco-Cabal, L., Grigoriadis, D.E., Rivier, J., Vale, W.W., Shinnick-Gallagher, P. and Gallagher, J.P. (2005) Chronic cocaine administration switches corticotropin-releasing factor2 receptor-mediated depression to facilitation of glutamatergic transmission in the lateral septum. J. Neurosci., 25: 577–583.

Lynch, M.A. (2004) Long-term potentiation and memory. Physiol. Rev., 84: 87–136.

Mesches, M.H., Fleshner, M., Heman, K.L., Rose, G.M. and Diamond, D.M. (1999) Exposing rats to a predator blocks primed burst potentiation in the hippocampus in vitro. J. Neurosci., 19: RC18.

Morilak, D.A., Barrera, G., Echevarria, D.J., Garcia, A.C., Hernandez, A., Ma, S. and Petre, C.O. (2005) Role of brain norepinephrine in the behavioral response to stress. Prog. Neuropsychopharmacol. Biol. Psychiatry, 29: 1214–1224.

Morsink, M.C., Steenbergen, P.J., Vos, J.B., Karst, H., Joëls, M., de Kloet, E.R. and Datson, N.A. (2006) Acute activation of hippocampal glucocorticoid receptors results in different waves of gene expression throughout time. J. Neuroendocrinol., 18: 239–252.

Orozco-Cabal, L., Pollandt, S., Liu, J., Shinnick-Gallagher, P. and Gallagher, J.P. (2006) Regulation of synaptic transmission by CRF receptors. Rev. Neurosci., 17: 279–307.

Pascual-Le Tallec, L. and Lombes, M. (2005) The mineralocorticoid receptor: a journey exploring its diversity and specificity of action. Mol. Endocrinol., 19: 2211–2221.

Pavlides, C., Ogawa, S., Kimura, A. and McEwen, B.S. (1996) Role of adrenal steroid mineralocorticoid and glucocorticoid receptors in long-term potentiation in the CA1 field of hippocampal slices. Brain Res., 738: 229–235.

Pu, Z., Krugers, H.J. and Joëls, M. (2007) Corticosterone time-dependently modulates beta-adrenergic effects on long-term potentiation in the hippocampal dentate gyrus. Learn. Mem., 14: 359–367.

Rainnie, D.G., Fernhout, B.J. and Shinnick-Gallagher, P. (1992) Differential actions of corticotropin releasing factor on basolateral and central amygdaloid neurones, in vitro. J. Pharmacol. Exp. Ther., 263: 846–858.

Rey, M., Carlier, E. and Soumireu-Mourat, B. (1987) Effects of corticosterone on hippocampal slice electrophysiology in normal and adrenalectomized BALB/c mice. Neuroendocrinology, 46: 424–429.

Rong, X.W., Chen, X.F. and Du, Y.C. (1993) Potentiation of synaptic transmission by neuropeptide AVP4-8 (ZNC(C)PR) in rat hippocampal slices. Neuroreport, 4: 1135–1138.

Roosevelt, R.W., Smith, D.C., Clough, R.W., Jensen, R.A. and Browning, R.A. (2006) Increased extracellular concentrations of norepinephrine in cortex and hippocampus following vagus nerve stimulation in the rat. Brain Res., 1119: 124–132.

Roozendaal, B. (2003) Systems mediating acute glucocorticoid effects on memory consolidation and retrieval. Prog. Neuropsychopharmacol. Biol. Psychiatry, 27: 1213–1223.

Rosenmund, C., Feltz, A. and Westbrook, G.L. (1995) Calcium-dependent inactivation of synaptic NMDA receptors in hippocampal neurons. J. Neurophysiol., 73: 427–430.

Sah, P. and Bekkers, J.M. (1996) Apical dendritic location of slow afterhyperpolarization current in hippocampal pyramidal neurons: implications for the integration of long-term potentiation. J. Neurosci., 16: 4537–4542.

Sidiropoulou, K., Joëls, M. and Poirazi, P. (2007) Modeling stress-induced adaptations in Ca^{++} dynamics. Neurocomputing, 70: 1640–1644.

Sigurdsson, T., Doyere, V., Cain, C.K. and Ledoux, J.E. (2007) Long-term potentiation in the amygdala: a cellular mechanism of fear learning and memory. Neuropharmacology, 52: 215–227.

Urban, I.J. and Killian, M.J. (1990) Two actions of vasopressin on neurons in the rat ventral hippocampus: a microiontophoretic study. Neuropeptides, 16: 83–90.

Valentino, R.J., Page, M.E. and Curtis, A.L. (1991) Activation of noradrenergic locus coeruleus neurons by hemodynamic stress is due to local release of corticotropin-releasing factor. Brain Res., 555: 25–34.

Van den Hooff, P. and Urban, I.J. (1990) Vasopressin facilitates excitatory transmission in slices of the rat dorso-lateral septum. Synapse, 5: 201–206.

Van den Hooff, P., Urban, I.J. and de Wied, D. (1989) Vasopressin maintains long-term potentiation in rat lateral septum slices. Brain Res., 505: 181–186.

Venero, C. and Borrell, J. (1999) Rapid glucocorticoid effects on excitatory amino acid levels in the hippocampus: a microdialysis study in freely moving rats. Eur. J. Neurosci., 11: 2465–2473.

Vidal, C., Jordan, W. and Zieglgansberger, W. (1986) Corticosterone reduces the excitability of hippocampal pyramidal cells in vitro. Brain Res., 383: 54–59.

Vouimba, R.M., Yaniv, D., Diamond, D. and Richter-Levin, G. (2004) Effects of inescapable stress on LTP in the amygdala versus the dentate gyrus of freely behaving rats. Eur. J. Neurosci., 19: 1887–1894.

Walling, S.G., Nutt, D.J., Lalies, M.D. and Harley, C.W. (2004) Orexin-A infusion in the locus ceruleus triggers norepinephrine (NE) release and NE-induced long-term potentiation in the dentate gyrus. J. Neurosci., 24: 7421–7426.

Wang, H.L., Tsai, L.Y. and Lee, E.H. (2000) Corticotropin-releasing factor produces a protein synthesis-dependent long-lasting potentiation in dentate gyrus neurons. J. Neurophysiol., 83: 343–349.

Wiegert, O., Joëls, M. and Krugers, H. (2006) Timing is essential for rapid effects of corticosterone on synaptic potentiation in the mouse hippocampus. Learn. Mem., 13: 110–113.

Wiegert, O., Pu, Z., Shors, S., Joëls, M. and Krugers, H. (2005) Glucocorticoid receptor activation selectively hampers N-methyl-D-aspartate receptor dependent hippocampal synaptic plasticity in vitro. Neuroscience, 135: 403–411.

Yehuda, R. (2006) Advances in understanding neuroendocrine alterations in PTSD and their therapeutic implications. Ann. N.Y. Acad. Sci., 1071: 137–166.

Zhou, J. and Cidlowski, J.A. (2005) The human glucocorticoid receptor: one gene, multiple proteins and diverse responses. Steroids, 70: 407–417.

Discussion: Chapter 1

DE RIJK: I was wondering if for the rapid effects, the balance of corticosteroids and noradrenalin is very important. If you look at the stress response of human individuals you see a huge variability in cortisol and noradrenalin levels. Is this a reflection of imbalance between these mediators?

JOËLS: More careful experiments in the future are necessary to figure out what a changing balance of these systems really means. In fact our preparation is ideal for that because we can give hormones in any known concentration and in any time frame. Also, I think it needs to be investigated whether it really matters that we have a strong adrenaline and little corticosterone response or the other way around, as well as the role of ACTH and vasopressin. There seems to be a certain degree of redundancy, but perhaps in reality these hormones each serve a different role.

DE KLOET: Could it be that these response patterns of the various stress hormones are in fact a clue toward understanding individual differences in vulnerability to disease?

JOËLS: That is possible. But again I say we know very little about this initial phase of the stress response and the rapid non-genomic effects. Maybe the clue to the individual differences is more in the delayed response. What we also shouldn't forget is that what I talked about concerns the CA1 area and may be different in another brain area. Also the wiring of the brain which is sensitive to early life experiences could be important, so that, e.g., there is much more activation of the amygdala compared to the CA1 area due to early life events.

GUNNAR: One of the things we see in humans is that we spend a lot of time making anticipatory responses to events. At least in the child development area children who have small elevations in cortisol and heart rate when things are new are often the more competent kids. The ones who seem to wait until bad stuff happens often look like they are not so competent. However, they are not stress level elevations — if you assume that stress level has to be above the early morning peak. Are these stress levels high enough for action in the brain?

JOËLS: With stress levels you mean cortisol levels?

GUNNAR: Right, cortisol will rise slightly, but it will not be anywhere near the morning peak.

JOËLS: I would say it would be really interesting to see if the sympathetic response influences the small effects due to cortisol.

GUNNAR: And the cortisol is just taking a ride with the sympathetic system and is not doing anything in the brain of these little infants?

JOËLS: That is one possibility. Another thing that may be relevant is that we've done some recent experiments on long-term potentiation and the influence of maternal care. There is a huge difference, not only under basal conditions but also with regard to the response to corticosterone.

LIBERZON: Is 1 h long enough for gene activation and protein interactions and all of that? You focus on a very narrow time-window.

JOËLS: If you use the exact same protocols and make slices as we do and then look at the gene expression with microarrays you see already after 1 h quite a number of genes being changed. After 3 h they are still changed and after 5 h it is more or less gone. You see waves of genes being differentially expressed. I should emphasize that with this microarray you cannot see the low abundance transcripts that probably are important for the neurophysiological responses we are looking at because, e.g., ion channel subunits or receptor subunits usually are less abundant. We found mostly suppression of gene expression after 1 h that would point to protein–protein interactions rather than transactivation. I agree with you that we have to be very careful how we interpret our data because it is a narrow time-window on which we focus.

E.R. de Kloet, M.S. Oitzl & E. Vermetten (Eds.)
Progress in Brain Research, Vol. 167
ISSN 0079-6123

CHAPTER 2

Glucocorticoids, developmental 'programming' and the risk of affective dysfunction

Jonathan R. Seckl*

Endocrinology Unit, Centre for Cardiovascular Science, Queen's Medical Research Institute, 47 Little France Crescent, Edinburgh EH16 4TJ, UK

Abstract: Early life environmental events have persisting effects on tissue structure and function, a phenomenon called 'developmental programming'. Exposure to stress and its glucocorticoid hormone mediators may underpin many such effects. Indeed, studies in animal models and observations in humans suggest that prenatal stress/glucocorticoid overexposure causes permanent cardiometabolic, neuroendocrine and behavioural effects in offspring. Such effects appear mediated via tissue-specific changes in gene expression. Underlying epigenetic changes in target gene promoters may ensure persistence of altered transcription long after the initial challenge. Posttraumatic stress disorder and other affective diseases may both act as environmental challenges if present in early life and may themselves be more likely in individuals made 'vulnerable' by early life stress.

Keywords: stress; glucocorticoid; 11β-hydroxysteroid dehydrogenase type 2; glucocorticoid receptor; epigenetic; intergenerational; hypothalamic-pituitary-adrenal; PTSD; anxiety

'Programming' and the developmental origins of brain disorders

In addition to the well-recognised effects of genes and the adult environment, it is broadly accepted that early life events have consequences that persist through the lifespan. Early life challenges impact upon 'peripheral' processes underpinning cardiovascular and metabolic disorders, a pathobiology illuminated by the pioneering epidemiological studies of David Barker and others. As the body's most complex and vulnerable organ, the brain is particularly susceptible to such early life influences. Many markers of an adverse early life environment (low birth weight, prematurity, exposure to toxins or pharmacological agents, postnatal illness or trauma) have been associated with persisting effects on brain structure and function. However, these findings reflect the 'disease diathesis' approach. This neglects the underlying biological subtleties of processes conserved through evolution. This consideration implies that permanent changes induced by transient effects in early life confer adaptive Darwinian advantages. Here, we briefly address the role of early life exposure to glucocorticoids and stress in programming of brain structure and function and explore this in relation to potential adaptations and the associated vulnerability to later disease.

Programming

The concept of early life physiological 'programming' has been advanced to explain the associations

*Corresponding author. Tel.: +44 (0) 131 242 6777/6769; Fax: +44 (0) 131 242 6779; E-mail: J.Seckl@ed.ac.uk

DOI: 10.1016/S0079-6123(07)67002-2

between perinatal environmental events, altered foetal/postnatal growth and development, and later pathophysiology (Barker et al., 1993b; Edwards et al., 1993; Seckl, 1998). Programming invokes an environmental 'factor' acting during a sensitive or vulnerable developmental period or 'window' to exert effects on the structure and function of tissues that persist throughout life. Different cells and tissues are sensitive at different developmental stages, so the effects of environmental challenges will have distinct effects depending not only on the challenge involved but also on its timing. The concept has been popularised in relationship to prenatal events, but development proceeds from preconception to puberty and programming has been demonstrated or suggested at all stages of this continuum.

Numerous human epidemiological studies have shown an association between lower, but still normal birth weight, and the subsequent development of the common cardiometabolic disorders of adult life, notably hypertension, type 2 diabetes and cardiovascular disease deaths (Barker, 1991a; Barker et al., 1993b; Fall et al., 1995; Yajnik et al., 1995; Curhan et al., 1996a, b; Leon et al., 1996; Lithell et al., 1996; Moore et al., 1996; Forsen et al., 1997; RichEdwards et al., 1997). The associations are largely independent of classical lifestyle risk factors that are merely additives to the effect of birth weight. The low birth weight–adult disease relationships are broadly continuous across the normal range (Barker, 1991a; Barker et al., 1993b; Curhan et al., 1996a, b), though premature babies also have increased cardiovascular risk in adult life (Irving et al., 2000). Additionally, postnatal catch-up growth amplifies the risk of adult cardiovascular disease (Barker, 1991a; Osmond et al., 1993; Levine et al., 1994; Leon et al., 1996; Forsen et al., 1997; Bavdekar et al., 1999; Law et al., 2002). Whilst such effects might reflect classical genetic effects, and genes in the insulin signaling pathway have been linked with both birth weight and adult disease, some work has suggested that the smaller of identical twins at birth has higher blood pressure in later life (Levine et al., 1994), though this has not been consistently reported (Baird et al., 2001). Whatever the limitations of human twin observations, the occurrence of associations between early life environmental manipulations and later physiology and disease risk in ostensibly inbred rodent models implicate at least an important role for early life environmental factors.

Birth weight and neuropsychiatric disorders

Birth weight is associated with affective disorders in adults and children, which effects apparently independent of maternal mental state and perhaps is more marked in female offspring (Thompson et al., 2001; Wiles et al., 2005; Alati et al., 2007; Costello et al., 2007). Similarly, birth weight has been linked with schizophrenia (Jones et al., 1998; Cannon et al., 2002) and indeed appears to compound genetic influences, at least of the catechol-*O*-methyltransferase gene association with anti-social behaviour and attention-deficit/hyperactivity disorder (Thapar et al., 2005). At least one report suggests a link between birth weight and subsequent vulnerability to post-traumatic stress disorder (PTSD, Famularo and Fenton, 1994). Such repeated, though by no means ubiquitous, reports of links between as relatively crude a measure of an adverse prenatal environment as low birth weight and later behavioural and psychiatric disorders merit some exploration.

Two major mechanistic hypotheses have been proposed to explain foetal programming: materno-foetal undernutrition and overexposure of the foetus to glucocorticoids/stress (Barker et al., 1993b; Edwards et al., 1993; Seckl, 1998). Here the focus is on glucocorticoids and their CNS targets. However, it should be recognised that maternal starvation and stress are difficult to separate, indeed starvation induces glucocorticoid hypersecretion and maternal stress, and glucocorticoid therapy reduces food intake. The offspring of mothers subject to starvation in the Dutch hunger winter of 1944–1945 and the Chinese famine of 1959–1961 have substantially increased rates of major psychiatric disorders (Susser and Lin, 1992; Susser et al., 1996; St Clair et al., 2005).

Glucocorticoid programming

Glucocorticoid treatment during pregnancy reduces birth weight in animal models, non-human primates (Reinisch et al., 1978; Ikegami et al., 1997; Nyirenda et al., 1998; French et al., 1999; Newnham et al., 1999; Newnham and Moss, 2001) and humans (French et al., 1999; Bloom et al., 2001). Birth weight reduction is most notable when glucocorticoids are administered in the later stages of pregnancy (Nyirenda et al., 1998). In human pregnancy, glucocorticoids are now used mainly in the management of women at risk of preterm delivery and in the antenatal treatment of foetuses at risk of congenital adrenal hyperplasia (CAH) due to 21-hydroxylase deficiency. In some reports antenatal glucocorticoids associate with reduced birth weight (French et al., 1999; Bloom et al., 2001), although normal birth weight has been reported in infants at risk of CAH whose mothers received relatively low dose of dexamethasone in utero from the first trimester (Forest et al., 1993; Mercado et al., 1995). Endogenous foetal cortisol levels are increased in intrauterine growth retardation and in preeclampsia, implicating endogenous glucocorticoids in foetal growth retardation (Goland et al., 1993, 1995).

Glucocorticoids have potent effects upon tissue development, accelerating maturation of the lung (Ward, 1994). Glucocorticoid receptors (GR) are expressed in most foetal tissues from early embryonal stages (Cole, 1995; Speirs et al., 2004). Expression of the higher affinity mineralocorticoid receptor (MR) has a more limited tissue distribution in development and is only present at later gestational stages, at least in rodents (Brown et al., 1996a). Additionally, GR are highly expressed in the placenta (Sun et al., 1997) where they mediate metabolic and anti-inflammatory effects. Clearly systems to transduce glucocorticoid actions upon the genome exist from early developmental stages, with complex cell-specific patterns of expression and presumably sensitivity to the steroid ligands (Speirs et al., 2004).

Physiology: placental 11β-hydroxysteroid dehydrogenase type 2 (11β-HSD-2)

11β-HSD-2, an NAD-dependent, 11β-dehydrogenase catalyses the rapid inactivation of physiological glucocorticoids (cortisol, corticosterone) to inert 11-keto forms (cortisone, 11-dehydrocorticosterone) (White et al., 1997). 11β-HSD-2 is highly expressed in the placenta. It is located at the interface between maternal and foetal circulations, in the syncytiotrophoblast in human (Brown et al., 1996a) and the labyrinthine zone in rodent placenta (Waddell and Atkinson, 1994; Waddell et al., 1998). Placental 11β-HSD-2 forms a potent barrier to maternal glucocorticoids (Lopez Bernal et al., 1980; Lopez Bernal and Craft, 1981; Benediktsson et al., 1997). This barrier is apparently incomplete since 10–20% of maternal cortisol crosses intact to the foetus. Indeed, in rodents the peak of the circadian rhythm of plasma corticosterone penetrates the 11β-HSD-2 barrier to an appreciable extent (Venihaki et al., 2000), presumably adding to the provision of glucocorticoids to the foetus for normal key developmental processes such as maturation of the lung. Dexamethasone and betamethasone are poor substrates for 11β-HSD-2 and therefore pass the placenta (Albiston et al., 1994; Brown et al., 1996a). In contrast, 11β-HSD-2 rapidly inactivates prednisolone to inert prednisone so it is unlikely to impact fully upon the foetus in vivo.

A relative deficiency of 11β-HSD-2, with consequent reduced placental inactivation of maternal steroids, is hypothesised to lead to overexposure of the foetus to glucocorticoids, retards foetal growth and programme responses leading to later disease (Edwards et al., 1993). In support of this idea, lower placental 11β-HSD-2 activity in rats associates with the smallest foetuses (Benediktsson et al., 1993). Similar associations have been reported in humans (Stewart et al., 1995; Shams et al., 1998; McTernan et al., 2001; Murphy et al., 2002), although not all studies have concurred (Rogerson et al., 1996, 1997). Additionally, markers of foetal exposure to glucocorticoids, such as cord blood levels of osteocalcin (a glucocorticoid-sensitive osteoblast product that does not cross the placenta), also correlate with placental 11β-HSD-2 activity (Benediktsson et al., 1995).

Humans homozygous (or compound heterozygous) for deleterious mutations of the 11β-HSD-2 gene have very low birth weight (Dave-Sharma et al., 1998), averaging 1.2 kg less than their heterozygous sibs. Though an initial report suggested that 11β-HSD-2 null mice have normal foetal weight in late gestation (Kotelevtsev et al., 1999), this appears to have reflected the 'genetic noise' of the crossed (129 × MF1) strain background of the original 11β-HSD-2 null mouse. When congenic mice on the C57Bl/6 strain background 11β-HSD-2, nullizygosity lowers birth weight (Holmes et al., 2006). Because maternal glucocorticoid levels are much higher than those of the foetus, subtle changes in placental 11β-HSD-2 activity may have profound effects on foetal glucocorticoid exposure (Lopez Bernal et al., 1980; Lopez Bernal and Craft, 1981).

A common mechanism may underlie foetal programming through maternal undernutrition and glucocorticoid exposure. Dietary protein restriction during rat pregnancy selectively attenuates 11β-HSD-2, but apparently not other placental enzymes (Langley-Evans et al., 1996b; Bertram et al., 2001; Lesage et al., 2001). Indeed in the maternal protein restriction model, offspring hypertension can be prevented by treating the pregnant dam with glucocorticoid synthesis inhibitors, and can be recreated by concurrent administration of corticosterone, at least in female offspring (Langley-Evans, 1997).

All these models of prenatal glucocorticoid exposure have persisting peripheral effects through the lifespan. Thus prenatal dexamethasone, stress or 11β-HSD-2 inhibition programmes higher adult blood pressure, glucose and insulin levels in rats, sheep and other models species (Benediktsson et al., 1993; Levitt et al., 1996; Lindsay et al., 1996a, b; Dodic et al., 1998, 1999, 2002a, b; Gatford et al., 2000; Sugden et al., 2001; Jensen et al., 2002). Importantly, recent data show that the glucocorticoid effects can be mediated directly on the foetus and its placenta rather than indirectly via alterations in maternal food intake or other aspects of her biology. Thus 'heterozygote' crosses of 11β-HSD-2$^{+/-}$ mice allow a single mother to bear wild-type, heterozygous and null offspring in the same pregnancy. Importantly, birth weight follows the foeto-placental genotype, as does offspring behaviour, with 11β-HSD-2 null offspring having the lowest birth weight and exhibiting the most 'anxious' adult phenotype compared with wild-type *littermates* (Holmes et al., 2006).

Stress and glucocorticoid programming of the brain

Maternal and/or foetal stressors alter developmental trajectories of specific brain structures with persistent effects (reviewed in Weinstock, 2001; Welberg and Seckl, 2001). Glucocorticoids are important for normal maturation in most regions of the developing CNS (Meaney et al., 1996; Korte, 2001), initiating terminal maturation, remodelling axons and dendrites, and the cell survival (Meyer, 1983). Prenatal glucocorticoid administration retards brain weight at birth in sheep (Huang et al., 1999), delaying maturation of neurons, myelination, glia and vasculature (Huang et al., 2001a, b). Exposure to glucocorticoids in utero has widespread acute effects upon neuronal structure and synapse formation (Antonow-Schlorke et al., 2003), and may permanently alter brain structure (Matthews, 2000). In rhesus monkeys, treatment with antenatal dexamethasone caused a dose-dependent neuronal degeneration of hippocampal neurones and reduced hippocampal volume which persisted at 20 months of age (Uno et al., 1990). Foetuses receiving multiple lower dose injections showed more severe damage than those receiving a single large injection. Human and animal studies have demonstrated that altered hippocampal structure may be associated with a number of consequences for memory and behaviour (Bremner et al., 1995; Sheline et al., 1996; Stein et al., 1997).

Given such widespread effects of glucocorticoids, it is unsurprising that GR and MR are highly expressed in the developing brain with complex ontogenies to allow selectivity of effects (Fuxe et al., 1985; Kitraki et al., 1997). Whether the receptors are occupied by endogenous glucocorticoids until late gestation is less certain as there is also plentiful 11β-HSD-2 in the CNS at midgestation (Brown et al., 1996a; Robson et al., 1998), which presumably 'protects' vulnerable developing cells from premature glucocorticoid action. 11β-HSD-2 expression is dramatically switched-off at the end of midgestation in the rat and mouse brain, coinciding with the terminal stage of neurogenesis (Brown et al., 1996b; Diaz et al., 1998). Similarly, in human foetal brain 11β-HSD-2 appears to be silenced between gestational weeks 19 and 26 (Stewart et al., 1994; Brown et al., 1996a). Postnatally, 11β-HSD-2 knockout mice are sensitive to cerebellar effects of exogenous corticosterone, whilst wild-type animals are not (Holmes et al., 2005). Thus, there appears to be an exquisitely timed system of protection and then exposure of developing brain regions to circulating glucocorticoids.

Programming the hypothalamic-pituitary-adrenal (HPA) axis

The hypothalamic-pituitary-adrenal (HPA) axis and its key limbic regulator, the hippocampus (Jacobson and Sapolsky, 1991), are particularly sensitive to glucocorticoids and their perinatal programming actions (Bohn, 1980; Gould et al., 1991a, b; Welberg and Seckl, 2001). Prenatal glucocorticoid exposure permanently increases basal plasma corticosterone levels in adult rats (Levitt et al., 1996; Welberg et al., 2001). This is apparently because the density of both types of corticosteroid receptors, GR and MR, is permanently reduced in the hippocampus, which are changes anticipated to attenuate HPA axis feedback sensitivity. Maternal undernutrition in rats (Langley-Evans et al., 1996a) and sheep (Hawkins et al., 2000) also affects adult HPA axis function, suggesting that HPA programming may be a common outcome of prenatal environmental challenge, perhaps acting in part via alterations in placental 11β-HSD-2 activity, which is selectively down-regulated by maternal dietary constraint (Langley-Evans et al., 1996b; Bertram et al., 2001). Consequent plasma glucocorticoid excess exacerbates hypertension and hyperglycaemia in such prenatal environmental programming models (Langley-Evans, 1997). Moreover, tissue glucocorticoid action is further increased by the documented elevations in hepatic and visceral adipose tissue glucocorticoid sensitivity (Nyirenda et al., 1998; Cleasby et al., 2003).

HPA axis programming also illustrates an important variable; it often differs between male and female offspring of the same litter. Sex-specific programming of the HPA axis has been reported for prenatal stress in rats (Weinstock et al., 1992; McCormick et al., 1995). In male guinea pigs, short-term prenatal exposure to dexamethasone significantly elevates subsequent basal plasma cortisol levels, whereas similarly exposed females have reduced HPA responses to stress. In contrast, males exposed to longer courses of prenatal glucocorticoids exhibit reduced plasma cortisol levels in adulthood, whilst females similarly exposed have higher plasma cortisol levels as adults in the follicular and early luteal phases of their oestrus cycles. In primates, offspring of mothers treated with dexamethasone during late pregnancy have elevated basal and stress-stimulated cortisol levels (Uno et al., 1994; de Vries et al., 2007).

Programming behaviour

Overexposure to glucocorticoids in utero leads to alterations in adult behaviour. Late gestational dexamethasone in rats apparently impairs coping in aversive situations later in life (Welberg et al., 2001). Prenatal glucocorticoid exposure also affects the developing dopaminergic system (Diaz et al., 1995, 1997) with implications for understanding of the developmental contributions to schizo-affective, attention-deficit hyperactivity and extrapyramidal disorders. Stressful events in the second trimester of human pregnancy associate with an increased incidence of offspring schizophrenia (Koenig et al., 2002). Prenatal exposure to dexamethasone may exert more widespread effects since it also increases the susceptibility of the cochlca to acoustic noise trauma in adulthood (Canlon et al., 2003).

Behavioural changes in adults exposed prenatally to glucocorticoids may be associated with altered functioning of the amygdala, a structural key to the expression of fear and anxiety. Intra-amygdala administration of corticotrophin-releasing hormone (CRH) is anxiogenic (Dunn and Berridge, 1990). Prenatal glucocorticoid exposure increases adult CRH levels specifically in the central nucleus of the amygdala, a key locus for its effects on fear and anxiety (Welberg et al., 2000, 2001). Prenatal stress similarly programmes increased anxiety-related behaviours with elevated CRH in the amygdala (Cratty et al., 1995).

Moreover, corticosteroids facilitate CRH mRNA expression in this nucleus (Hsu et al., 1998) and increase GR and/or MR in the amygdala (Welberg et al., 2000, 2001). The amygdala stimulates the HPA axis via a CRH signal (Feldman and Weidenfeld, 1998). So an elevated corticosteroid signal in the amygdala due to hypercorticosteronaemia in the adult offspring of dexamethasone-treated dams may produce the increased CRH levels in adulthood. A direct relationship between brain corticosteroid receptor levels and anxiety-like behaviour is supported by the phenotype of transgenic mice with selective loss of GR gene expression in the brain, which shows markedly reduced anxiety (Tronche et al., 1999).

CNS programming mechanisms

Indications of the molecular mechanisms by which early life environmental factors programme lifespan physiology come from the studies of the processes underpinning postnatal environmental programming of the HPA axis (Levine, 1957, 1962; Meaney et al., 1988, 1996). In these models, short daily handling of rat pups (Meaney et al., 1988) or merely more attentive maternal care to the pups (Liu et al., 1997) during the first 2 weeks of life permanently increases GR density in the hippocampus and prefrontal cortex, but not in other brain regions. This increase in receptor density potentiates the HPA axis sensitivity to glucocorticoid negative feedback and results in lower plasma glucocorticoid levels throughout life, a state compatible with a tighter HPA recovery from environmental stress (Meaney et al., 1989, 1992); note that this may be perceived as advantageous or disadvantageous depending on the actual postnatal challenges experienced. Neonatal glucocorticoid exposure may have similar effects (Catalani et al., 1993).

Such postnatal events act via ascending serotonergic pathways from the midbrain raphe nuclei to the hippocampus (Smythe et al., 1994). Activation of serotonin (5HT) induces GR gene expression in foetal hippocampal neurones in vitro (Mitchell et al., 1990) and in hippocampal neurones in vivo (Yau et al., 1997a). The induction of 5HT requires thyroid hormones that are elevated by appropriate maternal cues. At the postsynaptic hippocampal neuron, early postnatal events mediate their effects likely via the $5HT_7$ receptor subtype, which is regulated by glucocorticoids (Yau et al., 1997b) and positively coupled to cAMP generation (Meaney et al., 2000). The next step appears to involve stimulation of cAMP-associated and other transcription factors, most notably nerve growth factor-induced 1A (NGFI-A) and activator protein 2 (AP-2) (Meaney et al., 2000). NGFI-A and AP-2 may bind to the GR gene promoter (Encio and Detera-Wadleigh, 1991). This pathway might also be involved in some *prenatal* programming paradigms affecting the HPA axis since last trimester dexamethasone exposure increases 5HT transporter expression in the rat brain (Fumagalli et al., 1996; Slotkin et al., 1996), an effect predicted to reduce 5HT availability in the hippocampus and elsewhere. This may well induce a fall of GR and MR, the converse of postnatal higher level maternal care.

The GR promoter is complex, with multiple tissue-specific alternate untranslated first exons in rats (McCormick et al., 2000) and mice (Cole et al., 1995a, b, c), most within a transcriptionally active CpG island. All the GR mRNA species give rise to the same receptor protein encoded by exons 2–9. Tissue-specific exon 1 usage is regulated by perinatal environment manipulations. Indeed, neonatal handling permanently programmes increased expression of only one of the six alternate first exons (exon 1_7) utilised in the hippocampus (McCormick et al., 2000). Exon 1_7 contains sites which may bind AP-2 (Meaney et al., 2000) and have been documented to bind NGFI-A induced by perinatal events (Weaver et al., 2004).

The next question is how discrete perinatal events can permanently alter gene expression. Evidence has recently emerged for selective methylation/demethylation of specific CpG dinucleotides of the GR gene, notably around the putative NGFI-A site of exon 1_7. These sites are subject to differential and permanent demethylation, just after the time of birth, in association with the level maternal care. Thus transcription from this brain-enriched promoter is reduced with lower density maternal care (Weaver et al., 2004). Such changes affect NGFI-A binding to the promoter. Indeed,

GR under some circumstance can mediate differential demethylation of target gene promoters, which persists after steroid withdrawal, at least in liver-derived cells (Thomassin et al., 2001). This important novel mechanism of permanent determination of the set point of gene control by early life environmental events persisting throughout the lifespan remains to be assessed in other systems.

The GR gene: a common programming target?

Transgenic mice with a reduction of 30–50% in tissue levels of GR have striking neuroendocrine, metabolic and immunological abnormalities (Pepin et al., 1992), and whilst complete knockout is lethal (Cole et al., 1995a), tissue-specific knockout also has profound effects upon the manipulated cells and organs (Tronche et al., 1999; Bhattacharyya et al., 2007). The level of expression of GR is thus critical for cell function. GR gene expression is regulated in a tissue-specific manner. The GR promoter is complex, with multiple tissue-specific alternate untranslated first exons in rats (McCormick et al., 2000), mice (Cole et al., 1995c) and humans (Turner and Muller, 2005), most within a transcriptionally active 'CpG island'. All these mRNA species give rise to the same receptor protein, as only exons 2–9 encode the protein; the alternate untranslated first exons are spliced onto the common translated sequence beginning at exon 2. In the rat, two of the alternate exons are present in all tissues that have been studied, however others are tissue-specific (McCormick et al., 2000). This permits considerable complexity of tissue-specific variation in the control of GR expression without allowing any tissue to become GR deplete.

Neonatal handling, which increases aspects of maternal behaviour to her pups, notably licking and grooming, permanently programmes increased expression of only one of the six alternate first exons (exon 1_7) utilised in the hippocampus (McCormick et al., 2000). Exon 1_7 contains sites appropriate to bind the very third messenger/intermediate early gene transcription factors (AP-2, NGFI-A) themselves induced by the neonatal manipulation (Meaney et al., 2000). Indeed, NGFI-A induces transcriptional activity of the GR exon 1_7 promoter. The consensus NGFI-A site around exon 1_7 is subject to differential and permanent methylation/demethylation in association with variations in maternal care (Weaver et al., 2004), with methylation, as expected, inhibiting NGFI-A binding (Weaver et al., 2007). NGFI-A binding to its exon 1_7 promoter sequence is required for epigenetic reprogramming. Indeed, knockdown of NGFI-A in primary hippocampal cell culture prevents serotonin-induced demethylation of exon 1_7 and decreases its expression.

Such epigenetic effects are unlikely to be confined to the hippocampus or GR. Thus, GR in liver-derived cells mediates differential demethylation of target gene promoters, effects which persist after steroid withdrawal (Thomassin et al., 2001). During development, such target promoter demethylation occurs before birth and may fine-tune the promoter to 'remember' regulatory events occurring during development. Indeed, maternal malnutrition reduces methylation of the GR gene in postnatal liver (Lillycrop et al., 2005), perhaps underpinning the increased GR expression seen in adults following prenatal glucocorticoids (Nyirenda et al., 1998, 2006).

An analogous effect of maternal care has been reported for the estrogen receptor alpha (ERα) gene. One of its alternate first exons/promoters, ERα_{1b}, shows methylation differences. These associate with variations in the binding and effects of the transcription factor Stat5b upon expression of the receptor in the medial preoptic area and thus maternal behaviours (Champagne et al., 2006). Similarly, methylation of the PPARα gene in liver is reduced following prenatal calorie restriction (Lillycrop et al., 2005), though functional links are as yet unproven. In humans the biology may be conserved since low birth weight plus lower levels of maternal care associate with reduced adult hippocampal volume, a marker of vulnerability to affective and other neuropsychiatric disorders, albeit only in females (Buss et al., 2007).

Glucocorticoid programming in humans

Glucocorticoid treatment during pregnancy reduces birth weight (Seckl and Meaney, 2004), but there is a lack of long-term follow up studies addressing later effects of prenatal glucocorticoid exposure. Antenatal glucocorticoid administration has been linked with higher blood pressure in adolescence (Doyle et al., 2000). A number of studies aimed at establishing the long-term neurological and developmental effects of antenatal glucocorticoid exposure have been complicated by the fact that most of the children studied were born before term and were therefore already at risk of delayed neurological development. In a group of 6-year-old children, antenatal glucocorticoid exposure was associated with subtle effects on neurological function, including reduced visual closure and visual memory (MacArthur et al., 1982). Children exposed to dexamethasone in early pregnancy because they were at risk of congenital adrenal hyperplasia, and who were born at term, showed increased emotionality, unsociability, avoidance and behavioural problems (Trautman et al., 1995). These effects were seen in genetically unaffected glucocorticoid-exposed offspring. Furthermore, a recent study has shown that multiple doses of antenatal glucocorticoids given to women at risk of preterm delivery were associated with reduced head circumference in the offspring (French et al., 1999). There are also effects on behaviour; three or more courses of glucocorticoids associate with an increased risk of externalising behaviour problems, distractibility and inattention (French et al., 1999). Finally, a recent controlled trial of postnatal dexamethasone in premature babies showed associations with lower subsequent IQ and decrements of other cortical functions (Yeh et al., 2004).

As in other mammals, the humans HPA axis appears to be programmed by the early life environment. Higher plasma and urinary glucocorticoid levels are found in children and adults who were of lower birth weight (Clark et al., 1996; Phillips et al., 1998). This appears to occur in disparate populations (Phillips et al., 2000) and may precede overt adult disease (Levitt et al., 2000), at least in a socially disadvantaged South African population. Additionally, adult HPA responses to ACTH stimulation are exaggerated in those of low birth weight (Levitt et al., 2000; Reynolds et al., 2001), reflecting the stress axis biology elucidated in animal models. The HPA axis activation is associated with higher blood pressure, insulin resistance, glucose intolerance and hyperlipidaemia (Reynolds et al., 2001). Finally, the human GR gene promoter has multiple alternate untranslated first exons (Turner and Muller, 2005) analogous to the rat and mouse promoter structure. Whether these are subject to early life regulation and the molecular mechanisms by which this occurs remain to be determined.

Programming and PTSD

Reduced urinary, plasma and salivary cortisol levels in PTSD have been reported in several studies of trauma survivors (Yehuda, 2002), perhaps reflecting the increased tissue sensitivity to glucocorticoids in PTSD (Yehuda et al., 2004) and hence enhanced feedback. Recently, lower levels of salivary cortisol were reported in mothers who developed PTSD after being present at or near to the World Trade Centre atrocity on 11th Sept 2001 in New York than in mothers who did not develop PTSD. Crucially, the 1-year-old offspring of mothers with PTSD also had lower salivary cortisol levels (Yehuda et al., 2005). The changes were most apparent in babies born to mothers who were in the last 3 months of their pregnancies when the trauma occurred. Intriguingly there was no effect on birth weight, though the incidence of intrauterine growth retardation was increased (Berkowitz et al., 2003) and head circumference reduced (Engel et al., 2005). Monkeys exposed to low doses of dexamethasone in the second half of pregnancy also show persisting neuroendocrine changes, in this case hypercortisolaemia (as seen in the analogous rodent models), but no alteration in birth weight (de Vries et al., 2007). The direction of change in glucocorticoids may differ between perinatal challenges, their gestational timing and the species involved (some data imply that hypocortisolaemia with PTSD associates with increased tissue sensitivity to glucocorticoids so the target actions may be increased), but the principle of particular vulnerability of the HPA axis to developmental 'programming' seems consistent. The findings implicate in utero effects as major contributors to a possible biological risk factor for PTSD.

Intriguingly, offspring of Holocaust survivors with PTSD have significantly lower 24-h mean urinary cortisol excretion than offspring of Holocaust survivors without PTSD (Yehuda, 2002) suggesting similar intergenerational transmission as seen in glucocorticoid-exposed rats (Drake et al., 2005). Very recent data in Holocaust survivors suggest a link between the age of trauma exposure and the persisting phenotype, with younger trauma linking with changes in cortisol metabolism but less PTSD and metabolic syndrome-related pathologies in later life (Yehuda and Seckl, unpublished). Such effects may transmit into subsequent generations, since healthy adult children of Holocaust survivors with PTSD (and therefore lower plasma cortisol levels) themselves have lower cortisol levels though no PTSD. This appears to be confined to the children of Holocaust-exposed mothers with PTSD (Yehuda et al., 2007). The implication is that the marker of altered HPA axis functioning, itself perhaps a vulnerability factor, is transmitted into a subsequent generation, suggesting epigenetic 'inheritance' across at least one generation, as seen in rodent models.

Overview

Prenatal exposure to glucocorticoids may 'programme' a range of tissue-specific pathophysiologies. The foetus may be exposed to exogenous glucocorticoids, to active steroids of maternal origin or to its own adrenal products. The outcomes in a host of species and models are remarkably consistent with cardiometabolic and CNS effects predominating. Work on candidate mechanisms, GR gene programming, has illuminated a potential fundamental mechanism to underlie this rapidly emerging biology. Such fine-tuning of foetal physiology by the environment is conserved and therefore apparently important. Studies are unravelling the underlying processes, a prerequisite to rational treatments for the consequences of an adverse perinatal environment.

Abbreviations

AP-2	activator protein 2
CAH	congenital adrenal hyperplasia
CRH	corticotrophin-releasing hormone
GR	glucocorticoid receptor
HPA	hypothalamic-pituitary-adrenal
11β-HSD-2	11β-hydroxysteroid dehydrogenase type 2
5HT	serotonin
MR	mineralocorticoid receptor
NGFI-A	nerve growth factor-induced 1A
PTSD	posttraumatic stress disorder

Acknowledgements

Work in the author's laboratory is funded by grants from the Wellcome Trust, the Scottish Hospitals Endowments Research Trust, the Human Frontier Science Program, a European Union FP5 Award and the British Heart Foundation.

References

Alati, R., Lawlor, D.A., Al Mamun, A., Williams, G.M., Najman, J.M., O'Callaghan, M. and Bor, W. (2007) Is there a fetal origin of depression? Evidence from the Mater University Study of Pregnancy and its outcomes. Am. J. Epidemiol., 165(5): 575–582.

Albiston, A.L., Obeyesekere, V.R., Smith, R.E. and Krozowski, Z.S. (1994) Cloning and tissue distribution of the human 11β-hydroxysteroid dehydrogenase type 2 enzyme. Mol. Cell. Endocrinol., 105(2): R11–R17.

Antonow-Schlorke, I., Schwab, M., Li, C. and Nathanielsz, P.W. (2003) Glucocorticoid exposure at the dose used clinically alters cytoskeletal proteins and presynaptic terminals in the fetal baboon brain. J. Physiol. Lond., 547(1): 117–123.

Baird, J., Osmond, C., MacGregor, A., Snieder, H., Hales, C. and Phillips, D. (2001) Testing the fetal origins hypothesis in twins: the Birmingham twin study. Diabetologia, 44: 33–39.

Barker, D.J.P. (1991a) Fetal and infant origins of adult disease. BMJ, London.

Barker, D.J.P., Gluckman, P.D., Godfrey, K.M., Harding, J.E., Owens, J.A. and Robinson, J.S. (1993b) Fetal nutrition and cardiovascular disease in adult life. Lancet, 341: 938–941.

Barker, D.J.P., Hales, C.N., Fall, C.H.D., Osmond, C., Phipps, K. and Clarke, P.M.S. (1993a) Type 2 (non-insulin dependent) diabetes mellitus, hypertension and hyperlipidaemia (syndrome X): relation to reduced fetal growth. Diabetologia, 36: 62–67.

Bavdekar, A., Yajnik, C.S., Fall, C.H., Bapat, S., Pandit, A.N., Deshpande, V., Bhave, S., Kellingray, S.D. and Joglekar, C. (1999) Insulin resistance syndrome in 8-year-old Indian children: small at birth, big at 8 years, or both? Diabetes, 48(12): 2422–2429.

Benediktsson, R., Brennand, J., Tibi, L., Calder, A.A., Seckl, J.R. and Edwards, C.R.W. (1995) Fetal osteocalcin levels are related to placental 11β-hydroxysteroid dehydrogenase activity. Clin. Endocrinol., 42: 551–555.

Benediktsson, R., Calder, A.A., Edwards, C.R.W. and Seckl, J.R. (1997) Placental 11β-hydroxysteroid dehydrogenase type 2 is the placental barrier to maternal glucocorticoids: ex vivo studies. Clin. Endocrinol., 46: 161–166.

Benediktsson, R., Lindsay, R., Noble, J., Seckl, J.R. and Edwards, C.R.W. (1993) Glucocorticoid exposure in utero: a new model for adult hypertension. Lancet, 341: 339–341.

Berkowitz, G.S., Wolff, M.S., Janevic, T.M., Holzman, I.R., Yehuda, R. and Landrigan, P.J. (2003) The World Trade Center disaster and intrauterine growth restriction. JAMA, 290(5): 595–596.

Bertram, C., Trowern, A.R., Copin, N., Jackson, A.A. and Whorwood, C.B. (2001) The maternal diet during pregnancy programs altered expression of the glucocorticoid receptor and type 2 11beta-hydroxysteroid dehydrogenase: potential molecular mechanisms underlying the programming of hypertension in utero. Endocrinology, 142(7): 2841–2853.

Bhattacharyya, S., Brown, D., Brewer, J., Vogt, S. and Muglia, L. (2007) Macrophage glucocorticoid receptors regulate toll-like receptor-4-mediated inflammatory responses by selective inhibition of p38 MAP kinase. Blood (Epub: http://www.bloodjournal.org/cgi/reprint/blood-2006-10-048215v1).

Bloom, S.L., Sheffield, J.S., McIntire, D.D. and Leveno, K.J. (2001) Antenatal dexamethasone and decreased birth weight. Obstet. Gynecol., 97(4): 485–490.

Bohn, M.C. (1980) Granule cell genesis in the hippocampus of rats treated neonatally with hydrocortisone. Neuroscience, 5: 2003–2012.

Bremner, J.D., Randall, P., Scott, T.M., Bronen, R.A., Seibyl, J.P., Southwick, S.M., Delaney, R.C., McCarthy, G., Charney, D.S. and Innis, R.B. (1995) MRI-based measurement of hippocampal volume in patients with combat-related posttraumatic stress disorder [comment]. Am. J. Psychiatry, 152(7): 973–981.

Brown, R.W., Diaz, R., Robson, A.C., Kotelevtsev, Y., Mullins, J.J., Kaufman, M.H. and Seckl, J.R. (1996a) The ontogeny of 11β-hydroxysteroid dehydrogenase type 2 and mineralocorticoid receptor gene expression reveal intricate control of glucocorticoid action in development. Endocrinology, 137: 794–797.

Brown, R.W., Kotolevtsev, Y., Leckie, C., Lindsay, R.S., Lyons, V., Murad, P., Mullins, J.J., Chapman, K.E., Edwards, C.R.W. and Seckl, J.R. (1996b) Isolation and cloning of human placental 11β-hydroxysteroid dehydrogenase-2 cDNA. Biochem. J., 313: 1007–1017.

Buss, C., Lord, C., Wadiwalla, M., Hellhammer, D.H., Lupien, S.J., Meaney, M.J. and Pruessner, J.C. (2007) Maternal care modulates the relationship between prenatal risk and hippocampal volume in women but not in men. J. Neurosci., 27(10): 2592–2595.

Canlon, B., Erichsen, S., Nemlander, E., Chen, M., Hossain, A., Celsi, G. and Ceccatelli, S. (2003) Alterations in the intrauterine environment by glucocorticoids modifies the development programme of the auditory system. Eur. J. Neurosci., 17(10): 2035–2041.

Cannon, M., Jones, P.B. and Murray, R.M. (2002) Obstetric complications and schizophrenia: historical and meta-analytic review. Am. J. Psychiatry, 159(7): 1080–1092.

Catalani, A., Marinelli, M., Scaccianoce, S., Nicolai, R., Muscolo, L.A.A., Porcu, A., Koranyi, L., Piazza, P.V. and Angelucci, L. (1993) Progeny of mothers drinking corticosterone during lactation has lower stress-induced corticosterone secretion and better cognitive performance. Brain Res., 624: 209–215.

Champagne, F.A., Weaver, I.C.G., Diorio, J., Dymov, S., Szyf, M. and Meaney, M.J. (2006) Maternal care associated with methylation of the estrogen receptor-alpha 1b promoter and estrogen receptor-alpha expression in the medial preoptic area of female offspring. Endocrinology, 147(6): 2909–2915.

Clark, P.M., Hindmarsh, P.C., Shiell, A.W., Law, C.M., Honour, J.W. and Barker, D.J.P. (1996) Size at birth and adrenocortical function in childhood. Clin. Endocrinol., 45: 721–726.

Cleasby, M.E., Kelly, P.A.T., Walker, B.R. and Seckl, J.R. (2003) Programming of rat muscle and fat metabolism by in utero overexposure to glucocorticoids. Endocrinology, 144(3): 999–1007.

Cole, T., Blendy, J.A., Monaghan, A.P., Kriegelstein, K., Schmid, W., Fantuzzi, G., Hummler, E., Unsicker, K. and Schütz, G. (1995a) Targeted disruption of the glucocorticoid receptor blocks adrenergic chromaffin cell development and severely retards lung maturation. Genes Dev., 9: 1608–1621.

Cole, T.J. (1995) Cloning of the mouse 11beta-hydroxysteroid dehydrogenase type 2 gene: tissue specific expression and localization in distal convoluted tubules and collecting ducts of the kidney. Endocrinology, 136: 4693–4696.

Cole, T.J., Blendy, J.A., Monaghan, A.P., Krieglstein, K., Schmid, W., Aguzzi, A., Fantuzzi, G., Hummler, E., Unsicker, K. and Schutz, G. (1995b) Targeted disruption of the glucocorticoid receptor gene blocks adrenergic chromaffin cell-development and severely retards lung maturation. Genes Dev., 9(13): 1608–1621.

Cole, T.J., Blendy, J.A., Monaghan, A.P., Schmid, W., Aguzzi, A. and Schutz, G. (1995c) Molecular genetic analysis of glucocorticoid signaling during mouse development. Steroids, 60: 93–96.

Costello, E.J., Worthman, C., Erkanli, A. and Angold, A. (2007) Prediction from low birth weight to female adolescent depression — a test of competing hypotheses. Arch. Gen. Psychiatry, 64(3): 338–344.

Cratty, M.S., Ward, H.E., Johnson, E.A., Azzaro, A.J. and Birkle, D.L. (1995) Prenatal stress increases corticotropin-releasing factor (Crf) content and release in rat amygdala minces. Brain Res., 675(1–2): 297–302.

Curhan, G.C., Chertow, G.M., Willett, W.C., Spiegelman, D., Colditz, G.A., Manson, J.E., Speizer, F.E. and Stampfer, M.J. (1996a) Birth-weight and adult hypertension and obesity in women. Circulation, 94: 1310–1315.

Curhan, G.C., Willett, W.C., Rimm, E.B., Spiegelman, D., Ascherio, A.L. and Stampfer, M.J. (1996b) Birth weight and

adult hypertension, diabetes mellitus, and obesity in US men. Circulation, 94: 3246–3250.
Dave-Sharma, S., Wilson, R.C., Harbison, M.D., Newfield, R., Azar, M., Krozowski, Z.S., Funder, J.W., Shackleton, C.H.L., Bradlow, H.L., Wei, J.Q., Hertecant, J., Moran, A., Neiberger, R.E., Balfe, J.W., Fattah, A., Daneman, D., Akkurt, H.I., DeSantis, C. and New, M.I. (1998) Extensive personal experience — examination of genotype and phenotype relationships in 14 patients with apparent mineralocorticoid excess. J. Clin. Endocrinol. Metab., 83: 2244–2254.
Diaz, R., Brown, R.W. and Seckl, J.R. (1998) Ontogeny of mRNAs encoding glucocorticoid and mineralocorticoid receptors and 11β-hydroxysteroid dehydrogenases in prenatal rat brain development reveal complex control of glucocorticoid action. J. Neurosci., 18: 2570–2580.
Diaz, R., Fuxe, K. and Ogren, S.O. (1997) Prenatal corticosterone treatment induces long-term changes in spontaneous and apomorphine-mediated motor activity in male and female rats. Neuroscience, 81: 129–140.
Diaz, R., Ögren, S.O., Blum, M. and Fuxe, K. (1995) Prenatal corticosterone increases spontaneous and D-amphetamine induced locomotor activity and brain dopamine metabolism in prepubertal male and female rats. Neuroscience, 66: 467–473.
Dodic, M., Hantzis, V., Duncan, J., Rees, S., Koukoulas, I., Johnson, K., Wintour, E.M. and Moritz, K. (2002a) Programming effects of short prenatal exposure to cortisol. FASEB J., 16(9): 1017–1026.
Dodic, M., May, C.N., Wintour, E.M. and Coghlan, J.P. (1998) An early prenatal exposure to excess glucocorticoid leads to hypertensive offspring in sheep. Clin. Sci., 94: 149–155.
Dodic, M., Moritz, K., Koukoulas, I. and Wintour, E.M. (2002b) Programmed hypertension: kidney, brain or both? Trends Endocrinol. Metab., 13(9): 403–408.
Dodic, M., Peers, A., Coghlan, J., May, C., Lumbers, E., Yu, Z. and Wintour, E. (1999) Altered cardiovascular haemodynamics and baroreceptor-heart rate reflex in adult sheep after prenatal exposure to dexamethasone. Clin. Sci., 97: 103–109.
Doyle, L.W., Ford, G.W., Davis, N.M. and Callanan, C. (2000) Antenatal corticosteroid therapy and blood pressure at 14 years of age in preterm children. Clin. Sci., 98: 137–142.
Drake, A.J., Walker, B.R. and Seckl, J.R. (2005) Intergenerational consequences of fetal programming by in utero exposure to glucocorticoids in rats. Am. J. Physiol. Regul. Integr. Comp. Physiol., 288: R34–R38.
Dunn, A. and Berridge, C. (1990) Physiological and behavioral responses to corticotropin-releasing factor administration: is CRF a mediator of anxiety or stress responses. Brain Res. Rev., 15: 71–100.
Edwards, C.R.W., Benediktsson, R., Lindsay, R. and Seckl, J.R. (1993) Dysfunction of the placental glucocorticoid barrier: a link between the foetal environment and adult hypertension? Lancet, 341: 355–357.
Encio, I.J. and Detera-Wadleigh, S.D. (1991) The genomic structure of the human glucocorticoid receptor. J. Biol. Chem., 266: 7182–7188.
Engel, S.M., Berkowitz, G.S., Wolff, M.S. and Yehuda, R. (2005) Psychological trauma associated with the World Trade Center attacks and its effect on pregnancy outcome. Paediatr. Perinat. Epidemiol., 19(5): 334–341.
Fall, C.H.D., Osmond, C., Barker, D.J.P., Clark, P.M.S., Hales, C.N., Stirling, Y. and Meade, T.W. (1995) Fetal and infant growth and cardiovascular risk factors in women. Br. Med. J., 310: 428–432.
Famularo, R. and Fenton, T. (1994) Early developmental history and pediatric posttraumatic-stress-disorder. Arch. Pediatr. Adolesc. Med., 148(10): 1032–1038.
Feldman, S. and Weidenfeld, J. (1998) The excitatory effects of the amygdala on hypothalamo-pituitary-adrenocortical responses are mediated by hypothalamic norepinephrine, serotonin, and CRF-41. Brain Res. Bull., 45: 389–393.
Forest, M.G., David, M. and Morel, Y. (1993) Prenatal diagnosis and treatment of 21-hydroxylase deficiency. J. Steroid Biochem. Mol. Biol., 45: 75–82.
Forsen, T., Eriksson, J.G., Tuomilehto, J., Teramo, K., Osmond, C. and Barker, D.J.P. (1997) Mother's weight in pregnancy and coronary heart disease in a cohort of Finnish men: follow up study. Br. Med. J., 315: 837–840.
French, N.P., Hagan, R., Evans, S.F., Godfrey, M. and Newnham, J.P. (1999) Repeated antenatal corticosteroids: size at birth and subsequent development. Am. J. Obstet. Gynecol., 180(1 Pt 1): 114–121.
Fumagalli, F., Jones, S.R., Caron, M.G., Seidler, F.J. and Slotkin, T.A. (1996) Expression of mRNA coding for the serotonin transporter in aged vs. young rat brain: differential effects of glucocorticoids. Brain Res., 719: 225–228.
Fuxe, K., Wikstrom, A.-C., Okret, S., Agnati, L.F., Harfstrand, A., Yu, Z.-Y., Granholm, L., Zoli, M., Vale, W. and Gustafsson, J.-A. (1985) Mapping of glucocorticoid receptor immunoreactive neurons in the rat tel- and diencephalon using a monoclonal antibody against rat liver glucocorticoid receptor. Endocrinol., 117: 1803–1812.
Gatford, K.L., Wintour, E.M., de Blasio, M.J., Owens, J.A. and Dodic, M. (2000) Differential timing for programming of glucose homoeostasis, sensitivity to insulin and blood pressure by in utero exposure to dexamethasone in sheep. Clin. Sci., 98: 553–560.
Goland, R.S., Jozak, S., Warren, W.B., Conwell, I.M., Stark, R.I. and Tropper, P.J. (1993) Elevated levels of umbilical cord plasma corticotropin-releasing hormone in growth-retarded fetuses. J. Clin. Endocrinol. Metab., 77: 1174–1179.
Goland, R.S., Tropper, P.J., Warren, W.B., Stark, R.I., Jozak, S.M. and Conwell, I.M. (1995) Concentrations of corticotropin-releasing hormone in the umbilical-cord blood of pregnancies complicated by preeclampsia. Reprod. Fertil. Dev., 7: 1227–1230.
Gould, E., Woolley, C.S., Cameron, H.A., Daniels, D.C. and McEwen, B.S. (1991a) Adrenal steroids regulate postnatal development in the rat dentate gyrus: II. Effects of glucocorticoids on cell birth. J. Comp. Neurol., 313: 486–493.
Gould, E., Woolley, C.S. and McEwen, B.S. (1991b) Adrenal steroids regulate postnatal development in the rat dentate gyrus: I. Effects of glucocorticoids on cell death. J. Comp. Neurol., 313: 479–485.

Hawkins, P., Steyn, C., McGarrigle, H.H., Calder, N.A., Saito, T., Stratford, L.L., Noakes, D.E. and Hanson, M.A. (2000) Cardiovascular and hypothalamic-pituitary-adrenal axis development in late gestation fetal sheep and young lambs following modest maternal nutrient restriction in early gestation. Reprod. Fertil. Dev., 12(7–8): 443–456.

Holmes, M., Sangra, M., French, K., Whittle, I., Paterson, J., Mullins, J. and Seckl, J. (2005) 11beta-hydroxysteroid dehydrogenase type 2 protects the neonatal cerebellum from deleterious effects of glucocorticoids. Neurosci., 137: 865–873.

Holmes, M.C., Abrahamsen, C.T., French, K.L., Paterson, J.M., Mullins, J.J. and Seckl, J.R. (2006) The mother or the fetus? 11 beta-hydroxysteroid dehydrogenase type 2 null mice provide evidence for direct fetal programming of behavior by endogenous glucocorticoids. J. Neurosci., 26(14): 3840–3844.

Hsu, D., Chen, F., Takahashi, L. and Kalin, N. (1998) Rapid stress-induced elevations in corticotropin-releasing hormone mRNA in rat central amygdala nucleus and hypothalamic paraventricular nucleus: an in situ hybridization analysis. Brain Res., 788: 305–310.

Huang, W.L., Beazley, L.D., Quinlivan, J.A., Evans, S.F., Newnham, J.P. and Dunlop, S.A. (1999) Effect of corticosteroids on brain growth in fetal sheep. Obstet. Gynecol., 94(2): 213–218.

Huang, W.L., Harper, C.G., Evans, S.F., Newnham, J.P. and Dunlop, S.A. (2001a) Repeated prenatal corticosteroid administration delays astrocyte and capillary tight junction maturation in fetal sheep. Int. J. Dev. Neurosci., 19(5): 487–493.

Huang, W.L., Harper, C.G., Evans, S.F., Newnham, J.P. and Dunlop, S.A. (2001b) Repeated prenatal corticosteroid administration delays myelination of the corpus callosum in fetal sheep. Int. J. Dev. Neurosci., 19(4): 415–425.

Ikegami, M., Jobe, A.H., Newnham, J., Polk, D.H., Willet, K.E. and Sly, P. (1997) Repetitive prenatal glucocorticoids improve lung function and decrease growth in preterm lambs. Am. J. Respir. Crit. Care Med., 156(1): 178–184.

Irving, R.J., Belton, N.R., Elton, R.A. and Walker, B.R. (2000) Adult cardiovascular risk factors in premature babies. Lancet, 355: 2135–2136.

Jacobson, L. and Sapolsky, R. (1991) The role of the hippocampus in feedback regulation of the hypothalamic-pituitary-adrenal axis. Endocrinol. Rev., 12: 118–134.

Jensen, E.C., Gallaher, B.W., Breier, B.H. and Harding, J.E. (2002) The effect of a chronic maternal cortisol infusion on the late-gestation fetal sheep. J. Endocrinol., 174(1): 27–36.

Jones, P.B., Rantakallio, P., Hartikainen, A.L., Isohanni, M. and Sipila, P. (1998) Schizophrenia as a long-term outcome of pregnancy, delivery, and perinatal complications: a 28-year follow-up of the 1966 North Finland general population birth cohort. Am. J. Psychiatry, 155(3): 355–364.

Kitraki, E., Kittas, C. and Stylianopoulou, F. (1997) Glucocorticoid receptor gene expression during rat embryogenesis. An in situ hybridization study. Differentiation, 62: 21–31.

Koenig, J.I., Kirkpatrick, B. and Lee, P. (2002) Glucocorticoid hormones and early brain development in schizophrenia. Neuropsychopharmacology, 27(2): 309–318.

Korte, S.M. (2001) Corticosteroids in relation to fear, anxiety and psychopathology. Neurosci. Biobehav. Rev., 25(2): 117–142.

Kotelevtsev, Y., Brown, R.W., Fleming, S., Kenyon, C.J., Edwards, C.R.W., Seckl, J.R. and Mullins, J.J. (1999) Hypertension in mice lacking 11β-hydroxysteroid dehydrogenase type 2. J. Clin. Invest., 103: 683–689.

Langley-Evans, S.C. (1997) Hypertension induced by foetal exposure to a maternal low-protein diet, in the rat, is prevented by pharmacological blockade of maternal glucocorticoid synthesis. J. Hypertens., 15: 537–544.

Langley-Evans, S.C., Gardner, D.S. and Jackson, A.A. (1996a) Maternal protein restriction influences the programming of the rat hypothalamic-pituitary-adrenal axis. J. Nutr., 126: 1578–1585.

Langley-Evans, S.C., Philips, G., Benediktsson, R., Gardner, D., Edwards, C.R.W., Jackson, A.A. and Seckl, J.R. (1996b) Maternal dietary protein restriction, placental glucocorticoid metabolism and the programming of hypertension. Placenta, 17: 169–172.

Law, C.M., Shiell, A.W., Newsome, C.A., Syddall, H.E., Shinebourne, E.A., Fayers, P.M., Martyn, C.N. and de Swiet, M. (2002) Fetal, infant, and childhood growth and adult blood pressure: a longitudinal study from birth to 22 years of age. Circulation, 105(9): 1088–1092.

Leon, D.A., Koupilova, I., Lithell, H.O., Berglund, L., Mohsen, R., Vagero, D., Lithell, U.B. and McKeigue, P.M. (1996) Failure to realise growth potential in utero and adult obesity in relation to blood pressure in 50 year old Swedish men. Br. Med. J., 312: 401–406.

Lesage, J., Blondeau, B., Grino, M., Breant, B. and Dupouy, J.P. (2001) Maternal undernutrition during late gestation induces fetal overexposure to glucocorticoids and intrauterine growth retardation, and disturbs the hypothalamo-pituitary adrenal axis in the newborn rat. Endocrinology, 142(5): 1692–1702.

Levine, R.S., Hennekens, C.H. and Jesse, M.J. (1994) Blood pressure in prospective population based cohort of newborn and infant twins. Br. Med. J., 308: 298–302.

Levine, S. (1957) Infantile experience and resistance to physiological stress. Science, 126: 405–406.

Levine, S. (1962) Plasma-free corticosteroid response to electric shock in rats stimulated in infancy. Science, 135: 795–796.

Levitt, N., Lindsay, R.S., Holmes, M.C. and Seckl, J.R. (1996) Dexamethasone in the last week of pregnancy attenuates hippocampal glucocorticoid receptor gene expression and elevates blood pressure in the adult offspring in the rat. Neuroendocrinology, 64: 412–418.

Levitt, N.S., Lambert, E.V., Woods, D., Hales, C.N., Andrew, R. and Seckl, J.R. (2000) Impaired glucose tolerance and elevated blood pressure in low birth weight, non-obese young South African adults: early programming of the cortisol axis. J. Clin. Endocrinol. Metab., 85: 4611–4618.

Lillycrop, K.A., Phillips, E.S., Jackson, A.A., Hanson, M.A. and Burdge, G.C. (2005) Dietary protein restriction of pregnant rats induces and folic acid supplementation prevents epigenetic modification of hepatic gene expression in the offspring. J. Nutr., 135(6): 1382–1386.

Lindsay, R.S., Lindsay, R.M., Edwards, C.R.W. and Seckl, J.R. (1996a) Inhibition of 11β-hydroxysteroid dehydrogenase in pregnant rats and the programming of blood pressure in the offspring. Hypertension, 27: 1200–1204.

Lindsay, R.S., Lindsay, R.M., Waddell, B. and Seckl, J.R. (1996b) Programming of glucose tolerance in the rat: role of placental 11β-hydroxysteroid dehydrogenase. Diabetologia, 39: 1299–1305.

Lithell, H.O., McKeigue, P.M., Berglund, L., Mohsen, R., Lithell, U.B. and Leon, D.A. (1996) Relation of size at birth to non-insulin dependent diabetes and insulin concentrations in men aged 50–60 years. Br. Med. J., 312: 406–410.

Liu, D., Diorio, J., Tannenbaum, B., Caldji, C., Francis, D., Freedman, A., Sharma, S., Pearson, D., Plotsky, P.M. and Meaney, M.J. (1997) Maternal care, hippocampal glucocorticoid receptors, and hypothalamic-pituitary-adrenal responses to stress. Science, 277: 1659–1662.

Lopez Bernal, A. and Craft, I.L. (1981) Corticosteroid metabolism in vitro by human placenta, foetal membranes and decidua in early and late gestation. Placenta, 2: 279–285.

Lopez Bernal, A., Flint, A.P.F., Anderson, A.B.M. and Turnbull, A.C. (1980) 11β-hydroxysteroid dehydrogenase activity (E.C.1.1.1.146) in human placenta and decidua. J. Steroid Biochem., 13: 1081–1087.

MacArthur, B.A., Howie, R.N., Dezoete, J.A. and Elkins, J. (1982) School progress and cognitive development of 6-year-old children whose mothers were treated antenatally with betamethasone. Pediatrics, 70(1): 99–105.

Matthews, S.G. (2000) Antenatal glucocorticoids and programming of the developing CNS. Pediatr. Res., 47(3): 291–300.

McCormick, C.M., Smythe, J.W., Sharma, S. and Meaney, M.J. (1995) Sex-specific effects of prenatal stress on hypothalamic-pituitary-adrenal responses to stress and brain glucocorticoid receptor density in adult rats. Dev. Brain Res., 84: 55–61.

McCormick, J., Lyons, V., Jacobson, M., Diorio, J., Nyirenda, M., Weaver, S., J.L.W., Ester, W., Meaney, M.J., Seckl, J.R. and Chapman, K.E. (2000) 5′-heterogeneity of glucocorticoid receptor mRNA is tissue-specific: differential regulation of variant promoters by early life events. Mol. Endocrinol., 14: 506–517.

McTernan, C.L., Draper, N., Nicholson, H., Chalder, S.M., Driver, P., Hewison, M., Kilby, M.D. and Stewart, P.M. (2001) Reduced placental 11 beta-hydroxysteroid dehydrogenase type 2 mRNA levels in human pregnancies complicated by intrauterine growth restriction: an analysis of possible mechanisms. J. Clin. Endocrinol. Metab., 86(10): 4979–4983.

Meaney, M.J., Aitken, D.H., van Berkel, C., Bhatnagar, S. and Sapolsky, R.M. (1988) Effect of neonatal handling on age-related impairments associated with the hippocampus. Science, 239: 766–768.

Meaney, M.J., Aitken, D.H., Sharma, S. and Viau, V. (1992) Basal ACTH, corticosterone and corticosterone-binding globulin levels over the diurnal cycle, and hippocampal corticosteroid receptors in young and aged, handled and non-handled rats. Neuroendocrinology, 55: 204–213.

Meaney, M.J., Aitken, D.H., Viau, V., Sharma, S. and Sarrieau, A. (1989) Neonatal handling alters adrenocortical negative feedback sensitivity and hippocampal type II glucocorticoid receptor binding in the rat. Neuroendocrinol., 50: 597–604.

Meaney, M.J., Diorio, J., Francis, D., Weaver, S., Yau, J.L.W., Chapman, K.E. and Seckl, J.R. (2000) Postnatal handling increases the expression of cAMP-inducible transcription factors in the rat hippocampus: the effects of thyroid hormones and serotonin. J. Neurosci., 20: 3926–3935.

Meaney, M.J., Diorio, J., Francis, D., Widdowson, J., LaPlante, P., Caldji, C., Sharma, S., Seckl, J.R. and Plotsky, P.M. (1996) Early environmental regulation of forebrain glucocorticoid receptor gene expression: implications for adrenocortical responses to stress. Dev. Neurosci., 18: 49–72.

Mercado, A.B., Wilson, R.C., Cheng, K.C., Wei, J.Q. and New, M.I. (1995) Prenatal treatment and diagnosis of congential adrenal hyperlasia owing to 21-hydroxylase deficiency. J. Clin. Endocrinol. Metab., 80: 2014–2020.

Meyer, J.S. (1983) Early adrenalectomy stimulates subsequent growth and development of the rat brain. Exp. Neurol., 82: 432–446.

Mitchell, J.B., Rowe, W., Boksa, P. and Meaney, M.J. (1990) Serotonin regulates type II corticosteroid receptor binding in hippocampal cell cultures. J. Neurosci., 10: 1745–1752.

Moore, V.M., Miller, A.G., Boulton, T.J.C., Cockington, R.A., Craig, I.H., Magarey, A.M. and Robinson, J.S. (1996) Placental weight, birth measurements, and blood pressure at age 8 years. Arch. Dis. Child., 74: 538–541.

Murphy, V.E., Zakar, T., Smith, R., Giles, W.B., Gibson, P.G. and Clifton, V.L. (2002) Reduced 11beta-hydroxysteroid dehydrogenase type 2 activity is associated with decreased birth weight centile in pregnancies complicated by asthma. J. Clin. Endocrinol. Metab., 87(4): 1660–1668.

Newnham, J.P., Evans, S.F., Godfrey, M., Huang, W., Ikegami, M. and Jobe, A. (1999) Maternal, but not fetal, administration of corticosteroids restricts fetal growth. J. Matern. Fetal Med., 8(3): 81–87.

Newnham, J.P. and Moss, T.J. (2001) Antenatal glucocorticoids and growth: single versus multiple doses in animal and human studies. Semin. Neonatol., 6(4): 285–292.

Nyirenda, M., Dean, S., Lyons, V., Chapman, K. and Seckl, J. (2006) Prenatal programming of hepatocyte nuclear factor 4 alpha in the rat: a key mechanism in the 'fetal origins of hyperglycemia'? Diabetologia, 49: 1412–1420.

Nyirenda, M.J., Lindsay, R.S., Kenyon, C.J., Burchell, A. and Seckl, J.R. (1998) Glucocorticoid exposure in late gestation permanently programmes rat hepatic phosphoenolpyruvate carboxykinase and glucocorticoid receptor expression and causes glucose intolerance in adult offspring. J. Clin. Invest., 101: 2174–2181.

Osmond, C., Barker, D.J.P., Winter, P.D., Fall, C.H.D. and Simmonds, S.J. (1993) Early growth and death from cardiovascular disease in women. Br. Med. J., 307: 1524–1527.

Pepin, M.C., Pothier, F. and Barden, N. (1992) Impaired glucocorticoid receptor function in transgenic mice expressing antisense RNA. Nature, 355: 725–728.

Phillips, D.I., Fall, C.H.D., Whorwood, C.B., Seckl, J.R., Wood, P.J., Barker, D.J.P. and Walker, B.R. (1998) Elevated plasma cortisol concentrations: an explanation for the relationship between low birthweight and adult cardiovascular risk factors. J. Clin. Endocrinol. Metab., 83: 757–760.

Phillips, D.I.W., Walker, B.R., Reynolds, R.M., Flanagan, D.E.H., Wood, P.J., Osmond, C., Barker, D.J.P. and Whorwood, C.B. (2000) Low birth weight predicts elevated plasma cortisol concentrations in adults from 3 populations. Hypertension, 35(6): 1301–1306.

Reinisch, J.M., Simon, N.G., Karwo, W.G. and Gandelman, R. (1978) Prenatal exposure to prednisone in humans and animals retards intra-uterine growth. Science, 202: 436–438.

Reynolds, R.M., Walker, B.R., Syddall, H.E., Andrew, R., Wood, P.J., Whorwood, C.B. and Phillips, D.I.W. (2001) Altered control of cortisol secretion in adult men with low birth weight and cardiovascular risk factors. J. Clin. Endocrinol. Metab., 86: 245–250.

RichEdwards, J.W., Stampfer, M.J., Manson, J.E., Rosner, B., Hankinson, S.E., Colditz, G.A., Willett, W.C. and Hennekens, C.H. (1997) Birth weight and risk of cardiovascular disease in a cohort of women followed up since 1976. Br. Med. J., 315: 396–400.

Robson, A.C., Leckie, C., Seckl, J.R. and Holmes, M.C. (1998) Expression of 11β-hydroxysteroid dehydrogenase type 2 in the postnatal and adult rat brain. Mol. Brain Res., 61: 1–10.

Rogerson, F.M., Kayes, K. and White, P.C. (1996) No correlation in human placenta between activity or mRNA for the K (type 2) isozyme of 11β-hydroxysteroid dehydrogenase and fetal or placental weight. Tenth Int. Congr. Endocrinol. Abstr., P1–231: 193.

Rogerson, F.M., Kayes, K.M. and White, P.C. (1997) Variation in placental type 2 11beta-hydroxysteroid dehydrogenase activity is not related to birth weight or placental weight. Mol. Cell. Endocrinol., 128: 103–109.

Seckl, J.R. (1998) Physiologic programming of the fetus. Clin. Perinatol., 25: 939–964.

Seckl, J.R. and Meaney, M.J. (2004) Glucocorticoid programming. Biobehav. Stress Response Prot. Damag. Eff., 1032: 63–84.

Shams, M., Kilby, M.D., Somerset, D.A., Howie, A.J., Gupta, A., Wood, P.J., Afnan, M. and Stewart, P.M. (1998) 11beta hydroxysteroid dehydrogenase type 2 in human pregnancy and reduced expression in intrauterine growth retardation. Hum. Reprod., 13: 799–804.

Sheline, Y.I., Wang, P.W., Gado, M.H., Csernansky, J.G. and Vannier, M.W. (1996) Hippocampal atrophy in recurrent major depression. Proc. Natl. Acad. Sci. U.S.A., 93: 3908–3913.

Slotkin, T.A., Barnes, G.A., McCook, E.C. and Seidler, F.J. (1996) Programming of brainstem serotonin transporter development by prenatal glucocorticoids. Dev. Brain Res., 93: 155–161.

Smythe, J.W., Rowe, W.B. and Meaney, M.J. (1994) Neonatal handling alters serotonin (5-HT) turnover and 5-HT2 receptor binding in selected brain regions: relationship to the handling effect on glucocorticoid receptor expression. Dev. Brain Res., 80: 183–189.

Speirs, H., Seckl, J. and Brown, R. (2004) Ontogeny of glucocorticoid receptor and 11β-hydroxysteroid dehydrogenase type 1 gene expression identifies potential critical periods of glucocorticoid susceptibility during development. J. Endocrinol., 181: 105–116.

St Clair, D., Xu, M.Q., Wang, P., Yu, Y.Q., Fang, Y.R., Zhang, F., Zheng, X.Y., Gu, N.F., Feng, G.Y., Sham, P. and He, L. (2005) Rates of adult schizophrenia following prenatal exposure to the Chinese famine of 1959–1961. JAMA, 294(5): 557–562.

Stein, M.B., Koverola, C., Hanna, C., Torchia, M.G. and McClarty, B. (1997) Hippocampal volume in women victimized by childhood sexual abuse. Psychol. Med., 27(4): 951–959.

Stewart, P.M., Murry, B.A. and Mason, J.I. (1994) Type 2 11β-hydroxysteroid dehydrogenase in human fetal tissues. J. Clin. Endocrinol. Metab., 78: 1529–1532.

Stewart, P.M., Rogerson, F.M. and Mason, J.I. (1995) Type 2 11β-hydroxysteroid dehydrogenase messenger RNA and activity in human placenta and fetal membranes: its relationship to birth weight and putative role in fetal steroidogenesis. J. Clin. Endocrinol. Metab., 80: 885–890.

Sugden, M.C., Langdown, M.L., Munns, M.J. and Holness, M.J. (2001) Maternal glucocorticoid treatment modulates placental leptin and leptin receptor expression and materno-fetal leptin physiology during late pregnancy, and elicits hypertension associated with hyperleptinaemia in the early-growth-retarded adult offspring. Eur. J. Endocrinol., 145(4): 529–539.

Sun, K., Yang, K. and Challis, J.R.G. (1997) Differential expression of 11beta-hydroxysteroid dehydrogenase types 1 and 2 in human placenta and fetal membranes. J. Clin. Endocrinol. Metab., 82: 300–305.

Susser, E., Neugebauer, R., Hoek, H.W., Brown, A.S., Lin, S., Labovitz, D. and Gorman, J.M. (1996) Schizophrenia after prenatal famine. Arch. Gen. Psychiatry, 53: 25–31.

Susser, E.S. and Lin, S.P. (1992) Schizophrenia after prenatal exposure to the Dutch hunger winter of 1944–1945. Arch. Gen. Psychiatry, 49(12): 983–988.

Thapar, A., Langley, K., Fowler, T., Rice, F., Turic, D., Whittinger, N., Aggleton, J., Van den Bree, M., Owen, M. and O'Donovan, M. (2005) Catechol *O*-methyltransferase gene variant and birth weight predict early-onset antisocial behavior in children with attention-deficit/hyperactivity disorder. Arch. Gen. Psychiatry, 62(11): 1275–1278.

Thomassin, H., Flavin, M., Espinas, M. and Grange, T. (2001) Glucocorticoid-induced DNA demethylation and gene memory during development. EMBO J., 20: 1974–1983.

Thompson, C., Syddall, H., Rodin, I., Osmond, C. and Barker, D.J.P. (2001) Birth weight and the risk of depressive disorder in late life. Br. J. Psychiatry, 179: 450–455.

Trautman, P.D., Meyer-Bahlburg, H.F.L., Postelnek, J. and New, M.I. (1995) Effects of early prenatal dexamethasone on the cognitive and behavioral development of young children: results of a pilot study. Psychoneuroendocrinol., 20: 439–449.

Tronche, F., Kellendonk, C., Kretz, O., Gass, P., Anlag, K., Orban, P.C., Bock, R., Klein, R. and Schütz, G. (1999)

Disruption of the glucocorticoid receptor gene in the nervous system results in reduced anxiety. Nat. Genet., 23: 99–103.

Turner, J.D. and Muller, C.P. (2005) Structure of the glucocorticoid receptor (NR3C1) gene 5′ untranslated region: identification, and tissue distribution of multiple new human exon 1. J. Mol. Endocrinol., 35(2): 283–292.

Uno, H., Eisele, S., Sakai, A., Shelton, S., Baker, E., DeJesus, O. and Holden, J. (1994) Neurotoxicity of glucocorticoids in the primate brain. Horm. Behav., 28: 336–348.

Uno, H., Lohmiller, L., Thieme, C., Kemnitz, J.W., Engle, M.J., Roecker, E.B. and Farrell, P.M. (1990) Brain damage induced by prenatal exposure to dexamethasone in fetal rhesus macaques. I. Hippocampus. Dev. Brain Res., 53: 157–167.

Venihaki, M.A., Carrigan, P., Dikkes, P. and Majzoub, J. (2000) Circadian rise in maternal glucocorticoid prevents pulmonary dysplasia in fetal mice with adrenal insufficiency. Proc. Natl. Acad. Sci. U.S.A., 97: 7336–7341.

de Vries, A., Holmes, M., Heijnis, A., Seier, J., van Heerden, J., Louw, J., Wolfe-Coote, S., Meaney, M., Levitt, N. and Seckl, J. (2007) Prenatal dexamethasone exposure induces changes in offspring cardio-metabolic and hypothalamic-pituitary-adrenal axis function without alteration of birth weight in a non-human primate, the African vervet, *Chlorocebus aethiops*. J. Clin. Invest., 117: 1058–1067.

Waddell, B., Benediktsson, R., Brown, R. and Seckl, J.R. (1998) Tissue-specific mRNA expression of 11B-hydroxysteroid dehydrogenase types 1 and 2 and the glucocorticoid receptor within rat placenta suggest exquisite local control of glucocorticoid action. Endocrinology, 139: 1517–1523.

Waddell, B.J. and Atkinson, H.C. (1994) Production rate, metabolic clearance rate and uterine extraction of corticosterone during rat pregnancy. J. Endocrinol., 143: 183–190.

Ward, R.M. (1994) Pharmacologic enhancement of fetal lung maturation. Clin. Perinatol., 21: 523–542.

Weaver, I., Cervoni, N., Champagne, F., D'Alessio, A., Sharma, S., Seckl, J., Dymov, S., Szyf, M. and Meaney, M. (2004) Epigenetic programming by maternal behavior. Nat. Neurosci., 7: 847–854.

Weaver, I.C.G., D'Alessio, A.C., Brown, S.E., Hellstrom, I.C., Dymov, S., Sharma, S., Szyf, M. and Meaney, M.J. (2007) The transcription factor nerve growth factor-inducible protein A mediates epigenetic programming: altering epigenetic marks by immediate-early genes. J. Neurosci., 27(7): 1756–1768.

Weinstock, M. (2001) Alterations induced by gestational stress in brain morphology and behaviour of the offspring. Prog. Neurobiol., 65: 427–451.

Weinstock, M., Matlina, E., Maor, G.I., Rosen, H. and McEwen, B.S. (1992) Prenatal stress selectively alters the reactivity of the hypothalamic-pituitary adrenal system in the female rat. Brain Res., 595: 195–200.

Welberg, L.A.M. and Seckl, J.R. (2001) Prenatal stress, glucocorticoids and the programming of the brain. J. Neuroendocrinol., 13: 113–128.

Welberg, L.A.M., Seckl, J.R. and Holmes, M.C. (2000) Inhibition of 11β-hydroxysteroid dehydrogenase, the feto-placental barrier to maternal glucocorticoids, permanently programs amygdala glucocorticoid receptor mRNA expression and anxiety-like behavior in the offspring. Eur. J. Neurosci., 12: 1047–1054.

Welberg, L.A.M., Seckl, J.R. and Holmes, M.C. (2001) Prenatal glucocorticoid programming of brain corticosteroid receptors and corticotrophin-releasing hormone: possible implications for behaviour. Neuroscience, 104: 71–79.

White, P.C., Mune, T. and Agarwal, A.K. (1997) 11beta-Hydroxysteroid dehydrogenase and the syndrome of apparent mineralocorticoid excess. Endocrinol. Rev., 18: 135–156.

Wiles, N.J., Peters, T.J., Leon, D.A. and Lewis, G. (2005) Birth weight and psychological distress at age 45–51 years—results from the Aberdeen Children of the 1950s cohort study. Br. J. Psychiatry, 187: 21–28.

Yajnik, C.S., Fall, C.H.D., Vaidya, U., Pandit, A.N., Bavdekar, A., Bhat, D.S., Osmond, C., Hales, C.N. and Barker, D.J.P. (1995) Fetal growth and glucose and insulin metabolism in four-year- old Indian children. Diabet. Med., 12: 330–336.

Yau, J.L.W., Noble, J. and Seckl, J.R. (1997a) Site-specific regulation of corticosteroid and serotonin receptor subtype gene expression in the rat hippocampus following methylenedioxymethamphetamine: role of corticosterone and serotonin. Neuroscience, 78: 111–121.

Yau, J.L.W., Noble, J., Widdowson, J. and Seckl, J.R. (1997b) Impact of adrenalectomy on 5-HT6 and 5-HT7 receptor gene expression in the rat hippocampus. Mol. Brain Res., 45: 182–186.

Yeh, T., Lin, Y., Lin, H., Huang, C., Hsieh, W., Lin, C. and Tsai, C. (2004) Outcomes at school age after postnatal dexamethasone therapy for lung disease of prematurity. N. Engl. J. Med., 350: 1304–1313.

Yehuda, R. (2002) Current concepts: post-traumatic stress disorder. N. Engl. J. Med., 346(2): 108–114.

Yehuda, R., Engel, S.M., Brand, S.R., Seckl, J., Marcus, S.M. and Berkowitz, G.S. (2005) Transgenerational effects of posttraumatic stress disorder in babies of mothers exposed to the World Trade Center attacks during pregnancy. J. Clin. Endocrinol. Metab., 90(7): 4115–4118.

Yehuda, R., Golier, J.A., Yang, R.K. and Tischler, L. (2004) Enhanced sensitivity to glucocorticoids in peripheral mononuclear leukocytes in posttraumatic stress disorder. Biol. Psychiatry, 55(11): 1110–1116.

Yehuda, R., Teicher, M., Seckl, J., Grossman, R., Morris, A. and Bierer, L. (2007) Parental PTSD is a 'vulnerability' factor for low cortisol trait in offspring of Holocaust survivors. Arch. Gen. Psychiatry, 64(9): 1040–1048.

Discussion: Chapter 2

RICHTER-LEVIN: I am a little bit concerned about your terminology, the use of the term 'programming' to describe the long-term effects of early life environmental variables. Perhaps a programmed difference would be better. Such processes are dynamic, you program it one way or you program it the other, but you can't call some programmed and others non-programmed.

SECKL: You are right. I am saying that programming is a convenient jargon, a slang to describe what we recognise as persisting effects of early life, mapped onto the early life origins of disease observations to link with pathogenesis. Naturally, every animal has an exposure to early life variables, but the disease model has at least been helpful to recognise that some of these programmed pathways may associate with unfavourable late life outcomes and thus to stimulate research into mechanisms, both fundamental and pathogenic.

JOËLS: I wanted to ask about the study of the Holocaust survivors, what about the controls in that study, were they the relatives that for some reason were in the States?

SECKL: The controls were age and sex-matched Jewish people who had emigrated to the USA many years before. They had not suffered directly the Holocaust, but had typically moved to the USA a generation earlier. They came from the same European origin: they may have been exposed to low grade anti-Semitism in a variety of setting, but had not been through the 1930s in Europe.

JOËLS: But they were not like sisters or brothers, first-degree relatives.

SECKL: No, and unexposed first-degree relatives would be hard to find and perhaps difficult to show as untraumatised, even if vicariously.

JOËLS: So your argument that it is environment rather than gene effects, but this study is correlational.

SECKL: True, but we can only do retrospective observational studies in the context of Holocaust research for obvious reasons. Moreover, given that the events occurred more than 60 years ago, there are not enough survivors to do human genotyping for association studies. Our data show a correlation and this is suggestive of developmental programming. In the context of the animal work that I have shown you the rats and mice are isogenic, there is no genetic difference between them, and environmental manipulations in early life clearly have persisting effects that appear to cross into the next generation. So we can say that the data fit with an environmental hypothesis of causation and, whilst perhaps not proof, foetal programming is consistent with the observations. What is intriguing to me is that in the Holocaust survivors there is a graded age-associated vulnerability. The younger at exposure the greater the persisting steroid metabolism adjustments, but apparently the less risk for PTSD. Now one can turn these data around and say, 'Well if a subject was younger at exposure, they perceived events differently or had a different exposure or different experience'. That is fine and we don't have the data to support or gainsay this argument. It just seems perhaps a less likely explanation than being younger at exposure allows more developmental plasticity of responses, metabolic and others.

PITMAN: At one point you were saying that an apparently low functioning HPA axis may not necessarily reflect the tissue effect of the hormone.

SECKL: That is correct. Tissue responses depend on plasma hormone levels (plus binding by CBG in the circulation) and the sensitivity of the particular target tissue. The latter is determined by the density of the two intracellular receptors (GR and MR), pre-receptor metabolism by the isozymes of 11β-HSD and perhaps other enzymes such as 5alpha reductase, the expression of cell membrane pumps such as mdr1a and b and perhaps import processes as have recently been described for the analogous thyroid hormones. All these tissue factors are cell-specific and regulated, so it is a complex mix that determines how a tissue responds.

PITMAN: You were also saying that one of the reasons of the HPA axis activity might be depressed (in PTSD) was because the degradation of active cortisol is decreased and so it may be in fact that the apparently low HPA axis activity with increased feedback sensitivity could reflect an underlying increase state of the degree to which the target tissues are receiving glucocorticoids.

SECKL: That is the whole point of the emerging concept called 'intracrinology'. This is based on the realisation that metabolism of many steroids within cells is determining at the level of the individual cell what level of signal is 'perceived' at receptors and thus responses. However, there are two processes here. There is whole body aggregate metabolism of cortisol and local tissue metabolism. Whole body metabolism determines the half-life of cortisol, its clearance rate. If metabolism is decreased, a molecule of cortisol will on average persist longer. Therefore the HPA axis needs to be less active in order to maintain basal levels. The adrenal cortex can be hypotrophic without any change in blood levels. In contrast, in a tissue with significant levels of metabolic enzymes, some of which activate and others inactivate cortisol, levels of the steroid reaching receptors can be altered by such local metabolism. In PTSD we don't know what is happening at the level of local tissue sensitivity in the brain, but we have found in this particular Holocaust survivor population, altered cortisol metabolism (less degradation) by the major enzymes in liver and kidney. As the cortisol is being degraded less it is actually persisting more in these tissues. A consequence of lack of for example 11β-HSD-2 in kidney is hypertension, as indeed associated with the reduced activity of the enzyme in survivors.

PITMAN: So that when you get a situation in which the HPA axis is overactive or underactive which because of intracellular steroid metabolism results in the same end point physiological state.

SECKL: That is exactly right. In another animal we made in which we knocked out the 11β-HSD type 1 isozyme, which catalyses a reaction which goes the other way, an enzyme which makes glucocorticoids in target cells, and is expressed in sites of glucocorticoid negative feedback on the HPA axis. These mice have reduced feedback because they did not get the extra-amplified glucocorticoid boost in feedback tissues. In consequence, their corticosterone levels go up. Despite this the brain does well because it also normally expresses (and in the knockout mouse lacks) 11β-HSD-1. Inside neurons levels of corticosterone are lower, in the face of elevated plasma levels. This is all about what is within cells, and is perhaps not about the acute membrane effects Marian Joëls is describing. We have not looked at these, which would be an intriguing thing to do. Thus, metabolism in specific cells and organs is often a powerful determinant of steroid access to receptors. These are dynamic systems. So there is much to glucocorticoid effects in brain and elsewhere than merely point to measures of plasma, salivary or urinary levels.

E.R. de Kloet, M.S. Oitzl & E. Vermetten (Eds.)
Progress in Brain Research, Vol. 167
ISSN 0079-6123

CHAPTER 3

Amygdala modulation of memory-related processes in the hippocampus: potential relevance to PTSD

M.M. Tsoory, R.M. Vouimba, I. Akirav, A. Kavushansky, A. Avital and G. Richter-Levin*

Department of Psychology and the Brain and Behavior Research Center, University of Haifa, Haifa 31905, Israel

Abstract: A key assumption in the study of stress-induced cognitive and neurobiological modifications is that alterations in hippocampal functioning after stress are due to an excessive activity exerted by the amygdala on the hippocampus. Research so far focused on stress-induced impairment of hippocampal plasticity and memory but an exposure to stress may simultaneously also result in strong emotional memories. In fact, under normal conditions emotionally charged events are better remembered compared with neutral ones. Results indicate that under these conditions there is an increase in activity within the amygdala that may lead to memory of a different quality. Studying the way emotionality activates the amygdala and the functional impact of this activation we found that the amygdala modulates memory-related processes in other brain areas, such as the hippocampus. However, this modulation is complex, involving both enhancing and suppressing effects, depending on the way the amygdala is activated and the hippocampal subregion examined. The current review summarizes our findings and attempts to put them in context with the impact of an exposure to a traumatic experience, in which there is a mixture of a strong memory of some aspects of the experience but impaired memory of other aspects of that experience. Toward that end, we have recently developed an animal model for the induction of predisposition to stress-related disorders, focusing on the consequences of exposure to stressors during juvenility on the ability to cope with stress in adulthood. Exposing juvenile-stressed rats to an additional stressful challenge in adulthood revealed their impairment to cope with stress and resulted in significant elevation of the amygdala. Interestingly, and similar to our electrophysiological findings, differential effects were observed between the impact of the emotional challenge on CA1 and dentate gyrus subregions of the hippocampus. Taken together, the results indicate that long-term alterations within the amygdala contribute to stress-related mnemonic symptoms and suggest that elucidating further these intra-amygdala alterations and their effects on modulating other brain regions is likely to be beneficial for the development of novel approaches to treat stress-related disorders.

Keywords: amygdala; animal-model; anxiety; hippocampus; juvenile-stress; LTP; PTSD; stress

*Corresponding author. Tel.: +972-4-824-0962;
Fax: +972-4-828-8578; E-mail: gal.r-l@psy.haifa.ac.il

DOI: 10.1016/S0079-6123(07)67003-4

Amygdala modulation of synaptic plasticity in the hippocampus

The relationship between "stress" and "memory" is often conceived as one of "stress impairs memory." Indeed numerous findings support this concept. For example, exposing rats to a cat impaired spatial working memory in a Morris Water Maze task (Diamond et al., 1999), which depends on the integrity of the hippocampus (Morris et al., 1982). This "predator stress" procedure also impaired the induction of long-term potentiation (LTP) — a synaptic model of memory — in the hippocampus (Mesches et al., 1999; Vouimba et al., 2006). Stress-induced impairment of hippocampal-dependent memory and LTP has been observed for various stress procedure including tail-shock (Foy et al., 1987; Diamond and Rose, 1994; Garcia et al., 1997), forced exposure to brightly lit room (Xu et al., 1997), and platform stress (Maroun and Richter-Levin, 2003).

Since LTP is a model for activity-dependent plasticity assumed to be related to the formation of memories, these findings suggest that stress impede hippocampal-dependent learning and memory-related processes by disrupting plasticity in the hippocampus.

We have hypothesized that exposure to stressors impairs hippocampal functioning via the activation of the basolateral amygdala (BLA).

Stress increase BLA activity and synaptic plasticity

Examining the involvement of the BLA in stress modulation of learning and memory processes, we first evaluated the effects of exposure to stressors on BLA activity. We found that BLA response to the entorhinal cortex (EC) stimulation (Yaniv et al., 2000, 2003) was increased following exposure to a platform stress (Kavushansky and Richter-Levin, 2006). Moreover, injecting corticosterone (CORT) yielded a similar dose-dependent effect (Kavushansky and Richter-Levin, 2006). In addition, we have shown that platform stress enhanced amygdala synaptic plasticity (Vouimba et al., 2004), a finding also reported by others in both humans and rodents (Schaefer et al., 2002; Correll et al., 2005; McGaugh, 2005). Together, these findings suggest that stressful experiences indeed increase activity in the BLA, enabling it to disrupt hippocampal functioning.

Bidirectional effects of "stress" on memory-related processes: the involvement of the BLA

In contrast to the prevailing "stress impairs memory" concept there are several observations that suggest that "stress" or "emotionality" do not always impair memory formation, but rather they can also enhance hippocampal LTP and memory (Sapolsky, 2003; Kavushansky et al., 2006) and in some cases may drastically enhance some aspects of memory formation, as in the case of traumatic memories which haunt patients suffering from post-traumatic stress disorder (PTSD) (Van der Kolk and Fisler, 1995; Bower and Sivers, 1998).

Furthermore, studies suggest that the amygdala may also mediate stress-related enhancement of hippocampal memory processes and LTP (Richter-Levin and Akirav, 2003; Kim et al., 2001; McGaugh, 2002). Thus, the BLA may play a key role in both the impairing and enhancing effects of stress on hippocampal functioning (Liang et al., 1994; Akirav and Richter-Levin, 1999a, b; Kim et al., 2001) through a differential activation. Supporting this stance, we have found that stress effects on the BLA are not uniform, but may depend on stress characteristics, such as intensity, valence, duration, and controllability. For instance, we have shown (Fig. 1) that rats trained under "high-stress" conditions (cold water, 19°C) learnt faster to find the hidden platform in the Morris Water Maze than rats trained under "low-stress" conditions (warm water, 25°C) (Akirav et al., 2001). Moreover, in comparison with naïve rats, only rats that were trained under "high-stress" conditions exhibited significant increased extracellular signal-regulated kinases (ERK2) phosphorylation, indicative of activating mitogen-activated protein kinase (MAPK) signaling cascades in the BLA. No significant activation was evident among rats trained under "low-stress" conditions, or rats that did not learn the task well under "high-stress"conditions, nor among rats

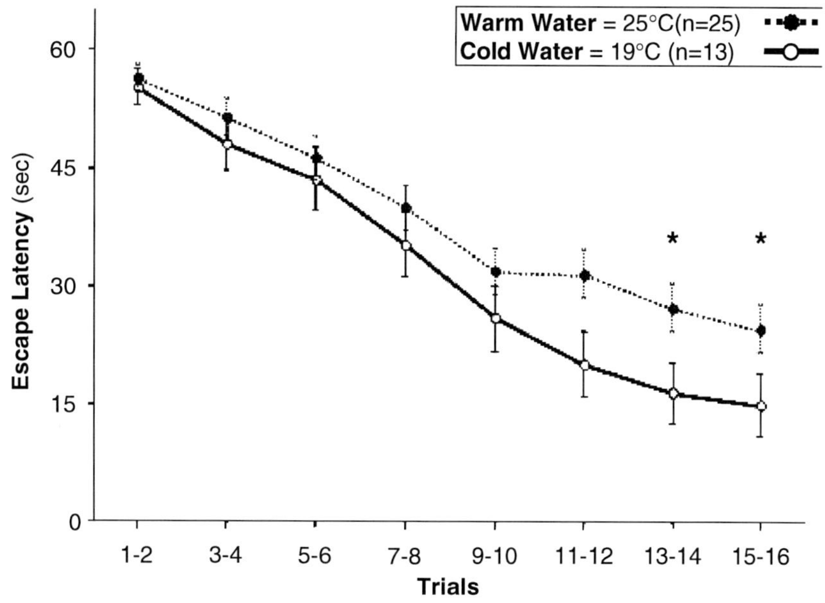

Fig. 1. Water temperature modulates stress levels and affects learning the Morris Water Maze. Rats trained under "high stress" conditions (cold water, 19°C) learnt faster to find the hidden platform in the Morris Water Maze than rats trained under "low stress" conditions (warm water, 25°C) (* $= p < 0.05$).

that had no platform to learn to find under "high-stress" conditions (Akirav et al., 2001). In the cornu ammonis field 1 (CA1), training was accompanied by increased phosphorylation of ERK2 only in animals that have acquired the task (irrespective of whether they were trained in cold or warm water). Thus, it is likely that the activation of the amygdala (as seen by the activation of ERK2) following an emotionally charged hippocampal-dependent learning experience led to the better performance of the cold water trained rats in the spatial task (Akirav et al., 2001).

Further investigating the involvement of the BLA in modulating learning processes, we have shown that while an acute exposure to platform stress facilitated LTP in the BLA, a repeated exposure suppressed long-lasting LTP in the BLA (Vouimba et al., 2004). Such changes were associated with normal or enhanced LTP in the hippocampal dentate gyrus (DG) for acute stress exposure (Vouimba et al., 2004; Kavushansky et al., 2006) and impaired DG LTP for repeated stress (Vouimba et al., 2004).

Thus, alteration of hippocampal functioning consecutive to stressful experiences may involve differential changes in the BLA activity and/or synaptic plasticity.

BLA influences memory by tagging important information

Both human and animal studies indicate that emotionality-induced enhanced memory formation involves the activation of the amygdala, but how may the BLA influence memory consolidation-related processes in the hippocampus remains to be studied.

The hippocampus, being involved in the transformation of short- into long-term memories should be able to sort out the more significant from the less relevant aspects of an experience in order to transform only the former into long-term memory. One mechanism that could contribute to this selection is the emotional significance of the experience. Emotionally significant events are likely to be important to remember and their emotional load could mark them as important — a function that we have termed "Emotional Tagging" (Richter-Levin and Akirav, 2003).

According to this proposed "Emotional Tagging" mechanism the activation of the amygdala in emotionally arousing events marks the experience as important and aids in enhancing synaptic plasticity in other brain regions (Akirav and Richter-Levin, 2002; Richter-Levin and Akirav, 2003). We have also proposed a potential neural mechanism that may underlie "Emotional Tagging." Long-term memory formation is considered to involve lasting alterations in synaptic efficacy, known as synaptic plasticity. Two factors were suggested as crucial for obtaining a synapse-specific long-term plasticity: (1) the successful activation of a synapse-specific, protein synthesis-independent tag (Frey and Morris, 1998) and (2) the activation of synapse-nonspecific protein synthesis (Matthies et al., 1990; Pittenger and Kandel, 1998). The activation of protein synthesis can then induce lasting plasticity only in those synapses marked by a tag. Interestingly and relevant to the "Emotional Tagging" hypothesis, Frey et al. (2001) demonstrated that the activation of the amygdala could transform "transient" (early-LTP) into "long-lasting" (late-LTP) plasticity. Thus, it seems reasonable to assume that the activation of the amygdala triggers neuromodulatory systems, which in turn reduce the threshold for the activation of the "Synaptic Tag," facilitating the transformation of early- into late-phase memory (Richter-Levin and Akirav, 2003).

BLA modulates hippocampal LTP

To further test this "Emotional Tagging" hypothesis we examined whether BLA activation can affect memory-related processes (LTP induction) in the hippocampus.

Priming the BLA before the induction of LTP in the DG by stimulating the perforant path (PP) enhanced DG LTP (Akirav and Richter-Levin, 1999b, 2002; Vouimba and Richter-Levin, 2005). Furthermore, this effect was found to be mediated by CORT and norepinephrine (NE), when administered either systemically or directly in the BLA (Akirav and Richter-Levin, 2002; Vouimba et al., 2007). Similar effects of BLA priming on DG LTP were reported by Ikegaya et al. (1995) demonstrating that BLA activation reduces the threshold for the induction of DG LTP.

Differential outcome of "stress" or BLA activation within the hippocampus: CA1 vs. DG

Taken together, these findings suggested us that "stress" affects hippocampal functioning by activating the BLA. However, we noted that while BLA activation was found to enhance LTP in the DG, "stress" was found to attenuate the induction of LTP at least in the CA1 (Maroun and Richter-Levin, 2003).

One possibility that could explain this discrepancy could be that, this premise needs some adjustments, i.e., that both "stress" and BLA activation affect the hippocampus but each in a different manner, the former impairing while the latter enhancing memory-related processes. Alternatively, the different effects of "stress" and BLA activation may derive from differential effects of both on the CA1 vs. the DG subregions of the hippocampus (Fig. 2).

The latter proposal warrants examination since the majority of research on the effects of "stress" on hippocampal functioning focused on the CA1 (e.g., Foy et al., 1987; Diamond and Rose, 1994; Maroun and Richter-Levin, 2003), while research on the effects of BLA activation on the hippocampus functioning has focused mainly on the DG subregion (e.g., Ikegaya et al., 1995; Akirav and Richter-Levin, 1999a, b, 2002; Frey et al., 2003).

Indeed, both "stress" and BLA activation impaired CA1 LTP (Vouimba and Richter-Levin, 2005; Kavushansky et al., 2006; Vouimba et al., 2006). However, while DG LTP enhancement appear to depend on CORT or NE transmission in the BLA, the attenuation of the CA1 LTP appear to be independent of these stress hormones (Vouimba et al., 2007). A similar pattern of differential outcomes between these subregions of the hippocampus was found for "stress." Exposure to uncontrollable swim stress enhanced DG LTP induction but impaired LTP in the CA1. Furthermore, these effects were moderated when controllable swim stress, presumed to represent lower

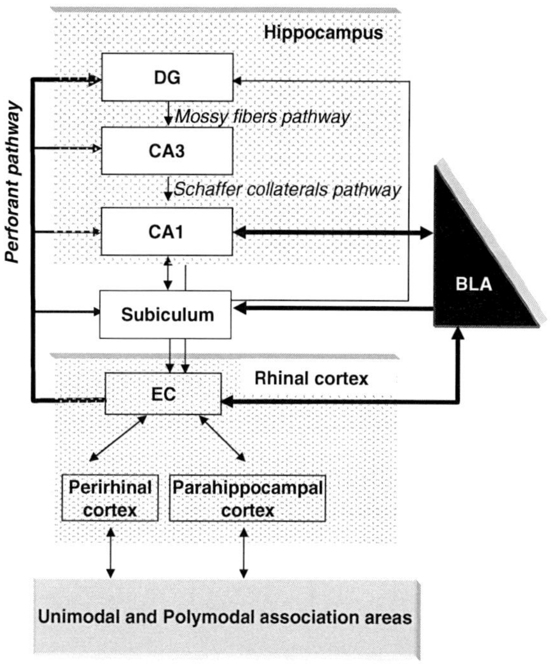

Fig. 2. A schematic diagram of hippocampal formation–amygdala connections. Unimodal and polymodal inputs from association areas reach the hippocampus (HPC) via the rhinal cortical regions (entorhinal, perirhinal, and parahippocampal cortices). Information flows from the entorhinal cortex (EC) to the HPC via the *perforant pathway*, the main afferent pathway to the HPC. Within the HPC, information coming into the dentate gyrus (DG) is processed and sent via the *mossy fiber pathway* to the CA3 subfield, which connects with the CA1 via the *Schaffer collateral pathway*. The CA1 sends significant output to the subiculum and both CA1 and subiculum project to the EC. The basolateral nucleus of the amygdala (BLA) sends monosynaptic projections to the CA1, the subiculum, and the EC. Given no monosynaptic interconnections between the BLA and DG, the BLA may modulate DG polysynaptically, i.e., via the EC or the subiculum.

stress levels, was applied (Kavushansky et al., 2006).

To summarize so far, both exposure to "stress" and BLA activation produced similar effects in the hippocampus, suppressing LTP in the CA1 while enhancing LTP induction in the DG. These results support the proposal that the effects of "stress" on hippocampal functioning are mediated to a large extent by the BLA differential modulation of hippocampal subregions. In support of this idea were the findings that exposure to "stress" activated the BLA and that lesion of the BLA suppressed the "stress" effects on CA1 LTP as well as on learning and memory (Akirav et al., 2001; Kim et al., 2001).

This novel realization that "stress" and amygdala activation exert a complex mixture of enhancing and suppressing effects on memory-related processes in different brain areas may suggest the following: under normal conditions emotionality may dictate the remembering of certain "important" (emotionally loaded) features of an event. However, if the same event is experienced under lower or higher levels of emotionality, altered memories may be formed, rendering certain features of the event as irrelevant while others as "unforgettable."

Post-traumatic stress disorder (PTSD)

Concomitant "stress" related impairment and enhancement of memories

Considering simultaneous opposite effects on memory processes brings to mind the PTSD, where the intense stress brought upon by the traumatic event confers a mixture of enhancing and suppressing effects on memory-related processes. On the one hand, as the diagnostic and statistical manual of mental disorders, 4th edition (DSM-IV) PTSD diagnosis requires, the "reexperiencing symptoms" like intrusive memories, recurrent dreams, flashbacks, and intense reactions in similar events (APA, 1994) indicate that there is enhancement of memories of certain features of the traumatic event. On the other hand, extensive research indicates that PTSD patients suffer also from impaired recall capacities that were related to altered hippocampal functioning (for review, see Nemeroff et al., 2006). We therefore speculated that the BLA response to stressful events, and its differential effects on the hippocampal subregions, may be of relevance to our understanding of the neurobiology of PTSD. We wanted to further examine this hypothesis but for that an animal model for PTSD was required. Unfortunately however, though some models were suggested, no animal model has gained a widespread consensus as a valid and suitable model for PTSD.

Most attempts to develop such a model of PTSD dealt with the question of what kind of a stress protocol should be employed; some studies dealt with the question of when a stressor becomes traumatic. For example, Cordero et al. (2002) proposed that, an electric foot shock is stressful at the intensity of 0.5 mA but becomes traumatic at the intensity of 1.0 mA. Others proposed employing ethologically relevant stressors, which might represent a relevant traumatic event (Cohen et al., 2003, 2006; Woodson et al., 2003; Adamec et al., 2004, 2006; Cohen and Zohar, 2004).

A novel animal model for PTSD

We were concerned with a different aspect of the PTSD phenomenon — the question of why do some people develop PTSD following a traumatic event while others do not. Clearly, understanding what underlies the predisposition to develop PTSD would be instrumental to our understanding of the disorder. Furthermore, it would help producing "affected" animals that could then be utilized in promoting the research into the neurobiology of the disorder.

Among the suggested factors that might contribute to the vulnerability to develop PTSD is the exposure to stressful events early in life. Early-life stress (ELS) is considered a significant risk factor, predisposing people to mal-adaptively respond to traumatic events later in life. Reports on ELS are most prevalent among those individuals who develop PTSD following a traumatic event and significantly less prevalent among those who did not develop the disorder (Nemeroff et al., 2006). To mimic these conditions our model consists of an exposure to "stress" early in life and a subsequent exposure to stress in adulthood.

A unique feature of our model is the age of the early exposure to stress. While most ELS rodent models focus on the perinatal to pre-weaning periods and involve some form of maternal deprivation or separation (for review, see Sanchez et al., 2001), we focused on another ELS-sensitive period in the rat ontogeny, namely the juvenile stage (~28 days), the earlier phase of the adolescent/post-weaning to pre-pubertal period (Avital and Richter-Levin, 2005; Avital et al., 2006; Tsoory and Richter-Levin, 2006; Tsoory et al., 2007a, b). During the adolescent period (21–42 days), substantial maturational processes occur in the rat limbic system, including in the hippocampus and amygdala-based neurocircuits (for review, see Spear, 2000).

During the juvenile period the hypothalamic-pituitary-adrenal (HPA) axis response reaches its developmental asymptote (Vazquez, 1998); however this response lasts considerably longer than in adults (Vazquez, 1998; Romeo et al., 2004). Romeo et al. (2004) suggested that this slower shutoff of the HPA axis during juvenility may derive from less centrally mediated feedback from various underdeveloped forebrain limbic regions at this age. Indeed, exposure to stressors during juvenility was reported to produce more pronounced effects than exposure at earlier or later ages, affecting object exploration in adulthood (Einon and Morgan, 1977), fluid intake (McGivern et al., 1996), and adulthood social and nonsocial behaviors associated with disregulation of endogenous opioid system development (Van den Berg et al., 1999a–c, 2000). Adult rats chronically exposed to variable stressors throughout juvenility had an enhanced acoustic startle response similar to patients with PTSD (Maslova et al., 2002).

Exposure to acute stressors during juvenility produces increased vulnerability to stressful events in adulthood (60 days of age), resulting in an augmented response to adverse experiences. Adult rats that were exposed to stress both during juvenility and adulthood exhibited enhanced startle response and reduced exploration in a novel setting not only in comparison with naïve unexposed rats but also in comparison with rats that were exposed to stress only during juvenility and in comparison with those exposed to stress only in adulthood (Avital and Richter-Levin, 2005). A similar pattern of effects was also evident at the age of 90 days (Fig. 3).

We have proceeded to compare the effects of recurrent exposure to stress during juvenility and in adulthood, with recurrent exposure to stress in adulthood. Indeed, adult rats exposed to stressors

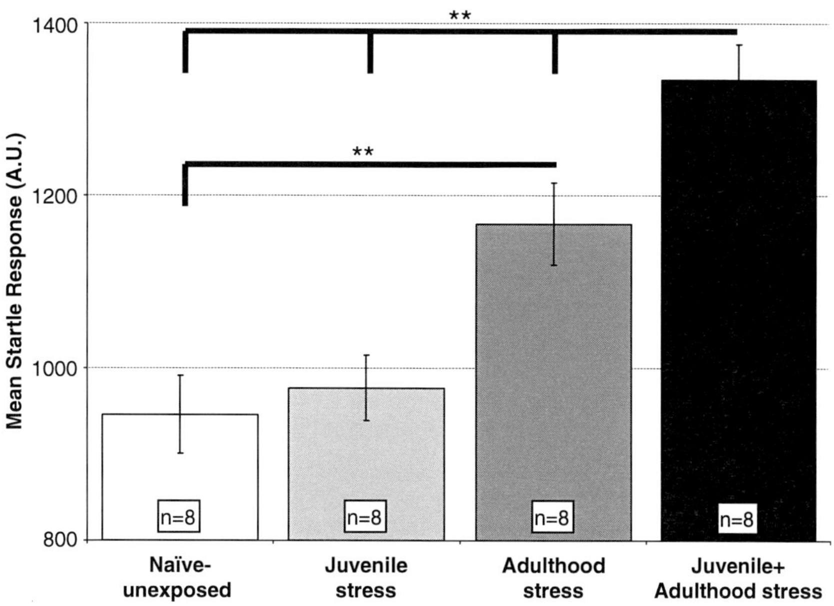

Fig. 3. The effect of exposure to stressors during juvenility, adulthood, or their recurrent combination on startle responses at the age of 90 days. Rats exposed to stress during juvenility and adulthood exhibited a significant higher startle response compared to all other groups. Rats exposed to stress only in adulthood had a significant increased startle response only in comparison with the naïve unexposed group (** $= p < 0.01$).

during juvenility and adulthood exhibited increased startle responses significantly greater than those of rats that were exposed to the stressors twice in adulthood (Avital and Richter-Levin, 2005). Rats exposed to stress twice in adulthood exhibited startle responses that did not differ from those of rats exposed to stress only once in adulthood.

It is noteworthy, that further examination of the model indicated that the effects of exposure to juvenile and adulthood stress are long-term (Avital and Richter-Levin, 2005). Comparing the startle responses of 60- and 80-days-old rats exposed either to stress only during juvenility, only in adulthood (59 days of age), or that underwent recurrent exposure to stress during juvenility and in adulthood, revealed that the effect of exposure to stress in adulthood alone diminished over time whereas the effect of recurrent exposure to stress during juvenility and adulthood did not diminish over time (Fig. 4).

We have moved on to examine the effects of a short-term juvenile exposure to variable stressors on adulthood coping responses by using stressful challenges, namely, novel-setting exploration and two-way shuttle avoidance learning. We chose to utilize the two-way shuttle avoidance task since learning and performance in this task are dependent on the hippocampus (Schwegler et al., 1981; Becker et al., 1997) and the amygdala (Savonenko et al., 2003) and poor two-way shuttle avoidance performance was observed following both negligible and high doses of injected CORT (Kademian et al., 2005). Selective breeding based on "high/low-avoidance" performance was suggested to relate to differences in "emotional" factors (state/trait anxiety) that influenced performance (Brush, 2003). Indeed, exposure to stressors during juvenility significantly reduced adulthood exploratory behavior and impaired adulthood learning under stressful conditions. It reduced the rates of avoidance responses and increased the rates of escape failures while learning the two-way shuttle avoidance task (Tsoory et al., 2007a). Furthermore, the effects of exposure to stressors during juvenility (27–29 days) were found to be stronger than those of exposure to the same stress protocol during

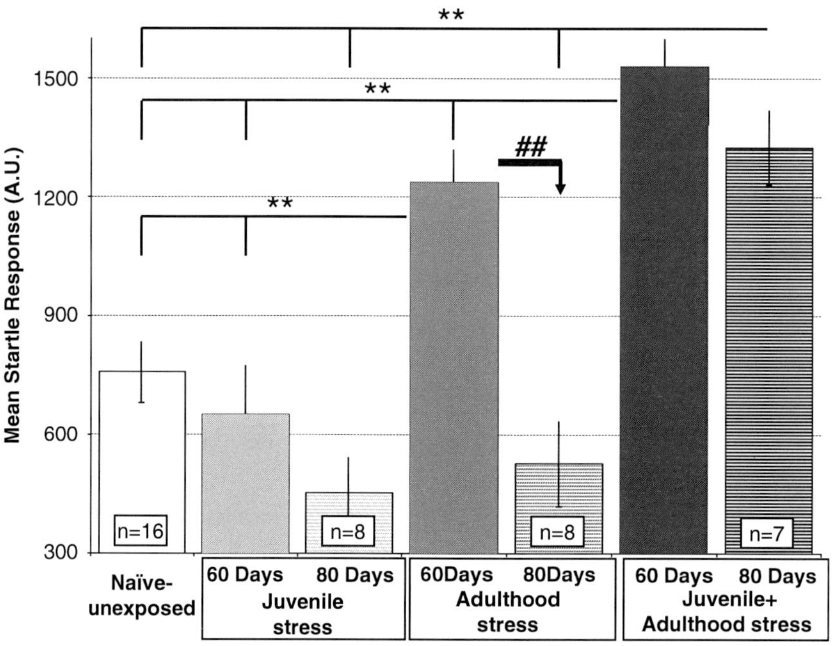

Fig. 4. The effect of exposure to stressors during juvenility, adulthood, or their recurrent combination on startle responses at the ages of 60 and 80 days. Recurrent exposure to stress (juvenility and adulthood) resulted in a significant higher startle response compared to all other groups at both 60 and 80 days ($^{**} = p<0.01$). The effect of exposure to stress only in adulthood diminished over time ($^{\#\#} = p<0.01$), while the effects of recurrent exposure did not.

mid-adolescence (33–35 days), indicating that within the post-weaning to pre-pubertal period the juvenile age (~28 days) is a stress-sensitive period (Tsoory and Richter-Levin, 2006).

Taken together, these results indicated "juvenility" (~28 days of age) as a stress-sensitive developmental period. Furthermore, the results strongly support the notion that "juvenile stress" may model the predisposing effect of ELS on stress responses later in life, which are related to PTSD (Appendix).

"Juvenile stress" affects neural cell adhesion molecules

Once a model was established, it could be utilized to study the neural consequences and correlates of an exposure to a trauma inducing experience. Among the candidate molecules to be studied in relation to pathological plasticity is the family of the neural cell adhesion molecules (NCAMs).

NCAMs are membrane-bound glycoproteins of the immunoglobulin superfamily of adhesion molecules, which mediate cell–cell interactions; by interacting with cytoskeletal components, they can activate specific intracellular signaling pathways (Cremer et al., 1997). The NCAM, polysialylated-NCAM (PSA-NCAM), and cell adhesion molecule L1 (CAM-L1) of this family play a pivotal role in neural development and regeneration, and are strongly implicated in synaptic plasticity and memory formation processes (Schachner, 1997; Kamiguchi, et al., 1998; Kiss et al., 2001; Welzl and Stork, 2003; Sandi, 2004; Gerrow and El-Husseini, 2006). The post-translational polysialylation of NCAM weakens its adhesive properties (Schachner, 1997) and thus it was suggested that PSA-NCAM acts as a plasticity promoter by decreasing overall cell adhesion, thereby allowing structural remodeling to occur (Rutishauser and Landmesser, 1996), whereas NCAM acts as a stability promoter (Ronn et al., 2000).

Alterations in the relative expression of PSA-NCAM to NCAM and in the expression of CAM-L1 were associated with development-related alterations (Edelman, 1984; Rutishauser and Jessell, 1988; Rutishauser, 1989; Kamiguchi, et al., 1998; Gerrow and El-Husseini, 2006). Alterations in the expression of these molecules were also found following learning, memory formation, and activity-dependent synaptic remodeling (Doyle et al., 1992a, b; Luthi et al., 1994; Ronn et al., 2000; Law et al., 2003; Welzl and Stork, 2003; Sandi, 2004).

A series of studies showed that chronic stress protocols known to produce cognitive and neural alterations (chronic restraint stress: 21 days × 6 h) markedly affected the expression of NCAMs, overall decreasing the expression of NCAM while increasing those of PSA-NCAM and CAM-L1 in the hippocampus and other brain areas (for review, see Sandi, 2004).

We have recently started to characterize the effects of exposure to stressors during juvenility on the expression levels of the NCAMs: NCAM, its polysialylated form PSA-NCAM, their expression ratio [PSA-NCAM/(NCAM + PSA-NCAM)], and CAM-L1 within the limbic system. Overall the results indicated that exposure to stressors during juvenility disrupts development-related alterations in the expression ratio of PSA-NCAM to NCAM in the BLA, CA1, DG, and EC (Tsoory et al., 2007b). It is noteworthy that differential effects were found between the CA1 and DG subregions of the hippocampus reminding us of the differences found between the effects of BLA activation on the DG and CA1 (Vouimba et al., 2007).

When the effects were examined soon after the "juvenile" stress exposure, 4 days following the last exposure to a stressor (at the age of 33 days) the expression ratio of PSA-NCAM to NCAM was increased among juvenile stressed rats in comparison with juvenile naïve rats in the CA1 but not in the DG (Fig. 5).

Similar differential effects were observed in adulthood for the expression of CAM-L1. Adult rats that were exposed to stress during juvenility exhibited increased levels of CAM-L1 expression in the CA1 compared with rats exposed to stress during juvenility and in adulthood. No such difference was found in the DG (Fig. 6).

Summary

Contrary to the prevailing concepts of "stress impairs memory-related processes" on the one hand and "stress promotes memory-related processes" on the other, our results indicate that an integrated view should be considered. The exposure to an emotionally or stressful experience modulates memory formation in a complex manner.

Some brain areas become more likely to process memories of certain aspects of the experience while memory formation in other brain areas may be suppressed. The result may not necessarily be "more" or "less" memory, but rather an "altered" memory. These alterations may relate to a range of features in the memory of the event, varying from differences with respect to which aspects of the experience are remembered, to how detailed the memory formed will be or how intense it will be, but also which brain areas will be recruited for its formation, maintenance, and recall.

In that respect, we were able to demonstrate that the amygdala, when activated, can concomitantly exert a mixture of plasticity-supporting and plasticity-suppressing influences. These are likely to contribute to the modification of the characteristics of the memory formed under emotional or stressful conditions.

It is easy to foresee that if amygdala functioning is altered due to a traumatic experience this would lead neither to the suppression of the traumatic event memory, nor to its enhancement, but rather to a complex mixture of both.

The "juvenile stress" protocol is suggested as an effective model for the induction of a predisposition and susceptibility to develop stress-related disorders in adulthood such as PTSD and post-traumatic depression.

The model can now be utilized to study how amygdala functioning is being modified following an exposure to a traumatic experience and how this affects the way the amygdala modulates memory-related processes in other brain regions.

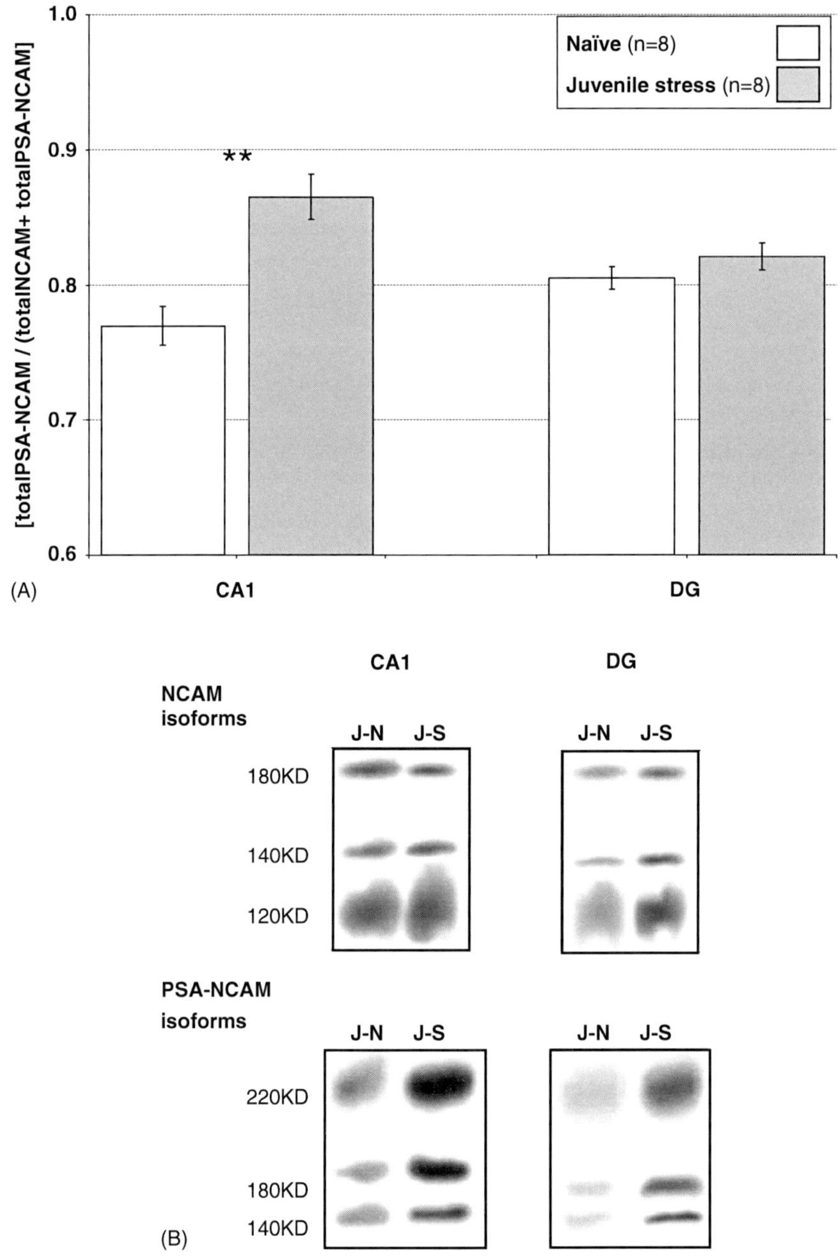

Fig. 5. The effect of exposure to stressors during juvenility on PSA-NCAM to NCAM expression ratio in the CA1 and DG subregions of the hippocampus 4 days after the exposure to "juvenile" stress. (A) Exposure to the "juvenile" stress protocol significantly increased the expression ratio of PSA-NCAM to NCAM in the CA1 but not in the DG (** $= p<0.01$). (B) The insets depict representative bands of NCAM isoforms (120, 140, 180 kDa); *upper panel*: PSA-NCAM isoforms (140, 180, 220 kDa); *lower panel*: across the J-N vs. J-S groups in the CA1 and DG.

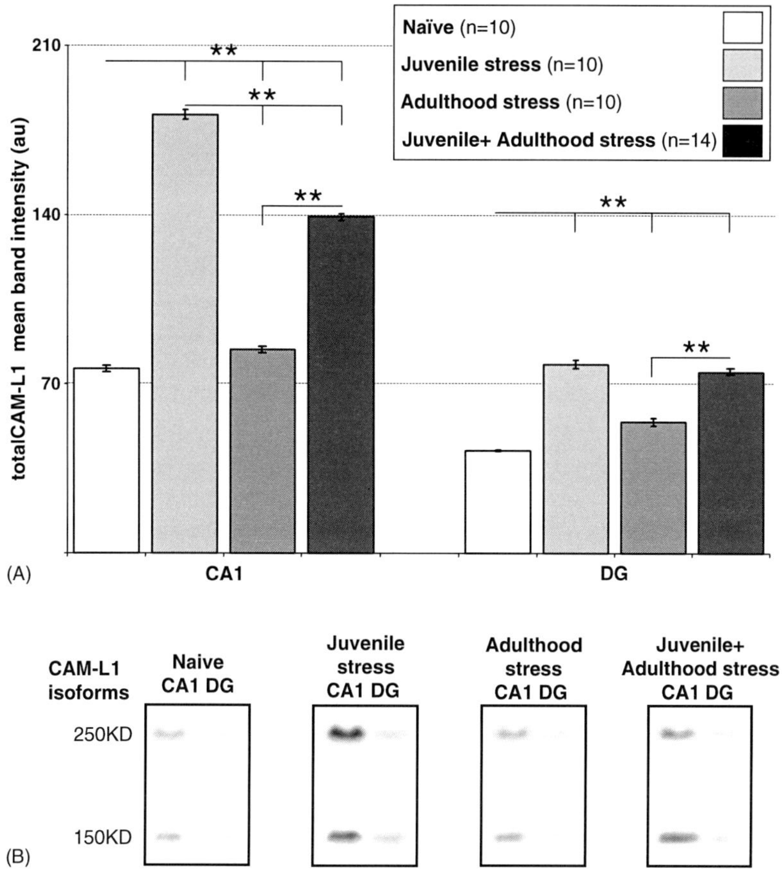

Fig. 6. The effects of exposure to stressors during juvenility, adulthood, or their recurrent combination on CAM-L1 expression in the CA1 and DG subregions of the hippocampus. (A) ANOVA of CAM-L1 expression levels across groups within each brain region indicated a significant main effect for groups in both the CA1 and DG ($p<0.01$). Significant post-hoc differences ($p<0.01$) between naïve rats and rats exposed to stress only in adulthood indicated that acute stress in adulthood significantly increased CAM-L1 expression levels in both CA1 and DG. Significant post-hoc differences between rats exposed only to juvenile stress and rats exposed to both juvenile and adulthood stress indicated that acute adulthood stress that followed a juvenile exposure to stress did not add to the juvenile stress-induced increase in the DG, but attenuated it in the CA1. (** – $p<0.01$). (B) The insets depict representative bands of CAM-L1 isoforms (150, 250 kDa) across groups in the CA1 and DG.

Appendix

Box 1: the "juvenile stress" model

On the basis of the observations in humans indicating early-life stress exposure as a significant risk factor for the emergence and persistence of PTSD (Nemeroff et al., 2006), our "juvenile stress" model consists of: (1) exposure to stressors early in life, during juvenility (~28 days of age) and (2) a subsequent exposure to a stressful challenge in adulthood (60 days of age at the earliest).

Juvenile stress protocols

We utilized either a repeated exposure to the "platform stress" — i.e., at the ages of 26–28 days rats were placed on an elevated platform for 30 min, three times a day; inter-trial interval (ITI), 60 min in home cage (Avital and Richter-Levin, 2005) — or a variable exposure, i.e., different inescapable stressors at the ages of 27–29 days: DAY1, forced swimming; DAY2, "platform stress;" DAY3, restraint or a short electric foot shock session (Tsoory and Richter-Levin, 2006; Tsoory et al., 2007a, b).

Adulthood stress protocols and behavioral assessments

We utilized in some studies a subsequent exposure to the "platform stress" (at either ~60 and/or 90 days of age), which was followed by the Open Field test, the Morris Water Maze task (spaced or massed training), and Acoustic Startle Response test (Avital and Richter-Levin, 2005). In other studies we employed the "two-way shuttle avoidance task" at 9 weeks of age (10 min free exploration in the apparatus; then one session comprising 100 trace conditioning trials); this task enabled us to simultaneously challenge the rats while assessing their ability to cope with learning under stressful conditions (Tsoory and Richter-Levin, 2006; Tsoory et al., 2007a, b).

The long-term consequences of exposure to "juvenile stress"

Adult rats exposed to stressors during juvenility and to a subsequent challenge in adulthood exhibited reduced exploration and increased avoidance from entering the arena's center in the Open Field, altered learning of the Morris Water Maze, and increased acoustic startle response (Avital and Richter-Levin, 2005). When challenged in the two-way shuttle avoidance task two "profiles" of altered coping with stress in adulthood were evident among adult juvenile stressed rats. (1) Anxious profile: low novel setting exploration, low rates of avoidance shuttles, moderate rates of escape shuttles, and low rates of escape failures; comprising ~40% of these rats. (2) Depressive profile: low novel setting exploration, low rates of avoidance shuttles, moderate rates of escape shuttles, and high rates of escape failures; comprising about a third of these rats. Less than a third of these rats appeared "unaffected" (Tsoory and Richter-Levin, 2006; Tsoory et al., 2007a). It is noteworthy that similar profiles and rates were evident when exposure to a predator scent was used as a stressor during juvenility and adulthood and behavioral profiling was based on altered behaviors in the elevated plus maze and startle responses (Tsoory et al., 2007a).

A substantial increase was observed in the expression ratio of PSA-NCAM to NCAM among adult juvenile stressed rats compared to adult juvenile stress-free rats in the BLA, CA1, DG, and EC (Tsoory et al., 2007b).

Abbreviations

BLA	basolateral amygdala
CA1	cornu ammonis field 1
CAM-L1	cell adhesion molecule L1
CORT	corticosterone
DG	dentate gyrus
DSM-IV	diagnostic and statistical manual of mental disorders, 4th edition
EC	entorhinal cortex
ELS	early-life stress
ERK2	extracellular signal-regulated kinases
HPA	hypothalamic-pituitary-adrenal
ITI	inter-trial interval
LTP	long-term potentiation
MAPK	mitogen-activated protein kinase
NCAM	neural cell adhesion molecule
NE	norepinephrine
PP	perforant path
PSA-NCAM	polysialylated neural cell adhesion molecule
PTSD	post-traumatic stress disorder

Acknowledgments

The first two authors contributed equally. This work was supported by a 2002 NARSAD Independent Investigator award to G.R.-L. and by the EU's PROMEMORIA grant number 512012 to G.R.-L.

References

Adamec, R., Head, D., Blundell, J., Burton, P. and Berton, O. (2006) Lasting anxiogenic effects of feline predator stress in mice: sex differences in vulnerability to stress and predicting severity of anxiogenic response from the stress experience. Physiol. Behav., 88(1–2): 12–29.

Adamec, R., Walling, S. and Burton, P. (2004) Long-lasting, selective, anxiogenic effects of feline predator stress in mice. Physiol. Behav., 83(3): 401–410.

Akirav, I. and Richter-Levin, G. (1999a) Biphasic modulation of hippocampal plasticity by behavioral stress and basolateral amygdala stimulation in the rat. J. Neurosci., 19(23): 10530–10535.

Akirav, I. and Richter-Levin, G. (1999b) Priming stimulation in the basolateral amygdala modulates synaptic plasticity in the rat dentate gyrus. Neurosci. Lett., 270(2): 83–86.

Akirav, I. and Richter-Levin, G. (2002) Mechanisms of amygdala modulation of hippocampal plasticity. J. Neurosci., 22(22): 9912–9921.

Akirav, I., Sandi, C. and Richter-Levin, G. (2001) Differential activation of hippocampus and amygdala following spatial learning under stress. Eur. J. Neurosci., 14(4): 719–725.

APA. (1994) Diagnostic and statistical manual of mental disorders (DSM-IV). American Psychiatric Association, Washington, DC.

Avital, A., Ram, E., Maayan, R., Weizman, A. and Richter-Levin, G. (2006) Effects of early-life stress on behavior and neurosteroid levels in the rat hypothalamus and entorhinal cortex. Brain Res. Bull., 68(6): 419–424.

Avital, A. and Richter-Levin, G. (2005) Exposure to juvenile stress exacerbates the behavioural consequences of exposure to stress in the adult rat. Int. J. Neuropsychopharmacol., 8(2): 163–173.

Becker, A., Letzel, K., Letzel, U. and Grecksch, G. (1997) Kindling of the dorsal and the ventral hippocampus: effects on learning performance in rats. Physiol. Behav., 62(6): 1265–1271.

Bower, G.H. and Sivers, H. (1998) Cognitive impact of traumatic events. Dev. Psychopathol., 10(4): 625–653.

Brush, F.R. (2003) Selection for differences in avoidance learning: the Syracuse strains differ in anxiety, not learning ability. Behav. Genet., 33(6): 677–696.

Cohen, H., Matar, M.A., Richter-Levin, G. and Zohar, J. (2006) The contribution of an animal model toward uncovering biological risk factors for PTSD. Ann. N.Y. Acad. Sci., 1071: 335–350.

Cohen, H. and Zohar, J. (2004) An animal model of posttraumatic stress disorder: the use of cut-off behavioral criteria. Ann. N.Y. Acad. Sci., 1032: 167–178.

Cohen, H., Zohar, J. and Matar, M. (2003) The relevance of differential response to trauma in an animal model of posttraumatic stress disorder. Biol. Psychiatry, 53(6): 463–473.

Cordero, M.I., Kruyt, N.D., Merino, J.J. and Sandi, C. (2002) Glucocorticoid involvement in memory formation in a rat model for traumatic memory. Stress, 5(1): 73–79.

Correll, C.M., Rosenkranz, J.A. and Grace, A.A. (2005) Chronic cold stress alters prefrontal cortical modulation of amygdala neuronal activity in rats. Biol. Psychiatry, 58(5): 382–391.

Cremer, H., Chazal, G., Goridis, C. and Represa, A. (1997) NCAM is essential for axonal growth and fasciculation in the hippocampus. Mol. Cell. Neurosci., 8(5): 323–335.

Diamond, D.M., Park, C.R., Heman, K.L. and Rose, G.M. (1999) Exposing rats to a predator impairs spatial working memory in the radial arm water maze. Hippocampus, 9(5): 542–552.

Diamond, D.M. and Rose, G.M. (1994) Stress impairs LTP and hippocampal-dependent memory. Ann. N.Y. Acad. Sci., 746: 411–414.

Doyle, E., Nolan, P.M., Bell, R. and Regan, C.M. (1992a) Hippocampal NCAM180 transiently increases sialylation during the acquisition and consolidation of a passive avoidance response in the adult rat. J. Neurosci. Res., 31(3): 513–523.

Doyle, E., Nolan, P.M., Bell, R. and Regan, C.M. (1992b) Intraventricular infusions of anti-neural cell adhesion molecules in a discrete posttraining period impair consolidation of a passive avoidance response in the rat. J. Neurochem., 59(4): 1570–1573.

Edelman, G.M. (1984) Modulation of cell adhesion during induction, histogenesis, and perinatal development of the nervous system. Annu. Rev. Neurosci., 7: 339–377.

Einon, D.F. and Morgan, M.J. (1977) A critical period for social isolation in the rat. Dev. Psychobiol., 10(2): 123–132.

Foy, M.R., Stanton, M.E., Levine, S. and Thompson, R.F. (1987) Behavioral stress impairs long-term potentiation in rodent hippocampus. Behav. Neural Biol., 48(1): 138–149.

Frey, S., Bergado-Rosado, J., Seidenbecher, T., Pape, H.C. and Frey, J.U. (2001) Reinforcement of early long-term potentiation (early-LTP) in dentate gyrus by stimulation of the basolateral amygdala: heterosynaptic induction mechanisms of late-LTP. J. Neurosci., 21(10): 3697–3703.

Frey, S., Bergado, J.A. and Frey, J.U. (2003) Modulation of late phases of long-term potentiation in rat dentate gyrus by stimulation of the medial septum. Neurosci., 118(4): 1055–1062.

Frey, U. and Morris, R.G. (1998) Synaptic tagging: implications for late maintenance of hippocampal long-term potentiation. Trends Neurosci., 21(5): 181–188.

Garcia, R., Musleh, W., Tocco, G., Thompson, R.F. and Baudry, M. (1997) Time-dependent blockade of STP and

LTP in hippocampal slices following acute stress in mice. Neurosci. Lett., 233(1): 41–44.
Gerrow, K. and El-Husseini, A. (2006) Cell adhesion molecules at the synapse. Front. Biosci., 11: 2400–2419.
Ikegaya, Y., Saito, H. and Abe, K. (1995) High-frequency stimulation of the basolateral amygdala facilitates the induction of long-term potentiation in the dentate gyrus in vivo. Neurosci. Res., 22(2): 203–207.
Kademian, S.M., Bignante, A.E., Lardone, P., McEwen, B.S. and Volosin, M. (2005) Biphasic effects of adrenal steroids on learned helplessness behavior induced by inescapable shock. Neuropsychopharmacol., 30(1): 58–66.
Kamiguchi, H., Hlavin, M.L. and Lemmon, V. (1998) Role of L1 in neural development: what the knockouts tell us. Mol. Cell. Neurosci., 12(1-2): 48–55.
Kavushansky, A. and Richter-Levin, G. (2006) Effects of stress and corticosterone on activity and plasticity in the amygdala. J. Neurosci. Res., 84(7): 1580–1587.
Kavushansky, A., Vouimba, R.M., Cohen, H. and Richter-Levin, G. (2006) Activity and plasticity in the CA1, the dentate gyrus, and the amygdala following controllable vs. uncontrollable water stress. Hippocampus, 16(1): 35–42.
Kim, J.J., Lee, H.J., Han, J.S. and Packard, M.G. (2001) Amygdala is critical for stress-induced modulation of hippocampal long-term potentiation and learning. J. Neurosci., 21(14): 5222–5228.
Kiss, J.Z., Troncoso, E., Djebbara, Z., Vutskits, L. and Muller, D. (2001) The role of neural cell adhesion molecules in plasticity and repair. Brain Res. Brain Res. Rev., 36(2-3): 175–184.
van der Kolk, B.A. and Fisler, R. (1995) Dissociation and the fragmentary nature of traumatic memories: overview and exploratory study. J. Trauma. Stress, 8(4): 505–525.
Law, J.W., Lee, A.Y., Sun, M., Nikonenko, A.G., Chung, S.K., Dityatev, A., Schachner, M. and Morellini, F. (2003) Decreased anxiety, altered place learning, and increased CA1 basal excitatory synaptic transmission in mice with conditional ablation of the neural cell adhesion molecule L1. J. Neurosci., 23(32): 10419–10432.
Liang, K.C., Hon, W. and Davis, M. (1994) Pre- and post-training infusion of *N*-methyl-D-aspartate receptor antagonists into the amygdala impair memory in an inhibitory avoidance task. Behav. Neurosci., 108(2): 241–253.
Luthi, A., Laurent, J.P., Figurov, A., Muller, D. and Schachner, M. (1994) Hippocampal long-term potentiation and neural cell adhesion molecules L1 and NCAM. Nature, 372(6508): 777–779.
Maroun, M. and Richter-Levin, G. (2003) Exposure to acute stress blocks the induction of long-term potentiation of the amygdala-prefrontal cortex pathway in vivo. J. Neurosci., 23(11): 4406–4409.
Maslova, L.N., Bulygina, V.V. and Markel, A.L. (2002) Chronic stress during prepubertal development: immediate and long-lasting effects on arterial blood pressure and anxiety-related behavior. Psychoneuroendocrinol., 27(5): 549–561.
Matthies, H., Frey, U., Reymann, K., Krug, M., Jork, R. and Schroeder, H. (1990) Different mechanisms and multiple stages of LTP. Adv. Exp. Med. Biol., 268: 359–368.
McGaugh, J.L. (2002) Memory consolidation and the amygdala: a systems perspective. Trends Neurosci., 25(9): 456.
McGaugh, J.L. (2005) Emotional arousal and enhanced amygdala activity: new evidence for the old perseveration-consolidation hypothesis. Learn. Mem., 12(2): 77–79.
McGivern, R.F., Henschel, D., Hutcheson, M. and Pangburn, T. (1996) Sex difference in daily water consumption of rats: effect of housing and hormones. Physiol. Behav., 59(4-5): 653–658.
Mesches, M.H., Fleshner, M., Heman, K.L., Rose, G.M. and Diamond, D.M. (1999) Exposing rats to a predator blocks primed burst potentiation in the hippocampus in vitro. J. Neurosci., 19(14): RC18.
Morris, R.G., Garrud, P., Rawlins, J.N. and O'Keefe, J. (1982) Place navigation impaired in rats with hippocampal lesions. Nature, 297(5868): 681–683.
Nemeroff, C.B., Bremner, J.D., Foa, E.B., Mayberg, H.S., North, C.S. and Stein, M.B. (2006) Posttraumatic stress disorder: a state-of-the-science review. J. Psychiatr. Res., 40(1): 1–21.
Pittenger, C. and Kandel, E. (1998) A genetic switch for long-term memory. C. R. Acad. Sci. III, 321(2-3): 91–96.
Richter-Levin, G. and Akirav, I. (2003) Emotional tagging of memory formation: in the search for neural mechanisms. Brain Res. Brain Res. Rev., 43(3): 247–256.
Romeo, R.D., Lee, S.J., Chhua, N., McPherson, C.R. and McEwen, B.S. (2004) Testosterone cannot activate an adult-like stress response in prepubertal male rats. Neuroendocrinology, 79(3): 125–132.
Ronn, L.C., Berezin, V. and Bock, E. (2000) The neural cell adhesion molecule in synaptic plasticity and ageing. Int. J. Dev. Neurosci., 18(2-3): 193–199.
Rutishauser, U. (1989) Neural cell-to-cell adhesion and recognition. Curr. Opin. Cell Biol., 1(5): 898–904.
Rutishauser, U. and Jessell, T.M. (1988) Cell adhesion molecules in vertebrate neural development. Physiol. Rev., 68(3): 819–857.
Rutishauser, U. and Landmesser, L. (1996) Polysialic acid in the vertebrate nervous system: a promoter of plasticity in cell–cell interactions. Trends Neurosci., 19(10): 422–427.
Sanchez, M.M., Ladd, C.O. and Plotsky, P.M. (2001) Early adverse experience as a developmental risk factor for later psychopathology: evidence from rodent and primate models. Dev. Psychopathol., 13(3): 419–449.
Sandi, C. (2004) Stress, cognitive impairment and cell adhesion molecules. Nat. Rev. Neurosci., 5(12): 917–930.
Sapolsky, R.M. (2003) Stress and plasticity in the limbic system. Neurochem. Res., 28(11): 1735–1742.
Savonenko, A., Werka, T., Nikolaev, E., Zielinski, K. and Kaczmarek, L. (2003) Complex effects of NMDA receptor antagonist APV in the basolateral amygdala on acquisition of two-way avoidance reaction and long-term fear memory. Learn. Mem., 10(4): 293–303.

Schachner, M. (1997) Neural recognition molecules and synaptic plasticity. Curr. Opin. Cell Biol., 9(5): 627–634.

Schaefer, S.M., Jackson, D.C., Davidson, R.J., Aguirre, G.K., Kimberg, D.Y. and Thompson-Schill, S.L. (2002) Modulation of amygdalar activity by the conscious regulation of negative emotion. J. Cogn. Neurosci., 14(6): 913–921.

Schwegler, H., Lipp, H.P., Van der Loos, H. and Buselmaier, W. (1981) Individual hippocampal mossy fiber distribution in mice correlates with two-way avoidance performance. Sci., 214(4522): 817–819.

Spear, L.P. (2000) The adolescent brain and age-related behavioral manifestations. Neurosci. Biobehav. Rev., 24(4): 417–463.

Tsoory, M., Cohen, H. and Richter-Levin, G. (2007a) Juvenile stress induces a predisposition to either anxiety or depressive-like symptoms following stress in adulthood. Eur. Neuropsychopharmacol., 17(4): 245–256.

Tsoory, M., Guterman, A. and Richter-Levin, G. (2007b) Exposure to stressors during juvenility disrupts development-related alterations in the PSA-NCAM to NCAM expression ratio: potential relevance for mood and anxiety disorders. Neuropsychopharmacology. http://www.nature.com/npp/journal/vaop/ncurrent/abs/1301397a.html

Tsoory, M. and Richter-Levin, G. (2006) Learning under stress in the adult rat is differentially affected by 'juvenile' or 'adolescent' stress. Int. J. Neuropsychopharmacol., 9(6): 713–728.

Van den Berg, C.L., Pijlman, F.T., Koning, H.A., Diergaarde, L., Van Ree, J.M. and Spruijt, B.M. (1999a) Isolation changes the incentive value of sucrose and social behaviour in juvenile and adult rats. Behav. Brain Res., 106(1-2): 133–142.

Van den Berg, C.L., Van Ree, J.M. and Spruijt, B.M. (1999b) Sequential analysis of juvenile isolation-induced decreased social behavior in the adult rat. Physiol. Behav., 67(4): 483–488.

Van den Berg, C.L., Van Ree, J.M. and Spruijt, B.M. (2000) Morphine attenuates the effects of juvenile isolation in rats. Neuropharmacol., 39(6): 969–976.

Van den Berg, C.L., Van Ree, J.M., Spruijt, B.M. and Kitchen, I. (1999c) Effects of juvenile isolation and morphine treatment on social interactions and opioid receptors in adult rats: behavioural and autoradiographic studies. Eur. J. Neurosci., 11(9): 3023–3032.

Vazquez, D.M. (1998) Stress and the developing limbic-hypothalamic-pituitary-adrenal axis. Psychoneuroendocrinol., 23(7): 663–700.

Vouimba, R.M., Munoz, C. and Diamond, D.M. (2006) Differential effects of predator stress and the antidepressant tianeptine on physiological plasticity in the hippocampus and basolateral amygdala. Stress, 9(1): 29–40.

Vouimba, R.M. and Richter-Levin, G. (2005) Physiological dissociation in hippocampal subregions in response to amygdala stimulation. Cereb. Cortex, 15(11): 1815–1821.

Vouimba, R.M., Yaniv, D., Diamond, D. and Richter-Levin, G. (2004) Effects of inescapable stress on LTP in the amygdala versus the dentate gyrus of freely behaving rats. Eur. J. Neurosci., 19(7): 1887–1894.

Vouimba, R.M., Yaniv, D. and Richter-Levin, G. (2007) Glucocorticoid receptors and beta-adrenoceptors in basolateral amygdala modulate synaptic plasticity in hippocampal dentate gyrus, but not in area CA1. Neuropharmacol., 52(1): 244–252.

Welzl, H. and Stork, O. (2003) Cell adhesion molecules: key players in memory consolidation? News Physiol. Sci., 18: 147–150.

Woodson, J.C., Macintosh, D., Fleshner, M. and Diamond, D.M. (2003) Emotion-induced amnesia in rats: working memory-specific impairment, corticosterone-memory correlation, and fear versus arousal effects on memory. Learn. Mem., 10(5): 326–336.

Xu, L., Anwyl, R. and Rowan, M.J. (1997) Behavioural stress facilitates the induction of long-term depression in the hippocampus. Nature, 387(6632): 497–500.

Yaniv, D., Schafe, G.E., LeDoux, J.E. and Richter-Levin, G. (2000) Perirhinal cortex and thalamic stimulation induces LTP in different areas of the amygdala. Ann. N.Y. Acad. Sci., 911: 474–476.

Yaniv, D., Vouimba, R.M., Diamond, D.M. and Richter-Levin, G. (2003) Simultaneous induction of long-term potentiation in the hippocampus and the amygdala by entorhinal cortex activation: mechanistic and temporal profiles. Neurosci., 120(4): 1125–1135.

Discussion: Chapter 3

OITZL: The developmental pattern of the HPA axis during early-life phase is well established. Can you comment on the characteristics of the HPA axis during the juvenile period?

RICHTER-LEVIN: On the one hand, in the juveniles, the HPA axis is beyond the stress hyporesponsive period and developed already. On the other hand, the stress response lingers when compared to the adult.

BUWALDA: What is the reason that you specifically choose 27–29 days as the juvenile age? Did you ever test at a later period, for instance at 50 days?

RICHTER-LEVIN: Yes, actually thank you for this question. We did test also at a later period. First, we found that in adulthood the juvenile stress population could be divided into two subgroups. A small subgroup showed symptoms of depression rather than of anxiety, but the majority of the animals were anxious. Exposed to stress a week later (around 36 days) the population with depressive symptoms disappeared and stressed rats only showed anxiety symptoms. This finding fits well with the human literature where we find indications that the exact age of exposure to trauma during childhood could make a difference in the potential risk for developing either depression or anxiety.

SCHMIDT: I wonder at what time you are weaning your animals and whether you think the stress of weaning will affect your paradigm.

RICHTER-LEVIN: This is a very good question. We have been dealing with this — because, of course, we want to be as close as possible to the natural weaning condition. We expected to find significant differences between animals that were weaned at age 21 or 28 days, but we did not. So these studies were done all with animals — the ones that I have shown here — that were weaned at age of 21. I believe it will be important to test effects in a population that was allowed weaning in a more natural setting (i.e., weaning as a process over several days).

E.R. de Kloet, M.S. Oitzl & E. Vermetten (Eds.)
Progress in Brain Research, Vol. 167
ISSN 0079-6123

CHAPTER 4

Commentary: neuroendocrine basis

E.R. de Kloet*

Medical Pharmacology, LACDR/LUMC, PO Box 9502, 2300 RA Leiden, The Netherlands

These first three contributions to the Colloquium on Stress and PTSD have provided highly interesting basic data that have important conceptual implications for the translation to clinical perspectives. I will first comment on the stress mediators, then address the dynamics of the glucocorticoid response and conclude with some remarks on the genetics and early life priming of stress reactivity in later life.

Stress mediators

Corticotrophin releasing hormone (CRH) is considered the central coordinator of the behavioural, autonomic and neuroendocrine stress response. The stress mediators activated by CRH are organized in the sympathetic nervous system and the hypothalamic-pituitary-adrenal (HPA) axis. The sympathicus comprises amines and co-localized neuropeptides such as neuropeptide Y. The HPA axis is driven by CRH, vasopressin and secretagogues of the pro-opiomelanocortin peptides, e.g. ACTH and the endorphins and of course the adrenal glucocorticoids cortisol and corticosterone, the latter steroid only in rodents. Over the past decades a wealth of data has become available on the action of these stress mediators on brain and behaviour (de Wied, 1997).

In recent years most attention was focused on glucocorticoids (CORT) as end product of the HPA axis. CORT is secreted in hourly pulses from the adrenal cortex; stressors cause secretory bursts at any time (Lightman, 2006). CORT feeds back on precisely those brain circuits that were initially activated by the stressor, which is usually a novel (either real or imagined) experience. Classically, this action of CORT on neural information processing is known to be mediated by two types of nuclear receptors that act as ligand-driven transcription factors involved in the regulation of genomic events, i.e. mineralocorticoid receptor (MR) and glucocorticoid receptor (GR) (de Kloet et al., 2005). MR and GR are both abundantly expressed in limbic neurons. The genomic control mediated by MR and GR modulates plasticity and metabolism of hippocampal circuits. New is that these nuclear receptors also reside in the membrane of limbic neurons. Recent data show that CORT can enhance excitability via these MR in a fast non-genomic fashion (Karst et al., 2005).

How do all these stress mediators cooperate in information processing? The model (Chapter 1 by Marian Joëls et al.) is based on dissection of the stress response in different phases, which depend on the time, place and context. In hippocampus the following four phases can be distinguished:

(1) In the initial alarm reaction stressful stimuli activate CRH neurons that organize the behavioural and physiological stress response; attention and vigilance are the initial

*Corresponding author. Tel.: +31-71-5276210/90; Fax: +31-71-5274715; E-mail: e.kloet@chem.leidenuniv.nl

DOI: 10.1016/S0079-6123(07)67004-6

behavioural manifestations together with perception and appraisal and other cognitive operations to assess the situation and to select the most appropriate coping response. To facilitate these emotional responses and cognitive operations the above mentioned 'stress' neurotransmitters and neuropeptides converge and synergize in critical targets of the limbic — midbrain, an excitatory reaction that now seems to be amplified and stabilized by CORT through membrane MR (Joëls et al., 2006).

(2) The stress mediators mobilized by the initial alarm reaction are essential, but if not controlled, they become harmful. Thus, these initial reactions are prevented from overshooting by inhibitory mechanisms (Sapolsky et al., 2000). In neuroendocrine realm, CORT can fulfil such a dampening role in the hippocampus. As occurs in many stress-activated cells and tissues (e.g. inflammatory, immune, metabolism) CORT attenuates the initial stress reaction through transrepression, a process that starts after several minutes and is mediated by GR.

(3) Then through GR recovery from the stressor is promoted and energy resources are replenished. Also the storage of the experience, i.e. coping with the experience, is promoted in the memory. These stimulatory actions of CORT on the storage of energy and information can be viewed as the organisms' ability to cope with similar challenges in the future. These adaptations, recently termed allostasis, all seem to depend on GR-mediated transactivation (Reichardt et al., 1998; Oitzl et al., 2001).

(4) Finally, what about the nuclear MR? This receptor was originally assigned a role in the threshold or sensitivity locally in cell and tissue reactions to stressors. We are now able to specify this role further; the neuroendocrine data suggest that the nuclear MR in limbic structures is of importance for the maintenance of the refractory period between the ultradian HPA pulses (Conway-Campbell et al., 2007).

Naturally, such a model raises more questions than answers. For instance, how do all these signals coordinate the initial stress reaction with the management of the later adaptive phase? And for this purpose, in which time domain and context do the individual mediators operate? Yet, the model is extremely helpful in the rational design of experiments. During the last decades the experiments of the groups of Joëls and de Kloet fortunately all are based on 20 min CORT exposure, which we now know mimics the CORT pulse under basal and stressful conditions (Joëls et al., 2007). Recently a study was completed to examine the temporal pattern of the genomic response to a CORT pulse; after 1 h we found mainly transrepression of CORT-responsive genes and after 3 h mainly transactivation (Morsink et al., 2006).

Dynamics of the CORT response

A rapid activation of the initial stress reaction by CRH is essential as long as its termination by CORT is efficient as well. If the stress reaction is inadequate, excessive or prolonged the management of the later adaptive phase may fail. One approach to study the role of stress mediators is by taking as leading principle the action mechanism of CORT. This is already a very complicated story, CORT action depends on CORT bio-availability, its receptor properties and the context in which CORT receptors operate (de Kloet et al., 1998) (Fig. 1).

(1) Bio-availability: determinants are binding to blood corticosteroid-binding globulin, penetration of the blood-brain barrier involving multidrug resistance P glycoprotein and intracellular metabolism of the naturally occurring glucocorticoids through 11β-hydroxy steroid dehydrogenase (11β-HSD) (Chapter 2 by Jonathan R. Seckl).

(2) Receptor properties: at the receptor level the CORT signal depends on an as yet unknown interaction with the membrane version of the nuclear receptors and the dissociation of this receptor from the initial heat shock protein — receptor complex.

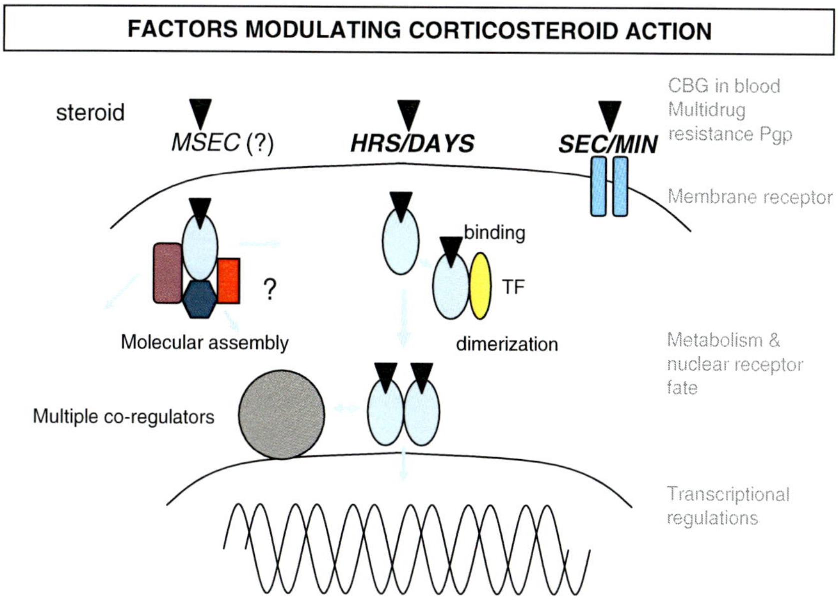

Fig. 1. Factors modulating corticosteroid action. The various levels of cellular organization are shown that can modulate corticosteroid action.

(3) Receptor polymorphisms are important and variants of the MR and GR gene have been identified which can modulate the CORT signal (DeRijk et al., 2006).
(4) Receptor interaction with other transcription factors determines how CORT exerts its transrepression effect on gene transcription dampening initial stress reactions.
(5) Receptors recruit patterns of co-regulators that may vary in different contexts (Meijer et al., 2003).
(6) The termination of the CORT receptor-mediated signal is not very well known, but at least the proteasome pathway is an important step (Conway-Campbell et al., 2007).

This brief overview of the various steps in CORT action is far for complete. Yet, it is important to realize that at the organismic level these factors can modulate the action of CORT in the coordination of cell and organ function. At the genomic level the recruitment of co-regulator cocktails provide CORT receptors the capacity to integrate various intracellular signalling pathways. Hence these inter- and intracellular mechanisms provide CORT with an enormous diversity in action, which is superimposed over its hourly ultradian rhythm. Hence patterns of CORT secretion matter.

Long-term control of stress reactivity: a question of balance

The balance hypothesis states that once the balance between onset and termination of the stress reaction by various stress mediators is disturbed the individual looses its ability to maintain homeostasis if challenged. This may lead to a condition of neuroendocrine dysregulation and impaired behavioural adaptation until a certain set point is surpassed, such that vulnerability is enhanced to stress-related diseases for which the individual is genetically predisposed. Leading principle in this hypothesis is the MR:GR balance (de Kloet et al., 2007) (Fig. 2).

Using the MR:GR balance as criterion has led to the notion that the set point of the HPA axis can be modulated by three defined inputs or ‘hits’:

(1) Genetic: a direct hit is the functional gene variants of MR and GR; indirect hits are

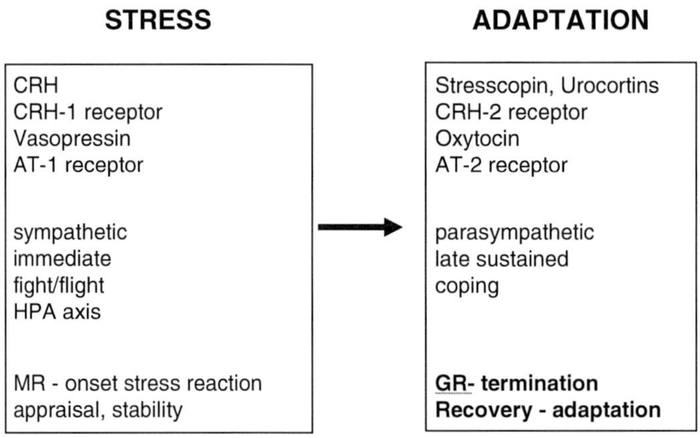

Fig. 2. The complementary action of mediators involved in the onset of the stress reaction and the management of later adaptations. CRH = corticotrophin-releasing hormone, AT = angiotensin II, HPA = hypothalamic-pituitary-adrenal axis, MR = mineralocorticoid receptor, GR = glucocorticoid receptor.

concerned with the genetic background in which MR and GR operate.

(2) Priming life experiences: these can be due to specific environmental challenges during pre- and early postnatal life of the pup (Chapter 2 by Seckl). Variations in maternal care have a lasting impact on development of the infant brain and may result in changes in gene expression through epigenetics, altered neuroendocrine regulations, emotional arousal and cognitive performance. The second hit not only includes perinatal life, but also relates to defined periods in juvenile life (Chapter 3 by Richter-Levin et al.).

(3) Later life stressors: these can be traumatic events and the recurrent memory to the event, or chronic daily stressors, which serve as trigger of a pathophysiological cascade in predisposed individuals.

The relevant question for this colloquium obviously is why some individuals become ill, but others actually gain strength from identical stressful situations. There are intriguing findings and there are paradoxes in this area of science. For instance, infants strongly attached to an overprotective mother usually display attenuated CORT reactivity in later life; a downregulated CORT secretion is also occurring in the aftermath of severe childhood trauma, and may be considered a fingerprint of PTSD. Is this a sign of inadequate CORT release unable to contain an exaggerated initial stress reaction? Does it mean that attenuated CORT reactivity is actually a predisposing factor for PTSD?

While maternal care reduces CORT reactivity in later life, animal models of maternal deprivation show enhanced CORT reactivity in later life. Is this bad or good? There is evidence that it is actually the nature of the stressor in later life, which may provide further understanding as to how to answer this question. Hence if an infant's CORT secretion is tuned to be hyporesponsive in later life because of enhanced maternal care, it is likely inadequate to cope with major stressors; for such conditions the highly reactive CORT responders exposed to adverse early life experience are expected to be better equipped. Thus, as a reductionist one could even view individual differences in coping with traumatic experiences by rephrasing the question: When does a CORT signal change from protective into harmful? What is the cause? What are the consequences?

Acknowledgement

The support from the Royal Netherlands Academy of Arts and Sciences is gratefully acknowledged.

References

Conway-Campbell, B.L., McKenna, M.A., Wiles, C.C., Atkinson, A.C., de Kloet, E.R. and Lightman, S.L. (2007) Proteasome-dependent downregulation of activated nuclear hippocampal glucocorticoid receptors determines responses to corticosterone. Endocrinology, August 9 [Epub ahead of press].

De Kloet, E.R., de Rijk, R.H. and Meijer, O.C. (2007) Therapy insight: is there an imbalanced response of mineralocorticoid and glucocorticoid receptors in depression? Nature Clin. Endocrinol. Metab., 3: 168–179.

DeRijk, R.H., Wüst, S., Meijer, O.C., Zennaro, M.C., Federenko, I.S., Hellhammer, D.H., Giacchetti, G., Vreugdenhil, E., Zitman, F.G. and de Kloet, E.R. (2006) A common polymorphism in the mineralocorticoid receptor modulates stress responsiveness. J. Clin. Endocrinol. Metab., 91: 5083–5089.

De Wied, D. (1997) The neuropeptide story. Geoffrey Harris Lecture, Budapest, Hungary, July 1994. Front. Neuroendocrinol., 18: 101–113.

Joëls, M., Karst, H., Krugers, H.J. and Lucassen, P.J. (2007) Chronic stress: implications for neuronal morphology, function and neurogenesis. Front. Neuroendocrinol., 28: 72–96.

Joëls, M., Pu, Z., Wiegert, O., Oitzl, M.S. and Krugers, H.J. (2006) Learning under stress: how does it work? Trends Cogn. Sci., 10: 152–158.

Karst, H., Berger, S., Turiault, M., Tronche, F., Schütz, G. and Joëls, M. (2005) Mineralocorticoid receptors are indispensable for nongenomic modulation of hippocampal glutamate transmission by corticosterone. Proc. Natl. Acad. Sci. U.S.A., 102: 19204–19207.

de Kloet, E.R., Joëls, M. and Holsboer, F. (2005) Stress and the brain: from adaptation to disease. Nat. Rev. Neurosci., 6: 6463–6475.

de Kloet, E.R., Vreugdenhil, E., Oitzl, M.S. and Joëls, M. (1998) Brain corticosteroid receptor balance in health and disease. Endocr. Rev., 19: 269–301.

Lightman, S.L. (2006) Patterns of exposure to glucocorticoid receptor ligand. Biochem. Soc. Trans., 34: 1117–1118.

Meijer, O.C., Karssen, A.M. and de Kloet, E.R. (2003) Cell- and tissue-specific effects of corticosteroids in relation to glucocorticoid resistance: examples from the brain. J. Endocrinol., 178: 13–18.

Morsink, M.C., Steenbergen, P.J., Vos, J.B., Karst, H., Joëls, M., de Kloet, E.R. and Datson, N.A. (2006) Acute activation of hippocampal glucocorticoid receptors results in different waves of gene expression throughout time. J. Neuroendocrinol., 18: 239–252.

Oitzl, M.S., Reichardt, H.M., Joëls, M. and de Kloet, E.R. (2001) Point mutation in the mouse glucocorticoid receptor preventing DNA binding impairs spatial memory. Proc. Natl. Acad. Sci. U.S.A., 98: 12790–12795.

Reichardt, H.M., Kaestner, K.H., Tuckermann, J., Kretz, O., Wessley, O., Bock, R., Gass, P., Schmid, W., Herrlich, P., Angel, P. and Schütz, G. (1998) DNA binding of the glucocorticoid receptor is not essential for survival. Cell, 93: 487–490.

Sapolsky, R.M., Romero, L.M. and Munck, A.U. (2000) How do glucocorticoids influence stress responses? Integrating permissive, suppressive, stimulatory, and preparative actions. Endocr. Rev., 21: 55–89.

General Discussion: Section I

LEVINE: I asked Jonathan (SECKL) something I would like to amplify on. CRH is excessively produced by the placenta and it is the only time in development that you can actually pick up CRH in the plasma. There is a lot of emphasis on the relationship of prenatal stress and CRH and I wonder if you can tie these systems together.

SECKL: Yes, Gig (Levine) that is a great question. It is an unusual CRH system in the human and the higher primate placenta, which produces large amounts of CRH, whereas the rodent placenta does not do this. Plasma CRH levels rise in consequence which might be expected to affect the pituitary corticotrophs, etc. However, the placenta also secretes into the maternal circulation a specific binding protein (CRHBP) which binds CRH, largely blocking any systemic action. So the thesis is that placental CRH is either working as a paracrine factor in the placenta and/or acts upon the foetus, which does not have the binding protein. CRH synthesis in the HPA axis system is 'normally' inhibited by glucocorticoids, but in the placenta glucocorticoids stimulate CRH. We hypothesized that if the mother has high glucocorticoid levels, perhaps if she is stressed, this might induce placental CRH, which spills over into the foetal circulation, thus stimulating the foetal HPA axis. This might produce a 'double hit' with both more maternal glucocorticoids passing intact across the placenta (the 11β-HSD2 'barrier' is not complete and a small but significant percentage of cortisol passes to the foetal blood) and increased placental CRH stimulating the foetal HPA system. Some unpublished data also suggest that in certain circumstances maternal stress may also lower (or fail to induce) the placental 11β-HSD2 barrier, perhaps a 'third hit'.

LEVINE: Is the placental 'feed-forward' of glucocorticoids on CRH unique?

SECKL: Glucocorticoids also stimulate CRH mRNA in the amygdala. Presumably this is all about the complexities of gene promoter control, which is site-specific and therefore complex.

SHALEV: Whereas in Holocaust survivors a correlation was between age of exposure and some of the endocrine outcomes, the scale is very different from the prenatal or perinatal stress models. I was wondering whether this is somehow you can complete the argument in the analogy in which it is an age-related effect for most survivors and the other factors you did describe.

SECKL: You quite correctly point out the problems of retrospective and historical human work such as the study of Holocaust survivors, especially studies 60 years after the events transpired. A more clearcut example of a single onset stressor was the 9/11 study which clearly had an effect on the HPA-axis of the offspring that was predominantly confined to those exposed in the third trimester of pregnancy. The holocaust data are purely correlative but what was very surprising to me, something we had not expected, was the strong correlations between 'age' at exposure and cortisol metabolism 60 years later. The trauma, which was both an extreme nutritional and psychological stress, was of course both extreme and prolonged. Nonetheless, the younger at exposure the greater the metabolic changes seen in old age and the less metabolic syndrome features found. My prejudice, my guess from the rodent literature is that the nutritional component of the stress is at least if not more important than the other modalities. One could speculate that the younger subjects made the greatest metabolic adjustments, which were a positive adaptation to their environment. Now they also got less PTSD in later life. Perhaps this is merely a reflection that they were younger at traumatization and therefore either less cognizant of what was going on, had less memories to compare at the time or were more plastic. But we should not dismiss the steroid metabolic changes as a 'protective' factor, though again this is pure speculation. In any case, the data are compatible with the concept of early life events having ongoing persisting cardiometabolic and neuropsychiatric effects, some of which will produce vulnerability to disease in certain adult circumstances others may associate with resistance to disease.

RICHTER-LEVIN: I wonder whether the age at which trauma is experienced also determines the outcome. Thus, if trauma is experienced at the different stages during development in infancy, childhood, adolescence would that enhance the vulnerability to PTSD or to depression or even more to resilience.

LIBERZON: Your comment reflects also my thoughts. I have actually two questions: one to you (Richter-Levin) about your data and regarding the second one I am interested to hear the response of the panel. So regarding your data, the reason I believe the juvenile or the neonatal period was investigated is because these are quiet periods in the development of the HPA axis. Now, the first week of rodent life when all separation studies were done, the HPA axis is different; it is quiescent because of the stress hyporesponsive period. My question is: what is special about the HPA axis of the 28 days of juvenile age, what is going on in the brain? What is a relevant circuitry in the brain that is specific to this age that might be involved?

RICHTER-LEVIN: First of all there is little known in the literature about the HPA axis and the brain circuits at the juvenile age, so the data I present are really the initial findings on this topic. In juveniles the hippocampus for example is still not yet fine-tuned, so the inhibitory influences of the interneurons on pyramidal cell excitability is not yet complete; the basic structure is complete but fine-tuning of the hippocampus is not yet complete. The amygdala is considered to be mature at day 28, but that cannot be said for the amygdala output to other regions.

GUNNAR: I was going to very briefly answer the question to Gal (Richter-Levin), because that is precisely one of the translational problems when interpreting rodent data to humans. The work that has been done in developmental psychopathology is focused on trying to ask this question about timing and resilience.

The best predictor of child outcome following a catastrophic event is a matter of their own behavioural performance prior to the catastrophic event. Of course that is also the case with adults as well. Of importance is also the social environment, the functioning of their families. So families that remain intact are capable to provide the function they are supposed to do. They tend to have the better functioning kids and the kids tend to be the ones that are resilient. But most often the trauma is abuse, right, then abuse is on top of families that are poorly functioning. In fact in the US 40–60% of the affected children that are identified as physically and sexually abused are also identified as neglected. Neglect begins early, so the model is a poorly functioning family. I will add one more thing and that humans have to learn a lot about how they need to function in the world based on their experiences with parents and other adults. So life in the world is not preprogrammed. If, the protective environment is lost one may become an incompetent organism that is more likely not be able to avoid threat

RICHTER-LEVIN: I agree. We actually try to focus on acute exposure of animals to stress under conditions they live normally before and normally after. I think in humans the question of age can be addressed because age would then interact with other factors.

PITMAN: Well I raised a question about the temporal patterns after the first presentation by Marian Joëls about the critical period in the aftermath of traumatic events. Is there a critical time window in which hormones can influence the pathogenesis of posttraumatic stress disorder? Does that critical period concern the excitatory action of cortisol or does it follow the subsequent inhibitory phase.

DE KLOET: Excellent. That is the point. When is the critical time window in the response pattern of the stress mediators in producing lasting changes in the brain that may enhance vulnerability to PTSD? So perhaps Marian (Joëls) would you like to comment?

JOËLS: Well, at this moment it is just guessing because it has not been addressed experimentally yet. I'd say that there is more redundancy in the initial phase of the stress response than in the later period. This is because there are more hormones that can do the same thing in the first phase than in the period that the slower genomic actions evolve. Also, we still know very little about these non-genomic effects of the steroids but we definitely know about the genomic effects that they are very sensitive to both the genetic and the developmental or environmental background of the animal. To give one example; the sensitivity to the effects of glucocorticoids depends very strongly on whether or not the animal has experienced a prolonged period of stress prior to the measurement. In the dentate, cells may not respond with a glutamate response after a single stress, but they do when they've had a background of stress several weeks before the experiment. The slow effects of corticosterone are very sensitive to both the genetic and environmental factors as you would expect if it has anything to do with the pathology.

PITMAN: Do you think processes underlying memory consolidation can be influenced by these manipulations with cortisol?

JOËLS: Yes, these studies have for instance been done by Melly Oitzl who has shown that the performance of animals in a Morris Maze task, which very much involves the CA1 hippocampal area, is impaired if the slow genomic effects are prevented by a glucocorticoid antagonist or if mutant mice are used in which dimerization and DNA binding of GRs is prevented. Such a blockade of GR-mediated genomic effects prevents consolidation of the learning in this task.

OITZL: I would like to add that it is not only the genomic effect of the glucocorticoid receptor, but that also the mineralocorticoid receptor is involved. In this respect the rapid membrane effect mediated by the mineralocorticoid receptors might affect the attention to the environmental circumstances in this task the animals have to learn. This rapid effect complemented by the slow genomic effects contribute to their performance.

RICHTER-LEVIN: I think we have to bear in mind that the stress system usually operates in a normal mode and rarely under conditions of extreme stress. In the normal mode it contributes to our daily life, helping in memorizing things and so on. The question we need to address is how the system works under relatively normal conditions and how its operation is modified during traumatic experiences. My second question is about the function of hippocampus that contains the dual CORT receptor types in relation to other brain regions that only express GR. To clarify myself, according to your scheme nuclear MR in hippocampus mediates the response to basal CORT, while both receptors respond under stress. The other brain regions are affected by CORT only through GR under stressful conditions. Would you like to comment on this?

JOËLS: It is one thing looking at corticosterone and cortisol levels and even the expression of receptors in particular tissues. Another is the function of the cell, which depends on many things. For instance, dentate cells do have GRs but normally we see little effect of this receptor; it is not understood why. However, with a background of chronic stress, GR activation induces strong effects. Actually this was also illustrated by Jonathan (Seckl): we can measure cortisol levels in the plasma but what is it telling you about what it is doing in the brain? And specifically what it is doing in specific parts of the brain. So, I think what I would like to promote here is that if you want to know what corticosterone elevations are doing in a particular brain area as they occur after stress look in that particular one cell to see what it's actually doing to its function, don't go for the expression of a gene or the receptor or the hormone changes itself. So look in the area you are interested in, the cell you are interested in, in the condition you are interested, and in the genetic and environmental background you are interested in and then you will get your answer. Because it is very difficult to predict just based on hormone levels and transcript levels what is going to happen. What they are actually doing depends on the network in which these cells function, depends on the cocktail of proteins that is locally present, all this determines what the hormone is going to do in that particular cell. I emphasized several times that I was talking about the CA1 area, that was the area we have investigated. It is very clear that there are different things happening in the dentate and the amygdala. This may be in fact very relevant for the condition we are talking about here, to what degree these areas are involved in behavioural tasks we study.

SECKL: Are we in the position that we can begin to put together the systems biology of what we think that some particular regions in the amygdala and the hippocampus contribute to how we can predict what the vulnerability of the system would be to one form of challenge or another or would that be far too soon in this biology.

JOËLS: At this moment it is pasting together little pieces of information. A lot of the information on the dentate and the prefrontal cortex and the basolateral amygdala is still lacking, there is clearly less than what is available about the CA1 area. But we already know that these areas respond to the same stimulus in a different way.

DE KLOET: This session is closed, thank you all for your comments.

SECTION II

The Vulnerable Phenotype: Circuits and Behavior

E.R. de Kloet, M.S. Oitzl & E. Vermetten (Eds.)
Progress in Brain Research, Vol. 167
ISSN 0079-6123

CHAPTER 5

Mice that under- or overexpress glucocorticoid receptors as models for depression or posttraumatic stress disorder

Sabine Chourbaji, Miriam A. Vogt and Peter Gass*

Central Institute of Mental Health, J 5, D-68159 Mannheim, Germany

Abstract: Modern molecular and pathophysiological concepts suggest that glucocorticoid receptors (GRs) play a crucial role for the pathogenesis, course and therapy of affective or emotional disorders. Specifically, an impairment of GR signaling has been associated with major depression, whereas overactivity or hyper-responsiveness of GRs have been conceptualized for posttraumatic stress disorder (PTSD). Recently, several research groups have generated transgenic mouse strains that under- or overexpress GRs, respectively. These animals seem to represent valuable tools for studying the foregoing hypotheses. Indeed, first results indicate that mice with a deficit in GR expression show a depression-like behavioral phenotype as well as characteristic neuroendocrinological changes observed in depressive patients. Particularly, GR heterozygous mice with a 50% reduction of GR expression represent a model for combined effects of both genetic and environmental manipulations, since their depression-like behavior becomes only manifest after stress-exposure. Thus, the phenotype of this strain mimics the human situation in depressive disorders, in which individuals at risk are predisposed to develop depressive episodes after stress. It is currently less clear whether, and in which way, mice that overexpress GRs can serve as models for PTSD, or mimic at least specific aspects of the clinical syndrome. The latter strains have still to be subjected to specific tests analyzing conditioning and sensitization processes in fearful situations. So far, mice with compromised GR expression seem to be a good tool to further study molecular, pathophysiological and cellular/structural alterations that underlie specific behavioral features such as despair or helplessness. A major challenge is to decipher which signs and symptoms in patients correspond to these animal behavioral constructs, and to elucidate whether it is possible to gain insights from the animals' response to specific treatments for human therapy.

Keywords: behavior; glucocorticoid receptor; HPA-system; learned helplessness; mice; stress

Introduction

Glucocorticoid receptors (GRs) and the hypothalamic-pituitary-adrenal (HPA)-system are thought to play an important role in depression, posttraumatic stress disorder (PSTD) and anxiety (Holsboer, 2000; Newport and Nemeroff, 2000; Korte, 2001; de Kloet, 2002; McEwen et al., 2002). Thus, patients with a severe major depression exhibit — as a group — hypersecretion of corticotrophin-releasing factor (CRF) and cortisol, while patients suffering

*Corresponding author. Tel.: +49 621 1703 2931;
Fax: +49 621 1703 2005; E-mail: peter.gass@zi-mannheim.de

DOI: 10.1016/S0079-6123(07)67005-8

from PTSD also hypersecrete CRH but paradoxically show normal or subnormal cortisol levels (Newport and Nemeroff, 2000; Gold et al., 2002). It is yet under intense debate whether these reported corticosteroid imbalances are causes or consequences of the respective behavioral syndromes (de Kloet et al., 1999; Holsboer, 2000; Lee et al., 2002). Beyond this "chicken or egg question" it has been speculated that major depression coincides with a hypoactivity of GR-regulated systems, whereas PTSD may be associated with a hyperactivity of GR (Holsboer, 2000; Newport and Nemeroff, 2000). Currently, it is not known, however, by which mechanisms GR might be dysregulated, nor how a potential dysregulation may subsequently influence neuronal gene expression, neural plasticity and behavior (Lee et al., 2002).

There are two types of intracellular receptors mediating the effects of circulating glucocorticoids, i.e. corticosterone in rodents and cortisol in humans: (i) the type I mineralocorticoid receptor (MR); and (ii) the type II GR (Beato et al., 1995; de Kloet et al., 1998). Both receptors are ligand-binding transcription factors that belong to the nuclear hormone receptor superfamily (Beato et al., 1995; Tronche et al., 1998). They modulate a wide range of neural functions including stress responsiveness and cognition (de Kloet et al., 1998; Holsboer, 2000; Reul et al., 2000). With respect to the former, MR and GR exert a negative feedback on the HPA-system. While the MR is thought to be responsible for the regulation of circadian glucocorticoid variations, the GR has been postulated to be involved in the modulation of stress effects with high levels of glucocorticoids, which also occur during depression (Heuser et al., 1994). According to current concepts, GR-mediated functions are disturbed particularly in patients with severe depressive episodes. In contrast, an increased responsiveness of GR seems to prevail in PTSD. So far, studies of the GR in the human brain mainly derive from indirect pharmacological challenge tests with GR agonists, like the dexamethasone (DEX) suppression test, which measures GR function in the pituitary and maybe in part also in the central nervous system (CNS) (Hatzinger et al., 1996).

Since in vivo expression and functional studies of GR are not feasible in the human brain, such analyses have to be done in experimental animals. Hence, there is an increasing demand to develop animal models of psychiatric disorders, in which the important facets of the disease state can be investigated, including analyses of brain tissues at time points of interest. Such animal models should preferably mimic specific features of the human condition with regard to its etiology, symptomatology, treatment or biological/pathophysiological basis — even if some of the clinical features cannot be modeled in animals, i.e., suicidal ideation, feelings of guilt, delusions, etc. However, it is not an aim of animal research to mimic the complexity of human nature but to model selected endophenotypes proposed to reflect core features of the respective psychiatric disorders (Siegmund and Wotjak, 2006). In mice, e.g., learned helplessness (LH) is a model based on the stress-diathesis-concept, which postulates unpredictability, uncontrollability and unavoidability of chronic stressors as crucial factors to induce depression-like states (Chourbaji et al., 2005). With respect to PTSD, an altered fear conditioning response and an elevated startle reaction have been suggested to represent psychobiological pathomechanisms (Morgan et al., 1995; Layton and Krikorian, 2002). Molecular, endocrinological and structural alterations detected in such models can represent pathogenic mechanisms of the disorder. Alternatively, they may only be consequences of the experimental protocol. For example, the so-called "depressive behavior" in mice is induced by stress-exposure, which by itself can modulate corticosteroid receptor expression or function. One way to support a causal relationship between the modeled disease and the molecular changes observed is the successful application of pharmacotherapy with reversal of both the molecular and the behavioral alterations.

Another approach to associate altered steroid signaling with behavioral changes involves the use of transgenic mouse models. Here, the alteration of corticosteroid receptor function is genetically induced (for review, see Urani et al., 2005). When tested in behavioral paradigms, it is possible to determine whether and how steroid receptors participate in the development of specific signs of modeled psychiatric disorders. For example, one can determine whether the absence or the alteration of a corticosteroid receptor renders animals more prone or more resistant to develop features of depression and PTSD. Molecular genetic manipulations allow generating "transgenic" and "knockout" lines. Classical transgenic mice carry additional copies of normal or abnormal genes in their genome, which usually results in a so-called "gain of function". However, transgenes can also be used to induce a "loss of function" when the inserted transgene is an antisense to a targeted gene. In classical knockout mice, specific target genes are disrupted leading to a "loss of function". However, knockin strategies, in which genetic material is knocked into a particular locus, e.g., knocking in tetracycline responsive elements downstream of a particular promoter, can also cause a gain of function. The main distinction between knockin/knockout techniques and transgenic techniques is that the former targets a specific locus, while the latter acts by random integration into the genome. The sophisticated combination of these techniques allows the generation of mutant mouse lines in which the genetic manipulation occurs only in selected cell populations and at a specific developmental stage, or in which the over- or underexpression of genes can even be turned on and switched off artificially at a specific time point determined by the researcher. However, aiming at modeling the pathophysiology of human disorders, one has to consider that most likely there is no psychiatric patient who completely lacks GR in specific brains areas. In this respect, more simple transgenic approaches, such as conventional GR heterozygous mice or animals overexpressing GR, may result in more naturalistic models for human disease states.

The role of glucocorticoid receptors in depression

The most consistent biological finding in patients with major depression is a hyperactivity of the HPA-system, postulated to be caused by impaired glucocorticoid signaling (Nemeroff, 1996; Holsboer, 2000). This hyperactivity is often reflected by a basal hypercortisolemia. The latter abnormality and its underlying pathophysiology, however, are sometimes masked and only become evident in challenge tests of the HPA-system. Thus, more often than elevated plasma cortisol one finds a typical non-suppression of cortisol levels after DEX treatment in depressive patients, as well as an increased adrenocorticotropic hormone (ACTH) and cortisol release in the combined DEX/corticotrophin-releasing hormone (CRH) test (von Bardeleben and Holsboer, 1991; Heuser et al., 1994). In these clinical tests, the synthetic glucocorticoid DEX is administrated to activate the negative feedback of the HPA-system, usually suppressing cortisol plasma levels in healthy control persons at the level of the pituitary (Barden et al., 1997).

The derived *Stress Hypothesis* postulates that stress (associated with high levels of circulating glucocorticoids) leads, via activation of GRs, to a plethora of pathophysiological and molecular alterations in different brain regions (de Kloet et al., 1998, 2002, 2005). One effect of glucocorticoids is the downregulation of brain-derived neurotrophic factor (BDNF) in the hippocampus, which has led to the *neurotrophin hypothesis of depression*. According to this concept the BDNF downregulation is crucial in the pathogenesis of depression (Smith et al., 1995; Duman et al., 1997; Wong and Licinio, 2001; Nestler et al., 2002; Cryan and Mombereau, 2004; Lang et al., 2004).

Despite an activation of GRs, the HPA-axis shows increasing levels of CRH, adrenocorticotrophic hormone and corticosterone, in depressive patients, which might be counterintuitive considering the negative feedback of glucocorticoids. In view of the fact that these hormones are dysregulated in depressive patients but normalized by antidepressive treatment, the HPA-axis is suggested to play a crucial role in the pathophysiology of depression. According to this hypothesis, mice with genetic alterations of the GR system would be expected to display changes of emotional behavior and depression-like states (Urani et al., 2003).

The role of glucocorticoid receptors in PTSD

Cortisol levels are often found to be below normal in PTSD, but may also be similar or greater than those of control subjects (for review see, Yehuda, 2006). Findings of changes in circadian rhythm suggest that there may be regulatory influences that result in a greater dynamic range of cortisol release over the diurnal cycle in PTSD. Studies using the DEX suppression test have convincingly demonstrated that there is an enhanced negative feedback inhibition of cortisol, at least at the level of the pituitary. An enhanced negative feedback inhibition can explain why ambient cortisol levels may be normal or even lower in the face of hypothalamic CRH hypersecretion, which has also been consistently described in PTSD. These effects have been postulated to result from increased responsiveness of central or peripheral GR. According to this hypothesis, mice overexpressing GR would be expected to display changes of emotional and cognitive behaviors that might resemble signs and symptoms of PTSD (Urani et al., 2003).

Mice with reduced GR expression

The availability of classical homozygous GR mutant mice ($GR^{-/-}$ mice) is restricted, since more than 90% of these animals die early due to impaired lung development. Only a small fraction of mice survive to adulthood and can be analyzed, exhibiting extreme elevations in plasma-ACTH (15-fold) as well as corticosterone (2.5-fold) (Cole et al., 1995, 2001). Since the molecular cause of survival might include alternative splice mechanisms resulting in uncontrollable residual GR activity, these surviving animals have not been used for behavioral analyses.

Therefore, due to the restricted accessibility of conventional GR knockout mice that do not survive to adulthood, nervous system specific knockout mice have been generated using the Cre/loxP recombination system under the control of the rat nestin promoter (Nes) (Tronche et al., 1999). These so-called GR^{NesCre} mice are viable and lack the GR in neurons and glial cells throughout the CNS. As one might have expected, the HPA-system of these animals is severely disinhibited/overactivated due to the lack of the negative feedback normally exerted at the level of the hypothalamus via GR, with drastically elevated levels of CRH, ACTH and corticosterone (Tronche et al., 1999). This results in a (neuro)endocrinological phenotype resembling Cushing's disease in man, with a redistribution of body fat, osteoporosis and immunological abnormalities. This phenotype of GR^{NesCre} mice is not surprising, because GR signaling outside the nervous system is intact in this strain.

Interestingly, Cushing patients have a very high risk to develop major depression. Therefore, it was unexpected that GR^{NesCre} mice do not show a depression-like phenotype in respective behavioral tests. They do not develop increased despair-like behavior in the Forced Swim Test (FST) as it would be expected in depressive mice (Tronche et al., 1999). Moreover, using standardized anxiety tests such as the Dark-Light Box (DLB) Test and the Elevated O-Maze (EOM) these animals display less anxiety-related behavior than wild-type littermate controls (Tronche et al., 1999). This paradoxical finding can be explained by the fact that the neurons of GR^{NesCre} mice do not express the GR. Consequently the hypercortisolism cannot affect neurons and cause subsequent changes in behavior. Thus, despite their hypercortisolism, these mice most likely represent a genetic model of resistance to depression. To test this hypothesis, they have to be subjected to stress-induced behavioral depression paradigms, such as a Chronic Stress Model of depression. Similar to GR^{NesCre} mice, also mice with a silencing of GR as a consequence of GR-antisense expression do not show a behavioral depression-like phenotype (Montkowski et al., 1995; Barden, 1996).

According to the *Stress Hypothesis of Depression*, mice that underexpress GR only to a certain extent due to a gene dosage alteration may mimic the situation of patients with affective disorders more closely. Such mice are GR heterozygous mice ($GR^{+/-}$), in which one GR allele has been ablated by a conventional knockout strategy (Tronche et al., 1998). This results in an ~50% reduction of GR expression on the protein level in the brain (Ridder et al., 2005). $GR^{+/-}$ mice — when subjected to a large standardized test battery for emotional behaviors including tests for depression-like signs such as despair (i.e., Porsolt FST and Tail Suspension Test (TST)) — surprisingly do not show a phenotype when compared to their wild-type littermates (Ridder et al., 2005). Furthermore, they reveal normal circadian levels of corticosterone (Ridder et al., 2005). However, when subjected to stress, $GR^{+/-}$ mice have a predisposition for depressive-like behaviors (LH) and depression-like neuroendocrinological abnormalities.

In the stress-induced Learned Helplessness Model, mice are exposed to a series of unpredictable, unavoidable and uncontrollable footshocks on 2 consecutive days. On the third day, ~30% of the animals demonstrate helplessness, i.e., coping deficits in a test situation (shuttle box) where aversive stimuli (footshocks) become avoidable, since they are announced by a light signal, signaling the mice to change the compartment and thereby avoiding the shock (Chourbaji et al., 2005). $GR^{+/-}$ mice demonstrate significantly increased helplessness compared to their wild-type littermates, measured by prolonged escape latencies and an increased number of failures in the shuttle box. The impaired coping of $GR^{+/-}$ mice may be interpreted as a correlate of depression-like behavior (Ridder et al., 2005).

An altered response to stress on the neuroendocrinological level is also found in this strain. Immobilization stress evokes a significant change in the response of the HPA-system in $GR^{+/-}$ animals. Since depression is a multigenetic/multifactorial disease, the dysregulation of a single system may not be sufficient to induce depression-like alterations in mice under basal conditions but render the affected individuals more sensitive to further environmental influences. In this context it is of interest that a selective impairment of GR function in hepatocytes does not affect gluconeogenesis under basal conditions, but evokes a gluconeogenesis deficit only under challenge conditions (by fasting the animals) (Opherk et al., 2004). In accordance with an altered stress sensitivity of the HPA-system, $GR^{+/-}$ mice exhibit a pathological DEX/CRH Test, currently the most relevant biological marker in patients for both, florid depression and the risk to develop a depressive episode. In many depressive patients, the alteration of the HPA-system is also not evident under baseline conditions (e.g. morning cortisol blood levels) but only becomes evident under challenge conditions. Such challenge conditions are artificially modeled by the DEX/CRH Test. In the $GR^{+/-}$ animal model, environmental stress factors provoke a behavioral as well as a neuroendocrinological depression-like phenotype (Ridder et al., 2005). This can be regarded as the strength of the model, because depression-vulnerable humans often develop a manifest depressive episode in response to external or internal stress factors. Another important environmental risk factor that may significantly modify emotional behavior in $GR^{+/-}$ mice could be an early childhood trauma (e.g. maternal separation). So far, respective experiments have not been performed with this strain.

Similar results as for $GR^{+/-}$ mice have also been obtained in a strain with a forebrain-specific GR knockout (FBGRKO): These mice with a forebrain-specific complete GR knockout induced via the calcium-calmodulin-dependent protein kinase II (CamKII) promoter have also a depressive-like phenotype (Boyle et al., 2005). They exhibit an impaired negative feedback regulation of the HPA-axis, as well as increased depression-like behaviors, which have been demonstrated in the Forced Swim and Tail Suspension Tests as well as in terms of anhedonic sucrose preference (Boyle et al., 2005). These findings indicate that the brain areas responsible for the phenotypic changes in $GR^{+/-}$ mice are located in the forebrain. In contrast to $GR^{+/-}$ mice, the depression-like phenotype became apparent at 4 months of age already under basal (unstressed) conditions (Boyle et al., 2005). Additionally, forebrain-specific complete GR knockout mice also exhibit decreased anxiety-like behaviors in the Elevated Plus-Maze (EPM) and the DLB (Boyle et al., 2006). While depression-like symptoms are reversible by chronic treatment with the tricyclic antidepressant imipramine, anxiety-related behaviors do not respond to the treatment (Boyle et al., 2005, 2006). These findings nicely demonstrate that by the use of conditional gene targeting strategies, alterations of GR expression or function in specific brain regions can be correlated with the occurrence of a depressive-like phenotype.

Mice with increased GR expression

As a proof of concept for their results obtained in $GR^{+/-}$ mice, the study by Ridder et al. (2005) has also examined a mouse strain that represents, from a molecular viewpoint, the opposite phenotype of $GR^{+/-}$ mice. These so-called YGR mice overexpress GR by a yeast artificial chromosome, resulting in a twofold increase of GR expression in the brain (Reichardt et al., 2000). One would expect that these animals show a phenotype opposite to the one observed in $GR^{+/-}$ mice. Indeed — in contrast to the increased stress sensitivity of $GR^{+/-}$ mice — YGR mice demonstrate stress resistance on both behavioral and neuroendocrinological levels. YGR mice exhibit less helplessness, reduced stress-induced corticosterone levels and oversuppression in the DEX suppression test (Ridder et al., 2005). Thus, YGR mice confirm the specificity of the findings in $GR^{+/-}$ mice. From a more general point of view, their phenotype is also in good agreement with the *Stress Hypothesis*, predicting a protective effect against stress by GR overexpression.

Using a CamKII-promoter construct, Wei et al. (2004) generated a mouse strain overexpressing GR (GRov) specifically in the forebrain (Wei et al., 2004). In contrast to the stress-resistant YGR mice with a general GR overexpression (see above), forebrain-specific overexpressing GR mice display increased immobility in the Porsolt FST and increased anxiety-like behavior in the EPM and in the DLB (Wei et al., 2004). These somewhat surprising results in light of similar findings obtained in FBGRKO mice have been interpreted by the authors as "increased emotional lability" (Wei et al., 2004). They related their mouse model to mixed or hypomanic states in bipolar disorder.

However, the criteria of face and construct validity favor the notion that the behavioral phenotype observed in GRov-mice rather corresponds to a PTSD-like behavioral syndrome than to bipolar disorder. Depression and anxiety are characteristic features of PTSD. However, several authors see the crucial cause for the development of PTSD in an extraordinarily strong memory of the aversive encounter, maintained, for instance, by resistance to extinction, by memory reinstatement, by overgeneralization or by disturbance of declarative memory due to a stronger automatic processing of trauma-related material (for review, see Siegmund and Wotjak, 2006). These constructs are based on conditioning and sensitization processes that can be investigated in established and standardized mouse models, in particular fear-conditioning paradigms. Today, the molecular and behavioral substrates of classical fear conditioning are well described, and knowledge concerning those of reconsolidation and extinction are advancing (Myers and Davis, 2002; Rodrigues et al., 2004). Furthermore, basic research has demonstrated that single exposure to a severe stressor may cause long-lasting increases in anxiety-like behavior, startle responses and electrical excitability of fear circuits (Adamec et al., 2001, 2005). YGR as well as GRov-mice should be subjected to this type of tests and paradigms in order to assess their validity as potential models of PTSD.

Currently it is not evident why YGR and GRov-mice are behaviorally different. A crucial developmental difference between the two strains is based on the fact that in YGR mice GR overexpression is present during complete development, while the CamKII-promoter driving overexpression of GR in GRov-mice becomes active only in postnatal weeks 2–3. Furthermore, YGR mice have a generalized GR overexpression throughout the brain, while the overexpression is restricted to forebrain areas in GRov-mice. Last but not least, both strains have a different genetic background. In particular the F1-hybrid background (C57BL6 and FVBN) of the YGR mice may be able to reduce the impact of genetic alterations compared to a mixed or inbred genetic backgrounds, a phenomenon that has been called hybrid vigor (Wolfer et al., 2002) (Table 1).

Implications for psychiatry: are results from GR mutant animals transferable?

In line with the concept that depression is a multifactorial disease (Claes, 2004), $GR^{+/-}$ mice represent a model for combined effects of both genetic and environmental manipulations. Their phenotype mimics the human situation in depressive disorders, in which individuals at risk are

Table 1. HPA dysregulation and behavioral symptoms in mice with targeted mutation of the GR

	Baseline HPA system				Challenged HPA system			Behavior			
	Hypothalamic CRH	Pituitary POMC/ ACTH	Plasma ACTH	Plasma cort	Stress-induced ACTH	Stress-induced cort	DEX/CRH test	Anxiety	Despair	Learned helplessness	Locomotion
Human depression	↑	n.d.	↑	↑	↓	↑	↑	↑	↑	↑	↓↑
$GR^{-/-}$	↑	↑	n.d.	n.d.	n.d.	n.d.	n.d.	n.d.	n.d.	n.d.	n.d.
GR^{NesCre}	↑	↑	↓	↑	↓	↔	n.d.	↓	↓	n.d.	↔
GR antisense	↓	↔	↔	↔	↑	↓	↑	↓	↓	n.d.	↔
$GR^{+/-}$	↑	n.d.	↔	↔	n.d.	↑	↑	↔	↔	↑	↔
FBGRKO	↔	n.d.	↔	↑	↑	↑	↑	↓	↑	n.d.	↔
YGR	↓	↓	↑	↔	n.d.	↓	↓	↔	↔	↓	↔
GRov	↔	↔	↔	↔	↔	↔	n.d.	↑	↑	n.d.	n.d.

Notes: Summary of changes in human and murine depression obtained from comparisons with control subjects/wild-type littermates. Mutant mice appear as they are mentioned in the text. Human CRH levels were determined after lumbar puncture in the cerebrospinal fluid; CRH expression in mice was derived from in situ hybridization or immunohistochemistry in the PVN. Challenges of the human HPA system done by CRH injection, not by stress. Parameters of locomotion were not applicable in humans. Despair behavior was assessed by the Forced Swim Test, Learned Helplessness in a Shuttle Box Paradigm. ↑↓ indicate increased and decreased depressive-like features, ↔ = not changed and n.d. = not done.

predisposed to develop depressive episodes after stress (Ridder et al., 2005). In this respect, $GR^{+/-}$ mice differ from mice with a forebrain-specific complete GR knockout (Boyle et al., 2005), which develop depression-like behavioral and neurochemical alterations during adulthood already under baseline conditions, i.e., without additional stress. However, the findings in both strains indicate that compromised GR function constitutes a crucial molecular risk factor in the pathophysiology of depression, indicating that both mouse lines are models with good face validity for further target-oriented physiological, biochemical and pharmacological investigations of GR function with regard to human depressive disorders. Interesting systems to investigate would therefore be the serotonergic and noradrenergic systems (e.g. monoamine tissue levels, monoamine transporter expression and function, pre- and postsynaptic receptor expression). Due to the availability of brain tissue, also neuroanatomical studies are feasible, e.g., with respect to neurogenesis and dendritic spine plasticity. Furthermore, GR mutant mice may be good models to better characterize potential diagnostic tools for depressive patients. Thus, GR deficient mice can be used to calibrate flow cytometry methods for analyses of GR expression of lymphocytes, or might be used to assess the quality of potential GR-ligands for positron emission tomography (PET) investigations (Wüst et al., 2003). With respect to effectiveness and potential mechanisms of pharmacotherapy, these models can be used to study which antidepressive substances work in mouse models of depression based on GR deficiency, and which depressive features are reversed by these substances. Thus, chronic treatment with the tricyclic antidepressant imipramine reverses the behavioral despair phenotype and influences the HPA-axis abnormalities in forebrain specific GR knockout mice (Boyle et al., 2005). In contrast, the anxiety-related behavioral signs observed in this model are not attenuated by treatment with imipramine (Boyle et al., 2006).

With respect to PTSD it seems to be too early to draw a conclusion whether and in which way GR mutant mice could serve as models for this disorder, or mimic at least specific aspects of the clinical syndrome. Both, constitutively GR-overexpressing YGR mice as well as forebrain specific, during adulthood GRov mice have to be studied more closely in specific tests and models that are thought to reflect core signs and symptoms of human PTDS. This includes analyses of conditioning and sensitization processes that can be investigated for instance in fear conditioning and fear extinction paradigms as well as in startle response test.

Conclusions

The findings in mouse strains with deficient glucocorticoid receptor signaling confirm the hypothesis that compromised GR function constitutes a crucial molecular risk factor in the development of depression-like behaviors. These mouse strains seem to be a good tool to further study molecular, pathophysiological and cellular/structural alterations that underlie specific behavioral features such as despair or helplessness. As a first potential molecular correlate for such changes we identified a downregulation of BDNF protein content in the hippocampus of $GR^{+/-}$ mice, which is in good agreement with the "neurotrophin hypothesis of depression". By using brain region or neuronal cell-type specific conditional transgenic strategies it might even be possible to attribute crucial GR functions to specific brain areas or neuronal networks. Of course, other genes apart from the GR may participate in the development of the characteristic alterations observed in GR compromised animals. Thus, preliminary data show alterations of MR expression in GR mutant animals, which should also have functional relevance with respect to the molecular action of corticosteroid receptors as dimers (de Kloet et al., 2007). It has yet to be shown whether such molecular alterations have further detrimental effect with respect to behavioral consequences, or are rather compensatory mechanisms of the organism. Technically, this question might be approached by the use of double-mutant animals, both for MR and GR. Since clinical data indicate that impairment of GR function defines a specific neuroendocrinological endophenotype of (severe) depression, GR mutant animals currently represent a model with good construct and face validity for this subgroup of depressed patients. Some of the patients' symptomatology might be attributable to the same brain regions identified in GR mutant mice. On the therapeutic side, it has been demonstrated that specific behavioral changes in GR mutant mice respond to specific drug treatments (e.g. imipramine), while other behavioral abnormalities remain unchanged (Boyle et al., 2005, 2006). A major challenge is to decipher which signs and symptoms in patients correspond to the animal behavioral constructs (e.g., despair, helplessness), and to elucidate whether it is possible to gain insights from the animals' response to specific treatments for human therapy.

Abbreviations

ACTH	adrenocorticotropic hormone
BDNF	brain-derived neurotrophic factor
CNS	central nervous system
CRE	cAMP response element
CRH	corticotrophin-releasing hormone
DEX	dexamethasone
DLB	Dark-Light Box
e.g.	exempli gratia
EOM	Elevated O-Maze
EPM	Elevated Plus-Maze
et al.	et alteri
FBGRKO	forebrain-specific glucocorticoid receptor knockout
FST	Forced Swim Test
GR	glucocorticoid receptor
GRov	glucocorticoid receptor overexpressing mice
$GR^{+/-}$	heterozygous glucocorticoid receptor knockout mice
HPA	hypothalamic-pituitary-adrenal
i.e.	id est
LH	learned helplessness
MR	mineralocorticoid receptor
Nes	nestin promoter
PET	positron emission tomography
PTSD	posttraumatic stress disorder
SFB	Sonderforschungsbereich
TST	Tail Suspension Test
YGR	glucocorticoid receptor overexpressing mice by a yeast artificial chromosome

Acknowledgments

This work was supported by a grant from the Deutsche Forschungsgemeinschaft to P.G. (SFB636/B3). M.A.V. has a scholarship from the Graduate College 791, University of Heidelberg.

References

Adamec, R.E., Blundell, J. and Burton, P. (2005) Neural circuit changes mediating lasting brain and behavioral response to predator stress. Neurosci. Biobehav. Rev., 29(8): 225–1241.

Adamec, R.E., Blundell, J. and Collins, A. (2001) Neural plasticity and stress induced changes in defense in the rat. Neurosci. Biobehav. Rev., 25(7–8): 721–744.

von Bardeleben, U. and Holsboer, F. (1991) Effect of age on the cortisol response to human corticotropin-releasing hormone in depressed patients pretreated with dexamethasone. Biol. Psychiatry, 29(10): 1042–1050.

Barden, N. (1996) Modulation of glucocorticoid receptor gene expression by antidepressant drugs. Pharmacopsychiatry, 29(1): 12–22.

Barden, N., Stec, I.S., Montkowski, A., Holsboer, F. and Reul, J.M. (1997) Endocrine profile and neuroendocrine challenge tests in transgenic mice expressing antisense RNA against the glucocorticoid receptor. Neuroendocrinology, 66(3): 212–220.

Beato, M., Herrlich, P. and Schütz, G. (1995) Steroid hormone receptors: many actors in search of a plot. Cell, 83: 851–857.

Boyle, M.P., Brewer, J.A., Funatsu, M., Wozniak, D.F., Tsien, J.Z., Izumi, Y., et al. (2005) Acquired deficit of forebrain glucocorticoid receptor produces depression like changes in adrenal axis regulation and behavior. Proc. Natl. Acad. Sci. U.S.A., 102(2): 473–478.

Boyle, M.P., Kolber, B.J., Vogt, S.K., Wozniak, D.F. and Muglia, L.J. (2006) Forebrain glucocorticoid receptors modulate anxiety-associated locomotor activation and adrenal responsiveness. J. Neurosci., 26(7): 1971–1978.

Chourbaji, S., Zacher, C., Sanchis-Segura, C., Dormann, C., Vollmayr, B. and Gass, P. (2005) Learned helplessness: validity and reliability of depressive-like states in mice. Brain Res. Brain Res. Protoc., 16(1–3): 70–78.

Claes, S.J. (2004) CRH, stress, and major depression: a psychobiological interplay. Vitam. Horm., 69: 117–150.

Cole, T.J., Blendy, J.A., Monaghan, A.P., Krieglstein, K., Schmid, W., Aguzzi, A., et al. (1995) Targeted disruption of the glucocorticoid receptor gene blocks adrenergic chromaffin cell development and severely retards lung maturation. Genes Dev., 9(13): 1608–1621.

Cole, T.J., Myles, K., Purton, J.F., Brereton, P.S., Solomon, N.M., Godfrey, D.I., et al. (2001) GRKO mice express an aberrant dexamethasone-binding glucocorticoid receptor, but are profoundly glucocorticoid resistant. Mol. Cell. Endocrinol., 173(1–2): 193–202.

Cryan, J.F. and Mombereau, C. (2004) In search of a depressed mouse: utility of models for studying depression-related behavior in genetically modified mice. Mol. Psychiatry, 9(4): 326–357.

Duman, R.S., Heninger, G.R. and Nestler, E.J. (1997) A molecular and cellular theory of depression. Arch. Gen. Psychiatry, 54(7): 597–606.

Gold, P.W., Drevets, W.C. and Charney, D.S. (2002) New insights into the role of cortisol and the glucocorticoid receptor in severe depression. Biol. Psychiatry, 52(5): 381–385.

Hatzinger, M., Reul, J.M.H.M., Landgraf, R., Holsboer, F. and Neumann, I. (1996) Combined dexamethasone/CRH test in rats: hypothalamo-pituitary-adrenocortical system alterations in aging. Neuroendocrinology, 64: 349–356.

Heuser, I., Yassouridis, A. and Holsboer, F. (1994) The combined dexamethasone/CRH test: a refined laboratory test for psychiatric disorders. J. Psychiatr. Res., 28(4): 341–356.

Holsboer, F. (2000) The corticosteroid receptor hypothesis of depression. Neuropsychopharmacology, 23(5): 477–501.

de Kloet, E.R. (2002) Stress in the brain: implications for treatment of depression. Acta Neuropsychiatrica, 14: 155–166.

de Kloet, E.R., Derijk, R.H. and Meijer, O.C. (2007) Therapy insight: is there an imbalanced response of mineralocorticoid and glucocorticoid receptors in depression? Nat. Clin. Pract. Endocrinol. Metab., 3(2): 168–179.

de Kloet, E.R., Joels, M. and Holsboer, F. (2005) Stress and the brain: from adaptation to disease. Nat. Rev. Neurosci., 6(6): 463–475.

de Kloet, E.R., Oitzl, M.S. and Joels, M. (1999) Stress and cognition: are corticosteroids good or bad guys? Trends Neurosci., 22(10): 422–426.

de Kloet, E.R., Vreugdenhil, E., Oitzl, M.S. and Joels, M. (1998) Brain corticosteroid receptor balance in health and disease. Endocrinol. Rev., 19(3): 269–301.

Korte, S.M. (2001) Corticoids in relation to fear, anxiety and psychopathology. Neurosci. Behav. Rev., 25: 117–142.

Lang, U.E., Hellweg, R. and Gallinat, J. (2004) BDNF serum concentrations in healthy volunteers are associated with depression-related personality traits. Neuropsychopharmacology, 29(4): 795–798.

Layton, B. and Krikorian, R. (2002) Memory mechanisms in posttraumatic stress disorder. J. Neuropsychiatry Clin. Neurosci., 14(3): 254–261.

Lee, A.L., Ogle, W.O. and Sapolsky, R.M. (2002) Stress and depression: possible links to neuron death in the hippocampus. Bipolar Disord., 4(2): 117–128.

McEwen, B.S., Magarinos, A.M. and Reagan, L.P. (2002) Studies of hormone action in the hippocampal formation: possible relevance to depression and diabetes. J. Psychosom. Res., 53(4): 883–890.

Montkowski, A., Barden, N., Wotjak, C., Stec, I., Ganster, J., Meaney, M., et al. (1995) Long-term antidepressant treatment reduces behavioural deficits in transgenic mice with impaired glucocorticoid receptor function. J. Neuroendocrinol., 7(11): 841–845.

Morgan III, C.A., Grillon, C., Southwick, S.M., Davis, M. and Charney, D.S. (1995) Fear-potentiated startle in posttraumatic stress disorder. Biol. Psychiatry, 38(6): 378–385.

Myers, K.M. and Davis, M. (2002) Behavioral and neural analysis of extinction. Neuron, 36(4): 567–584.

Nemeroff, C.B. (1996) The corticotropin-releasing factor (CRF) hypothesis of depression: new findings and new directions. Mol. Psychiatry, 1(4): 336–342.

Nestler, E.J., Barrot, M., DiLeone, R.J., Eisch, A.J., Gold, S.J. and Monteggia, L.M. (2002) Neurobiology of depression. Neuron, 34: 13–25.

Newport, D.J. and Nemeroff, C.B. (2000) Neurobiology of posttraumatic stress disorder. Curr. Opin. Neurobiol., 10: 211–218.

Opherk, C., Tronche, F., Kellendonk, C., Kohlmuller, D., Schulze, A., Schmid, W., et al. (2004) Inactivation of the glucocorticoid receptor in hepatocytes leads to fasting hypoglycemia and ameliorates hyperglycemia in streptozotocin-induced diabetes mellitus. Mol. Endocrinol., 18(6): 1346–1353.

Reichardt, H.M., Umland, T., Bauer, A., Kretz, O. and Schutz, G. (2000) Mice with an increased glucocorticoid receptor gene dosage show enhanced resistance to stress and endotoxic shock. Mol. Cell. Biol., 20(23): 9009–9017.

Reul, J.M., Gesing, A., Droste, S., Stec, I.S., Weber, A., Bachmann, C., et al. (2000) The brain mineralocorticoid receptor: greedy for ligand, mysterious in function. Eur. J. Pharmacol., 405(1–3): 235–249.

Ridder, S., Chourbaji, S., Hellweg, R., Urani, A., Zacher, C., Schmid, W., et al. (2005) Mice with genetically altered glucocorticoid receptor expression show altered sensitivity for stress-induced depressive reactions. J. Neurosci., 25(26): 6243–6250.

Rodrigues, S.M., Schafe, G.E. and LeDoux, J.E. (2004) Molecular mechanisms underlying emotional learning and memory in the lateral amygdala. Neuron, 44(1): 75–91.

Siegmund, A. and Wotjak, C.T. (2006) Toward an animal model of posttraumatic stress disorder. Ann. N.Y. Acad. Sci., 1071: 324–334.

Smith, M.A., Makino, S., Kvetnansky, R. and Post, R.M. (1995) Effects of stress on neurotrophic factor expression in the rat brain. Ann. N.Y. Acad. Sci., 771: 234–239.

Tronche, F., Kellendonk, C., Kretz, O., Gass, P., Anlag, K., Orban, P.C., et al. (1999) Disruption of the glucocorticoid receptor gene in the nervous system results in reduced anxiety. Nat. Genet., 23(1): 99–103.

Tronche, F., Kellendonk, C., Reichardt, H.M. and Schutz, G. (1998) Genetic dissection of glucocorticoid receptor function in mice. Curr. Opin. Genet. Dev., 8(5): 532–538.

Urani, A., Chourbaji, S. and Gass, P. (2005) Mutant mouse models of depression: candidate genes and current mouse lines. Neurosci. Biobehav. Rev., 29(4–5): 805–828.

Urani, A., Chourbaji, S., Henn, F. and Gass, P. (2003) The neurotrophin hypothesis of depression revisited by transgenic mice. Clin. Neurosci. Res., 3: 263–269.

Wei, Q., Lu, X.Y., Liu, L., Schafer, G., Shieh, K.R., Burke, S., et al. (2004) Glucocorticoid receptor overexpression in forebrain: a mouse model of increased emotional lability. Proc. Natl. Acad. Sci. U.S.A., 101(32): 11851–11856.

Wolfer, D.P., Crusio, W.E. and Lipp, H.P. (2002) Knockout mice: simple solutions to the problems of genetic background and flanking genes. Trends Neurosci., 25(7): 336–340.

Wong, M.L. and Licinio, J. (2001) Research and treatment approaches to depression. Nat. Rev. Neurosci., 2(5): 343–351.

Wüst, F., Carlson, K.E. and Katzenellenbogen, J.A. (2003) Synthesis of novel arylpyrazolo corticosteroids as potential ligands for imaging brain glucocorticoid receptors. Steroids, 68: 177–191.

Yehuda, R. (2006) Advances in understanding neuroendocrine alterations in PTSD and their therapeutic implications. Ann. N.Y. Acad. Sci., 1071: 137–166.

Discussion: Chapter 5

SCHMIDT: In your talk you mentioned the heterozygous GR knockouts. Did you also study mice with a heterozygous brain-specific knockout? Do they have a different phenotype compared to the controls?

GASS: That has not been done so far and will probably not be done, because we will have mice with conditional brain region specific deletions of the GR using specific neuronal promoters that drive the CRE-recombinase. I do not believe that there would be much difference between heterozygous brain-specific knockouts compared to the general heterozygous mice we used.

SCHMIDT: And also the pituitary, do you think that if you skip the pituitary in the knockout paradigm, would you still see the same effects?

GASS: I think that the hypothalamus as center of control of the HPA-axis and as area for negative feedback of GR signaling is more important than the pituitary.

DE KLOET: If the heterozygous GR knockout animal with a strongly reduced expression of GR is considered an animal model for depression, would a GR antagonist qualify as an antidepressant?

GASS: This is a complicated story. We should treat our GR heterozygous mice with such compounds and then see what happens. We have not done that.

DE KLOET: I am just wondering about the rationale. The heterozygous GR mutants have reduced GR expression rather than an altered sensitivity of these receptors.

GASS: GR antagonists are effective in rather high doses. When you consider our complete knockout mouse with no GR left in the brain as a model for stress-resistance, a rather complete blockade of GR in the brain may be better for the affected individual, at least at certain times of the disease, than a 50% reduction in function or expression. Additional experiments are necessary to prove that.

JOËLS: If you have these beautiful tools that make this mouse knock out — how would you use these to make your PTSD mouse?

GASS: Originally I was expecting that the learned helplessness model would allow to model PTSD in GR mutant mice, now it has turned out to be a good depression model. I think good models for PTSD will be fear conditioning and fear extinction, and I would rather expect a PTSD-like phenotype in GR overexpressing mice than in GR compromised animals.

E.R. de Kloet, M.S. Oitzl & E. Vermetten (Eds.)
Progress in Brain Research, Vol. 167
ISSN 0079-6123

CHAPTER 6

Adrenal stress hormones, amygdala activation, and memory for emotionally arousing experiences

Benno Roozendaal*, Areg Barsegyan and Sangkwan Lee

Center for the Neurobiology of Learning and Memory, Department of Neurobiology and Behavior, University of California, Irvine CA 92697-3800, USA

Abstract: Extensive evidence indicates that stress hormones released from the adrenal glands are critically involved in memory consolidation of emotionally arousing experiences. Epinephrine or glucocorticoids administered after exposure to emotionally arousing experiences enhance the consolidation of long-term memories of these experiences. Our findings indicate that adrenal stress hormones influence memory consolidation via interactions with arousal-induced activation of noradrenergic mechanisms within the amygdala. In turn, the amygdala regulates memory consolidation via its efferent projections to many other brain regions. In contrast to the enhancing effects on consolidation, high circulating levels of stress hormones impair memory retrieval and working memory. Such effects also require noradrenergic activation of the amygdala and interactions with other brain regions.

Keywords: basolateral amygdala; corticosterone; cortisol; epinephrine; glucocorticoids; memory consolidation; memory retrieval; norepinephrine; stress

Introduction

Emotionally significant experiences typically leave lasting and vivid memories (Bohannon, 1988; Neisser et al., 1996). And, it certainly seems highly adaptive that we record and retain lasting memories of our significant experiences. However, intensely emotional experiences such as automobile accidents, fires, muggings, rapes, wartime battles or terrorists' bombings can also create maladaptive traumatic memories and result in the development of posttraumatic stress disorder (PTSD). The scope of this paper is to summarize recent findings of animal and human experiments on the effects of emotional arousal on brain activity in influencing memory. These findings may provide some understanding of the neurobiological processes that underlie traumatic memories and PTSD as well as some possible implications for therapeutic intervention.

There is extensive evidence that stress hormones released from the adrenal glands are critically involved in regulating memory consolidation of emotionally arousing experiences. Epinephrine or glucocorticoids, as well as specific agonists for their receptors, administered after exposure to emotionally arousing experiences enhance the consolidation of long-term memories of these experiences (McGaugh and Roozendaal, 2002). As is discussed below, the findings indicate that adrenal stress hormone effects on the enhancement of memory consolidation depend on interactions with arousal-induced noradrenergic activation of the

*Corresponding author. Tel.: +1 (949) 824-5250; Fax: +1 (949) 824 2952; E-mail: broozend@uci.edu

DOI: 10.1016/S0079-6123(07)67006-X

amygdala. Such amygdala activation strengthens the storage of different kinds of information via its projections to many other brain regions. However, stress and emotional arousal do not only induce strong memories of new information; they can also impair our remembering. The evidence from many animal and human studies indicates that the mechanisms that enhance the consolidation of new, emotionally arousing experiences are also responsible for the stress-induced impairment of memory retrieval and working memory.

Adrenal stress hormone effects on memory consolidation require emotional arousal

It is well established that exposure to stressful or emotionally arousing events leads to the activation of the autonomic nervous system and hypothalamic-pituitary-adrenocortical axis (HPA axis). Activation of the autonomic nervous system results in the release of catecholamines from the adrenal medulla and sympathetic nerve endings, whereas activation of the HPA axis culminates in the release of cortisol (corticosterone in rodents) into general circulation. The degree to which these hormonal systems are activated depends on a variety of parameters, including age and gender of the subjects as well as the severity and type of stressor employed (Kopin, 1995; Korte, 2001). As removal of these hormones by adrenalectomy or synthesis blockade impairs memory consolidation (Oitzl and de Kloet, 1992; Roozendaal et al., 1996a, c; Pugh et al., 1997; Liu et al., 1999), such evidence indicates that stress hormones may act as endogenous modulators of memory consolidation.

Facilitatory effects of epinephrine and glucocorticoid administration on memory consolidation have been found in animal and human subjects with many different types of training (Gold and van Buskirk, 1975; Flood et al., 1978; Introini-Collison and McGaugh, 1986; Liang et al., 1986; Lupien and McEwen, 1997; Costa-Miserachs et al., 1994; Sandi and Rose, 1997; Roozendaal, 2000; Buchanan and Lovallo, 2001; Cahill and Alkire, 2003). Glucocorticoid administration also facilitates the consolidation of memory of fear extinction whereas suppression of glucocorticoid function impairs fear extinction memory consolidation (Bohus and Lissak, 1968; Barrett and Gonzalez-Lima, 2004; Cai et al., 2006; Yang et al., 2006). The memory-enhancing effects of epinephrine and glucocorticoid administration are dose- and time-dependent. Moderate doses of posttraining epinephrine or glucocorticoids enhance retention performance, whereas lower or higher doses are less effective. And, consistent with the evidence that these hormones facilitate time-dependent processes underlying long-term consolidation of the memory trace, memory enhancement is greatest when the injections are administered shortly after training (Gold and van Buskirk, 1975).

Evidence from several kinds of studies indicates that these hormones do not act in isolation but, rather, that synergistic actions of epinephrine and corticosterone may be essential in mediating stress effects on memory enhancement. Borrell et al. (1983, 1984) reported that glucocorticoids alter the sensitivity of epinephrine in influencing memory consolidation in adrenalectomized rats. Further, in adrenally intact rats, administration of metyrapone, a corticosterone-synthesis inhibitor that reduces the elevation of circulating corticosterone induced by aversive stimulation, attenuates the memory-enhancing effects of epinephrine administered posttraining (Roozendaal et al., 1996b).

Importantly, recent evidence suggests that adrenal stress hormones do not enhance memory consolidation of all kinds of training, but preferentially modulate memory consolidation of emotionally arousing experiences. Learning tasks in animal experiments are often emotionally arousing because of the motivational stimulation necessary to elicit changes in behavior. It is obvious that with the use of such experimental conditions it is not possible to determine whether emotional arousal is a prerequisite in regulating stress hormone influences on memory processes. We recently investigated the importance of emotional arousal in mediating glucocorticoid effects on memory consolidation in rats trained on an object recognition task (Okuda et al., 2004). Although no rewarding or aversive stimulation is used during object recognition training, such training induces modest novelty-induced stress or arousal (de Boer et al., 1990). However, extensive habituation of rats to the training apparatus (in the absence of any objects) prior to the training reduces the arousal level induced by object recognition training. We found that corticosterone administered systemically immediately after training enhanced 24-h retention performance of rats that were not previously habituated to the experimental context (i.e., emotionally aroused rats). In contrast, corticosterone did not affect 24-h retention of rats that had received extensive prior habituation to the experimental context and, thus, had decreased novelty-induced emotional arousal during training (Okuda et al., 2004).

An intimate link between the level of emotional arousal at encoding and the efficacy of adrenal stress hormones in influencing memory consolidation has also been demonstrated in humans. Cortisol administration selectively enhanced long-term memory of emotionally arousing, but not emotionally neutral, items (Buchanan and Lovallo, 2001; Kuhlmann and Wolf, 2006). Consistent with these findings, Abercrombie et al. (2006) reported that levels of endogenous cortisol correlated with enhanced memory consolidation only in individuals who were emotionally aroused. Moreover, epinephrine or cold pressor stress (known to induce the release of epinephrine and cortisol) enhanced memory consolidation only when associated with emotional arousal at encoding (Cahill and Alkire, 2003; Cahill et al., 2003). Collectively, these findings indicate that at least some degree of training-associated endogenous emotional arousal is essential for enabling stress hormone effects on memory consolidation.

Role of the amygdala in mediating stress hormone effects on memory consolidation

A critical question raised by this evidence is why stress hormones may selectively enhance memory consolidation of emotionally arousing experiences. Our findings suggest that interactions between stress hormones and amygdala activity may be key in determining this selectivity. It is well established that emotional experiences that induce the release of adrenal stress hormones also activate the amygdala (Campeau et al., 1991; Pelletier et al., 2005). Extensive evidence from our as well as other laboratories indicates that the enhancing effects of stress hormone administration on the consolidation of memory of emotionally arousing experiences involve the amygdala. Lesions of the basolateral complex of the amygdala (BLA; consisting of the lateral, basal and accessory basal nuclei) block the memory-modulatory effects induced by posttraining systemic injections of epinephrine or glucocorticoids as well as that by drugs affecting a variety of other neuromodulatory systems (Cahill and McGaugh, 1991; Roozendaal and McGaugh, 1996a; McGaugh, 2004). As lesions of the adjacent central nucleus of the amygdala (CEA) are ineffective, the BLA appears to be the critical region of the amygdala in mediating stress and arousal effects on memory consolidation. Moreover, and in support of this view, posttraining infusions of a specific glucocorticoid receptor (GR) agonist administered into the BLA, but not the CEA, enhance memory consolidation, whereas intra-BLA infusions of a GR antagonist impair memory consolidation or block the facilitatory effects of stress exposure (Roozendaal and McGaugh, 1997a; Conrad et al., 2004; Donley et al., 2005). These findings thus indicate that the BLA may be a general, but critical, gateway in mediating stress hormone effects on memory consolidation.

The enhancing effects of adrenal stress hormones on memory consolidation depend on the integrity of the amygdala noradrenergic system (McGaugh, 2000). Studies using in vivo microdialysis and HPLC have shown that emotional arousal induces the release of norepinephrine in the amygdala (Galvez et al., 1996; Quirarte et al., 1998). Furthermore, amygdala norepinephrine levels assessed following aversively motivated inhibitory avoidance training correlate with retention latencies tested 24 h later (McIntyre et al., 2002), whereas posttraining infusions of norepinephrine or β-adrenoceptor agonists into the amygdala (or selectively into the BLA) produce dose-dependent enhancement of memory consolidation (Liang et al., 1986; Ferry et al., 1999; Hatfield and McGaugh, 1999; LaLumiere et al., 2003). Williams et al. (1998) reported that epinephrine administered systemically immediately after inhibitory avoidance training increased norepinephrine levels in the amygdala. As epinephrine is a polar substance that does not readily cross the blood-brain barrier, a peripheral-central pathway is most likely involved in mediating epinephrine effects on amygdala activity in modulating memory consolidation (McGaugh et al., 1996; Williams and Clayton, 2001). It is now well established that systemic epinephrine can activate peripheral β-adrenoceptors located on vagal afferents terminating in the nucleus of the solitary tract (NTS). In turn, noradrenergic cell groups in the NTS send direct projections to the amygdala (Fallon and Ciofi, 1992). In addition, the NTS regulates noradrenergic activity of the forebrain via indirect projections to noradrenergic cell groups in the locus coeruleus (Williams and Clayton, 2001). In support of the view that epinephrine effects on memory consolidation require noradrenergic activity within the BLA, infusions of a β-adrenoceptor antagonist administered into the amygdala block the memory-enhancing effects of peripherally administered epinephrine (Liang et al., 1986).

Glucocorticoids also require noradrenergic activation within the BLA to influence memory for emotionally arousing training. A β-adrenoceptor antagonist infused into the BLA blocks the memory-enhancing effect of systemically administered glucocorticoids (Quirarte et al., 1997; Roozendaal et al., 2006a, b). Unlike catecholamines, glucocorticoid hormones readily enter the brain and bind directly to adrenal steroid receptors in the BLA and other brain regions (Reul and de Kloet, 1985; de Kloet, 1991). Extensive evidence indicates that glucocorticoid effects on memory consolidation involve a selective activation of the high-affinity GR (Oitzl and de Kloet, 1992; Roozendaal et al., 1996c). Glucocorticoids are known to act through intracellular and intranuclear receptors and can affect gene transcription by direct binding of receptor homodimers to DNA or via protein–protein interactions with other transcription factors (Beato et al., 1995; Datson et al., 2001; Oitzl et al., 2001). However, glucocorticoids may also act more rapidly by interacting with membrane-associated receptors (Orchinik et al., 1991; Karst et al., 2005; Tasker et al., 2006). In line with this evidence, we reported that GR activation in the BLA may facilitate memory consolidation via a rapid potentiation of the norepinephrine signal cascade through an interaction with G-protein-mediated effects (Roozendaal et al., 2002). Activation of β-adrenoceptors in the BLA enhances memory consolidation via stimulation of the adenosine 3′,5′-cyclic monophosphate (cAMP)/protein kinase A pathway (Ferry et al., 1999). Intra-BLA infusions of the GR antagonist RU 38486 (mifepristone) attenuated the dose-response effects of a β-adrenoceptor agonist on retention enhancement for inhibitory avoidance training (Roozendaal et al., 2002). As the GR antagonist had no effect on memory enhancement induced by posttraining intra-BLA infusions of the synthetic cAMP analog 8-Br-cAMP, these findings indicate that cAMP activation is downstream from the interaction of glucocorticoids with the noradrenergic system. In addition to such postsynaptic actions, glucocorticoid administration may increase the availability of norepinephrine in the BLA via an activation of GRs located in brainstem noradrenergic cell groups. Posttraining infusions of the GR agonist RU 28362 into the NTS dose-dependently enhanced memory consolidation of inhibitory avoidance training and the memory enhancement was blocked by intra-BLA infusions of a β-adrenoceptor antagonist (Roozendaal et al., 1999b). The findings of a recent in vivo microdialysis experiment support the view that glucocorticoids may facilitate the training-induced release of norepinephrine in the amygdala (McIntyre et al., 2004). Figure 1 summarizes the interactions of adrenal stress hormones with the noradrenergic system in the BLA that are involved in regulating memory consolidation. Extensive evidence indicates that noradrenergic activity within the BLA also plays a critical role in mediating the modulatory effects of other hormones and neurotransmitters, including corticotropin-releasing hormone, vasopressin, opioid peptides and adrenocorticotropin, on memory consolidation (Roozendaal, 2007).

Based on the evidence summarized above, it may be hypothesized that emotional arousal-induced increases in noradrenergic activity within the BLA are essential in enabling stress hormone effects on memory consolidation. Such a mechanism may then provide a direct explanation of the findings that stress hormones selectively enhance memory consolidation of emotionally arousing

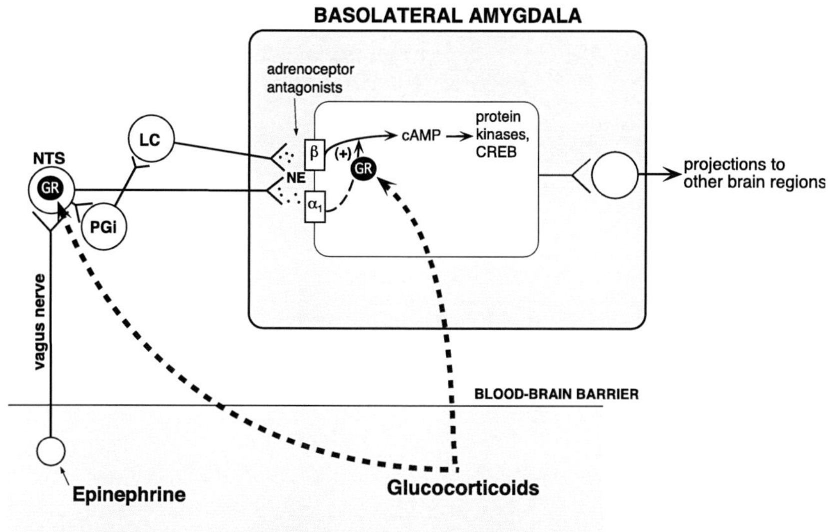

Fig. 1. Interactions of adrenal stress hormones with the noradrenergic system in the BLA in modulating memory consolidation. Adrenal stress hormones are released during training in emotionally arousing tasks and are known to enhance memory consolidation. Epinephrine, which does not cross the blood-brain barrier, induces the release of norepinephrine (NE) in the BLA by activating vagal afferents to the nucleus of the solitary tract (NTS). Noradrenergic neurons in the NTS project directly to the BLA, and indirectly via the locus coeruleus (LC). Norepinephrine binds to both β-adrenoceptors and α_1-adrenoceptors at postsynaptic sites and activates cAMP formation. Glucocorticoids freely enter the brain and bind to glucocorticoid receptors (GRs) in brainstem noradrenergic neurons to potentiate norepinephrine release in the BLA, as well as postsynaptically in BLA neurons to facilitate the norepinephrine signaling cascade. Glucocorticoids may influence the β-adrenoceptor-cAMP system via a coupling with α_1-adrenoceptors. These stress hormone effects on noradrenergic activation in the BLA are required for regulating memory consolidation in other brain regions. $\alpha_1 = \alpha_1$-adrenoceptor; β = β-adrenoceptor; cAMP = adenosine 3′,5′-cyclic monophosphate; CREB, cAMP response-element binding; PGi, nucleus paragigantocellularis. (Reprinted from Roozendaal (2000), with permission.)

experiences. We recently investigated this issue in rats trained on an object recognition task. As is discussed above, corticosterone enhances memory of object recognition training when administered to naïve rats, but is ineffective in rats that have reduced training-associated emotional arousal because of prior habituation to the experimental context (Okuda et al., 2004). As is shown in Fig. 2A, we found that, in non-habituated (i.e., emotionally aroused) rats, the β-adrenoceptor antagonist propranolol blocked the corticosterone-induced memory enhancement (Roozendaal et al., 2006b). Propranolol infused directly into the BLA also blocked the enhancing effects of corticosterone on object recognition memory. To determine whether the failure of corticosterone to enhance memory consolidation under low-arousing conditions is due to insufficient training-induced noradrenergic activation, low doses of the α_2-adrenoceptor antagonist yohimbine, which increases norepinephrine levels in the brain, were co-administered with the corticosterone to well-habituated rats immediately after object recognition training. The critical finding of this study was that such an augmented noradrenergic tone was sufficient to mimic the effects of emotional arousal in that simultaneously administered corticosterone-enhanced memory consolidation (Fig. 2B, Roozendaal et al., 2006b). Further, in habituated rats, corticosterone activated BLA neurons, as assessed by phosphorylated cAMP response-element binding (pCREB) protein immunoreactivity levels, only in animals also given yohimbine. Such observations strongly suggest that because glucocorticoid (and epinephrine) effects on memory consolidation require noradrenergic activation within the BLA, they only modulate memory under emotionally arousing conditions that induce the release of norepinephrine.

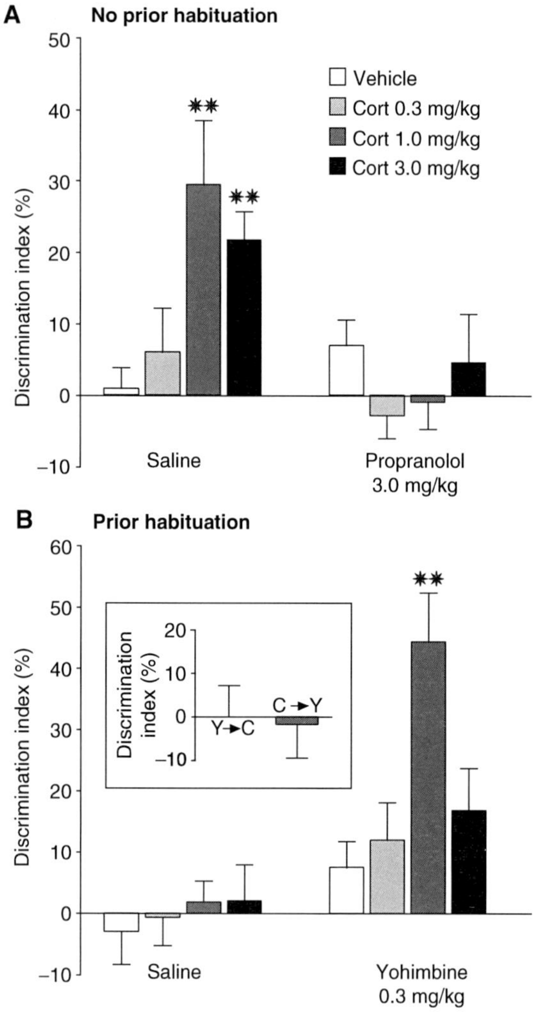

Fig. 2. Glucocorticoid effects on memory consolidation for object recognition training require noradrenergic activation. Data represent discrimination index (mean ± SEM) in percent on a 24-h retention trial. The discrimination index was calculated as the difference in time spent exploring the novel and familiar objects, expressed as the ratio of the total time spent exploring both objects. A: Effects of immediate posttraining administration of the β-adrenoceptor antagonist propranolol (3.0 mg/kg, s.c.) on corticosterone-induced enhancement of object recognition memory in naïve rats. B: Effect of co-administration of the α_2-adrenoceptor antagonist yohimbine (0.3 mg/kg, s.c.) with corticosterone on object recognition memory in habituated rats. $**P<0.01$ compared with the corresponding vehicle group. B (Inset) Effect of posttraining injections of yohimbine (0.3 mg/kg, s.c.) and corticosterone (1.0 mg/kg, s.c.) separated by a 4-h delay. Y→C, yohimbine administered immediately after training and corticosterone 4 h later; C→Y, corticosterone administered immediately after training and yohimbine 4 h later. (Reprinted from Roozendaal et al. (2006b), with permission.)

Human studies have also provided evidence that stress hormone effects on memory enhancement for emotionally arousing experiences require amygdala activity. Memory for emotionally arousing material is not enhanced in human subjects with selective bilateral damage to the amygdala, as it is in normal subjects (Cahill et al., 1995; Adolphs et al., 1997). Studies using brain positron emission tomography (PET) and functional magnetic resonance imaging techniques have provided evidence that amygdala activity during viewing of emotionally arousing stimuli correlates highly with the subjects' recall of the material assessed weeks later (Cahill et al., 1996; Hamann et al., 1999, 2002; Canli et al., 2000). Furthermore, the relationship between amygdala activity during encoding and subsequent memory was greatest for the material rated as being the most emotionally intense. In a recent study, van Stegeren et al. (2007) showed that amygdala activity during viewing of emotionally arousing pictures was greatest for those subjects who responded with a large increase in endogenous cortisol. And, importantly, a β-adrenoceptor antagonist blocked the increase in amygdala activity and enhanced retention induced by either emotional arousal or endogenous cortisol (Cahill et al., 1994; Strange and Dolan, 2004; van Stegeren et al., 2005, 2007).

Interactions of the amygdala with other brain regions

Adrenal stress hormone effects on memory consolidation have been obtained in experiments using many different kinds of emotionally arousing training, including inhibitory avoidance, contextual and cued fear conditioning, water-maze spatial and cued training, object recognition and conditioned taste aversion. And, as these different training experiences are known to engage different brain systems during both training and the consolidation occurring after training (Izquierdo et al., 1997; Packard and Knowlton, 2002; Gold, 2004), the stress hormone-induced modulation no doubt involves receptor activation in a variety of brain regions. In support of this view, corticosterone enhances memory consolidation of conditioned taste aversion when infused posttraining into the insular cortex or BLA, but is ineffective when administered into the hippocampus, a region that does not play a significant role in learning and memory of conditioned taste aversion (Miranda et al., unpublished observation). However, consistent with a role of the hippocampus in spatial/contextual learning and memory (Morris et al., 1982; Maren and Fanselow, 1997; Sacchetti et al., 1999), glucocorticoids administered into the hippocampus influence memory of water-maze spatial training (Roozendaal and McGaugh, 1997b). Glucocorticoids infused into the hippocampus also enhance memory of inhibitory avoidance training (Roozendaal and McGaugh, 1997b). In fear conditioning tasks, including inhibitory avoidance, the rats learn that footshock occurs in a specific context. Further, that information can be learned if rats are first exposed to the context and then, on a subsequent day, given a brief footshock (Malin and McGaugh, 2006). As is shown in Fig. 3, we found that infusions of a GR agonist administered into the hippocampus after context exposure enhanced the subsequent conditioning whereas infusions administered after the footshock training were ineffective (Lee and Roozendaal, unpublished observation). In contrast, the GR agonist infused into the BLA enhanced retention when administered after either the context or footshock training, consistent with extensive evidence that BLA activity modulates memory for many different kinds of experiences. Other findings suggest that corticosterone infusions into the caudate nucleus may selectively modulate the consolidation of memory of implicit (i.e., stimulus-response) forms of learning (Medina et al., 2007). Thus, these recent findings indicate that glucocorticoids act in different brain regions to enhance the consolidation of different aspects of information acquired during the training.

Although the BLA plays a critical role in regulating stress hormone effects on memory consolidation of many different types of training, extensive evidence indicates that it is not a permanent storage site of such memory traces (McGaugh, 2004). The findings of many recent studies indicate that the BLA interacts with other brain regions in regulating memory consolidation of different kinds of information (McGaugh, 2002). The BLA interacts with the hippocampus in regulating stress effectson memory of spatial/contextual information (Packard et al., 1994). The BLA projects both directly and indirectly (via the entorhinal cortex) to the hippocampus (Pikkarainen et al., 1999; Petrovich et al., 2001). As indicated, infusion of the GR agonist RU 28362 into the hippocampus enhances memory consolidation of inhibitory avoidance training. Importantly, lesions of the BLA or infusions of a

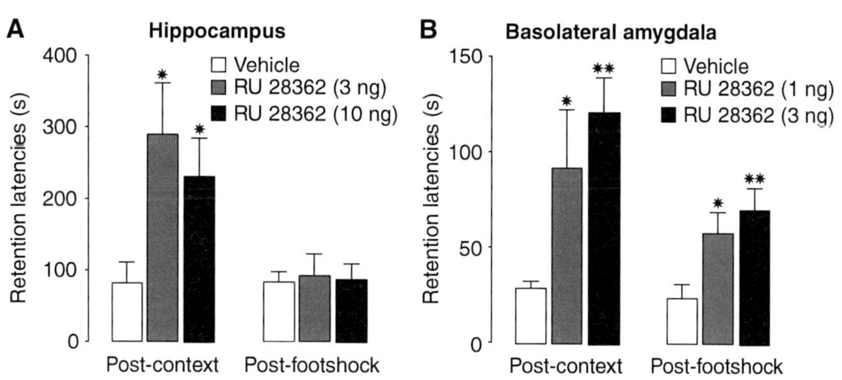

Fig. 3. Differential effect of glucocorticoid administration into the hippocampus and BLA on memory for context and footshock. Rats explored an inhibitory avoidance apparatus for 3 min and then, on a subsequent day, given a brief footshock in that context. Data represent 48-h retention latencies (mean + SEM) in seconds. A: Posttraining infusions of the GR agonist RU 28362 (3 or 10 ng in 0.5 µl) into the dorsal hippocampus enhanced inhibitory avoidance retention latencies when administered after context exposure but not after the shock exposure. B: Posttraining infusions of RU 28362 (1 or 3 ng in 0.2 µl) into the BLA enhanced inhibitory avoidance retention latencies when administered after either the context exposure or the shock experience. $*p<0.05$; $**p<0.01$ compared with the corresponding vehicle group (Lee and Roozendaal, unpublished observation).

β-adrenoceptor antagonist into the BLA block the enhancing effect of posttraining intra-hippocampal infusions of the GR agonist (Roozendaal and McGaugh, 1997b; Roozendaal et al., 1999a). BLA lesions also block the impairing effect of a GR antagonist infused into the hippocampus on spatial memory in a water maze (Roozendaal and McGaugh, 1997b). Thus, these findings indicate that an intact and functional BLA is required for enabling memory modulation of spatial/contextual information induced by manipulation of GR activity in the hippocampus. Similarly, electrophysiological findings indicate that BLA activity influences stress- or perforant path stimulation-induced long-term potentiation in the hippocampus (Ikegaya et al., 1994, 1997; Akirav and Richter-Levin, 1999; Frey et al., 2001; Nakao et al., 2004; Pape et al., 2005). Norepinephrine and corticosterone both influence the effects of BLA stimulation on synaptic plasticity in the dentate gyrus, but not CA1 area (Akirav and Richter-Levin, 2002; Vouimba et al., 2007). Findings of human imaging studies also indicate that amygdala activation influences memory processing involving the hippocampus. Activity of the amygdala and (para)hippocampal region is correlated during emotional arousal (Hamann et al., 1999; Richardson et al., 2004) and such activation is correlated with subsequent retention (Dolcos et al., 2004). The findings of a "path analysis" (structural equation modeling) study (Kilpatrick and Cahill, 2003) of amygdala activity scanned, using PET, while subjects viewed neutral or emotionally arousing films (Cahill et al., 1996) suggest that emotional arousal increased amygdala influences on activity of the ipsilateral parahippocampal gyrus (and ventrolateral prefrontal cortex). Thus, these studies of emotionally influenced memory in human subjects are consistent with findings of animal experiments and indicate that emotional arousal-induced amygdala (BLA) activation may be a critical step in enabling stress hormone effects in modulating memory processes involving other brain regions, including hippocampus-dependent explicit/declarative memory (Fig. 4).

Although BLA influences on hippocampus-dependent memory processes may depend on direct connections between both brain regions, other evidence indicates that converging influences of the BLA and hippocampus onto the nucleus accumbens are also involved. The BLA projects to the nucleus accumbens primarily via the stria terminalis (Kelley et al., 1982; Wright et al., 1996). The possible involvement of the BLA-nucleus accumbens pathway in modulating memory consolidation was suggested by the finding that lesions of the stria terminalis or the nucleus accumbens, like lesions of the BLA, blocked the enhancing effects of systemically administered dexamethasone, a synthetic glucocorticoid, on inhibitory avoidance memory (Roozendaal and McGaugh, 1996b; Setlow et al., 2000). Furthermore, the finding that unilateral lesions of the BLA combined with contralateral (unilateral) lesions of the nucleus accumbens also blocked the dexamethasone effect indicates that these two structures interact in influencing memory consolidation (Setlow et al., 2000). Hippocampal glucocorticoid influences on memory also require a functional BLA-nucleus accumbens pathway. Like BLA lesions, bilateral lesions of either the stria terminalis or nucleus accumbens block the memory-enhancing effects of posttraining intra-hippocampal infusions of RU 28362 (Roozendaal et al., 2001). As both the BLA and the hippocampus are known to project to the nucleus accumbens, these findings suggest that this brain region may be a critical locus of converging BLA and hippocampal modulatory influences on memory consolidation (Mulder et al., 1998). It is not known whether the BLA-nucleus accumbens pathway is also involved in mediating BLA neuromodulatory influences on memory consolidation involving other brain regions.

The BLA also interacts with the medial prefrontal cortex (mPFC) in mediating stress hormone effects on memory consolidation. However, the findings suggest that this interaction may serve quite a different role in memory than that described between the BLA and hippocampus. The mPFC is primarily involved in higher cognitive functions such as thought, decision-making and working memory (Ramnani and Owen, 2004; Fellows and Farah, 2007) and exerts a strong inhibitory (i.e., fear-reducing) influence on behaviors (Amat et al., 2005). Stress exposure is known to impair this mPFC-dependent inhibitory control (Lyons et al., 2000). We found that posttraining infusions of the GR agonist RU 28362

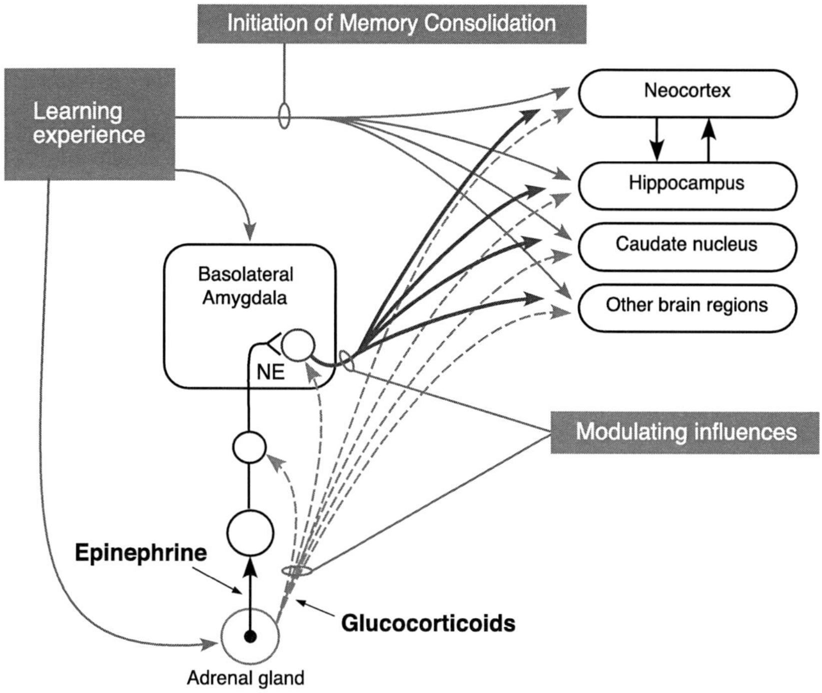

Fig. 4. Emotional arousal-induced modulation of memory consolidation. Experiences initiate memory storage in many brain regions involved in the forms of memory represented. Emotionally arousing experiences also release adrenal epinephrine and glucocorticoids and activate the release of norepinephrine in the BLA. The BLA modulates memory consolidation by influencing neuroplasticity in other brain regions. (Reprinted from McGaugh (2000), with permission.)

administered into the mPFC enhanced memory consolidation of inhibitory avoidance training (Roozendaal et al., unpublished observation; cf. Liang, 2001; Runyan et al., 2004). However, consistent with evidence of inhibitory influences between the mPFC and BLA (McDonald, 1991; Perez-Jaranay and Vives, 1991; Rosenkranz and Grace, 2002), RU 28362 infused into the mPFC after inhibitory avoidance training, but not in non-trained animals, also increased BLA activity, as assessed with phosphorylation levels of extracellular-regulated kinase type 1/2 (ERK1/2), a member of the mitogen-activated protein kinase family (Roozendaal et al., unpublished observation). Furthermore and importantly, blockade of this increase in phosphorylated ERK1/2 levels in the BLA with the MEK inhibitor PD98059 or lesions of the BLA blocked the memory enhancement induced by intra-mPFC infusions of the GR agonist. Thus, these findings suggest that a GR agonist infused into the mPFC may enhance memory consolidation via a stimulatory influence on BLA activity, i.e., a loss of inhibitory control.

Role of the amygdala in mediating adrenal stress hormone effects on memory retrieval and working memory

Most studies investigating adrenal stress hormone effects on memory have focused on the neurobiological mechanisms underlying the consolidation of recent experiences. However, evidence indicates that adrenal stress hormones also influence memory retrieval and working memory. Most of these studies investigated the effects of either peripherally administered stress hormones or examined the effects of direct infusions into the hippocampus and mPFC, brain regions that are critically involved in regulating memory retrieval and working memory. However, consistent with its role in memory consolidation, recent findings indicate that the

BLA, via its projections to these brain regions, plays an important modulatory role in regulating stress hormone effects on these memory functions.

Memory retrieval

Several studies investigated the effects of peripherally administered stress hormones on memory retrieval in rats. Stress exposure or the glucocorticoid corticosterone administered systemically shortly before retention testing on inhibitory avoidance and water-maze spatial tasks, 24 h after training, induce retention impairment (de Quervain et al., 1998; Yang et al., 2003; Rashidy-Pour et al., 2004; Roozendaal et al., 2004c; Sajadi et al., 2006). Extensive evidence indicates that glucocorticoids can affect such retention performance by impairing the retrieval of previously learned information. Likewise, stress-level doses of cortisol or cortisone administered to human subjects impair delayed recall on episodic tasks (de Quervain et al., 2000; Wolf et al., 2001; Buss et al., 2004; Het et al., 2005; Kuhlmann et al., 2005a, b; Buchanan et al., 2006). In contrast to stress hormone effects on memory consolidation, the impairing effects of glucocorticoids on memory retrieval are temporary and subside when the hormone levels return to baseline. Additionally, as protein synthesis blockade does not prevent glucocorticoid influences on memory retrieval (Sajadi et al., 2006), the effects are likely mediated through a non-genomic, possibly membrane receptor-mediated, mechanism. Glucocorticoid-induced memory retrieval impairment depends, in part, on GR activation in the hippocampus, a brain region importantly involved in memory retrieval. The GR agonist RU 28362 administered into the hippocampus shortly before retention testing impairs the retrieval of spatial memory (Roozendaal et al., 2003, 2004a). Moreover, findings from a PET study in human subjects indicate that cortisone administration reduces activity of the parahippocampal gyrus during the retrieval of episodic information (de Quervain et al., 2003). The effects of epinephrine on memory retrieval have not been investigated.

Glucocorticoid effects on memory retrieval are highly comparable to those of studies obtained in memory consolidation in that the effects depend critically on an interaction with noradrenergic mechanisms. Systemic administration of the β-adrenoceptor antagonist propranolol blocks the memory retrieval impairment induced by concurrent injections of corticosterone (Roozendaal et al., 2004c). A β-adrenoceptor antagonist infused into the hippocampus also prevents the retrieval-impairing effect of a GR agonist administered concurrently (Roozendaal et al., 2004a). As stimulation of β_1-adrenoceptors with systemic injections of the selective agonist xamoterol induces memory retrieval impairment comparable to that seen after corticosterone administration (Roozendaal et al., 2004a), the findings suggest that glucocorticoid effects on memory retrieval impairment involve a facilitation of noradrenergic mechanisms. Other evidence from animal studies indicates that the BLA interacts with the hippocampus in mediating glucocorticoid effects on memory retrieval impairment. Lesions of the BLA or, as is shown in Fig. 5, infusions of a β-adrenoceptor antagonist into the BLA block the impairing effect of a GR agonist infused into

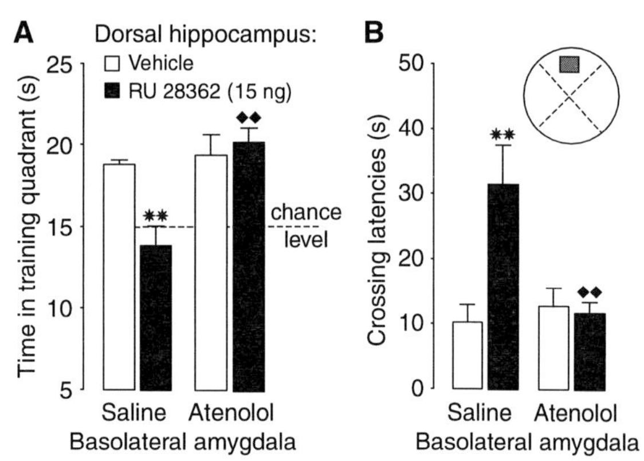

Fig. 5. Interaction between the BLA and hippocampus in mediating glucocorticoid effects on memory retrieval. The β_1-adrenoceptor antagonist atenolol (0.5 μg in 0.2 μl) administered into the BLA 60 min before retention testing blocked the impairing effects induced by the GR agonist RU 28362 (15 ng in 0.5 μl) infused into the hippocampus 60 min before retention testing on probe-trial retention performance in a water maze. A: Time spent in the training quadrant (mean + SEM) in seconds during the 60-s probe trial. B: Latencies (mean + SEM) in seconds to cross the former platform location. B (Inset) Location of the platform and the four quadrants in the maze. $**p<0.01$ compared with the corresponding vehicle group; ◆◆$p<0.01$ compared with the saline-RU 28362 group. (Reprinted from Roozendaal et al. (2004a), with permission.)

the hippocampus on memory retrieval (Roozendaal et al., 2003, 2004a). Such evidence is consistent with that from studies examining stress hormone effects on memory retrieval in humans. Glucocorticoids or psychosocial stress only impair retrieval of emotionally arousing information or during emotionally arousing test conditions (Kuhlmann et al., 2005a, b; Buchanan et al., 2006; de Quervain et al., 2007). Further, the β-adrenoceptor antagonist propranolol blocks the glucocorticoid-induced impairment of memory retrieval in humans (de Quervain et al., 2007). Findings of imaging studies in human subjects indicate that successful retrieval of emotionally arousing information induces greater activity in and connectivity between the amygdala and hippocampus than retrieval of emotionally neutral information (Dolan, 2000; Dolcos et al., 2005; Greenberg et al., 2005; Smith et al., 2006).

Working memory

Evidence from lesion, pharmacological, imaging and clinical studies indicates that working memory, a dynamic processes whereby information is updated continuously, depends on the integrity of the mPFC (Brito et al., 1982; Fuster, 1991). Stress exposure impairs performance of rats on a delayed alternation task, a task commonly used to assess working memory in rodents (Arnsten and Goldman-Rakic, 1998). Basal levels of endogenous glucocorticoids are required to maintain prefrontal cortical function (Mizoguchi et al., 2004), but systemic injections of stress doses of corticosterone or intra-mPFC administration of the GR agonist RU 28362 impair delayed alternation performance in rats (Roozendaal et al., 2004b). As similar GR agonist infusions into the mPFC do not impair delayed alternation performance on non-mnemonic control tasks that have similar motivational and motor demands (Barsegyan and Roozendaal, unpublished observation), these findings strongly suggest that glucocorticoids, via GR activation, impair working memory. Additionally, stress-level cortisol treatment is known to impair working memory performance in human subjects (Lupien et al., 1999; Young et al., 1999; Wolf et al., 2001). Importantly, glucocorticoids interact with noradrenergic mechanisms in inducing working memory impairment. A β-adrenoceptor antagonist administered systemically blocks the impairing effect of corticosterone on working memory in rats (Roozendaal et al., 2004b). Furthermore, as is shown in Fig. 6, a β-adrenoceptor antagonist or cAMP blocker infused into the mPFC also blocks working memory impairment induced by a GR agonist administered concurrently (Barsegyan and Roozendaal, unpublished observation). Animal studies have shown that glucocorticoid effects on working memory also depend on functional interactions between the BLA and the mPFC. Disruption of BLA activity blocks the effect on working memory of a GR agonist administered into the mPFC (Roozendaal et al., 2004b). This evidence provides strong support for the hypothesis that BLA activity modulates stress and emotional

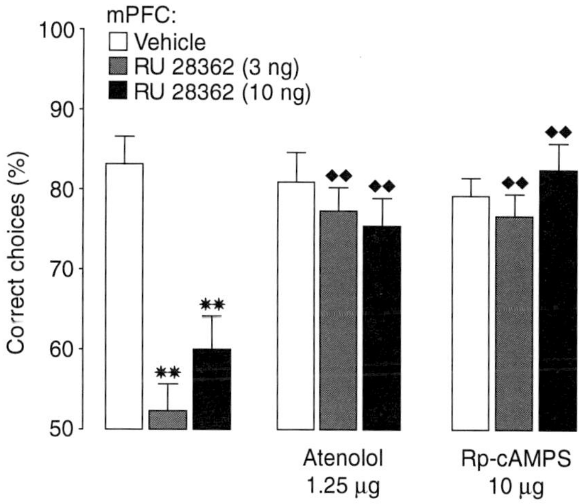

Fig. 6. Glucocorticoid–noradrenergic interactions in the mPFC in working memory. The β_1-adrenoceptor antagonist atenolol (1.25 μg in 0.5 μl) or cAMP blocker Rp-cAMPS (10 μg in 0.5 μl) administered into the mPFC 60 min before testing blocked the impairing effects induced by the GR agonist RU 28362 (3 or 10 ng) administered concurrently on delayed alternation performance in rats. Data represent correct choices (mean + SEM) in percent. $**p<0.01$ compared with the corresponding vehicle group; ◆ ◆$p<0.01$ compared with the saline-RU 28362 group (Barsegyan and Roozendaal, unpublished observation).

arousal effects on working memory in other brain regions.

Concluding remarks

The evidence summarized indicates that adrenal stress hormones influence memory processes in various animal and human memory tasks. Acutely administered or released epinephrine or glucocorticoids enhance the consolidation of long-term memory but impair memory retrieval and working memory. Stress hormone effects on these different memory functions depend on the arousal state and noradrenergic activation of the BLA. Our findings suggest that exposure to stress hormones or emotionally arousing training experiences, evoking noradrenergic activation, may induce BLA activation, in concert with inhibitory effects on the mPFC and some other brain regions, to create a brain state that promotes the long-term storage of these emotionally arousing events and, thus, preserve significant information. Although more experimentation is needed to elucidate the functional relevance of the temporary impairment of memory retrieval and working memory observed concurrently by this pattern of brain activity (de Kloet et al., 1999; Roozendaal, 2002), it is important to stress that these effects should not a priori be considered as maladaptive but that they may actually aid to an accurate storage of this new information (e.g., by blocking retroactive interference).

These neurobiological processes may play a critical role in the development and maintenance of traumatic memories and PTSD. Many patients with stress-associated diseases, such as PTSD, show sustained neuroendocrine abnormalities, which include increased catecholaminergic activity and impairment in glucocorticoid signaling (Kosten et al., 1987; Yehuda et al., 1992; Yehuda, 2002). In this respect it is noteworthy that, consistent with the extensive evidence that a β-adrenoceptor antagonist impairs memory consolidation, propranolol administered to recently traumatized patients attenuated the development of PTSD (Pitman et al., 2002; Vaiva et al., 2003). The etiology of PTSD, however, is complex and may depend not only on an over-consolidation of memory of traumatic experiences, but the symptoms, e.g., intrusive recollections or nightmares, may also be kept alive by an excessive memory retrieval or inability to extinguish. In line with the evidence that glucocorticoids impair memory retrieval and facilitate memory extinction, it has been reported that sustained administration of stress doses of glucocorticoids after a massive stress exposure can reduce the development of PTSD (Schelling et al., 2004) and might also be useful for the treatment of established PTSD (Aerni et al., 2004).

Abbreviations

BLA	basolateral complex of the amygdala
cAMP	adenosine 3′,5′-cyclic monophosphate
CEA	central nucleus of the amygdala
ERK1/2	extracellular-regulated kinase type 1/2
GR	glucocorticoid receptor
HPA axis	hypothalamic-pituitary-adrenocortical axis
mPFC	medial prefrontal cortex
NTS	nucleus of the solitary tract
pCREB	phosphorylated cAMP response-element binding
PET	positron emission tomography
PTSD	posttraumatic stress disorder

Acknowledgments

Research was supported by NSF grant IOB-0618211 and NIMH grant MH12526 to Benno Roozendaal.

References

Abercrombie, H.C., Speck, N.S. and Monticelli, R.M. (2006) Endogenous cortisol elevations are related to memory facilitation only in individuals who are emotionally aroused. Psychoneuroendocrinology, 31: 187–196.

Adolphs, R., Cahill, L., Schul, R. and Babinsky, R. (1997) Impaired declarative memory for emotional material following bilateral amygdala damage in humans. Learn. Mem., 4: 51–54.

Aerni, A., Traber, R., Hock, C., Roozendaal, B., Schelling, G., Papassotiropoulos, A., Nitsch, R.M., Snyder, U. and de Quervain, D.J.-F. (2004) Low-dose cortisol for symptoms of post-traumatic stress disorder. Am. J. Psychiatry, 161: 1488–1490.

Akirav, I. and Richter-Levin, G. (1999) Biphasic modulation of hippocampal plasticity by behavioral stress and basolateral amygdala stimulation in the rat. J. Neurosci., 19: 10530–10535.

Akirav, I. and Richter-Levin, G. (2002) Mechanisms of amygdala modulation of hippocampal plasticity. J. Neurosci., 22: 9912–9921.

Amat, J., Baratta, M.V., Paul, E., Bland, S.T., Watkins, L.R. and Maier, S.F. (2005) Medial prefrontal cortex determines how stressor controllability affects behavior and dorsal raphe nucleus. Nat. Neurosci., 8: 365–371.

Arnsten, A.F.T. and Goldman-Rakic, P.S. (1998) Noise stress impairs prefrontal cortical cognitive function in monkeys: evidence for a hyperdopaminergic mechanism. Arch. Gen. Psychiatry, 55: 362–369.

Barrett, D. and Gonzalez-Lima, F. (2004) Behavioral effects of metyrapone on Pavlovian extinction. Neurosci. Lett., 37: 91–96.

Beato, M., Herrlich, P. and Schutz, G. (1995) Steroid hormone receptors: many actors in search of a plot. Cell, 83: 851–857.

de Boer, S.F., Koopmans, S.J., Slangen, J.L. and van der Gugten, J. (1990) Plasma catecholamine, corticosterone and glucose responses to repeated stress in rats: effect of interstressor interval length. Physiol. Behav., 47: 1117–1124.

Bohannon III, J.N. (1988) Flashbulb memories for the space shuttle disaster: a tale of two theories. Cognition, 29: 179–196.

Bohus, B. and Lissak, K. (1968) Adrenocortical hormones and avoidance behaviour of rats. Int. J. Neuropharmacol., 7: 301–306.

Borrell, J., de Kloet, E.R. and Bohus, B. (1984) Corticosterone decreases the efficacy of adrenaline to affect passive avoidance retention of adrenalectomized rats. Life Sci., 34: 99–104.

Borrell, J., de Kloet, E.R., Versteeg, D.H. and Bohus, B. (1983) Inhibitory avoidance deficit following short-term adrenalectomy in the rat: the role of adrenal catecholamines. Behav. Neural Biol., 39. 241–258.

Brito, G.N.O., Thomas, G.J., Davis, B.J. and Gingold, S.I. (1982) Prelimbic cortex, mediodorsal thalamus, septum and delayed alternation in rats. Exp. Brain Res., 46: 52–58.

Buchanan, T.W. and Lovallo, W.R. (2001) Enhanced memory for emotional material following stress-level cortisol treatment in humans. Psychoneuroendocrinology, 26: 307–317.

Buchanan, T.W., Tranel, D. and Adolphs, R. (2006) Impaired memory retrieval correlates with individual differences in cortisol response but not autonomic response. Learn. Mem., 13: 382–387.

Buss, C., Wolf, O.T., Witt, J. and Hellhammer, D.H. (2004) Autobiographic memory impairment following acute cortisol administration. Psychoneuroendocrinology, 29: 1093–1096.

Cahill, L. and Alkire, M.T. (2003) Epinephrine enhancement of human memory consolidation: interaction with arousal at encoding. Neurobiol. Learn. Mem., 79: 194–198.

Cahill, L., Babinsky, R., Markowitsch, H.J. and McGaugh, J.L. (1995) The amygdala and emotional memory. Nature, 377: 295–296.

Cahill, L., Gorski, L. and Le, K. (2003) Enhanced human memory consolidation with post-learning stress: interaction with the degree of arousal at encoding. Learn. Mem., 10: 270–274.

Cahill, L., Haier, R.J., Fallon, J., Alkire, M.T., Tang, C., Keator, D., Wu, J. and McGaugh, J.L. (1996) Amygdala activity at encoding correlated with long-term, free recall of emotional information. Proc. Natl. Acad. Sci. U.S.A., 93: 8016–8021.

Cahill, L. and McGaugh, J.L. (1991) NMDA-induced lesions of the amygdaloid complex block the retention enhancing effect of posttraining epinephrine. Psychobiology, 19: 206–210.

Cahill, L., Prins, B., Weber, M. and McGaugh, J.L. (1994) Beta-adrenergic activation and memory for emotional events. Nature, 371: 702–704.

Cai, W.H., Blundell, J., Han, J., Greene, R.W. and Powell, C.M. (2006) Postreactivation glucocorticoids impair recall of established fear memory. J. Neurosci., 26: 9560–9566.

Campeau, S., Hayward, M.D., Hope, B.T., Rosen, J.B., Nestler, E.J. and Davis, M. (1991) Induction of the c-fos proto-oncogene in rat amygdala during unconditioned and conditioned fear. Brain Res., 565: 349–352.

Canli, T., Zhao, Z., Brewer, J., Gabrieli, J.D. and Cahill, L. (2000) Event-related activation in the human amygdala associates with later memory for individual emotional experience. J. Neurosci., 20: RC99.

Conrad, C.D., MacMillan II, D.D., Tsekhanov, S., Wright, R.L., Baran, S.E. and Fuchs, R.A. (2004) Influence of chronic corticosterone and glucocorticoid receptor antagonism in the amygdala on fear conditioning. Neurobiol. Learn. Mem., 81: 185–199.

Costa-Miserachs, D., Portell-Cortes, I., Aldavert-Vera, L., Torras-Garcia, M. and Morgado-Bernal, I. (1994) Long-term memory facilitation in rats by posttraining epinephrine. Behav. Neurosci., 108: 469–474.

Datson, N.A., van der Perk, J., de Kloet, E.R. and Vreugdenhil, E. (2001) Identification of corticosteroid-responsive genes in rat hippocampus using serial analysis of gene expression. Eur. J. Neurosci., 14: 675–689.

Dolan, R.J. (2000) Functional neuroimaging of the human amygdala during emotional processing and learning. In: Aggleton J.P. (Ed.), The Amygdala (2nd ed.). Oxford University, New York, pp. 631–653.

Dolcos, F., LaBar, K.S. and Cabeza, R. (2004) Interaction between the amygdala and the medial temporal lobe memory system predicts better memory for emotional events. Neuron, 42: 855–863.

Dolcos, F., LaBar, K.S. and Cabeza, R. (2005) Remembering one year later: role of the amygdala and the medial temporal lobe memory system in retrieving emotional memories. Proc. Natl. Acad. Sci. U.S.A., 102: 2626–2631.

Donley, M.P., Schulkin, J. and Rosen, J.B. (2005) Glucocorticoid receptor antagonism in the basolateral amygdala and ventral hippocampus interferes with long-term memory of contextual fear. Behav. Brain Res.: 197–205.

Fallon, J.H. and Ciofi, P. (1992) Distribution of monoamines within the amygdala. In: Aggleton J.P. (Ed.), The Amygdala: Neurobiological Aspects of Emotion, Memory, and Mental Dysfunction. Wiley-Liss, New York, pp. 84–114.

Fellows, L.K. and Farah, M.J. (2007) The role of ventromedial prefrontal cortex in decision making: judgment under uncertainty or judgment per se? Cereb. Cortex, (in press).

Ferry, B., Roozendaal, B. and McGaugh, J.L. (1999) Basolateral amygdala noradrenergic influences on memory storage are mediated by an interaction between β- and α_1-adrenoceptors. J. Neurosci., 19: 5119–5123.

Flood, J.F., Vidal, D., Bennett, E.L., Orme, A.E., Vasquez, S. and Jarvik, M.E. (1978) Memory facilitating and anti-amnestic effects of corticosteroids. Pharmacol. Biochem. Behav., 8: 81–87.

Frey, S., Bergado-Rosado, J., Seidenbecher, T., Pape, H.C. and Frey, J.U. (2001) Reinforcement of early long-term potentiation (early-LTP) in dentate gyrus by stimulation of the basolateral amygdala: heterosynaptic induction mechanisms of late-LTP. J. Neurosci., 21: 3697–3703.

Fuster, J.M. (1991) The prefrontal cortex and its relation to behavior. Prog. Brain Res., 87: 201–211.

Galvez, R., Mesches, M. and McGaugh, J.L. (1996) Norepinephrine release in the amygdala in response to footshock stimulation. Neurobiol. Learn. Mem., 66: 253–257.

Gold, P.E. (2004) Coordination of multiple memory systems. Neurobiol. Learn. Mem., 82: 230–242.

Gold, P.E. and van Buskirk, R. (1975) Facilitation of time-dependent memory processes with posttrial epinephrine injections. Behav. Biol., 13: 145–153.

Greenberg, D.L., Rice, H.J., Cooper, J.J., Cabeza, R., Rubin, D.C. and LaBar, K.S. (2005) Co-activation of the amygdala, hippocampus and inferior frontal gyrus during autobiographical memory retrieval. Neuropsychologia, 43: 659–674.

Hamann, S.B., Eli, T.D., Grafton, S.T. and Kilts, C.D. (1999) Amygdala activity related to enhanced memory for pleasant and aversive stimuli. Nat. Neurosci., 2: 289–303.

Hamann, S.B., Eli, T.D., Hoffman, J.M. and Kilts, C.D. (2002) Ecstasy and agony: activation of the human amygdala in positive and negative emotions. Psychol. Sci., 13: 135–141.

Hatfield, T. and McGaugh, J.L. (1999) Norepinephrine infused into the basolateral amygdala posttraining enhances retention in a spatial water maze task. Neurobiol. Learn. Mem., 91: 232–239.

Het, S., Ramlow, G. and Wolf, O.T. (2005) A meta-analytic review of the effects of acute cortisol administration on human memory. Psychoneuroendocrinology, 30: 771–784.

Ikegaya, Y., Nakanishi, K., Saito, H. and Abe, K. (1997) Amygdala ß-noradrenergic influence on hippocampal long-term potentiation in vivo. NeuroReport, 8: 3143–3146.

Ikegaya, Y., Saito, H. and Abe, K. (1994) Attenuated hippocampal long-term potentiation in basolateral amygdala-lesioned rats. Brain Res., 656: 157–174.

Introini-Collison, I.B. and McGaugh, J.L. (1986) Epinephrine modulates long-term retention of an aversively motivated discrimination. Behav. Neural Biol., 45: 358–365.

Izquierdo, I., Quillfeldt, J.A., Zanatta, M.S., Quevedo, J., Schaeffer, E., Schmitz, P.K. and Medina, J.H. (1997) Sequential role of hippocampus and amygdala, entorhinal cortex and parietal cortex in formation and retrieval of memory for inhibitory avoidance in rats. Eur. J. Neurosci., 9: 786–793.

Karst, H., Berger, S., Turiault, M., Tronche, F., Schütz, G. and Joëls, M. (2005) Mineralocorticoid receptors are indispensable for nongenomic modulation of hippocampal glutamate transmission by corticosterone. Proc. Natl. Acad. Sci. U.S.A., 102: 19204–19207.

Kelley, A.E., Domesick, V.B. and Nauta, W.J. (1982) The amygdalostriatal projection in the rat — an anatomical study by anterograde and retrograde tracing methods. Neuroscience, 7: 615–630.

Kilpatrick, L. and Cahill, L. (2003) Amygdala modulation of parahippocampal and frontal regions during emotionally influenced memory storage. Neuroimage, 20: 2091–2099.

de Kloet, E.R. (1991) Brain corticosteroid receptor balance and homeostatic control. Front. Neuroendocrinol., 12: 95–164.

de Kloet, E.R., Oitzl, M.S. and Joëls, M. (1999) Stress and cognition: are corticosteroids good or bad guys? Trends Neurosci., 22: 422–426.

Kopin, I.J. (1995) Definitions of stress and sympathetic neuronal responses. Ann. N.Y. Acad. Sci., 771: 19–30.

Korte, S.M. (2001) Corticosteroids in relation to fear, anxiety and psychopathology. Neurosci. Biobehav. Rev., 25: 117–142.

Kosten, T.R., Mason, J.W., Giller, E.L., Ostroff, R.B. and Harkness, L. (1987) Sustained urinary norepinephrine and epinephrine elevation in post-traumatic stress disorder. Psychoneuroendocrinology, 12: 13–20.

Kuhlmann, S., Kirschbaum, C. and Wolf, O.T. (2005a) Effects of oral cortisol treatment in healthy young women on memory retrieval of negative and neutral words. Neurobiol. Learn. Mem., 83: 158–162.

Kuhlmann, S., Piel, M. and Wolf, O.T. (2005b) Impaired memory retrieval after psychosocial stress in healthy young men. J. Neurosci., 25: 2977–2982.

Kuhlmann, S. and Wolf, O.T. (2006) Arousal and cortisol interact in modulating memory consolidation in healthy young men. Behav. Neurosci., 120: 217–223.

LaLumiere, R.T., Buen, T.V. and McGaugh, J.L. (2003) Post-training intra-basolateral amygdala infusions of norepinephrine enhance consolidation of memory for contextual fear conditioning. J. Neurosci., 23: 6754–6758.

Liang, K.C. (2001) Epinephrine modulation of memory: amygdala activation and regulation of long-term memory storage. In: Gold P.E. and Greenough W.T. (Eds.), Memory consolidation: Essays in Honor of James L. McGaugh, pp. 165–183.

Liang, K.C., Juler, R.G. and McGaugh, J.L. (1986) Modulating effects of posttraining epinephrine on memory: involvement of the amygdala noradrenergic system. Brain Res., 368: 125–133.

Liu, L., Tsuji, M., Takeda, H., Takada, K. and Matsumiya, T. (1999) Adrenocortical suppression blocks the enhancement of memory storage produced by exposure to psychological stress in rats. Brain Res., 821: 134–140.

Lupien, S.J., Gillin, C.J. and Hauger, R.L. (1999) Working memory is more sensitive than declarative memory to the acute effects of corticosteroids: a dose-response study in humans. Behav. Neurosci., 113: 420–430.

Lupien, S.J. and McEwen, B.S. (1997) The acute effects of corticosteroids on cognition: integration of animal and human model studies. Brain Res. Rev., 24: 1–27.

Lyons, D.M., Lopez, J.M., Yang, C. and Schatzberg, A.F. (2000) Stress-level cortisol treatment impairs inhibitory control of behavior in monkeys. J. Neurosci., 20: 7816–7821.

Malin, E.L. and McGaugh, J.L. (2006) Differential involvement of the hippocampus, anterior cingulate cortex, and basolateral amygdala in memory for context and footshock. Proc. Natl. Acad. Sci. U.S.A., 103: 1959–1963.

Maren, S. and Fanselow, M.S. (1997) Electrolytic lesions of the fimbria/fornix, dorsal hippocampus, or entorhinal cortex produce anterograde deficits in contextual fear conditioning in rats. Neurobiol. Learn. Mem., 67: 142–149.

McDonald, A.J. (1991) Organization of amygdaloid projections to the prefrontal cortex and associated striatum in the rat. Neuroscience, 44: 1–14.

McGaugh, J.L. (2000) Memory: a century of consolidation. Science, 287: 248–251.

McGaugh, J.L. (2002) Memory consolidation and the amygdala: a systems perspective. Trends Neurosci., 25: 456–461.

McGaugh, J.L. (2004) The amygdala modulates the consolidation of memories of emotionally arousing experiences. Annu. Rev. Neurosci., 27: 1–28.

McGaugh, J.L., Cahill, L. and Roozendaal, B. (1996) Involvement of the amygdala in memory storage: interaction with other brain systems. Proc. Natl. Acad. Sci. U.S.A., 93: 13508–13514.

McGaugh, J.L. and Roozendaal, B. (2002) Role of adrenal stress hormones in forming lasting memories in the brain. Curr. Opin. Neurobiol., 12: 205–210.

McIntyre, C.K., Hatfield, T. and McGaugh, J.L. (2002) Norepinephrine levels in the amygdala following inhibitory avoidance training predict retention score. Eur. J. Neurosci., 16: 1223–1226.

McIntyre, C.K., Roozendaal, B. and McGaugh, J.L. (2004) Glucocorticoid treatment enhances training-induced norepinephrine release in the amygdala. Soc. Neurosci. Abstr. 772.12 online.

Medina, A.C., Charles, J.R., Espinoza-González, V., Sanchez-Resendis, O., Prado-Alcalá, R.A., Roozendaal, B. and Quirarte, G.L. (2007) Glucocorticoid administration into the dorsal striatum facilitates memory consolidation of inhibitory avoidance training but not of the context or footshock components. Learn. Mem., in press.

Mizoguchi, K., Ishige, A., Takeda, S., Aburada, M. and Tabita, T. (2004) Endogenous glucocorticoids are essential for maintaining prefrontal cortical cognitive function. J. Neurosci., 24: 5492–5499.

Morris, R.G.M., Garrud, P., Rawlins, J.N.P. and O'Keefe, J. (1982) Place navigation impaired in rats with hippocampal lesions. Nature, 297: 681–683.

Mulder, A.B., Hodenpijl, M.G. and Lopes da Silva, F.H. (1998) Electrophysiology of the hippocampal and amygdaloid projections to the nucleus accumbens of the rat: convergence, segregation, and interaction of inputs. J. Neurosci., 18: 5095–5102.

Nakao, K., Matsuyama, K., Matsuki, N. and Ikegaya, Y. (2004) Amygdala stimulation modulates hippocampal synaptic plasticity. Proc. Natl. Acad. Sci. U.S.A., 101: 14270–14275.

Neisser, U., Winograd, E., Bergman, E.T., Schreiber, C.A., Palmer, S.E. and Weldon, M.S. (1996) Remembering the earthquake: direct experience vs. hearing the news. Memory, 4: 337–357.

Oitzl, M.S. and de Kloet, E.R. (1992) Selective corticosteroid antagonists modulate specific aspects of spatial orientation learning. Behav. Neurosci., 108: 62–71.

Oitzl, M.S., Reichardt, H.M., Joëls, M. and de Kloet, E.R. (2001) Point mutation in the mouse glucocorticoid receptor preventing DNA-binding impairs spatial memory. Proc. Natl. Acad. Sci. U.S.A., 98: 12790–12795.

Okuda, S., Roozendaal, B. and McGaugh, J.L. (2004) Glucocorticoid effects on object recognition memory require training-associated emotional arousal. Proc. Natl. Acad. Sci. U.S.A., 101: 853–858.

Orchinik, M., Murray, T.F. and Moore, F.L. (1991) A corticosteroid receptor in neuronal membranes. Science, 252: 1848–1851.

Packard, M.G., Cahill, L. and McGaugh, J.L. (1994) Amygdala modulation of hippocampal-dependent and caudate nucleus-dependent memory processes. Proc. Natl. Acad. Sci. U.S.A., 91: 8477–8481.

Packard, M.G. and Knowlton, B.J. (2002) Learning and memory functions of the basal ganglia. Annu. Rev. Neurosci., 25: 563–593.

Pape, H.C., Narayanan, R.T., Smid, J., Stork, O. and Seidenbecher, T. (2005) Theta activity in neurons and networks of the amygdala related to long-term fear memory. Hippocampus, 15: 874–880.

Pelletier, J.G., Likhtik, E., Filali, M. and Paré, D. (2005) Lasting increases in basolateral amygdala activity after emotional arousal: implications for facilitated consolidation of emotional memories. Learn. Mem., 12: 96–102.

Perez-Jaranay, J.M. and Vives, F. (1991) Electrophysiological study of the response of medial prefrontal cortex neurons to stimulation of the basolateral nucleus of the amygdala in the rat. Brain Res., 564: 97–101.

Petrovich, G.D., Canteras, N.S. and Swanson, L.W. (2001) Combinatorial amygdalar inputs to hippocampal domains and hypothalamic behavior systems. Brain Res. Rev., 38: 247–289.

Pikkarainen, M., Ronkko, S., Savander, V., Insausti, R. and Pitkanen, A. (1999) Projections from the lateral, basal, and accessory basal nuclei of the amygdala to the hippocampal formation in rat. J. Comp. Neurol., 403: 229–260.

Pitman, R.K., Sanders, K.M., Zusman, R.M., Healy, A.R., Cheema, F., Lasko, N.B., Cahill, L. and Orr, S.P. (2002) Pilot study of secondary prevention of posttraumatic stress disorder with propranolol. Biol. Psychiatry, 51: 189–192.

Pugh, C.R., Tremblay, D., Fleshner, M. and Rudy, J.W. (1997) A selective role for corticosterone in contextual-fear conditioning. Behav. Neurosci., 111: 503–511.

de Quervain, D.J.-F., Aerni, A. and Roozendaal, B. (2007) Preventive effect of β-adrenoceptor blockade on glucocorticoid-induced memory retrieval deficits. Am. J. Psychiatry, 164: 967–969.

de Quervain, D.J.-F., Henke, K., Aerni, A., Treyer, V., McGaugh, J.L., Berthold, T., Nitsch, R.M., Buck, A., Roozendaal, B. and Hock, C. (2003) Glucocorticoid-induced impairment of declarative memory retrieval is associated with reduced blood flow in the medial temporal lobe. Eur. J. Neurosci., 17: 1296–1302.

de Quervain, D.J.-F., Roozendaal, B. and McGaugh, J.L. (1998) Stress and glucocorticoids impair retrieval of long-term spatial memory. Nature, 394: 787–790.

de Quervain, D.J.-F., Roozendaal, B., Nitsch, R.M., McGaugh, J.L. and Hock, C. (2000) Acute cortisone administration impairs retrieval of long-term declarative memory in humans. Nat. Neurosci., 3: 313–314.

Quirarte, G.L., Galvez, R., Roozendaal, B. and McGaugh, J.L. (1998) Norepinephrine release in the amygdala in response to footshock and opioid peptidergic drugs. Brain Res., 808: 134–140.

Quirarte, G.L., Roozendaal, B. and McGaugh, J.L. (1997) Glucocorticoid enhancement of memory storage involves noradrenergic activation in the basolateral amygdala. Proc. Natl. Acad. Sci. U.S.A., 94: 14048–14053.

Ramnani, N. and Owen, A.M. (2004) Anterior prefrontal cortex insights into function from anatomy and neuroimaging. Nat. Rev. Neurosci., 5: 184–194.

Rashidy-Pour, A., Sadeghi, H., Taherain, A.A., Vafaei, A.A. and Fathollahi, Y. (2004) The effects of acute restraint stress and dexamethasone on retrieval of long-term memory in rats: an interaction with opiate system. Behav. Brain Res., 154: 193–198.

Reul, J.M.H.M. and de Kloet, E.R. (1985) Two receptor systems for corticosterone in the rat brain: microdistribution and differential occupation. Endocrinology, 117: 2505–2512.

Richardson, M.P., Strange, B.A. and Dolan, R.J. (2004) Encoding of emotional memories depends on amygdala and hippocampus and their interactions. Nat. Neurosci., 7: 278–285.

Roozendaal, B. (2000) Glucocorticoids and the regulation of memory consolidation. Psychoneuroendocrinology, 25: 213–238.

Roozendaal, B. (2002) Stress and memory: opposing effects of glucocorticoids on memory consolidation and memory retrieval. Neurobiol. Learn. Mem., 78: 578–595.

Roozendaal, B. (2007) Norepinephrine and long-term memory function. In: Ordway G.A., Schwartz M.A. and Frazer A. (Eds.), Brain Norepinephrine: Neurobiology and Therapeutics. Cambridge University Press, pp. 236–274.

Roozendaal, B., Bohus, B. and McGaugh, J.L. (1996a) Dose-dependent suppression of adreno-cortical activity with metyrapone: effects on emotion and learning. Psychoneuroendocrinology, 21: 681–693.

Roozendaal, B., Carmi, O. and McGaugh, J.L. (1996b) Adrenocortical suppression blocks the memory-enhancing effects of amphetamine and epinephrine. Proc. Natl. Acad. Sci. U.S.A., 93: 1429–1433.

Roozendaal, B., Griffith, Q.K., Buranday, J., de Quervain, D.J.-F. and McGaugh, J.L. (2003) The hippocampus mediates glucocorticoid-induced impairment of spatial memory retrieval: dependence on the basolateral amygdala. Proc. Natl. Acad. Sci. U.S.A., 100: 1328–1333.

Roozendaal, B., Hahn, E.L., Nathan, S.V., de Quervain, D.J.-F. and McGaugh, J.L. (2004a) Glucocorticoid effects on memory retrieval require concurrent noradrenergic activity in the hippocampus and basolateral amygdala. J. Neurosci., 24: 8161–8169.

Roozendaal, B., Hui, G.K., Hui, I.R., Berlau, D.J., McGaugh, J.L. and Weinberger, N.M. (2006a) Basolateral amygdala noradrenergic activity mediates corticosterone-induced enhancement of auditory fear conditioning. Neurobiol. Learn. Mem., 86: 249–255.

Roozendaal, B. and McGaugh, J.L. (1996a) Amygdaloid nuclei lesions differentially affect glucocorticoid-induced memory enhancement in an inhibitory avoidance task. Neurobiol. Learn. Mem., 65: 1–8.

Roozendaal, B. and McGaugh, J.L. (1996b) The memory-modulatory effects of glucocorticoids depend on an intact stria terminalis. Brain Res., 709: 243–250.

Roozendaal, B. and McGaugh, J.L. (1997a) Glucocorticoid receptor agonist and antagonist administration into the basolateral but not central amygdala modulates memory storage. Neurobiol. Learn. Mem., 67: 176–179.

Roozendaal, B. and McGaugh, J.L. (1997b) Basolateral amygdala lesions block the memory-enhancing effect of glucocorticoid administration in the dorsal hippocampus of rats. Eur. J. Neurosci., 9: 76–83.

Roozendaal, B., McReynolds, J.R. and McGaugh, J.L. (2004b) The basolateral amygdala interacts with the medial prefrontal cortex in regulating glucocorticoid effects on working memory impairment. J. Neurosci., 24: 1385–1392.

Roozendaal, B., Nguyen, B.T., Power, A.E. and McGaugh, J.L. (1999a) Basolateral amygdala noradrenergic influence enables enhancement of memory consolidation induced by hippocampal glucocorticoid receptor activation. Proc. Natl. Acad. Sci. U.S.A., 96: 11642–11647.

Roozendaal, B., Okuda, S., Van der Zee, E.A. and McGaugh, J.L. (2006b) Glucocorticoid enhancement of memory requires arousal-induced noradrenergic activation in the basolateral amygdala. Proc. Natl. Acad. Sci. U.S.A., 103: 6741–6746.

Roozendaal, B., Portillo-Marquez, G. and McGaugh, J.L. (1996c) Basolateral amygdala lesions block glucocorticoid-induced modulation of memory for spatial learning. Behav. Neurosci., 110: 1074–1083.

Roozendaal, B., de Quervain, D.J.-F., Ferry, B., Setlow, B. and McGaugh, J.L. (2001) Basolateral amygdala-nucleus accumbens interactions in mediating glucocorticoid effects on memory consolidation. J. Neurosci., 21: 2518–2525.

Roozendaal, B., de Quervain, D.J.-F., Schelling, G. and McGaugh, J.L. (2004c) A systemically administered β-adrenoceptor antagonist blocks corticosterone-induced impairment of contextual memory retrieval in rats. Neurobiol. Learn. Mem., 81: 150–154.

Roozendaal, B., Quirarte, G.L. and McGaugh, J.L. (2002) Glucocorticoids interact with the basolateral amygdala β-adrenoceptor-cAMP/PKA system in influencing memory consolidation. Eur. J. Neurosci., 15: 553–560.

Roozendaal, B., Williams, C.L. and McGaugh, J.L. (1999b) Glucocorticoid receptor activation in the rat nucleus of the solitary tract facilitates memory consolidation: involvement of the basolateral amygdala. Eur. J. Neurosci., 11: 1317–1323.

Rosenkranz, J.A. and Grace, A.A. (2002) Cellular mechanisms of infralimbic and prelimbic prefrontal cortical inhibition and dopaminergic modulation of basolateral amygdala neurons in vivo. J. Neurosci., 22: 324–337.

Runyan, J.D., Moore, A.N. and Dash, P.K. (2004) A role for prefrontal cortex in memory storage for trace fear conditioning. J. Neurosci., 24: 1288–1295.

Sacchetti, B., Lorenzini, C.A., Baldi, E., Tassoni, G. and Bucherelli, C. (1999) Auditory thalamus, dorsal hippocampus, basolateral amygdala, and perirhinal cortex role in the consolidation of conditioned freezing to context and to acoustic conditioned stimulus in the rat. J. Neurosci., 19: 9570–9578.

Sajadi, A.A., Samaei, S.A. and Rashidy-Pour, A. (2006) Intra-hippocampal microinjections of anisomycin did not block glucocorticoid-induced impairment of memory retrieval in rats: an evidence for non-genomic effects of glucocorticoids. Behav. Brain Res., 173: 158–162.

Sandi, C. and Rose, S.P.R. (1997) Training-dependent biphasic effects of corticosterone in memory formation for a passive avoidance task in chicks. Psychopharmacology, 133: 152–160.

Schelling, G., Roozendaal, B. and de Quervain, D.J.-F. (2004) Can posttraumatic stress disorder be prevented with glucocorticoids? Ann. N.Y. Acad. Sci., 1032: 158–166.

Setlow, B., Roozendaal, B. and McGaugh, J.L. (2000) Involvement of a basolateral amygdala complex — nucleus accumbens pathway in glucocorticoid-induced modulation of memory storage. Eur. J. Neurosci., 12: 367–375.

Smith, A.P., Stephen, K.E., Rugg, M.D. and Dolan, R.J. (2006) Task and content modulate amygdala-hippocampal connectivity in emotional retrieval. Neuron, 49: 631–638.

van Stegeren, A.H., Goekoop, R., Everaerd, W., Scheltens, P., Barkhof, F., Kuijer, J.P.A. and Rombouts, S.A.R.B. (2005) Noradrenalin mediates amygdala activation in men and women during encoding of emotional material. Neuroimage, 24: 898–909.

van Stegeren, A.H., Wolf, O.T., Everaerd, W., Scheltens, P., Barkhof, F. and Rombouts, S.A.R.B. (2007) Endogenous cortisol level interacts with noradrenergic activation in the human amygdala. Neurobiol. Learn. Mem., 87: 57–66.

Strange, B.A. and Dolan, R.J. (2004) β-Adrenergic modulation of emotional-evoked human amygdala and hippocampal responses. Proc. Natl. Acad. Sci. U.S.A., 101: 11454–11458.

Tasker, J.G., Di, S. and Malcher-Lopes, R. (2006) Rapid glucocorticoid signaling via membrane-associated receptors. Endocrinology, 147: 5549–5556.

Vaiva, G., Ducrocq, F., Jezequel, K., Averland, B., Lestavel, P., Brunet, A. and Marmar, C.R. (2003) Immediate treatment with propranolol decreases posttraumatic stress disorder two months after trauma. Biol. Psychiatry, 54: 947–949.

Vouimba, R.-M., Yaniv, D. and Richter-Levin, G. (2007) Glucocorticoid receptors and β-adrenoceptors in basolateral amygdala modulate synaptic plasticity in hippocampal dentate gyrus, but not in area CA1. Neuropharmacology, 52: 244–252.

Williams, C.L. and Clayton, E.C. (2001) Contribution of brainstem structures in modulating memory storage processes. In: Gold P.E. and Greenough W.T. (Eds.), Memory Consolidation: Essays in Honor of James L. McGaugh. American Psychological Association, Washington, DC, pp. 141–163.

Williams, C.L., Men, D., Clayton, E.C. and Gold, P.E. (1998) Norepinephrine release in the amygdala after systemic injections of epinephrine or inescapable footshock: contribution of the nucleus of the solitary tract. Behav. Neurosci., 112: 1414–1422.

Wolf, O.T., Convit, A., McHugh, P.F., Kandil, E., Thorn, E.L., De Santi, S., McEwen, B.S. and de Leon, M.J. (2001) Cortisol differentially affects memory in young and elderly men. Behav. Neurosci., 115: 1002–1011.

Wright, C.I., Beijer, A.V. and Groenewegen, H.J. (1996) Basal amygdaloid complex afferents to the rat nucleus accumbens are compartmentally organized. J. Neurosci., 16: 1877–1893.

Yang, Y., Cao, J., Xiong, W., Zhang, J., Zhou, Q., Wei, H., Liang, C., Deng, J., Li, T., Yang, S., Xu, L. and Xu, L. (2003) Both stress experience and age determine the impairment or enhancement effect of stress on spatial memory retrieval. J. Endocrinol., 178: 45–54.

Yang, Y.L., Chao, P.K. and Lu, K.T. (2006) Systemic and intra-amygdala administration of glucocorticoid agonist and antagonist modulate extinction of conditioned fear. Neuropsychopharmacology, 31: 912–924.

Yehuda, R. (2002) Current status of cortisol findings in posttraumatic stress disorder. Psychiatr. Clin. North Am., 25: 341–368 vii.

Yehuda, R., Southwick, S., Giller, E.L., Ma, X. and Mason, J.W. (1992) Urinary catecholamine excretion and severity of PTSD symptoms in Vietnam combat veterans. J. Nerv. Ment. Dis., 180: 321–325.

Young, A.H., Sahakian, B.J., Robbins, T.W. and Cowen, P.J. (1999) The effects of chronic administration of cortisone on cognitive function in normal male volunteers. Psychopharmacology, 145: 260–266.

Discussion: Chapter 6

RICHTER-LEVIN: You indicated, actually you showed very nicely, that you define your studies in the normal learning and memory domain. Within this, you find that corticosterone enhances memory for the tasks that you have been looking at. So you may be in a good position to ask when corticosterone becomes traumatic. Do you have any experience at which dosage this may occur?

ROOZENDAAL: Then we would need to be able to define what makes an experience traumatic for a rat. This may be very difficult to do. For example, do rats experience a footshock as being traumatic? Rats normally live in their cages, deprived of much sensory stimulation. Thus, it is a completely unnatural, and perhaps traumatic, experience for a rat to receive even a mild footshock. But, we have not really looked at this issue. I do not think that the dose of corticosterone determines whether an experience becomes traumatic or not. If a rat by itself becomes highly aroused by the training experience, it is very difficult with a corticosterone injection to further enhance memory consolidation. The same dose of corticosterone that normally enhances memory may then impair the consolidation.

DE KLOET: Could you elaborate further on the question how corticosterone acts in learning and memory processes given its suppressive and permissive actions, and its genomic vs. non-genomic actions?

ROOZENDAAL: In our studies, we found that corticosterone enhances memory consolidation via a permissive interaction with arousal-induced noradrenergic activation. In conditions with low emotional arousal, corticosterone injections do not enhance memory consolidation. Thus, corticosterone requires such noradrenergic activation. This is not true the other way around: injecting low-aroused rats with a high dose of yohimbine to induce noradrenergic stimulation is sufficient to enhance memory consolidation. This corticosterone effect on the noradrenergic system in enhancing memory consolidation appears to involve a non-genomic GR action. However, as shown by Marian Joëls GR activation can also induce long-lasting suppression, but this effect was obtained in experimental conditions with low levels of emotional arousal and noradrenergic activation. The situation may also be different when you work with levels of glucocorticoids that are below baseline. Then you might also expect an MR involvement. I do not know if such effects are suppressive or permissive.

JOËLS: What did you say, the genomic, or the non-genomic effect?

ROOZENDAAL: Glucocorticoid effects on noradrenergic activity are rapidly induced and, thus, likely non-genomic. But to affect consolidation, ultimately some kind of genomic process must be involved. Non-genomic actions of glucocorticoids on the noradrenergic system may be sufficient to impair memory retrieval and working memory.

E.R. de Kloet, M.S. Oitzl & E. Vermetten (Eds.)
Progress in Brain Research, Vol. 167
ISSN 0079-6123

CHAPTER 7

Adult neurogenesis and systemic adaptation: animal experiments and clinical perspectives for PTSD

Noriko U. Takemura* and Nobumasa Kato

Department of Neuropsychiatry, Graduate School of Medicine, The University of Tokyo, 7-3-1 Hongo, Bunkyo-ku, Tokyo 113-8655, Japan

Abstract: The life-long persistence of neuron production in the adult mammalian central nervous system was established at the end of the 20th century and since then, intensive studies have been carried out to determine the biological role of neuronal turnover in the mature brain. To date, evidence has been found of involvement in learning/memory function and stress-related mental disorders. With a discussion of speculative link between impaired amygdala-relevant neurogenesis and PTSD in an animal model, we here review across species the functional significance of adult neurogenesis from the point of view of systemic adaptation.

Keywords: adult neurogenesis; species; learning and memory; adaptation; stress; depression

Historical overview

It was once a universal belief that neurogenesis does not occur in the adult mammalian central nervous system. However, through the technological advancement of recent decades this tenet of neuroscience has been overturned by evidence to the contrary; evidence that neurogenesis does indeed occur in adult mammals, including humans. The first evidence for adult mammalian neurogenesis was reported by Joseph Altman and colleagues in the 1960s who showed continual neuron production in the subventricular zone (SVZ) lining the anterior lateral ventricle (LV) wall (Altman, 1969) and the subgranular zone (SGZ) of the dentate gyrus (DG) of the hippocampus (Altman and Das, 1965) in rodents. Surprisingly, at that early stage of such research, Altman predicted that newborn neurons could be derived from a sort of stem/progenitor cells; "if the general observation is valid that mitotic figures are absent in the brain of adult mammals, these findings might suggest that the labeled (newborn) neurons were formed from undifferentiated cells which divided mitotically during the period at which the administered thymidine-H^3 was available" (Altman, 1962). Although electron microscopy clearly visualized the ultrastructure of [^{3}H]thymidine-labeled newborn neurons in the SVZ and SGZ (Kaplan and Hinds, 1977), the idea of adult neurogenesis in mammals was still regarded with great skepticism at that time.

In the late 1990s, immunohistochemical double staining combined with confocal microscopy definitively identified new neurons generated in the adult brain, and the introduction of stereology, an unbiased stereological cell counting technique, permitted in vivo quantitative analysis of these neurons (Kuhn et al., 1996, 1997; Kempermann

*Corresponding author. Tel.: +81 3 5800 9263;
Fax: +81 3 5800 6894; E-mail: nutakemura@yahoo.co.jp

DOI: 10.1016/S0079-6123(07)67007-1

et al., 1998). In parallel, in the adult brain, the presence of neural stem cells, defined by the characteristics of long-term self-renewal and multipotentiality, has been identified (Reynolds and Weiss, 1992; Lois and Alvarez-Buylla, 1993; McKay, 1997; reviewed in Gage 2000). Since then, functional incorporation of newborn neurons into preexisting neural networks have been shown with characteristics of their developmental stage-typical physiological properties having been observed (van Praag et al., 2002; Belluzzi et al., 2003; Carleton et al., 2003; Fukuda et al., 2003; Overstreet et al., 2004; Schmidt-Hieber et al., 2004).

One of the central issues of recent debate is what biological contribution is made by neurogenesis in the mammalian central nervous system? Biological experiments using molecular genetics and irradiation techniques, and also theoretical neural network models, suggest it has functional implications in learning and memory (Snyder et al., 2005; reviewed in Meltzer et al., 2005; Lledo et al., 2006) and in mood disorders (reviewed in Sapolsky, 2004).

It may be worth mentioning here why irradiation is useful in this field of research. Compared to other methods, the irradiation paradigm is particularly advantageous because of its relatively uninvasive nature, dose-dependent cell type-selective elimination performance, and easy and quick operation. Radiation sensitivity varies among cell types, and low-dose irradiation can achieve a proliferating cell type-specific ablation (reviewed in Wojtowicz, 2006). Proliferating cells and immature neurons are known to be the most sensitive cell populations in the adult brain; a dose above 1 Gy reduces proliferating progenitor cells and immature neurons in rat SVZ (Shinohara et al., 1997). In contrast, matured cells are quite resistant to radiation exposure; in rodents, 8 Gy dose has no effect on astrocytes and microglia (Chiang et al., 1993); 10 Gy causes no effect on mature neurons (Peissner et al., 1999) and is also well below the threshold for frank vascular changes, demyelination, or radionecrosis (Monje et al., 2002). The evidence of normal synaptic physiology of 10 Gy irradiated hippocampal neurons (Snyder et al., 2001) further support the above observations. There are, however, also some disadvantages; a 10 Gy dose induces an inflammation response that disrupts the neurogenic potential of the rodent hippocampus (Monje et al., 2002), and causes a significant loss of radiation-sensitive subpopulation of endothelial cells (Lyubimova et al., 2001). This disrupted microenvironment possibly induces unknown dysfunction in matured cells that may depend on that environment. Thus, although imperfect, irradiation is one of the most advantageous tools for adult neurogenesis research.

Biological significance of adult neurogenesis

There are two possible lines of research aimed at understanding the functional significance of adult neurogenesis: comparative study across species and that within a model animal. Adult neurogenesis occurs under physiological conditions with varied topological identities across species. Knowledge of interactions between the regionality (both the occurrence and absence) and species-typical behaviors would tell us its common biological significance in the adult brain (reviewed in Zupanc, 2001). Another approach is detecting how it behaves within and beyond a physiological range in a given animal. Although various physical/mental life events impact on living things, the system usually returns to a basic homeostatic state within hours or days by means of intrinsic adaptation mechanisms.

Extreme acute/chronic extrinsic events, however, can force the system beyond its adaptation limit, and in such cases there may occasionally be permanent traces of damage. It is known that such physiological and pathological events impact on the magnitude of adult neurogenesis. Efforts to understand such reactive aspects of neurogenesis may provide clues of its roles in the adult brain. Indeed, both cross-species and within-species approaches have identified the consequences outlined below.

Adult neurogenesis under physiological conditions

Physiological post-developmental neurogenesis is evolutionarily conserved in many biological systems. In the invertebrates: insects (Scotto-Lomassese et al., 2003) and crustaceans (reviewed in Schmidt and Demuth, 1998; Harzsch et al., 1999), and in the vertebrates: fishes (reviewed in Zupanc, 2001), reptiles (Font et al., 2001), avians (reviewed in Nottebohm, 2004), and mammals (reviewed in Ming and Song, 2005; Lledo et al., 2006). Beyond its regional and phenotypic diversity among species, accumulated evidence has suggested a common biological significance of adult neurogenesis in learning and memory function relevant to species-typical life strategies.

Insects

In the adult house cricket, a cluster of mitotically active cells continually differentiates into interneurons of the mushroom body (Cayre et al., 1994), which plays a dominant role for olfactory associative learning (reviewed in Davis, 2005; Margulies et al., 2005). It is shown that sensory experiences influence adult neurogenesis (Scotto-Lomassese et al., 2000, 2002), and radiation-induced disruption of neurogenesis resulted in impairment of associative learning/memory for odor (Scotto-Lomassese et al., 2003).

Birds

Probably the most comprehensive evidence of interaction between adult neurogenesis and brain function has been shown in songbirds (reviewed in Nottebohm, 2002, 2004). In adult canaries, neurogenesis persists in the high vocal center (HVC), which is a key nucleus in the song learning system connecting an ascending auditory pathway with descending motor pathways to produce learned song (reviewed in Nottebohm, 2002). Neural precursors in the nearby lateral ventricular zone migrate into HVC where half or more differentiate into projection neurons innervating to the robust nucleus of the archistriatum (RA), one of the systems for learned vocal behavior (Goldman and Nottebohm, 1983). Another, smaller, group of cells become HVC interneurons. In vivo auditory responsiveness of the newborn interneurons definitively showed their functional incorporation into the matured neural networks (Paton and Nottebohm, 1984).

Functional roles of neurogenesis have been implicated in song learning/memory and phonation. The magnitude of the turnover of HVC neurons is well correlated with seasonal periods of song instability and restabilization (Alvarez-Buylla and Kirn, 1997). Strikingly, in adult zebra finches, a targeted death of HVC-RA neurons caused song deterioration, and the behavioral impairment was restored accompanying endogenously activated HVC neurogenesis (Scharff et al., 2000). Furthermore, neuronal turnover is activity-dependent as deprivation of either song input (Wang et al., 1999) or song output (Li et al., 2000) both reduce the survival of newborn neurons. There are also some inconsistent observations; adult neurogenesis occurs in other seasonal species that do not change their songs (Tramontin and Brenowitz, 1999), and in canaries at times when song is stable (Alvarez-Buylla et al., 1990). These results might infer an additional role of neurogenesis other than that in learning/memory and motor coordination functions.

Rodents

In rodents, neural stem/progenitor cells in the SVZ migrate along the rostral migratory stream (RMS) and differentiate into granule cells and periglomerular interneurons in the olfactory bulb (OB) (reviewed in Lledo et al., 2006). Functional significance of OB neurogenesis has been implicated in odor discrimination and memory formation. Enhanced OB neurogenesis improved olfactory memory (Rochefort et al., 2002), and the reduction in genetically modified mice resulted in an impairment in odor discrimination (Gheusi et al., 2000; Enwere et al., 2004). Activity-dependent incorporation of newborn neurons into preexisting OB networks has also been shown; odor-enriched environment (Rochefort et al., 2002) and olfactory discrimination learning (Alonso et al., 2006) both enhanced survival of newly generated neurons (Rochefort et al., 2002), whereas in anosmic mice, it was significantly attenuated (Petreanu and Alvarez-Buylla, 2002).

In the hippocampus, the ventral edge of the granule cell layer (or SGZ) harbors neural stem/progenitor cells. Most of these cells differentiate into excitatory granule neurons. The hippocampus is classically characterized as a structure that acts as a gateway for passing information into memory, and one of the least controversial theoretical roles of DG is in pattern separation. DG granule cells provide distinct codes to the downstream CA3 creating highly sparse representations of entorhinal inputs (reviewed in Aimone et al., 2006). Activity-dependent regulation of DG neurogenesis has also been detected; environmental complexity (Kempermann et al., 1997, 1998) and physical activity (van Praag et al., 1999) both enhance survival of newborn neurons. Despite extensive research, however, the functional significance of hippocampal neurogenesis is not yet well understood (reviewed in Greenough et al., 1999; Ming and Song, 2005; Lledo et al., 2006). One of the reasons is the lack of an in vivo assay system that can disrupt neurogenesis specifically.

Computational models

One way to manipulate adult neurogenesis completely is to use neural network simulation models. Although results obtained from a model network does not reflect exactly what happens in the biological network, it can, nevertheless, help to generate hypotheses just as well as can biological experiments. To date, multilayer neural networks predict quite consistently that the occurrence of neuronal turnover acts to endow an already plastic neural network with a higher order form of plasticity (reviewed in Meltzer et al., 2005; Lledo et al., 2006). Accordingly, in the case of OB and DG circuitries, the mechanism appears to be beneficial to both new information learning and old memory clearance (Cecchi et al., 2001; Chambers et al., 2004; Deisseroth et al., 2004). It is also hypothesized that neuronal turnover may optimize temporal associations in memory (reviewed in Aimone et al., 2006), and that it can improve recall of information facilitating precise segregation among stored similar items (Becker, 2005). Overall, under physiological conditions, ongoing neurogenesis appears to be beneficial to the biological system, improving the reliability of its computational machinery for information evaluation.

Primates

Evidence has also accumulated for the occurrence of DG neurogenesis in postnatal nonhuman primates (Gould et al., 1997, 1998; Kornack and Rakic, 1999) and humans (Eriksson et al., 1998). Perhaps in contrast to expectations, it has been shown that a substantially lower magnitude of neurogenesis occurs in humans and macaque monkeys compared to that in rodents. Daily production value rates as one new neuron per 24,000 and 2000 existing granule neurons, in monkeys and mice, respectively (Kornack, 2000). These results may infer that the functional role of neurogenesis in primates is less crucial than that in rodents. Furthermore, while OB neurogenesis is evident in macaque monkeys (Pencea et al., 2001), it is absent in humans (Sanai et al., 2004). The authors have hypothesized that the lack of human OB neurogenesis may be explained by our relatively micro-osmotic capabilities.

Adult neurogenesis under pathological conditions

Depression

It is known that the stress response of the hypothalamic-pituitary-adrenal (HPA) axis-mediated glucocorticoid negative feedback is essential to hold the biological system within a homeostatic state. Although various life events impact on it, the organism usually returns to the basal state within hours or days. Under excessive situations, however, it may be pushed over a "homeostasis boundary" and be unable to recover fully within a reasonable period of time. In humans, this homeostasis breakdown often leads to mental illnesses such as anxiety and mood disorders. Structural and/or functional abnormalities of the hippocampus have often been observed in patients with such conditions, and agents that can ameliorate the symptoms often normalize changes in hippocampus structure and function. The hippocampus is, therefore, the most extensively studied brain substructure in this context. In parallel, adult neurogenesis research has proposed that systemic adaptation could be facilitated by virtue of neurogenesis. Thus, these two lines of research lead to the hypothesis that there may be a causal link between DG neurogenesis and anxiety/mood disorders. On this point, however, both consistent and inconsistent observations have been made (reviewed in Sapolsky, 2004; Rosenbrock et al., 2005; Reif et al., 2006), so that we must continue to seek for definitive evidence.

Epilepsy

Activation of adult neurogenesis within a physiological range seems to be advantageous to the biological system. The question remains, however, as to whether the phenomenon of nonphysiological reactive neurogenesis (e.g., injury-induced neurogenesis) has biological consequences, and if so whether they are beneficial or adverse for the organism. In human temporal lobe epilepsy, the dentate granule cells give rise to abnormal mossy fiber sprouting into the molecular layer of the DG and plausibly form granule cell–granule cell recurrent excitatory synaptic connections. In addition the mossy fiber-CA3 pyramidal cell projections are aberrantly reorganized (Sutula et al., 1989; Houser et al., 1990). These phenomena have also been described in animal models of temporal lobe epilepsy (Tauck and Nadler, 1985; Sutula et al., 1988). The ectopic granule-like neurons express their abnormal physiological properties in synchrony with the CA3 pyramidal cell epileptiform bursts (Scharfman et al., 2000). A causal link has been suggested, therefore, between the histological abnormality and hippocampal hyperexcitability in relation to the pathogenesis of recurrent spontaneous seizures (Mathern et al., 1996).

In the field of adult neurogenesis research, recent evidence has shown that experimentally induced epileptic events dramatically increase DG neurogenesis in rodents. These newborn cells are normal granule neurons and ectopic granule-like cells (in the hilus and DG inner molecular layer), and axons of the latter project aberrantly to both the CA3 pyramidal cell region and the dentate inner molecular layer (Parent et al., 1997). These phenotypic consequences have led to the suggestion that there may be an interaction between reactive neurogenesis and epileptogenesis. Nevertheless, there is also evidence inconsistent with this hypothesis. It is shown that aberrant granule cell axonal projections stabilize the network by preferentially innervating inhibitory neurons and thereby restoring recurrent inhibition (Sloviter, 1992). In accordance with this observation, cessation of reactive neurogenesis was observed following a prolonged stimulation that developed secondary generalized motor seizures (Nakagawa et al., 2000). These outcomes suggest that the reactive neurogenesis is a type of self-protection mechanism. Furthermore, evidence for seizure-induced network reorganization without neurogenesis has been shown (Parent et al., 1999), indicating a minor contribution of newborn neurons to the process. Further investigations may answer the issue of whether reactive adult neurogenesis is a cause or result, is a contributor to, or an adaptive response to psychiatric illnesses.

Adult white matter neurogenesis and PTSD

In addition to the well-described rodent adult neurogenic regions, SVZ and SGZ, we recently identified a novel adult neurogenic region, the temporal germinal layer (TGL), within the white matter adjacent to the caudal edge of the lateral nucleus of the amygdala (LA) of rats (Fig. 1; Takemura, 2005). TGL neurogenesis occurs within the fiber pathway connecting LA to other regions. An impaired TGL neurogenesis caused an increased amygdala excitation, and has led animals to hold a long-term fear memory (Takemura and Kato, unpublished observation). This phenotype in an animal model reminds us of symptoms of human post-traumatic stress disorder (PTSD). A single exposure to a distressing physical or psychological event results, for a proportion of such victims, in prolonged/repetitive physical/mental responses due to an intrusive recollection of the distressing memory (reviewed in Pitman et al., 2006; Vieweg et al., 2006). Because one of the underlying mechanisms for internal vulnerability

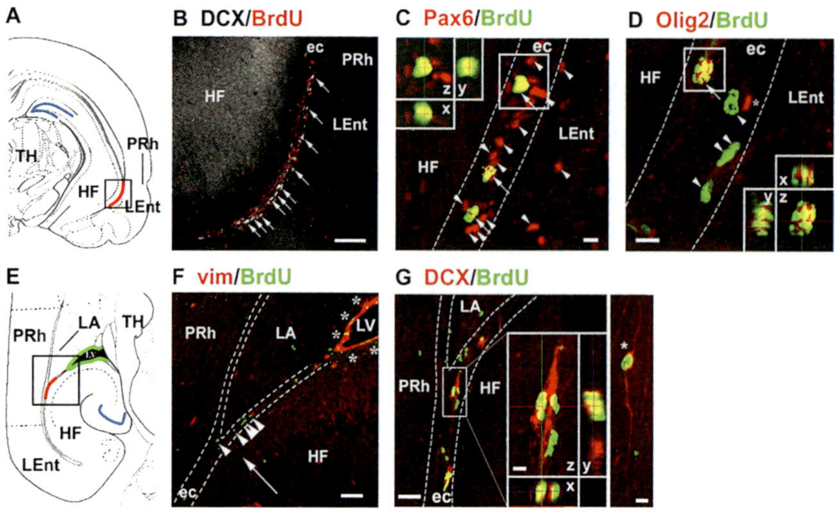

Fig. 1. (A) Representative schematic drawings, including the TGL (indicated in red) and SGZ (blue). Coronal plane at AP −4.8 mm from bregma. (B) A coronal brain section at around AP −4.8 mm from bregma. DCX^+ (white stain, arrows) and $BrdU^+$ cells (red) within the external capsule. Scale bar: 200 μm. (C) BrdU (green) and Pax6 (red) double positive cells (arrows). Pax6 single positive cells (arrowheads). Broken lines represent the border of the external capsule (ec). Scale bar: 10 μm. (D) $BrdU^+$ cells (green) and Olig2 (red) double positive cells (arrow). $BrdU^+$ cells (arrowheads). An Olig2 single positive cell (asterisk). Scale bar: 10 μm. (E) Representative schematic drawings, including the TGL (red), SGZ (blue), and SVZ (green). Horizontal plane at DV −7.6 mm from bregma. (F) Horizontal sections at approximately −7.6 mm from bregma. Arrows indicate the central area of the TGL. $BrdU^+$ cells (arrowheads) in the TGL locate at a distance from the lateral ventricle wall, visualized by vimentin (vim) positive ependymal cells (asterisks). Broken lines represent the border of the external capsule (ec). Scale bar: 100 μm. (G) $BrdU^+$ cells (green) coexpressed DCX (red) in the adjacent section of (F). Scale bars: 10 μm (right panel, inset), 50 μm (left panel). Abbreviations: DCX, doublecortin; ec, external capsule; HF, hippocampal formation; LA, lateral nucleus of the amygdala; LEnt, lateral entorhinal cortex; LV, lateral ventricle; NeuN, neuronal nuclei; PRh, perirhinal cortex; TH, thalamus; vim, vimentin. Adapted with permission from Takemura (2005).

factors to PTSD has been linked to anatomical/functional abnormality in the amygdala, it is intriguing to speculate that an amygdala-relevant neurogenesis in humans might function as TGL neurogenesis in rodents. As it has not yet been examined whether corresponding amygdala-relevant neurogenesis occurs in the human brain, this first clue may lead to future investigations.

Conclusion

Post-developmental neurogenesis is conserved across evolutionary boundaries, and its regionality and magnitude have been linked to species-typical adaptation demands. Amygdala-associated adult neurogenesis has been shown in rats (Takemura, 2005) and monkeys (Bernier et al., 2002), but has not yet been looked for in other species. Future investigation into the degree and location of amygdala-relevant neurogenesis across species may add insights into the differing roles neurogenesis fulfils in support of different species-typical brain functions.

Abbreviations

DG	dentate gyrus
HPA axis	hypothalamic–pituitary–adrenal axis
HVC	high vocal center
LV	lateral ventricle
OB	olfactory bulb
PTSD	post-traumatic stress disorder
RA	archistriatum
RMS	rostral migratory stream
SGZ	subgranular zone
SVZ	subventricular zone
TGL	temporal germinal layer

References

Aimone, J.B., Wiles, J. and Gage, F.H. (2006) Potential role for adult neurogenesis in the encoding of time in new memories. Nat. Neurosci., 9(6): 723–727.

Alonso, M., Viollet, C., Gabellec, M.M., Meas-Yedid, V., Olivo-Marin, J.C. and Lledo, P.M. (2006) Olfactory discrimination learning increases the survival of adult-born neurons in the olfactory bulb. J. Neurosci., 26(41): 10508–10513.

Altman, J. (1962) Are new neurons formed in the brains of adult mammals? Science, 135: 1127–1128.

Altman, J. (1969) Autoradiographic and histological studies of postnatal neurogenesis: IV. Cell proliferation and migration in the anterior forebrain, with special reference to persisting neurogenesis in the olfactory bulb. J. Comp. Neurol., 137(4): 433–457.

Altman, J. and Das, G.D. (1965) Autoradiographic and histological evidence of postnatal hippocampal neurogenesis in rats. J. Comp. Neurol., 124(3): 319–335.

Alvarez-Buylla, A. and Kirn, J.R. (1997) Birth, migration, incorporation, and death of vocal control neurons in adult songbirds. J. Neurobiol., 33(5): 585–601.

Alvarez-Buylla, A., Kirn, J.R. and Nottebohm, F. (1990) Birth of projection neurons in adult avian brain may be related to perceptual or motor learning. Science, 249(4975): 1444–1446.

Becker, S. (2005) A computational principle for hippocampal learning and neurogenesis. Hippocampus, 15(6): 722–738.

Belluzzi, O., Benedusi, M., Ackman, J. and LoTurco, J.J. (2003) Electrophysiological differentiation of new neurons in the olfactory bulb. J. Neurosci., 23(32): 10411–10418.

Bernier, P.J., Bedard, A., Vinet, J., Levesque, M. and Parent, A. (2002) Newly generated neurons in the amygdala and adjoining cortex of adult primates. Proc. Natl. Acad. Sci. U.S.A., 99(17): 11464–11469.

Carleton, A., Petreanu, L.T., Lansford, R., Alvarez-Buylla, A. and Lledo, P.M. (2003) Becoming a new neuron in the adult olfactory bulb. Nat. Neurosci., 6(5): 507–518.

Cayre, M., Strambi, C. and Strambi, A. (1994) Neurogenesis in an adult insect brain and its hormonal control. Nature, 368: 57–59.

Cecchi, G.A., Petreanu, L.T., Alvarez-Buylla, A. and Magnasco, M.O. (2001) Unsupervised learning and adaptation in a model of adult neurogenesis. J. Comput. Neurosci., 11(2): 175–182.

Chambers, R.A., Potenza, M.N., Hoffman, R.E. and Miranker, W. (2004) Simulated apoptosis/neurogenesis regulates learning and memory capabilities of adaptive neural networks. Neuropsychopharmacology, 29(4): 747–758.

Chiang, C.S., McBride, W.H. and Withers, H.R. (1993) Radiation-induced astrocytic and microglial responses in mouse brain. Radiother. Oncol., 29(1): 60–68.

Davis, R.L. (2005) Olfactory memory formation in *Drosophila*: from molecular to systems neuroscience. Annu. Rev. Neurosci., 28: 275–302.

Deisseroth, K., Singla, S., Toda, H., Monje, M., Palmer, T.D. and Malenka, R.C. (2004) Excitation-neurogenesis coupling in adult neural stem/progenitor cells. Neuron, 42(4): 535–552.

Enwere, E., Shingo, T., Gregg, C., Fujikawa, H., Ohta, S. and Weiss, S. (2004) Aging results in reduced epidermal growth factor receptor signaling, diminished olfactory neurogenesis, and deficits in fine olfactory discrimination. J. Neurosci., 24(38): 8354–8365.

Eriksson, P.S., Perfilieva, E., Bjork-Eriksson, T., Alborn, A.M., Nordborg, C., Peterson, D.A. and Gage, F.H. (1998) Neurogenesis in the adult human hippocampus. Nat. Med., 4(11): 1313–1317.

Font, E., Desfilis, E., Perez-Canellas, M.M. and Garcia-Verdugo, J.M. (2001) Neurogenesis and neuronal regeneration in the adult reptilian brain. Brain Behav. Evol., 58(5): 276–295.

Fukuda, S., Kato, F., Tozuka, Y., Yamaguchi, M., Miyamoto, Y. and Hisatsune, T. (2003) Two distinct subpopulations of nestin-positive cells in adult mouse dentate gyrus. J. Neurosci., 23(28): 9357–9366.

Gage, F.H. (2000) Mammalian neural stem cells. Science, 287(5457): 1433–1438.

Gheusi, G., Cremer, H., McLean, H., Chazal, G., Vincent, J.D. and Lledo, P.M. (2000) Importance of newly generated neurons in the adult olfactory bulb for odor discrimination. Proc. Natl. Acad. Sci. U.S.A., 97(4): 1823–1828.

Goldman, S.A. and Nottebohm, F. (1983) Neuronal production, migration, and differentiation in a vocal control nucleus of the adult female canary brain. Proc. Natl. Acad. Sci. U.S.A., 80(8): 2390–2394.

Gould, E., McEwen, B.S., Tanapat, P., Galea, L.A. and Fuchs, E. (1997) Neurogenesis in the dentate gyrus of the adult tree shrew is regulated by psychosocial stress and NMDA receptor activation. J. Neurosci., 17(7): 2492–2498.

Gould, E., Tanapat, P., McEwen, B.S., Flugge, G. and Fuchs, E. (1998) Proliferation of granule cell precursors in the dentate gyrus of adult monkeys is diminished by stress. Proc. Natl. Acad. Sci. U.S.A., 95(6): 3168–3171.

Greenough, W.T., Cohen, N.J. and Juraska, J.M. (1999) New neurons in old brains: learning to survive? Nat. Neurosci., 2(3): 203–205.

Harzsch, S., Miller, J., Benton, J. and Beltz, B. (1999) From embryo to adult: persistent neurogenesis and apoptotic cell death shape the lobster deutocerebrum. J. Neurosci., 19(9): 3472–3485.

Houser, C.R., Miyashiro, J.E., Swartz, B.E., Walsh, G.O., Rich, J.R. and Delgado-Escueta, A.V. (1990) Altered patterns of dynorphin immunoreactivity suggest mossy fiber reorganization in human hippocampal epilepsy. J. Neurosci., 10(1): 267–282.

Kaplan, M.S. and Hinds, J.W. (1977) Neurogenesis in the adult rat: electron microscopic analysis of light radioautographs. Science, 197(4308): 1092–1094.

Kempermann, G., Kuhn, H.G. and Gage, F.H. (1997) More hippocampal neurons in adult mice living in an enriched environment. Nature, 386(6624): 493–495.

Kempermann, G., Kuhn, H.G. and Gage, F.H. (1998) Experience-induced neurogenesis in the senescent dentate gyrus. J. Neurosci., 18(9): 3206–3212.

Kornack, D.R. (2000) Neurogenesis and the evolution of cortical diversity: mode, tempo, and partitioning during development and persistence in adulthood. Brain Behav. Evol., 55(6): 336–344.

Kornack, D.R. and Rakic, P. (1999) Continuation of neurogenesis in the hippocampus of the adult macaque monkey. Proc. Natl. Acad. Sci. U.S.A., 96(10): 5768–5773.

Kuhn, H.G., Dickinson-Anson, H. and Gage, F.H. (1996) Neurogenesis in the dentate gyrus of the adult rat: age-related decrease of neuronal progenitor proliferation. J. Neurosci., 16(6): 2027–2033.

Kuhn, H.G., Winkler, J., Kempermann, G., Thal, L.J. and Gage, F.H. (1997) Epidermal growth factor and fibroblast growth factor-2 have different effects on neural progenitors in the adult rat brain. J. Neurosci., 17(15): 5820–5829.

Li, X.C., Jarvis, E.D., Alvarez-Borda, B., Lim, D.A. and Nottebohm, F. (2000) A relationship between behavior, neurotrophin expression, and new neuron survival. Proc. Natl. Acad. Sci. U.S.A., 97(15): 8584–8589.

Lledo, P.M., Alonso, M. and Grubb, M.S. (2006) Adult neurogenesis and functional plasticity in neuronal circuits. Nat. Rev. Neurosci., 7(3): 179–193.

Lois, C. and Alvarez-Buylla, A. (1993) Proliferating subventricular zone cells in the adult mammalian forebrain can differentiate to neurons and glia. Proc. Natl. Acad. Sci. U.S.A., 90(5): 2074–2077.

Lyubimova, N.V., Coultas, P.G., Yuen, K. and Martin, R.F. (2001) In vivo radioprotection of mouse brain endothelial cells by Hoechst 33342. Br. J. Radiol., 74(877): 77–82.

Margulies, C., Tully, T. and Dubnau, J. (2005) Deconstructing memory in *Drosophila*. Curr. Biol., 15(17): R700–R713.

Mathern, G.W., Babb, T.L., Leite, J.P., Pretorius, K., Yeoman, K.M. and Kuhlman, P.A. (1996) The pathogenic and progressive features of chronic human hippocampal epilepsy. Epilepsy Res., 26(1): 151–161.

McKay, R. (1997) Stem cells in the central nervous system. Science, 276(5309): 66–71.

Meltzer, L.A., Yabaluri, R. and Deisseroth, K. (2005) A role for circuit homeostasis in adult neurogenesis. Trends Neurosci., 28(12): 653–660.

Ming, G.L. and Song, H. (2005) Adult neurogenesis in the mammalian central nervous system. Annu. Rev. Neurosci., 28: 223–250.

Monje, M.L., Mizumatsu, S., Fike, J.R. and Palmer, T.D. (2002) Irradiation induces neural precursor-cell dysfunction. Nat. Med., 8(9): 955–962.

Nakagawa, E., Aimi, Y., Yasuhara, O., Tooyama, I., Shimada, M., McGeer, P.L. and Kimura, H. (2000) Enhancement of progenitor cell division in the dentate gyrus triggered by initial limbic seizures in rat models of epilepsy. Epilepsia, 41(1): 10–18.

Nottebohm, F. (2002) Neuronal replacement in adult brain. Brain Res. Bull., 57(6): 737–749.

Nottebohm, F. (2004) The road we travelled: discovery, choreography, and significance of brain replaceable neurons. Ann. N.Y. Acad. Sci., 1016: 628–658.

Overstreet, L.S., Hentges, S.T., Bumaschny, V.F., de Souza, F.S., Smart, J.L., Santangelo, A.M., Low, M.J., Westbrook, G.L. and Rubinstein, M. (2004) A transgenic marker for newly born granule cells in dentate gyrus. J. Neurosci., 24(13): 3251–3259.

Parent, J.M., Tada, E., Fike, J.R. and Lowenstein, D.H. (1999) Inhibition of dentate granule cell neurogenesis with brain irradiation does not prevent seizure-induced mossy fiber synaptic reorganization in the rat. J. Neurosci., 19(11): 4508–4519.

Parent, J.M., Yu, T.W., Leibowitz, R.T., Geschwind, D.H., Sloviter, R.S. and Lowenstein, D.H. (1997) Dentate granule cell neurogenesis is increased by seizures and contributes to aberrant network reorganization in the adult rat hippocampus. J. Neurosci., 17(10): 3727–3738.

Paton, J.A. and Nottebohm, F. (1984) Neurons generated in the adult brain are recruited into functional circuits. Science, 255(4666): 1046–1048.

Peissner, W., Kocher, M., Treuer, H. and Gillardon, F. (1999) Ionizing radiation-induced apoptosis of proliferating stem cells in the dentate gyrus of the adult rat hippocampus. Brain Res. Mol. Brain Res., 71(1): 61–68.

Pencea, V., Bingaman, K.D., Freedman, L.J. and Luskin, M.B. (2001) Neurogenesis in the subventricular zone and rostral migratory stream of the neonatal and adult primate forebrain. Exp. Neurol., 172(1): 1–16.

Petreanu, L. and Alvarez-Buylla, A. (2002) Maturation and death of adult-born olfactory bulb granule neurons: role of olfaction. J. Neurosci., 22(14): 6106–6113.

Pitman, R.K., Gilbertson, M.W., Gurvits, T.V., May, F.S., Lasko, N.B., Metzger, L.J., Shenton, M.E., Yehuda, R. and Orr, S.P. (2006) Clarifying the origin of biological abnormalities in PTSD through the study of identical twins discordant for combat exposure. Ann. N.Y. Acad. Sci., 1071: 242–254.

van Praag, H., Kempermann, G. and Gage, F.H. (1999) Running increases cell proliferation and neurogenesis in the adult mouse dentate gyrus. Nat. Neurosci., 2(3): 266–270.

van Praag, H., Schinder, A.F., Christie, B.R., Toni, N., Palmer, T.D. and Gage, F.H. (2002) Functional neurogenesis in the adult hippocampus. Nature, 415(6875): 1030–1034.

Reif, A., Fritzen, S., Finger, M., Strobel, A., Lauer, M., Schmitt, A. and Lesch, K.P. (2006) Neural stem cell proliferation is decreased in schizophrenia, but not in depression. Mol. Psychiatry, 11(5): 514–522.

Reynolds, B.A. and Weiss, S. (1992) Generation of neurons and astrocytes from isolated cells of the adult mammalian central nervous system. Science, 255(5052): 1707–1710.

Rochefort, C., Gheusi, G., Vincent, J.D. and Lledo, P.M. (2002) Enriched odor exposure increases the number of newborn neurons in the adult olfactory bulb and improves odor memory. J. Neurosci., 22(7): 2679–2689.

Rosenbrock, H., Bloching, A., Weiss, C. and Borsini, F. (2005) Partial serotonergic denervation decreases progenitor cell proliferation in the adult rat hippocampus, but has no effect on rat behavior in the forced swimming test. Pharmacol. Biochem. Behav., 80(4): 549–556.

Sanai, N., Tramontin, A.D., Quinones-Hinojosa, A., Barbaro, N.M., Gupta, N., Kunwar, S., Lawton, M.T., McDermott,

M.W., Parsa, A.T., Manuel-Garcia Verdugo, J., Berger, M.S. and Alvarez-Buylla, A. (2004) Unique astrocyte ribbon in adult human brain contains neural stem cells but lacks chain migration. Nature, 427(6976): 740–744.

Sapolsky, R.M. (2004) Is impaired neurogenesis relevant to the affective symptoms of depression? Biol. Psychiatry, 56(3): 137–139.

Scharff, C., Kirn, J.R., Grossman, M., Macklis, J.D. and Nottebohm, F. (2000) Targeted neuronal death affects neuronal replacement and vocal behavior in adult songbirds. Neuron, 25(2): 481–492.

Scharfman, H.E., Goodman, J.H. and Sollas, A.L. (2000) Granule-like neurons at the hilar/CA3 border after status epilepticus and their synchrony with area CA3 pyramidal cells: functional implications of seizure-induced neurogenesis. J. Neurosci., 20(16): 6144–6158.

Schmidt, M. and Demuth, S. (1998) Neurogenesis in the central olfactory pathway of adult decapod crustaceans. Ann. N.Y. Acad. Sci., 855: 277–280.

Schmidt-Hieber, C., Jonas, P. and Bischofberger, J. (2004) Enhanced synaptic plasticity in newly generated granule cells of the adult hippocampus. Nature, 429(6988): 184–187.

Scotto-Lomassese, S., Strambi, C., Aouane, A., Strambi, A. and Cayre, M. (2002) Sensory inputs stimulate progenitor cell proliferation in an adult insect brain. Curr. Biol., 12(12): 1001–1005.

Scotto-Lomassese, S., Strambi, C., Strambi, A., Aouane, A., Augier, R., Rougon, G. and Cayre, M. (2003) Suppression of adult neurogenesis impairs olfactory learning and memory in an adult insect. J. Neurosci., 23(28): 9289–9296.

Scotto-Lomassese, S., Strambi, C., Strambi, A., Charpin, P., Augier, R., Aouane, A. and Cayre, M. (2000) Influence of environmental stimulation on neurogenesis in the adult insect brain. J. Neurobiol., 45(3): 162–171.

Shinohara, C., Gobbel, G.T., Lamborn, K.R., Tada, E. and Fike, J.R. (1997) Apoptosis in the subependyma of young adult rats after single and fractionated doses of X-rays. Cancer Res., 57(13): 2694–2702.

Sloviter, R.S. (1992) Possible functional consequences of synaptic reorganization in the dentate gyrus of kainate-treated rats. Neurosci. Lett., 137(1): 91–96.

Snyder, J.S., Kee, N. and Wojtowicz, J.M. (2001) Effects of adult neurogenesis on synaptic plasticity in the rat dentate gyrus. J. Neurophysiol., 85(6): 2423–2431.

Snyder, J.S., Hong, N.S., McDonald, R.J. and Wojtowicz, J.M. (2005) A role for adult neurogenesis in spatial long-term memory. Neuroscience, 130(4): 843–852.

Sutula, T., Cascino, G., Cavazos, J., Parada, I. and Ramirez, L. (1989) Mossy fiber synaptic reorganization in the epileptic human temporal lobe. Ann. Neurol., 26(3): 321–330.

Sutula, T., He, X.X., Cavazos, J. and Scott, G. (1988) Synaptic reorganization in the hippocampus induced by abnormal functional activity. Science, 239(4844): 1147–1150.

Takemura, N.U. (2005) Evidence for neurogenesis within the white matter beneath the temporal neocortex of the adult rat brain. Neuroscience, 134(1): 121–132.

Tauck, D.L. and Nadler, J.V. (1985) Evidence of functional mossy fiber sprouting in hippocampal formation of kainic acid-treated rats. J. Neurosci., 5(4): 1016–1022.

Tramontin, A.D. and Brenowitz, E.A. (1999) A field study of seasonal neuronal incorporation into the song control system of a songbird that lacks adult song learning. J. Neurobiol., 40(3): 316–326.

Vieweg, W.V., Julius, D.A., Fernandez, A., Beatty-Brooks, M., Hettema, J.M. and Pandurangi, A.K. (2006) Posttraumatic stress disorder: clinical features, pathophysiology, and treatment. Am. J. Med., 119(5): 383–390.

Wang, N., Aviram, R. and Kirn, J.R. (1999) Deafening alters neuron turnover within the telencephalic motor pathway for song control in adult zebra finches. J. Neurosci., 19(23): 10554–10561.

Wojtowicz, J.M. (2006) Irradiation as an experimental tool in studies of adult neurogenesis. Hippocampus, 16(3): 261–266.

Zupanc, G.K. (2001) A comparative approach towards the understanding of adult neurogenesis. Brain Behav. Evol., 58(5): 246–249.

Discussion: Chapter 7

PITMAN: Do you think the decreased production of neurones adjacent to the amygdala of the animals reflects decreased production of GABA interneurons?

KATO: It is an important point. A reduced TGL neurogenesis increased amygdala neuron firing. This result suggests that TGL produces inhibitory neurons. Currently, we have tried to find out the evidence.

SECKL: I enjoyed your talk. Data from Meaney and colleagues have suggested that thyroid hormones are important in the short-term maternal separation paradigm. Do you think your finding of transthyretin in the choroid plexus might mediate the increased passage of T3 to the neonatal brain? Have you looked at other components of thyroid hormonal signaling?

KATO: In our own experiment, we have studied the change in gene expression by means of microarray and found no change at all after maternal separation except for transthyretin. Actually we are doing similar experiments in the neonatal mild hypothyroid rat that shows in later life hyperactivity and learning impairment. It could be related to a model of neurodevelopmental disorder and perhaps somehow to PTSD as well.

SANDI: I do not know if I understood you correctly about the flight responses. Do you mean the testing or training or the flight response, or does it concern the conditioning phase?

KATO: We employed a fear conditioning paradigm using ultrasound (USS), which evokes innate fear in rodents. Because USS itself can induce fear in animals, the behavior in response to the initial USS can be considered not only conditioning but also testing phase. We divided the behavior into freezing and flight, and found the novel neurogenic area near the amygdala was only related to flight but not to freezing behavior.

E.R. de Kloet, M.S. Oitzl & E. Vermetten (Eds.)
Progress in Brain Research, Vol. 167
ISSN 0079-6123

CHAPTER 8

Commentary: behavioral phenotype

Melly S. Oitzl*

Division of Medical Pharmacology, LACDR/LUMC, University of Leiden, P.O. Box 9502, 2300 RA Leiden, The Netherlands

Abstract: PTSD arises by definition as a direct consequence of the experience of an acute severe stressor. The formation of traumatic memory and its extinction, sympathetic and adrenocortical stress systems activity in relation to individual vulnerability form the core of animal models for PTSD.

Keywords: animal model; validity; resilience; conditioned fear; fear extinction; mineralocorticoïd receptor; glucocorticoid receptor

The previous presentations showed challenging data from preclinical research in rats and mice that offer quite some opportunities to develop potential animal models for PTSD. Carmen Sandi underlined the relevance of personality traits in inter-individual differences; the role of neural cell adhesion molecules in various brain areas was addressed for understanding neural substrates in the formation of fear memories. Peter Gass, using GR mutant mice, specifically addressed the involvement of the GR in stress-related diseases. Benno Roozendaal showed compelling evidence supporting glucocorticoid and catecholamine interactions that may underlie the facilitation as well as the impairment of memory consolidation and retrieval; these interactions occurred in the amygdala and the hippocampus, and could be modulated by inputs of the prefrontal cortex. Nobumasa Kato described correlations between neurogenesis in the amygdala and periventricular areas with fear conditioning and traumatic early life experience.

Overall, the impact of the glucocorticoids and catecholamines and the role of their receptors in regulation of conditioned fear memories was the central theme of the presentations that were concerned with animal models for PTSD. The actions exerted by the stress hormones were examined in the amygdala, hippocampus and prefrontal cortex, which are brain areas thought to be predominantly involved in the processing of fearful stimuli. At the same time the output of these regions can modulate the behavioral stress response. Therefore, I will focus my commentary on the validity of animal models of conditioned fear and how these models relate to PTSD.

In classical fear conditioning tasks, the subjects very quickly learn to associate a sensory stimulus [conditioned stimulus (CS), such as a tone, light or odor] or environmental context with a coinciding aversive stimulus [unconditioned stimulus (US), such as an electric shock]. If a lasting memory is formed, the subsequent re-exposure to the CS or context will elicit a conditioned fear response (CR). This conditioned response is accompanied by autonomic reactions, which manifest itself as tachycardia, pressor response and hypoalgesia. Also neuroendocrine activation occurs, which

*Corresponding author. Tel.: +31 71 5276289;
Fax: +31 71 5274715; E-mail: m.oitzl@lacdr.leidenuniv.nl

DOI: 10.1016/S0079-6123(07)67008-3

results in the release of the hormones of the hypothalamic-pituitary-adrenal (HPA) axis as well as species-specific defensive behavior characterized by freezing, risk assessment and flight (LeDoux, 2000). This response pattern is rapid and robust, and would allow for a precise control of the different phases of fear memory including the modulation of fear memory through stimulus predictability and level of emotional arousal.

Since the quality of an animal model depends largely on detailed knowledge of the cause(s), course and symptoms of the disorders, the attempts to model PTSD have a rather exceptional "advantage": (i) PTSD arises by definition as a direct consequence of the experience of an acute severe stressor; (ii) in addition to a variety of autonomic, neuroendocrine and behavioral symptoms, fear is central to PTSD. There is a large individual difference, however, in the outcome of animals exposed to severe stressors. This individual difference is based on previous adverse experiences in interaction with genetic background and personality traits. Specific examples of animal models based on this reasoning are given in the contributions by J.R. Seckl, M.M. Tsory et al., N.U. Takemura and N. Kato and S. Chourbaji et al.

Fear is a protective mechanism and aimed to promote an adaptive response to danger. It is well known that an emotionally arousing event triggers an acute stress response, and that the convergence of the various stress mediators in time and space can strengthen the memory for that event (Joëls et al., 2006). Given this knowledge, fear memories preferably are analyzed in relation to the various phases of memory processing: acquisition, consolidation and retrieval. The following possible causes of PTSD are addressed: (1) An exceptional strong encoding process of fearful memories of the adverse event. (2) A dysfunction of processes that facilitate extinction of fear memory. (3) The role of glucocorticoids and catecholamines (and their interaction) in distinct memory phases.

Glucocorticoids secreted or administered at the right time and site in the brain will facilitate ongoing behavior, normalize maladaptive behavior and eliminate behavioral responses that are of no more relevance. Besides these memory-strengthening effects the modulation of fear memory by glucocorticoids and catecholamines also can have apparent memory impairing effects if the hormones are administered out of context. These aspects of dosage and timing for the action of these two hormones have been documented in great detail over the past decennia (for review, see de Kloet and Oitzl, 2006). Recently, the action of the classical stress hormones came again in focus, particularly with respect to their interaction in distinct brain areas (Chapter 6 by B. Roozendaal et al.). Furthermore, the functional significance of the two corticosteroid receptors mineralocorticoid (MR) and glucocorticoid receptor (GR), respectively (de Kloet et al., 1999) has been validated using mutant mice that have these receptors deleted or overexpressed (Chapter 5 by S. Chourbaji et al.). Moreover, recently a fast non-genomic action mediated by MR was discovered in the hippocampus that might contribute to the expression of initial fear responses (Karst et al., 2005). Much new data for the PTSD animal model can be expected to be generated from the recently developed MR and GR mutant mouse models that are based on brain site specific inducible knock-out, knock-in or downregulated genes.

Animal models for PTSD usually are based on rather short-term observations of fear-motivated behavior. Less attention has been paid to more remote fear memories and this is surprising because characteristic features of PTSD are the long-lasting recurring responses to the original stressful event. Although a distinct acute stressor is the trigger for PTSD, the long-lasting traumatic consequences also represent the chronic experience of stressful conditions. Recurrent fear and stress may have important implications for the extinction of fear. Reinstatement, renewal, spontaneous recovery and reconsolidation are other aspects of fear memory that need to be integrated in future animal models for PTSD.

Conditioned and unconditioned fear-related behaviors require the integrated activity of neuronal networks. The questions that arise are: Does the fear-conditioning event in the laboratory represent a "traumatic" memory for the animal? Is a physiological or a pathological fear process under study? Although it is likely, we do not know whether the same brain structures (amygdala, hippocampus, prefrontal cortex) and molecular mechanisms, e.g., corticosteroids, catecholamines, and their targets such as neural cell adhesion molecules — NCAMs, BDNF (S. Chourbaji et al.), neurogenesis (Chapter 7 by N.U. Takemura and N. Kato) are involved in physiological as well as pathological processing of fear. For example, does overgeneralization or sensitization of fear-motivated behavior indicate an imbalance or dysfunction in one of the fear controlling brain sites or does it point to the involvement of another brain area? Does an impairment of contextual fear conditioning or lack of contextual fear extinction indicate a mismatch of hippocampal/cortical/amygdala integration? As a new approach towards these questions behavior analyzed with simultaneous tetrode recordings already has provided links to these limbic brain circuits (Quirk et al., 2006). Of course it can be expected that more answers and questions will be provided in human brain imaging studies.

Given its rich symptomatology and comorbidity, PTSD is likely not a "one gene disorder". Stress hormone signaling pathways undoubtedly contribute to the development of PTSD as the modulations of these hormones on fear memory proved to be useful in animal models for stress-related brain diseases in general. Large-scale phenotyping using behavioral and neuroendocrine selection criteria such as high anxiety, social avoidance, impaired extinction, arousal, sleep disturbances, might render more appropriate animal models for PTSD. While currently the focus is mainly on the face validity of the negative symptoms of the disorder, it is as challenging to find out why a large group of individuals is resilient to the adverse effects of stress. To characterize their behavioural phenotype in relation to genetic analysis and manipulation may offer an alternative approach to generate a novel animal model for PTSD.

References

Joëls, M., Pu, Z., Wiegert, O., Oitzl, M.S. and Krugers, H.J. (2006) Learning under stress: how does it work? Trends Cogn. Sci., 10(4): 152–158.

Karst, H., Berger, S., Turiault, M., Tronche, F., Schutz, G. and Joëls, M. (2005) Mineralocorticoid receptors are indispensable for nongenomic modulation of hippocampal glutamate transmission by corticosterone. Proc. Natl. Acad. Sci. U.S.A., 102(52): 19204–19207.

de Kloet, E.R. and Oitzl, M.S. (2006) Cortisol and PTSD: animal experiments and clinical perspective. In: Kato N., Kawata M. and Pitman R.K. (Eds.), PTSD: Brain Mechanisms and Clinical Implications. Springer-Verlag, Tokyo, Japan, pp. 13–27.

de Kloet, E.R., Oitzl, M.S. and Joëls, M. (1999) Stress and cognition: are corticosteroids good or bad guys? Trends Neurosci., 22: 422–426.

LeDoux, J.E. (2000) The amygdala and emotion: a view through fear. In: Aggleton J.P. (Ed.), The Amygdala. Oxford University Press, Oxford, pp. 289–310.

Quirk, G.J., Garcia, R. and Gonzalez-Lima, F. (2006) Prefrontal mechanisms in extinction of conditioned fear. Biol. Psychiatry, 60(4): 337–343.

General Discussion: Section II

OITZL: To underline the relevance of context, time-dependency and convergence of stress and stress hormone activity for behavior and especially fear memories, I would like to raise the following questions for the general discussion.

Firstly, do the effects of stress through glucocorticoids and catecholamines affect specific stages of memory processes in such a way that fear conditioning tasks will be affected in animal models representative for PTSD?

Secondly, does the current knowledge on mechanisms and predictors for PTSD allow the design of a novel animal model?

ROOZENDAAL: In response to the first question: adrenalectomy removes only the peripheral but not the central responses of the stress system. Of course it is true that peripheral catecholamines cannot penetrate the blood-brain barrier, while glucocorticoids do. These facts complicate the extrapolation of peripheral administration of catecholamines and glucocorticoids towards their interactions in the amygdala and hippocampus. However, in conditions of extreme stress, things might be different. In answer to the second question: 'If our work can provide a model for PTSD'. I do not think so. But, I want to underline that our work has a large impact on the treatment of PTSD. For example, Roger Pitman is using beta-blockers based on the animal and human studies of Larry Cahill. Moreover, Gustav Schelling is using glucocorticoid treatment of PTSD patients, also based on our findings. Thus, we might consider our avoidance paradigm and the understanding of interactions of glucocorticoids and catecholamines in different brain areas as learning under traumatic conditions.

BREMNER: I have a general question. Why are we talking about memory facilitating effects of glucocorticoids if dexamethasone and hydrocortisone inhibit declarative memory?

ROOZENDAAL: It depends on the timing of glucocorticoid treatment. In the laboratory, individuals are experiencing stressors or receive glucocorticoid treatment just before they have to learn a task. Retrieval is usually 20–60 min later, which means that glucocorticoids are still elevated. For acquisition and short-term memory retrieval most studies report impairment after such prior exposures to glucocorticoids or stress. Long-term consolidation, upon which glucocorticoids have a facilitating effect, has not taken place yet. Studies that use longer time intervals report an enhancement of memory specifically for the arousal (stress) condition.

LIBERZON: Before we go into such details, I want to ask a more provocative question. How do we really know that fear memory formation has anything to do with PTSD? Is there any evidence that PTSD patients have enhanced fear memories?

ROOZENDAAL: Actually, the work of Gustav Schelling using glucocorticoids to suppress PTSD symptoms indicates that it is rather a memory retrieval problem.

LIBERZON: I am looking for mechanistic arguments. I am asking: Do we have any evidence that fear-related memory is actually involved in the pathophysiology of PTSD?

PITMAN: That's too general. Are we talking about declarative, implicit or hippocampal-associated memory, which are completely different? Is it the ability to elicit a physiological response or the ability to give facts of the traumatic event? Depending on which aspect, there is plenty of evidence.

LIBERZON: I am talking about specific fear-related memories. Which one would you like to address specifically?

PITMAN: About 20 studies report larger physiological responses in traumatized people with PTSD compared to traumatized people without PTSD. This kind of conditioned response is one part of the memory. However, there is little evidence for the ability to narrate about the facts of the traumatic event, which is another kind of memory. We know that.

LIBERZON: Thanks. How do we separate that from simply physiological reactivity to a particular stimulus which we remember? Exaggerated physiological reactivity could be driven on the efferent part of the response, not in the position of the memory trace, not even in the retrieval of the trace.

PITMAN: One piece of evidence against that is: It is pretty selective for the traumatic memory of people with PTSD rather than for their other memories.

BREMNER: But — no one has ever shown that people with PTSD remember emotional traumatic material more. They don't look at the results from emotional word lists.

SHALEV: This is a general discussion — Do you mean implicit memory or data from Stroop tests?

BREMNER: We have done this using combat slides followed by a forced recall. Did PTSD patients remember combat slides better — no, they didn't! This never gets published because it is not a positive finding.

OITZL: Given the way memory systems are discussed up to now, is it becoming more clear that animal models (did) contribute to our understanding of PTSD?

BREMNER: The point is that declarative memory impairment is not a necessary feature of PTSD. The core is the traumatic memory. And there arises the question: did PTSD patients have an enhanced traumatic memory formation. The answer is — we don't have evidence for that.

LIBERZON: I'd like a moment of reflection: Do we all assume that there is a real source of pathology in PTSD? This is different from having some symptoms. But is it so that fear memory consolidation, you may even expand it to fear memory reconsolidation and extinction is a central part of the pathophysiological process, and from there, it leads to exaggerated responses? — One may argue the other way round and state that the emotional responses are totally dysregulated. We have hyperactivation that is not related to memory and therefore, we should not focus on processes of fear memory but more specific to anxiety responses to contextual cues.

SANDI: I do not see why these are two alternatives and fear memories might not play a role. I think we have to distinguish between the fear memories of the traumatic experience and the ability to acquire new emotional memories that are not related to the core of the pathology. In animal experiments, very strong fear conditioning might relate to inhibition of molecular reprocessing that may also prevent the integration of the experience in a memory trace, while other degrees of stress facilitate fear memories, as it is expressed in the well-known inverted U-shape of stress and acquisition of new information. I am rather convinced that both, emotional fear memories and the more declarative memory processes interact in this kind of pathology.

SHALEV: I think we have to be serious about the lack of translational relevance of the preclinical findings for PTSD. The data that were presented today indicate that PTSD may not be the only human condition that fits the profile. PTSD is most probably a more complex condition than we considered up to now. True, it includes an element of fear responses and a narrowing of facts, but it also includes elements like redundancy of intrusive recollections — a phenomenon, that is totally unpredicted by the fear memory model. Elements of depression, the numbing of feelings, hypervigilance are somewhat predicted by fear conditioning, also by other models. As clinicians, we have to define these different elements. We may also think about avoidance, surrender, numbing, depression and hypervigilance in relation to territorial defense. However, we are very bad in disentangling these conditions into the different components, which would benefit biological research.

OITZL: To summarize — we should expand from fear models to a variety of other experimental designs that will characterize or reflect certain elements of the human disorder. Meta-analysis might help to figure out which condition of the animal fits most closely the human situation. A first step would be to shift the focus from fear to other elements.

SHALEV: Yes, thinking of territory and imminence (the threat of something to happen) related to my PTSD patients: they have lost the ability to consider themselves living in a safe territory. One major element in animal research could be to explain why PTSD patients are unable to territorialize fear. This ability might not involve hippocampal or amygdala functions, but other brain systems that recognize spatial and time relations in a different way.

LIBERZON: I think this is a big challenge for the clinicians. That's actually why I started my arguments. I do not think that there is an animal model for PTSD, as there is no animal model for depression or schizophrenia. We all know that these are multifactorial diseases. What we can try is, to focus on the following: what subsets of symptoms associate with subsets of neurobiological changes, subsets of predisposing factors — and try to model these subsets. However, we should not try to apply that to, e.g., numbing, avoidance behavior. It might be a component of intrusive memories or other PTSD symptoms. For example, fear might be the connection between several symptoms and predispose for pathophysiological processes.

KATO: It is not necessary that an animal model presents the complete picture of PTSD in humans; this is in fact, impossible. It depends which symptoms might be critical in terms of development or treatment of PTSD. I suppose impairment of extinction of fear memories may be a central feature, because intrusive memories are long-lasting and form a most discrete trait. There is a relation to memory in early life. Concerning neural stimulation and neurogenesis, we might consider stressful early life events like maternal deprivation, to dampen the activity of the lateral amygdala, which might be of significance for the development of PTSD.

BREMNER: From brain imaging studies in humans we know that so-called fear-inducing pictures and reading of scripts of a trauma often fail to activate the amygdala. However, classical fear conditioning always activates the amygdala. What I see is that PTSD patients overgeneralize. They respond to conditioned but also to all other kinds of stimuli. Is PTSD a disorder of excessive conditioned fear responses, or is it a specific kind of the stimuli that elicits fear responses, is it a failure of selective acquisition of fear?

SHALEV: The acquisition of fear is a necessary but non-sufficient condition for PTSD.

STAM: There are animal models that address the generalization of fear, which might be called a problem of generalized sensitization. These models do not look only at conditioned responses but test generalization to different test conditions. Animals do become hyperresponsive. So far, the discussion has been concentrated on conditioned fear, but I think unconditioned hyperresponsivity has to be considered as an important component in animal models.

YEHUDA: I would like us to start answering these questions by saying: it depends — instead of yes or no. PTSD looks different in different people. Animal models can help us to understand these individual differences in fear acquisition, consolidation and extinction. That would be very helpful to clinicians, because everybody who is traumatized becomes afraid. You pointed out that it is not a bad thing to become afraid. The idea is that for some people the mechanism that becomes engaged in healing is not engaged in the activation of the memory. The memory is distressing also because it contains so much physiological kick that it is very reminiscent of somebody who is just experiencing the trauma (and not long ago). I think, we the people who know the PTSD patients can articulate exactly what the problems are. It is not expected that one animal model will cover the whole scope of PTSD. Another valuable aspect is the developmental perspective. Here, I do not mean childhood experiences — I want to address the fact of age, specifically in relation to the bidirectional effects of glucocorticoids, the inverted U-shape. We have data that glucocorticoid treatment of young PTSD patients worsens their state, while treatment of older PTSD patients alleviates the symptoms. These findings also relate to the question you raised about the context: again it is not the yes or no, but — under what condition. You know that in response to treatment the animal work is helpful, but what we have to figure out is: who definitely can be helped with propranolol and who definitely can be helped with glucocorticoids — and — who cannot.

SECTION III

The Clinical Perspective: Conceptualization of PTSD and Related Disorders

E.R. de Kloet, M.S. Oitzl & E. Vermetten (Eds.)
Progress in Brain Research, Vol. 167
ISSN 0079-6123

CHAPTER 9

Transgenerational transmission of cortisol and PTSD risk

Rachel Yehuda* and Linda M. Bierer

The Traumatic Stress Studies Program, Department of Psychiatry, Mount Sinai School of Medicine and Bronx Veterans Affairs, James J. Peters VAMC, 116-A, OOMH-PTSD, 130 West Kingsbridge Road, Bronx, NY 10468, USA

Abstract: Parental posttraumatic stress disorder (PTSD) appears to be a relevant risk factor for the development of PTSD, as evidenced by a greater prevalence of PTSD, but not trauma exposure, in adult offspring of Holocaust survivors with PTSD, compared to children of Holocaust-exposed parents without PTSD. This paper summarizes recent neuroendocrine studies in offspring of parents with PTSD. Offspring of trauma survivors with PTSD show significantly lower 24-h mean urinary cortisol excretion and salivary cortisol levels as well as enhanced plasma cortisol suppression in response to low dose dexamethasone administration than offspring of survivors without PTSD. In all cases, neuroendocrine measures were negatively correlated with severity of parental PTSD symptoms, even after controlling for PTSD and even other symptoms in offspring. Though the majority of our work has focused on adult offspring of Holocaust survivors, recent observations in infants born to mothers who were pregnant on 9/11 demonstrate that low cortisol in relation to parental PTSD appears to be present early in the course of development and may be influenced by in utero factors such as glucocorticoid programming. Since low cortisol levels are particularly associated with the presence of maternal PTSD the findings suggest the involvement of epigenetic mechanisms.

Keywords: posttraumatic stress disorder; glucocorticoid programming; epigenetics; cortisol; intergenerational effects; risk factors

Introduction

Biological alterations in posttraumatic stress disorder (PTSD) were initially conceptualized as being caused by trauma exposure. With the realization that only a proportion of trauma-exposed persons develop PTSD, it became clear that trauma exposure alone could not account for either the existence of PTSD, or its concomitant neurobiology (Yehuda and McFarlane, 1995). This recognition has prompted the search to identify risk factors that influence the development of PTSD following trauma, and subsequently, elucidate their biological basis.

Some studies have suggested that dissociation or panic attacks during trauma exposure, lack of social support subsequent to it, avoidant coping behaviors, or the development of maladaptive cognitions, increase the likelihood of developing PTSD (Brewin et al., 2000). Far more provocative is the idea that there are pre-traumatic risk factors that contribute to, or even possibly predict PTSD. Several studies have suggested that pre-existing

*Corresponding author. Tel.: +1 718 584 9000 x 6964; Fax: +1 718 741 4775; E-mail: Rachel.yehuda@va.gov

DOI: 10.1016/S0079-6123(07)67009-5

personality, behaviors, and psychological traits contribute to the development of PTSD (McFarlane, 1989; Breslau et al., 1998). Other pre-traumatic risk factors include exposure to adversity prior to the "focal" trauma (i.e., the event that immediately precipitated the PTSD (Bremner et al., 1993) or family history of psychiatric disorder (McFarlane, 1988; Davidson et al., 1998). The latter finding, in particular, presents a gateway to detection of longstanding molecular, genetic, or epigenetic phenomena that might be informative about biological mechanisms of PTSD and/or risk for this disorder.

To date there are no established biological risk factors for PTSD. However, about a decade ago, our group reported an association between parental PTSD and the subsequent development of PTSD in adult offspring of Holocaust survivors (Yehuda et al., 1998a). There was a higher prevalence of PTSD in adult offspring of Holocaust survivors compared to demographically similar comparison subjects even though there were no group differences in trauma exposure (Yehuda et al., 1998b). This paper reviews findings that support the idea that parental PTSD may be associated with biological changes that render offspring more vulnerable to the effects of trauma exposure, and discusses the implications of these findings.

The evidence for parental PTSD as a risk factor for PTSD

The finding that trauma survivors who develop PTSD following exposure to adversity are more likely to have a family history of psychopathology was first made in 1918 (Oppenheimer and Rothschild, 1918; Wolfsohn, 1918). These initial observations were replicated by other investigators who found similar associations in World War I and World War II veterans and their families (Swan, 1921; Cohen et al., 1948). More contemporary community-based studies confirmed that respondents with PTSD are on average three times more likely than trauma survivors without PTSD to report family mental illness of anxiety, depression, psychosis, and antisocial behavior. This has been demonstrated in studies of PTSD following rape (Davidson et al., 1998), disaster (McFarlane, 1988), and combat exposure (Reich et al., 1996). Though parental trauma and PTSD were not measured specifically in the above studies, the findings are consistent with the idea that a trauma survivor's risk for PTSD might be related to family history of PTSD-like symptoms.

The link between PTSD in offspring and trauma exposure in parents was first made by Solomon et al. (1988), who reported that war veterans who were children of Holocaust survivors had higher rates of PTSD, and a different clinical picture, compared to Lebanon War veterans whose parents were not survivors. Our previous observations extended this finding by demonstrating not only that Holocaust offspring were more likely to have PTSD following their own traumatic events than offspring of demographically comparable Jewish controls (Yehuda et al., 1998a), but that PTSD in Holocaust offspring was specifically related to the presence of PTSD in the Holocaust survivor (Yehuda et al., 1998b).

Is parental PTSD a genetic risk factor?

The finding of a familial association with PTSD is not, per se, evidence of genetics. However, this possibility becomes more plausible in considering findings of an increased prevalence of PTSD among trauma survivors who also had a twin with PTSD compared to trauma survivors whose exposed twin did not develop PTSD. An increased risk for the development of PTSD in monozygotic than dizygotic twins has been demonstrated in combat veterans (True et al., 1993), and in population-based studies of civilians (Koenen et al., 2002; Stein et al., 2002).

Nonetheless, to date, very few genes have been identified as associated with PTSD. Significant associations were found with a variable number tandem repeat (VNTR) polymorphism in an untranslated region of the dopamine transporter gene in a relatively small convenience sample of subjects with PTSD compared to those without PTSD (Segman and Shalev, 2003). In another small study, no association was found between two glucocorticoid receptor (GR) polymorphisms, N363S and BclI, and the diagnosis of PTSD although PTSD patients homozygous for the BclI GG genotype tended to be more responsive to a peripheral test of glucocorticoid sensitivity and displayed more severe PTSD symptoms (Bachmann et al., 2005).

The limited number of gene-related findings in PTSD may reflect the complexity involved in executing research in this area. Alternatively, the failure to find susceptibility loci for PTSD may suggest that PTSD risk is not related to differences in genetic polymorphisms. Indeed, there are alternative explanations even for the twin findings. For example, though PTSD was more likely to be concordant in monozygotic than dizygotic twins, there was also a genetic association with exposure to interpersonal violence (Stein et al., 2002). In contrast, the concordance rate of exposure to accidents or disaster among identical and fraternal twins was not different. Thus, genetic risk for assault trauma reflected personality traits — anger or irritability — that increased the likelihood for assault, which then increases the likelihood of PTSD (Stein et al., 2002). By linking genetic predisposition to the risk of assaultive trauma the findings imply the biological factors that predict exposure characteristics may be distinct from those that are associated with the development of PTSD. In considering Holocaust survivors, for example, the primary risk for exposure to the Holocaust was being Jewish and living in Nazi-occupied Europe in the 1930s. While it may be that persons who developed PTSD in response to Holocaust exposure did so based on specific pre-trauma risk factors, it is implausible that the same genetic predisposition that explains Holocaust exposure also explains the development of PTSD.

Alternative explanations for the association between parental PTSD and PTSD in offspring

There are several potential explanations for the finding of a greater prevalence rate of PTSD among offspring of Holocaust survivors that do not necessarily implicate genetic mechanisms. The development of PTSD following exposure to a traumatic event requires a subjective perception of threat, which includes feelings of helplessness, or lack of competence for survival from such experiences. Individual differences in the presence of cognitive schemas such as, whether the world is fundamentally safe or not, or whether persons are to blame for traumatic events that befall them, have been demonstrated to underlie difference in PTSD prevalence according to trauma type. Thus, a rape victim who feels she may have provoked her assailant, or is in some other way responsible for what happened, may be more likely to develop PTSD than another victim who does not have this perception. Arguably, prior experience is a major contributor to such subjective interpretations.

The presence of a parental trauma — particularly one as life altering as was exposure to the Nazi Holocaust — can also influence an offspring's response to trauma in that perspective about the world, such as whether it is fundamentally a dangerous place and/or whether individuals are competent to deal with adversity, can be transmitted from parent to child. Children can also "learn" symptoms from seeing them in parents. The increased prevalence of PTSD in Holocaust offspring may also reflect deficits in child rearing that are experienced as abuse or neglect, and increase the likelihood of exposure to subsequent events, as well as the vulnerability to develop PTSD symptoms, in the same way that prior traumatization has been described as a risk factor for subsequent trauma and for PTSD (Clancy et al., 2006; Koenen et al., 2007). Children can also become traumatized by Holocaust-related stories themselves, or by visualizing what their parents experienced. Such forms of vicarious traumatization have been known to occur even among therapists exposed to traumatic material of their patients, and may be particularly salient sources of vulnerability in children. Thus, even before entertaining specific models of biological transmission of vulnerability, including heritable or developmental genetic alterations, there are several possible mechanisms to account for the observation of increased risk for PTSD in relation to parental exposure and/or PTSD.

There are two broad conceptions that may be applicable to biological transmission of risk from parent to child. The first is that the same biological alteration that explains symptoms in parents (e.g., a genetic polymorphism) is transmitted to the offspring. The second is that the response in the offspring is a consequence of a change in the parents, such as "programming" due to early environmental modifications, in either the parent or child, which may produce permanent and transmissible changes. An example is the type of glucocorticoid programming, recently demonstrated to be influenced by maternal care that is determined by epigenetic mechanisms. It is this latter possibility of programming that has become particularly interesting based on the convergence of findings concerning hypothalamic-pituitary-adrenal (HPA) axis alterations in PTSD. This possibility can be discussed in light of findings relating to the association between HPA measures and parental PTSD.

Biological correlates of parental PTSD: studies of Holocaust offspring

Though observations of low cortisol levels in PTSD (Mason et al., 1986) were initially interpreted as reflecting pathophysiology of the disorder, cortisol-related alterations in PTSD may reflect pre-existing vulnerability factors that increase the probability of developing PTSD following trauma exposure. In several prospective, longitudinal studies, lower cortisol levels in the acute aftermath of trauma were associated with either the subsequent development of PTSD, or with the well-established risk factor of prior trauma exposure (Resnick et al., 1995; Delahanty et al., 2003). These findings suggested that reduced cortisol levels at the time of a trauma may compromise the inhibition of stress-induced biological responses (e.g., during and following a traumatic event), resulting in a prolonged physiological/emotional distress which would then facilitate the development of PTSD in at least some at risk persons (Yehuda, 2002). The question that arose from the above findings, however, concerns the origin of low cortisol levels in at risk persons. In consideration of the above observations pertaining to reduced cortisol, and in light of the findings that parental PTSD is a risk factor for PTSD, we set out to examine HPA axis parameters in adult children of Holocaust survivors.

In an initial study, we examined mean 24-h urinary cortisol excretion in 11 Holocaust offspring with no parental PTSD and 24 offspring with parental PTSD, compared to 15 demographically similar offspring of Jews born to parents who did not undergo the trauma of the Holocaust (Yehuda et al., 2000). Holocaust offspring were born after World War II, or after their parents had escaped to safety during the War, and were raised through adolescence by at least one biological parent who had been interned in a Nazi concentration camp during World War II or faced comparably severe threats in hiding. These offspring subjects were further subdivided based on whether at least one parent met the diagnostic criteria for lifetime PTSD according to the Parental PTSD Questionnaire (PPQ), completed by the offspring. This scale has recently been validated against 58 clinical interviews of the parents using the Clinician Administered PTSD Scale (CAPS) (Blake et al., 1995) and found to have extremely high predictive strength for PTSD diagnosis, with 15 out of 17 items showing significant correlations between offspring-rated and clinician-rated parental symptoms based on direct interview of the parents (Yehuda et al., 2006). Offspring with parental PTSD were further subdivided into those with and without PTSD (based on their own life events). The offspring with PTSD (and parental PTSD) had significantly lower cortisol levels than the two groups without parental PTSD. However, offspring without any lifetime PTSD who had the risk factor of parental PTSD also had significantly lower cortisol levels than the latter two groups, and showed cortisol levels that were only slightly (but not significantly) higher than offspring with PTSD. We concluded from this study that low cortisol levels were associated with the risk factor of parental PTSD.

Findings of studies examining urinary cortisol excretion in PTSD and trauma survivors have been highly variable, possibly because of the room for methodological error in procedures that place responsibility for complete collections in the subjects' hands. We were therefore interested in replicating the observation of low cortisol in relation to parental PTSD using a different subject group, but also a different collection procedure, in which we would examine cortisol output over the diurnal cycle. We hypothesized that alterations in plasma cortisol levels and chronobiological parameters would be observed only in the offspring with parental PTSD, and that offspring without parental PTSD would demonstrate chronobiological parameters similar to the comparison sample (Yehuda et al., 2007b).

A total of 23 men and 26 women (33 offspring, 16 comparison subjects) were examined. In this study, as well, we subdivided offspring on the basis of presence or absence of parental PTSD, using the PPQ. However, we did not include any subjects with current or lifetime PTSD so that we could focus exclusively on biological signals resulting from parental Holocaust exposure and/or PTSD. The procedures of the study involved admission to the general clinical research center on the evening before the study, insertion of an i.v. at 5:30 a.m., and, after a stabilization period of approximately 60 min, collection of blood samples every 30 min for a 24-h period. As described in Yehuda et al. (2007b), lower cortisol values were observed for offspring with parental PTSD than for offspring of survivors without PTSD or comparison subjects.

We also examined responsivity of the HPA axis using the low dose (0.50 mg) dexamethasone (DEX) suppression test. The study group was comprised of 25 offspring (16 with, 12 without, parental PTSD) and 16 comparison subjects, with similar inclusion/exclusions described above (Yehuda et al., 2007a). There were significant group differences in pre- and post-DEX cortisol values. Moreover, greater cortisol suppression was found in offspring with parental (but not their own) PTSD than in offspring without parental PTSD or comparison subjects.

Cortisol levels in Holocaust offspring: cause or effect of vulnerability conferred by parental PTSD?

Not only do offspring of Holocaust survivors with PTSD demonstrate a greater frequency of PTSD than offspring of survivors without PTSD and controls, they have also, in some studies, endorsed more childhood traumatic antecedents (Yehuda et al., 1998a). Accordingly, it has not been possible to rule out that cortisol levels may reflect responses of offspring to their own early experiences of adversity. In all the above studies, however, we attempted to examine the relationship between cortisol and parental PTSD by controlling for either endorsement of childhood trauma [scores on the Childhood Trauma Questionnaire (CTQ), Bernstein et al., 1994] and/or current symptoms in the offspring. In all three cohorts studied above using mean 24-h urinary cortisol excretion, 24-h plasma cortisol mean, and cortisol suppression in response to DEX, the relationships with parental PTSD were maintained or strengthened after considering the contribution of childhood trauma.

Since in all three reports, the attribution of parental PTSD was made by the offspring who rated parental symptoms, a concern arises regarding the contribution of offspring attribution of parental PTSD to cortisol measures we obtained. Although we have previously established that there is strong agreement between offspring ratings of parental PTSD and independent clinician ratings of the parent, if offspring cortisol levels were also associated with their perception of parental PTSD, this could lead to a circularity of inference. For example, if having low cortisol reflected offspring characteristics, such as irritability or emotional withdrawal, or psychopathology, this might result in a greater projection of (or sensitivity to) analogous parental characteristics so that "parental" PTSD would partly reflect offspring characteristics. Controlling for offspring PTSD symptoms and diagnosis of other mood and anxiety disorders in the analysis of the relationship between cortisol and parental PTSD only partially addresses this concern. However, in the aftermath of 9/11, a unique opportunity arose to examine the relationship between maternal PTSD and PTSD in the infant offspring, addressing the above issue of attribution, since the group assignment of parental (maternal PTSD) could now be made independently from the offspring's assessment.

Cortisol levels in mothers and babies

It was not possible in our studies of Holocaust offspring, who were already adults at assessment to know whether the phenomenon we had observed would have also been present early in life (Yehuda et al., 2005). Measuring cortisol levels in infants who had recently been born to mothers experiencing trauma and PTSD in pregnancy and in the early post-partum period therefore provided a unique opportunity to determine the origin of low cortisol in offspring.

Examination of the relationship between maternal PTSD symptoms and salivary cortisol levels obtained at awakening and at bedtime in 38 infants (ranging in age from ~9 months to 1 year) and their mothers who were directly exposed to the World Trade Center (WTC) collapse on 9/11 during their pregnancy revealed that salivary cortisol levels were significantly lower in the offspring of women with PTSD than in those whose mothers had not developed PTSD. Interestingly, there was a significant correlation between severity of maternal PTSD symptoms and cortisol levels in infants, but no correlations between infant cortisol and maternal depression severity. The finding of lowered cortisol among infants with maternal PTSD suggested that the similar findings observed in adult offspring of Holocaust survivors were probably unrelated to their own "trauma" history or parental consequences (e.g., poor parenting) that occur well after the immediate postnatal period. Rather, the findings likely implicated earlier and possibly more subtle influences relating to mother–infant attachments or early social regulation.

When data were examined including trimester of maternal exposure to 9/11 a significant effect of maternal PTSD status in infants born to mothers pregnant in the third trimester was observed. Cortisol levels in these mothers with PTSD were also significantly lower than in mothers without PTSD. These findings suggest the possibility of in utero transmission of this effect. This was plausible since the human HPA axis appears to be programmed by early life influences (Francis et al., 1999; Weaver et al., 2002; Seckl, 2004). In both animal and human studies, maternal exposure to glucocorticoids during pregnancy can result in lower birth weight and higher glucocorticoid levels in offspring, leading ultimately, to adult disease (e.g., hypertension, insulin resistance, and hyperlipidemia) (O'Regan et al., 2001) and depression (Halligan et al., 2004).

Are cortisol levels more related to maternal than paternal PTSD?

Because the above-findings implicated a role for in utero factors in that associations between low cortisol and maternal PTSD were greatest in those exposed in the third trimester, we reconsidered the Holocaust offspring finding in relation to parental gender (Yehuda et al., 2007b). This was accomplished by subdividing data from offspring reported above on the basis of whether the father only, mother only, both, or neither parent had PTSD, compared to controls. The analysis demonstrated that offspring with only paternal PTSD ($n = 6$) were not significantly different in mean cortisol release than offspring with no parental PTSD or comparison subjects ($n = 16$). Mean cortisol release was similar for offspring with PTSD in both parents ($n = 9$) and those with only maternal PTSD ($n = 8$), whereas both of these groups differed from offspring with no parental PTSD and comparison subjects (Yehuda et al., 2007b). Even after controlling for Body Mass Index (BMI), current depression or anxiety diagnosis, and CAPS lifetime total score in regression analysis, there were significant negative partial correlations between mean cortisol level and the presence of both maternal and paternal PTSD. However, in a regression analysis, after additionally controlling for the presence of PTSD in the other parent, only maternal PTSD retained its significant negative association with offspring mean cortisol. The results here must be interpreted with caution since obviously the number of subjects in some subgroups is very small, but when considered in the context of the findings of infants from mothers exposed to 9/11 while pregnant, they suggest that there may be unique contributions of maternal PTSD to lowered cortisol levels in offspring.

Potential mechanisms for transgenerational transmission

The above findings of lower cortisol levels in offspring of persons with PTSD could be explained by early social regulation in that babies being raised under conditions of neglect or abusive care have lower ambient cortisol levels than normal (Gunnar and Vazquez, 2001), presumably mirroring findings from non-human primates. Marmoset monkeys exposed to early maternal separations also show reduced basal cortisol 3–5 months later (Dettling et al., 2002), as do monkeys who were exposed to the more stressful peer vs. mother rearing condition (Clarke et al., 1998). In rodents as well, results of cross-fostering studies demonstrate that even relatively brief exposures to alterations in postnatal maternal care during a critical period can have permanent behavioral and neuroendocrine alterations in offspring (Meaney, 2001). Therefore, it may be that PTSD symptoms in mothers — that may lead to the display of stress-reactive behavior toward their offspring during a critical developmental window(s) — may have long-lived effects on glucocorticoid regulation in offspring.

However, glucocorticoid programming in utero (Drake and Walker, 2004; Seckl, 2004) may also be important. In that alterations in GR sensitivity are known to occur very early during the postnatal period, possibly coinciding with the period in which normal maternal infant attachments are developing, it is possible that low cortisol levels in offspring is mediated through alterations in GR sensitivity (Mathews, 2002). In animal models, maternal behaviors, which alter the infants "environment," in turn result in long-lived changes in hippocampal GR expression and HPA function that are subsequently transmitted intergenerationally (Meaney, 2001). The hippocampal GR gene is particularly sensitive to postnatal "programming" because it has complex tissue-specific promoters that are susceptible to epigenetic modification (Sutherland and Costa, 2003). Modifications of these promoters during development have been shown to alter the "set-point" of receptor function in feedback sites and hence glucocorticoid secretion (McCormick et al., 2000). It is also possible that changes in GR sensitivity have even earlier origins than the postnatal period. It is important to consider such mechanisms based on the effects of PTSD on cortisol in mothers exposed in the third trimester of pregnancy, where stress induced increases in maternal glucocorticoids may have begun a process of glucocorticoid programming. This hypothesis is supported by the well known influences of maternal stress (glucocorticoids) on fetal brain development, which are, in part, dependent on the gestational age of the fetus (Drake and Walker, 2004; Seckl, 2004).

Early handling in rats and vulnerability to psychopathology in offspring: a crisis in translation?

One of the issues that has arisen in the "translation" of the early handling work to mental health-related issues is the interpretation of a manipulation of maternal behaviors induced by "early handling." The consequence of early handling in mothers is to increase maternal licking and grooming. This behavior has been widely interpreted as beneficial, as are associated biological outcomes in the adult offspring rat. As a result, it is has been challenging to draw the appropriate inferences from the early handling model to psychopathology, or even consider the nature of the relationship (if any) between these observations and transgenerational effects that might be operational with respect to cortisol and PTSD. However, insofar as early handling in rodents results in HPA alterations in the same direction as those described in PTSD [e.g., increased GR sensitivity, enhanced cortisol (corticosterone) suppression following DEX administration, lower ambient glucocorticoid levels (Meaney et al., 1985, 1991)], it is compelling to attempt to do so.

The presence of a similar behaviorally transmitted effect that is deemed positive in the rat, but might be problematic for humans, is not a translational crisis. The essential point of linking the early handling phenomenon in rats with transgenerational vulnerability in humans is to offer proof of principle that environmental exposures can result in persisting alterations in GR expression that underlie individual differences in endocrine function (Macri and Wurbel, 2006). That these differences may be associated with increased vulnerability to the development of PTSD following trauma exposure in humans is noteworthy, and underscores the necessity of highlighting that consequences to the individual of glucocorticoid programming may be context-dependent. That is, maternal PTSD may confer risk for PTSD in the offspring by influencing the predisposition to a modification that may later influence the response to a traumatic event, though not necessarily in an adverse manner. What gets "programmed" may simply be the set point of glucocorticoid secretion and an enhanced capacity for responsiveness of the HPA axis, which might be more adaptive to offspring who continue to live in a similar (dangerous) environment as their parents. For example, it would be adaptive for a Holocaust survivor parent to "transmit" the tendency or capacity for hypervigilance to seemingly "neutral" stimuli (e.g., a uniformed police officer) to an offspring that might be reared in the same environment as the parent (i.e., Nazi-occupied Europe), yet easy to see how this transmitted tendency for hypervigilance might interfere with functioning in a different environment (e.g., urban township during peacetime).

It is also possible that specific aspects of apparently similar behaviors in rodent and human mothers have different consequences. That increased "licking and grooming" of the handled mother is interpreted to result in positive effects is a reflection of anthropomorphic notions of more infant contact being associated with "better" parenting, the value placed on lower cortisol responses to stress as necessarily indicating less disruption, and our assumptions that superior cognitive performance on memory tasks are of greater advantage than higher endocrine responses to stress and less good cognitive performance. As a result, this paradigm has become a heuristic model for environmentally induced developmental effects that are positive, whereas in reality, the measurable alterations interpreted as positive may not be uniformly so, or may be occurring at the expense of other more subtle effects that still require delineation.

For example, observing Holocaust mothers and babies, one might similarly conclude that mothers with PTSD form attachment to infants marked by enhanced attention and contact (i.e., rather than relative emotional neglect or physical abuse). Though we have previously demonstrated increased emotional abuse and neglect, there were in fact no differences in Holocaust offspring with or without parental PTSD with respect to physical or sexual abuse, compared to controls (Yehuda et al., 2001). Thus, increased attention by the mother and possibly a greater reluctance to undergo voluntary separations from such offspring is not a contradiction with the finding of greater emotional abuse and neglect. Quite the opposite may be true. The strong need to protect offspring within the context of a great sense of loss or treat, and fear of loss of one's offspring, may compromise the mother's ability to create the space that is critical for the offspring's ultimate separation and individuation. Moreover, the clinical characteristics of adult Holocaust offspring who present for psychotherapy are consistent with issues relating to attachment. Such patients typically complain of difficulties in physically and emotionally separating from parents, implying strong attachments, though not attachments that protect from the risk of developing psychopathology. Thus, their clinical complaints are not incompatible with mothers who are excessively "licking and grooming," but unlike rodents who cannot be asked their opinion regarding the impact of such behavior on their ability to form social relationships or on other areas of adaptive functioning, human offspring can be quite clear that such contacts can lead to impairments in circumscribed domains (though not necessarily with regard to subsequent cognitive development, which is the usual outcome measured in early handled rats to demonstrate positive long-term effects of this manipulation).

Characteristics of maternal attachment related to ambient cortisol levels in adult offspring

In order to explore further the possibility that the quality of maternal attachment might influence offspring cortisol levels, we examined responses on the Parental Bonding Scale (Parker et al., 1979) among offspring of Holocaust survivors with ($n = 23$) and without ($n = 18$) maternal PTSD and comparison subjects ($n = 19$). This scale identifies both positive parental bonding (as reflected by scores for maternal and paternal care), as well the presence of more insecure parental attachments (as reflected by scores for maternal and paternal overprotection). As in previous studies of this high-risk group, we chose subjects without their own PTSD to avoid confounding the interpretation of associations with cortisol findings. The sample characteristics are shown in Table 1. Groups differed significantly in age, with comparison subjects being significantly younger, respectively, than offspring groups with and without PTSD, which did not differ from each other [post-hoc tests by Least Significant Difference (LSD)]. Gender distribution and BMI did not differ significantly among the groups, however the offspring groups were preponderantly composed of women, while the control group was predominantly male. Groups differed in scores on the Mississippi PTSD Scale (Keane et al., 1988), which provides a continuous measure of the effects of trauma exposure, including trauma-related or PTSD symptomatology, on a person's life, with offspring with maternal PTSD demonstrating significantly higher scores than offspring without maternal PTSD ($p = .014$) and controls ($p = .008$). Scores for the Beck Depression Inventory (Beck et al., 1961) also differed significantly among the three groups, but in this case scores for offspring with maternal PTSD differed only from those of controls in post-hoc testing ($p = .001$), with depression scores for offspring of mothers exposed to the Holocaust but without PTSD showing scores non-significantly lower than offspring with maternal PTSD but greater than controls. With respect to urinary cortisol (Table 1), values for offspring with maternal PTSD were significantly lower than controls ($p = .021$), and lower than offspring without maternal PTSD at a trend level of significance ($p = .081$). After co-varying for age and BMI, the results were essentially unchanged.

As had been previously noted for offspring with parental PTSD, ratings of childhood traumatic antecedents with the CTQ differed among the groups, but only for emotional abuse and neglect and for physical neglect, with scores in each of

Table 1. Childhood abuse, parental care and overprotection, and current symptoms in offspring with and without maternal PTSD and comparison subjects

Variable	Comparison subjects ($n = 19$)	Offspring without maternal PTSD ($n = 18$)	Offspring with maternal PTSD ($n = 23$)	ANOVA F (df) p
	Mean ± SD	Mean ± SD	Mean ± SD	
Age	44.4±9.5	49.8±6.5	50.4±7.3	3.45 (2,57) .039
Gender (M/F)	12 M/7 F	6 M/12 F	7 M/16 F	$\chi^2 = 5.32$ (2) .070
BMI	26.0±4.3	25.0±5.5	26.0±4.4	.28 (2,57) ns[a]
Paternal PTSD, Y/N (%)	na[b]	4 Y/14 N (22%)	9 Y/14 N (39%)	$\chi^2 = 1.33$ (1) ns
Beck Depression Scale	3.2±3.3	7.1±6.2	9.4±6.5	6.18 (2,56) .004
Mississippi PTSD Scale	65.9±15.2	67.1±13.0	80.5±20.3	4.89 (2,56) .011
Childhood Trauma Questionnaire (CTQ)				
Emotional abuse	6.9±2.3	7.0±3.4	10.1±5.2	4.36 (2,55) .018
Physical abuse	6.0±2.2	5.2±.4	6.1±2.3	1.26 (2,55) ns
Sexual abuse	5.6±2.1	5.1±.47	6.6±3.9	1.48 (2,55) ns
Emotional neglect	8.8±4.0	7.9±3.2	11.4±5.2	3.67 (2,55) .032
Physical neglect	5.6±.9	5.6±1.4	7.3±2.4	6.29 (2,55) .003
Total score	32.9±8.1	30.9±6.8	41.5±13.4	6.20 (2,55) .004
Parental Bonding Instrument (PBI)				
Maternal care	26.2±8.9	28.3±7.7	19.4±10.9	5.12 (2,57) .009
Maternal overprotection	11.3±7.4	13.1±9.8	18.1±11.0	2.76 (2,56) .072
Paternal care	24.7±8.2	21.5±8.2	23.1±10.0	.56 (2,55) ns
Paternal overprotection	9.8±6.3	14.0±11.0	13.5±8.2	1.24 (2,55) ns
Urinary cortisol (μg/24 h)	71.31±34.71	66.13±27.30	49.46±27.15	3.13 (2,57) .052

[a]ns: non-significant at $p \geq 1.0$.
[b]na: not applicable.

these subscales for offspring with maternal PTSD being significantly higher than those of offspring without maternal PTSD or controls (Table 1). Offspring with maternal PTSD showed significantly lower scores than either of the other groups for maternal care, but higher scores for maternal overprotection. There were no group differences in either paternal care or paternal overprotection ratings.

Figure 1 shows that maternal overprotection scores were significantly negatively correlated with urinary cortisol in offspring with maternal PTSD ($r = -537$, $p = .008$), but this relationship was not apparent in the other two groups, who demonstrated positive (though non-significant) associations, respectively (for offspring without parental PTSD, $r = .164$, ns; and for controls, $r = .249$, ns). In fact, the inverse correlation for cortisol and maternal overprotection in offspring with maternal PTSD differed significantly from the weakly positive associations of these variables in offspring without PTSD ($p = .012$) and controls ($p = .022$). These relationships for cortisol and maternal overprotection scores, as well as the significance of the differences among the associations, were retained when all correlations were controlled for age and BMI (e.g., for offspring with maternal PTSD, $r = -.538$, $\mathrm{df} = 19$, $p = .012$). When paternal overprotection was examined, there were no significant associations with cortisol. Furthermore, when the above correlations were additionally controlled for paternal PTSD, the negative relationship of urinary cortisol with maternal overprotection was also unchanged ($r = -.546$,

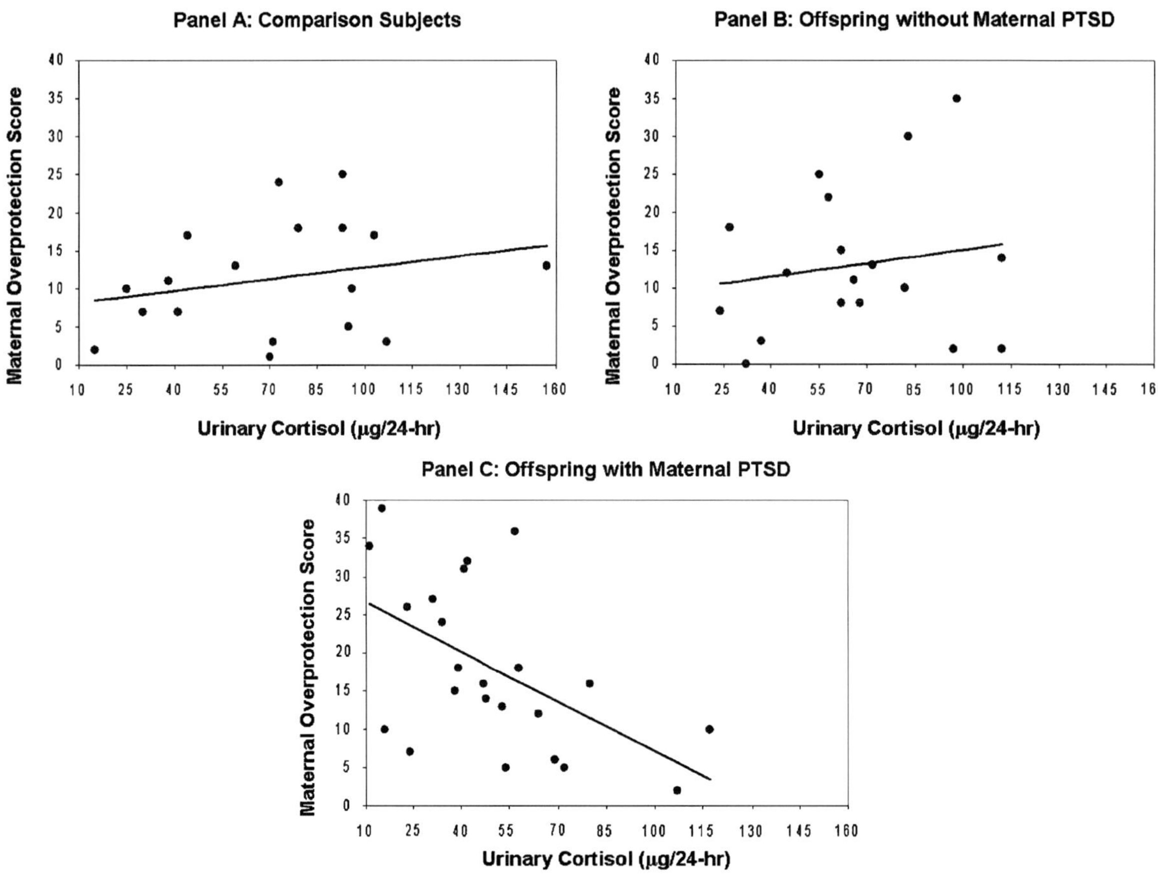

Fig. 1. Relationship of 24-h cortisol to maternal overprotection scores among offspring with and without PTSD and controls.

df = 18, $p = .013$), indicating that the association is appropriately ascribed to an influence of maternal PTSD.

Finally, among offspring with maternal PTSD, maternal overprotection scores were not substantially related to those of childhood abuse or neglect, indicating that this subscale describes an aspect of maternal attachment at least somewhat independent of childhood trauma that is negatively associated with cortisol.

What has been learned?

PTSD is associated with a distinct profile in that corticotropin releasing factor (CRF) levels appear to be increased while ambient urinary and plasma levels of cortisol have been found to be lower than in normals in many studies (Yehuda, 2002). PTSD is also characterized by increased cortisol-suppression in response to DEX administration in most studies, likely resulting from increased responsiveness of GR. This profile is different from that observed in studies of acute and chronic stress and depression, which have been associated with increased CRF and cortisol levels and reduced cortisol suppression to DEX, and GR responsiveness. The above findings may in part offer an explanation for the increased cortisol signaling capacity in PTSD that results in a more efficient suppression of the HPA axis despite what appears to be an attenuated glucocorticoid production from the adrenal glands.

We have previously suggested that alterations in GR sensitivity represent primary alterations in PTSD, but the relationship between this measure and ambient cortisol levels has remained difficult to pin down. This is because glucocorticoids such as cortisol both regulate GR and are further regulated by them. Furthermore, glucocorticoids exert their action not only through GR, but also through the mineralocorticoid receptor (MR). Both these receptor types are involved in the regulation of the HPA axis. Of particular relevance to PTSD is that the MR affinity for cortisol is high, such that even very low ambient hormone concentrations can result in receptor occupancy. This allows MR to mediate basal glucocorticoid functions such as circadian regulation, and may contribute to lower ambient hormone production, thus affecting the number of available GR binding sites, and further affecting GR responsiveness (de Kloet et al., 2005). The phenomenon of increased GR responsiveness has most recently been demonstrated in PTSD by examining the differential effects of DEX on the inhibition of lysozyme synthesis in lymphocytes (Yehuda et al., 2004). The presence of detectable differences in a gene product related to the activity of GR (i.e., lysozyme activity) in any target tissue (i.e., the lymphocyte) suggests the possibility of an alteration related to the expression of GR genes, whether it be a naturally occurring variant in GR polymorphisms that results in increased GR sensitivity or an epigenetic modification, such as cytosine methylation of a GR promoter. Were such changes present in the brain of vulnerable persons, one could postulate that under certain environmental conditions, such as in response to a traumatic event, increased cortisol signaling would alter or interact with psychological and biological risk factors to result in the PTSD clinical phenotype. Such possibilities are plausible in the context of the transgenerational findings.

Relevance of epigenetic analyses to PTSD risk

The study of epigenetic modifications may provide important insights into PTSD risk and pathophysiology since it provides a mechanism for transgenerationally transmissible functional change in genomic activity that can be induced by environmental events. In particular, methylation of polymerase II promoters (Sutherland and Costa, 2003), which provide an efficient way of gene silencing, provides a concrete molecular mechanism through which genetic–environmental interactions occur in PTSD, since this mechanism that appears to be operative in programming the activity of hippocampal genes regulating HPA activity by early life events (i.e., differences in maternal care).

Indeed, when considering that PTSD is fundamentally a response to an environmental event that is likely formed, not so much by the objective characteristics of the event, but by subjective interpretations of its meaning — it becomes obvious that neither genetic analyses nor an understanding of the normative biological responses to stress or fear, can provide the information needed to understand why PTSD results in only a proportion of those exposed. Clearly genetic analyses alone will simply not detect environment–gene activity connections, and though endocrine or other biological markers examining stress effects can in principle detect them, the response of a person to trauma exposure in adulthood is often determined by a lifelong pattern of responding to life events that may obfuscate the impact of earlier events. The study of epigenetic modifications may provide a relatively stable measure that reflects early life events, rather than the cumulative effects of stress, that can help delineate developmental influences on biological alterations in PTSD from those reflecting pathophysiology.

Abbreviations

BMI	Body Mass Index
CAPS	Clinician Administered PTSD Scale
CRF	corticotropin releasing factor
CTQ	Childhood Trauma Questionnaire
DEX	dexamethasone
GR	glucocorticoid receptor
HPA	hypothalamic-pituitary-adrenal
LSD	Least Significant Difference
MR	mineralocorticoid receptor
PPQ	Parental PTSD Questionnaire
PTSD	posttraumatic stress disorder
VNTR	variable number tandem repeat
WTC	World Trade Center

Acknowledgments

This work was supported by an NIMH R01 MH64675-01 entitled "Biology of Risk and PTSD in Holocaust Survivor Offspring, and, in part by a grant (5 M01 RR00071) for the Mount Sinai General Clinical Research Center from the National Institute of Health. The authors acknowledge Dr. James Schmeidler for statistical consultation and Amanda Bell for her assistance in the preparation of this manuscript.

References

Bachmann, A.W., Sedgley, T.L., Jackson, R.V., Gibson, J.N., Young, R.M. and Torpy, D.J. (2005) Glucocorticoid receptor polymorphisms and post-traumatic stress disorder. Psychoneuroendocrinology, 30(3): 297–306.

Beck, A.T., Ward, C.H., Mendelson, M., Mock, J. and Erbaugh, J. (1961) An inventory for measuring depression. Arch. Gen. Psychiatry, 4: 561–571.

Bernstein, D.P., Fink, L., Handelsman, L., Foote, J., Lovejoy, M., Wenzel, K., Sapareto, E. and Ruggiero, J. (1994) Initial reliability and validity of a new retrospective measure of child abuse and neglect. Am. J. Psychiatry, 151(8): 1132–1136.

Blake, D.D., Weathers, F.W., Nagy, L.M., Kaloupek, D.G., Klauminzer, G., Charney, D.S. and Keane, T.M. (1995) A clinician rating scale for assessing current and lifetime PTSD: the CAPS-1. Behav. Ther., 13: 187–188.

Bremner, J.D., Southwick, S.M., Johnson, D.R., Yehuda, R. and Charney, D.S. (1993) Childhood physical abuse and combat-related posttraumatic stress disorder in Vietnam veterans. Am. J. Psychiatry, 150(2): 235–239.

Breslau, N., Kessler, R.C., Chilcoat, H.D., Schultz, L.R., Davis, G.C. and Andreski, P. (1998) Trauma and posttraumatic stress disorder in the community: the 1996 Detroit Area Survey of Trauma. Arch. Gen. Psychiatry, 55(7): 626–632.

Brewin, C.R., Andrews, B. and Valentine, J.D. (2000) Meta-analysis of risk factors for posttraumatic stress disorder in trauma-exposed adults. J. Consult. Clin. Psychol., 68(5): 748–766.

Clancy, C.P., Graybeal, A., Tompson, W.P., Badgett, K.S., Feldman, M.E., Calhoun, P.S., Erkanli, A., Hertzberg, M.A. and Beckham, J.C. (2006) Lifetime trauma exposure in veterans with military-related posttraumatic stress disorder: association with current symptomatology. J. Clin. Psychiatry, 67(9): 1346–1353.

Clarke, A.S., Kraemer, G.W. and Kupfer, D.J. (1998) Effects of rearing condition on HPA axis response to fluoxetine and desipramine treatment over repeated social separations in young rhesus monkeys. Psychiatry Res., 79(2): 91–104.

Cohen, M.E., White, P.D. and Johnson, R.E. (1948) Neurocirculatory asthenia, anxiety neurosis or the effort syndrome. Arch. Int. Med., 81: 260–281.

Davidson, J.R., Tupler, L.A., Wilson, W.H. and Connor, K.M. (1998) A family study of chronic post-traumatic stress disorder following rape trauma. J. Psychiatr. Res., 32(5): 301–309.

Delahanty, D.L., Raimonde, A.J., Spoonster, E. and Cullado, M. (2003) Injury severity, prior trauma history, urinary cortisol levels, and acute PTSD in motor vehicle accident victims. J. Anxiety Disord., 17(2): 149–164.

Dettling, A.C., Feldon, J. and Pryce, C.R. (2002) Early deprivation and behavioral and physiological responses to social separation/novelty in the marmoset. Pharmacol. Biochem. Behav., 73(1): 259–269.

Drake, A.J. and Walker, B.R. (2004) The intergenerational effects of fetal programming: non-genomic mechanisms for the inheritance of low birth weight and cardiovascular risk. J. Endocrinol., 180(1): 1–16.

Francis, D.D., Caldji, C., Champagne, F., Plotsky, P.M. and Meaney, M.J. (1999) The role of corticotropin-releasing factor–norepinephrine systems in mediating the effects of early experience on the development of behavioral and endocrine responses to stress. Biol. Psychiatry, 46(9): 1153–1166.

Gunnar, M.R. and Vazquez, D.M. (2001) Low cortisol and a flattening of expected daytime rhythm: potential indices of risk in human development. Dev. Psychopathol., 13(3): 515–538.

Halligan, S.L., Herbert, J., Goodyer, I.M. and Murray, L. (2004) Exposure to postnatal depression predicts elevated cortisol in adolescent offspring. Biol. Psychiatry, 55(4): 376–381.

Keane, T.M., Caddell, J.M. and Taylor, K.L. (1988) Mississippi Scale for Combat-Related Posttraumatic Stress Disorder: three studies in reliability and validity. J. Consult. Clin. Psychol., 56(1): 85–90.

de Kloet, E.R., Joels, M. and Holsboer, F. (2005) Stress and the brain: from adaptation to disease. Nat. Rev. Neurosci., 6(6): 463–475. Review

Koenen, K.C., Harley, R., Lyons, M.J., Wolfe, J., Simpson, J.C., Goldberg, J., Eisen, S.A. and Tsuang, M. (2002) A twin registry study of familial and individual risk factors for trauma exposure and posttraumatic stress disorder. J. Nerv. Ment. Dis., 190(4): 209–218.

Koenen, K.C., Moffitt, T.E., Poulton, R., Martin, J. and Caspi, A. (2007) Early childhood factors associated with the development of post-traumatic stress disorder: results from a longitudinal birth cohort. Psychol. Med., 37(2): 181–192.

Macri, S. and Wurbel, H. (2006) Developmental plasticity of HPA and fear responses in rats: a critical review of the maternal mediation hypothesis. Horm. Behav., 50(5): 667–680.

Mason, J.W., Giller, E.L., Kosten, T.R., Ostroff, R.B. and Podd, L. (1986) Urinary free-cortisol levels in posttraumatic stress disorder patients. J. Nerv. Ment. Dis., 174(3): 145–149.

Mathews, S.G. (2002) Early programming of the hypothalamo-pituitary-adrenal axis. Trends Endocrinol. Metab., 13(9): 373–380.

McCormick, J.A., Lyons, V., Jacobson, M.D., Noble, J., Diorio, J., Nyirenda, M., Weaver, S., Ester, W., Yau, J.L., Meaney, M.J., Seckl, J.R. and Chapman, K.E. (2000) 5′-Heterogeneity of glucocorticoid receptor messenger RNA is tissue specific: differential regulation of variant transcripts by early-life events. Mol. Endocrinol., 14(4): 506–517.

McFarlane, A.C. (1988) The aetiology of post-traumatic stress disorders following a natural disaster. Br. J. Psychiatry J. Ment. Sci., 152: 116–121.

McFarlane, A.C. (1989) The aetiology of post-traumatic morbidity: predisposing, precipitating and perpetuating factors. Br. J. Psychiatry J. Ment. Sci., 154: 221–228.

Meaney, M.J. (2001) Maternal care, gene expression, and the transmission of individual differences in stress reactivity across generations. Annu. Rev. Neurosci., 24: 1161–1192.

Meaney, M.J., Aitken, D.H., Bhatnagar, S. and Sapolsky, R.M. (1991) Postnatal handling attenuates certain neuroendocrine, anatomical, and cognitive dysfunctions associated with aging in female rats. Neurobiol. Aging, 12(1): 31–38.

Meaney, M.J., Aitken, D.H., Bodnoff, S.R., Iny, L.J. and Sapolsky, R.M. (1985) The effects of postnatal handling on the development of the glucocorticoid receptor systems and stress recovery in the rat. Prog. Neuropsychopharmacol. Biol. Psychiatry, 9(5–6): 731–734.

Oppenheimer, B.S. and Rothschild, M.A. (1918) The psychoneurotic factor in the irritable heart of soldiers. JAMA, 70: 1919–1922.

O'Regan, D., Welberg, L.L., Holmes, M.C. and Seckl, J.R. (2001) Glucocorticoid programming of pituitary-adrenal function: mechanisms and physiological consequences. Semin. Neonatol., 6(4): 319–329.

Parker, G., Tupling, H. and Brown, L.B. (1979) A parental bonding instrument. Br. J. Med. Psychol., 52: 1–10.

Reich, J., Lyons, M. and Cai, B. (1996) Familial vulnerability factors to post-traumatic stress disorder in male military veterans. Acta Psychiatr. Scand., 93(2): 105–112.

Resnick, H.S., Yehuda, R., Pitman, R.K. and Foy, D.W. (1995) Effect of previous trauma on acute plasma cortisol level following rape. Am. J. Psychiatry, 152(11): 1675–1677.

Seckl, J.R. (2004) Prenatal glucocorticoids and long-term programming. Eur. J. Endocrinol., 151(Suppl 3): U49–U62.

Segman, R.H. and Shalev, A.Y. (2003) Genetics of posttraumatic stress disorder. CNS Spectr., 8(9): 693–698.

Solomon, Z., Mikulincer, M. and Avitzur, E. (1988) Coping, locus of control, social support, and combat-related posttraumatic stress disorder: a prospective study. J. Pers. Soc. Psychol., 55(2): 279–285.

Stein, M.B., Jang, K.L., Taylor, S., Vernon, P.A. and Livesley, W.J. (2002) Genetic and environmental influences on trauma exposure and posttraumatic stress disorder symptoms: a twin study. Am. J. Psychiatry, 159(10): 1675–1681.

Sutherland, J.E. and Costa, M. (2003) Epigenetics and the environment. Ann. N.Y. Acad. Sci., 983: 151–160.

Swan, J.M. (1921) An analysis of ninety cases of functional disease in soldiers. Arch. Intern. Med., 28: 586–602.

True, W.R., Rice, J., Eisen, S.A., Heath, A.C., Goldberg, J., Lyons, M.J. and Nowak, J. (1993) A twin study of genetic and environmental contributions to liability for posttraumatic stress symptoms. Arch. Gen. Psychiatry, 50(4): 257–264.

Weaver, I.C., Szyf, M. and Meaney, M.J. (2002) From maternal care to gene expression: DNA methylation and the maternal programming of stress responses. Endocr. Res., 28(4): 699.

Wolfsohn, J.M. (1918) The predisposing factors of war psychoneuroses. Lancet, 1: 177–180.

Yehuda, R. (2002) Current status of cortisol findings in posttraumatic stress disorder. Psychiatr. Clin. North Am., 25: 341–368.

Yehuda, R., Bierer, L.M., Schmeidler, J., Aferiat, D.H., Breslau, I. and Dolan, S. (2000) Low cortisol and risk for PTSD in adult offspring of holocaust survivors. Am. J. Psychiatry, 157(8): 1252–1259.

Yehuda, R., Blair, W., Labinsky, E. and Bierer, L.M. (2007a) Effects of parental PTSD on the cortisol response to dexamethasone administration in their adult offspring. Am. J. Psychiatry, 164(1): 163–166.

Yehuda, R., Engel, S.M., Brand, S.R., Seckl, J., Marcus, S.M. and Berkowitz, G.S. (2005) Transgenerational effects of posttraumatic stress disorder in babies of mothers exposed to the World Trade Center attacks during pregnancy. J. Clin. Endocrinol. Metab., 90(7): 4115–4118.

Yehuda, R., Golier, J.A., Yang, R.K. and Tischler, L. (2004) Enhanced sensitivity to glucocorticoids in peripheral mononuclear leukocytes in posttraumatic stress disorder. Biol. Psychiatry, 55(11): 1110–1116.

Yehuda, R., Halligan, S.L. and Bierer, L.M. (2001) Relationship of parental trauma exposure and PTSD to PTSD, depressive and anxiety disorders in offspring. J. Psychiatr. Res., 35(5): 261–270.

Yehuda, R., Labinsky, E., Tischler, L., Brand, S.R., Lavin, Y., Blair, W., Bierer, L.M., Goodman, R.Z. and Grossman, R.A. (2006) Are adult offspring reliable informants about parental PTSD? A validation study. Ann. N.Y. Acad. Sci., 1071: 484–487.

Yehuda, R. and McFarlane, A.C. (1995) Conflict between current knowledge about posttraumatic stress disorder and its original conceptual basis. Am. J. Psychiatry, 152(12): 1705–1713.

Yehuda, R., Schmeidler, J., Giller Jr., E.L., Siever, L.J. and Binder-Brynes, K. (1998a) Relationship between posttraumatic stress disorder characteristics of Holocaust survivors and their adult offspring. Am. J. Psychiatry, 155(6): 841–843.

Yehuda, R., Schmeidler, J., Wainberg, M., Binder-Brynes, K. and Duvdevani, T. (1998b) Vulnerability to posttraumatic stress disorder in adult offspring of Holocaust survivors. Am. J. Psychiatry, 155(9): 1163–1171.

Yehuda, R., Teicher, M.H., Seckl, J.R., Grossman, R.A., Morris, A. and Bierer, L.M. (2007b) Maternal but not paternal PTSD is a 'vulnerability' factor for low cortisol trait in offspring of Holocaust survivors. Arch. Gen. Psychiatry, 64(9): 1040–1048.

Discussion: Chapter 9

RICHTER-LEVIN: I wanted to ask whether it is not risky to include also this group under the definition of PTSD. Is it not an additional disorder to the definition of PTSD that does not have the same roots, the same mechanism, and so on?

YEHUDA: Right, well that's a good point, glad that you brought that up. I don't think that all comes from this kind of risk. This is an example, one type of group that is vulnerable and for that reason that makes them vulnerable and their history and their background might explain why they might catastrophize the response to trauma. But not all traumas, practically traumas that involve long separation and personal violence, right? It doesn't necessarily explain why somebody else would get PTSD. These are not the low IQ, cognitively impaired PTSD people. With the paper that came out from Mark Gibertson and Roger Pitman very recently showed cognitive deficits in the non-exposed co-twins in PTSD Vietnam veterans with PTSD are (end of tape).

RICHTER-LEVIN: Is it a good translation of what you presented that actually we don't have PTSD (posttraumatic). That is, you have a kind of a syndrome; one of the ways to get to the syndrome is trauma (and then it would be posttraumatic), but that there are other ways to get to this same syndrome (without the exposure to a well-defined trauma), so that the name PTSD is not a good name for this syndrome.

YEHUDA: In the offspring, or in anyone?

RICHTER-LEVIN: If this population of offspring has the same syndrome as someone who is exposed to trauma but they were not exposed to a well-defined trauma, then you have a syndrome, but only one of the ways to get to the syndrome is posttrauma.

YEHUDA: OK I mean we can keep the initials and think of it as posttraumatic sensitization disorder. You know that kind of work, but that is why I raised earlier whether there is a final common pathway and one biology that fits all or whether what we are looking at are different kinds of trajectories that look similar and for us to call it all PTSD. The political history of PTSD was that prior to the establishment of the diagnosis there was a diagnosis like concentration camp syndrome, and other s; the idea was to bring it all together under an umbrella, and you are saying that is maybe not such a good idea, maybe not.

RICHTER-LEVIN: I am asking, not saying.

YEHUDA: You are asking, but it is a testable hypothesis that should be asked for an answer. With lot of clinical experience I have been in a position to really study the typical cultural differences. I can tell you that holocaust survivors are different from Vietnam veterans but they share some PTSD symptoms, but in terms of a biological basis do I want to go after the differences or do I want to go after the common you know PTSD.

E.R. de Kloet, M.S. Oitzl & E. Vermetten (Eds.)
Progress in Brain Research, Vol. 167
ISSN 0079-6123

CHAPTER 10

Early care experiences and HPA axis regulation in children: a mechanism for later trauma vulnerability

Megan R. Gunnar* and Karina M. Quevedo

Institute of Child Development, University of Minnesota, Minneapolis, MN 55455, USA

Abstract: Post-traumatic stress disorder (PTSD) is associated with functional abnormalities of the hypothalamic-pituitary-adrenocortical (HPA) axis. Emerging evidence suggests that failures in social regulation of the HPA axis in young children manifested as neglectful or abusive care may play a role in shaping cortico-limbic circuits involved in processing experiences threatening experiences encountered later in life. Low cortisol levels, particularly near the peak of the diurnal rhythm, have been reported in abused, neglected and deprived children. Thus early imprinting effects of parenting quality on the HPA system regulation may be one of the mechanisms causing heightened risk of PTSD in responses to later trauma. However there is also evidence that the altered patterns of cortisol production seen in the context of early adverse care are not permanent, and remit once the care children receive improves. What awaits study is whether periods of atypical cortisol levels and altered HPA function early in life, even if transient, impact brain development in ways that heighten vulnerability to PTSD in response to traumas experienced later.

Keywords: stress; early experience; cortisol; children

Post-traumatic stress disorder (PTSD) develops in only a subset of individuals who experience trauma. Hyper-arousal of threat-responsive neurobiological systems, including the sympathetic nervous system, and hypo-arousal of the hypothalamic-pituitary-adrenocortical (HPA) system at the time of trauma exposure are hypothesized to represent pre-existing risk factors for the development of PTSD (for review, see Delahanty and Nugent, 2006). Hypo-activity of the HPA axis is believed to enhance risk of PTSD perhaps through failing to counter-regulate hyper-arousal in other trauma-sensitive neurobiological systems (Yehuda, 2001). In adults, prior trauma increases both the risk of PTSD and the risk of blunted cortisol responses to trauma (Delahanty and Nugent, 2006). Exposure to trauma during childhood is believed to be particularly critical in increasing the risk of PTSD in response to later traumatic experiences (Teicher et al., 2002). In studies predicting PTSD from early childhood experiences, however, measures of childhood trauma include a range of experiences from sexual abuse to more subtle forms of emotional abuse (Yehuda, 2002). Furthermore, the animal models used to support arguments that childhood abuse shapes vulnerability to developing PTSD do not clearly point to childhood trauma as a necessary antecedent (Heim et al., 1997). Rather, they point to lack or loss of expectable parental

*Corresponding author. Tel.: +1-612-624-2846;
Fax: +1-612-624-6373; E-mail: gunnar@umn.edu

DOI: 10.1016/S0079-6123(07)67010-1

stimulation as critical in altering the development of the HPA axis and neurobiological systems involved in regulating reactions to stressful and traumatic events (Gunnar and Fisher, 2006).

In the following review we will examine the impact of variations in parental care on activity of the HPA axis in animal models and young children. Our focus on the HPA axis is based on the putative role of this neuroendocrine system in the etiology of PTSD (Yehuda, 2001). We will argue that parental care provides a social regulator of activity of the HPA axis in infants and young children (Gunnar and Donzella, 2002). Disturbances in parental care, thus, disrupt normal regulation of this neuroendocrine system. However, the effects of these disruptions are varied. In some instances, disturbances in the typical circadian rhythm of the axis are observed, in others the presence of the caregiver fails to buffer activation of the axis in response to events that produce fearful behavior, while in still others the parent's presence and interaction with the child stimulates elevations in cortisol. Whether these different effects on HPA axis activity reflect common or varied elements in the parent–offspring relationships are not yet known. However, there is now substantial evidence that parental care impacts activity of the HPA axis in the young of many species, including our own. As such, this evidence supports arguments that failures in the caregiving system may increase the risk of PTSD through altering the development of stress- and threat-responsive neurobiological systems. This review assumes the reader's knowledge of the neuroana tomy and physiology of the HPA axis (for review, see de Kloet, 1991; Herman et al., 2005) and mechanisms through which altered activity of this axis may impact neurodevelopment (for review, see Gunnar and Vazquez, 2006).

Animal models: parental care and HPA activity

Much of the animal work has been conducted on rats. During infancy (postnatal days 4–14) rats exhibit low basal levels of corticosterone (the predominant steroid hormone produced by the HPA axis in rodents) and hypo-responsiveness of the HPA axis to many stressors (Rosenfeld et al., 1992). This hypo-responsive period in rats is maintained by specific components of maternal care: licking and grooming and the delivery of milk into the gut (Suchecki et al., 1993a, b, 1995). Both manipulations that alter licking and grooming and normal variations in this aspect of maternal care produce long-term alterations in the HPA axis and in emotional behavior. Short (3–5 min) repeated daily separations in infancy result in the development of attenuated HPA reactivity and fearfulness to stressors encountered later in life. In contrast, longer separations (e.g., 180 min daily) produce the development of a more hyper-responsive HPA axis and a more fearful animal (for review, see Sanchez et al., 2001). Repeated, prolonged separations also produce numerous other alterations that support vulnerability to stressors throughout life (Cirulli et al., 2000, 2003; Roceri et al., 2004). Early, repeated separations are believed to produce their effects, in part, through altering maternal care: increasing maternal licking and grooming in the case of brief, daily separations and reducing maternal licking and grooming in the case of prolonged daily separations (Heim and Nemeroff, 2001; Denenberg, 1999; Tang et al., 2006).

The argument that parental care mediates the impact of early separations is enhanced by evidence that normal variations in maternal licking and grooming produce the same effects as early separations. When the offspring of high licking and grooming dams are followed, they exhibit better regulation of the HPA axis and reduced fearful behavior in adulthood (Caldji et al., 1998). These effects are not due to common maternal and offspring genetics, as they can be produced in pups cross-fostered from low-to-high licking and grooming dams (Francis and Meaney, 1999). Furthermore, there is now good evidence that maternal care in rats produces long-term impacts on stress and emotional behavior through affecting methylation of the glucocorticoid receptor (GR) gene in the hippocampus. This gene produces the GR, the receptor that is critically involved in negative feedback of the HPA axis in response to psychosocial stressors (De Kloet, 1991). In response to low maternal care (i.e., low licking and grooming), there is greater methylation of the GR gene which results in fewer glucocorticoid receptors in the hippocampus and poorer negative feedback regulation of the HPA axis (Meaney and Szyf, 2005). Pharmacological intervention can correct these impacts on GR methylation. Thus offsprings of low licking and grooming dams who were centrally infused with methionine showed a reversal of the effects of maternal behavior on the fearful behavior, HPA axis regulation and GR methylation (Weaver et al., 2005). Postnatal experiences can also alter some, but not all of these effects. Thus exposure of juvenile rats to an enriched environment prevents expression of the heightened fearfulness and elevated corticosterone in response to stressors, but it does not alter patterns of GR methylation, suggesting that these animals may maintain a vulnerability to traumatic stimulation later in life (Francis et al., 2002).

In non-human primates, the work on early parental care has progressed through the lens of attachment theory. Through this lens, the mother–infant relationship is viewed as a stress buffer (Bowlby, 1969; Suomi, 1995). The natural ecology of most primate species is social. Mother and infant live in troupes that include other mothers and infants, adult males and the infant's older siblings. Species differ in how often the infant is parented by other members of the troupe — a phenomenon termed alloparenting or aunting. Although the availability of alloparents does not prevent elevations in cortisol to maternal separation, it does result in a more rapid termination of the HPA response and a reduced likelihood that despair behavior will develop during prolonged separations (Levine and Wiener, 1988). Thus, in primate infants maternal buffering is a function that can be provided to some extent by other nurturing conspecifics.

While repeated prolonged separations in rats may produce hyper-cortisolism, in monkeys they may reduce HPA activity (Levine et al., 1997; Dettling et al., 2002; Sanchez et al., 2005). In rhesus macaques, patterns of low basal output of cortisol following repeated separations have been accompanied by evidence of heightened startle responses (Sanchez et al., 2005), a pattern reminiscent of PTSD. Similarly, manipulations like variable foraging that produced disturbed mother–infant relations do not necessarily produce hyper-activation of the HPA axis in adulthood. Nevertheless, animals reared in these paradigms do exhibit elevated corticotropin-releasing hormone (CRH) in cerebrospinal fluid, altered serotonergic and noradrenergic activity and fearful behavior (Rosenblum and Andrews, 1994). Likewise, animals reared under conditions of parental deprivation as adults do not exhibit heightened HPA responses to stressors or neurochemical and structural changes in the HPA system (for review, see Sanchez et al., 2001), but do exhibit changes in monoamine systems (Kraemer et al., 1989), hippocampal development (Siegel et al., 1993) and in CRH receptors in the prefrontal cortex and amygdala (Sanchez et al., 1999). These differences from the rodent model likely reflect the greater maturity of the HPA axis at birth relative to its maturity in the rat (Gunnar and Vazquez, 2006). To summarize, in non-human primates parental care is a potent regulator of activity of the infant's HPA axis; however, hyper-reactivity of the axis as a consequence of adverse early care has typically not been observed. Despite a general failure to observer hyper-reactivity of the axis following adverse early care experiences, in non-human primates many of the anxiety-associated behavioral and neurobiological sequelae of disturbances in early parental care have been observed.

Human development and the hypothalamic-pituitary axis

With these animal data in mind, we turn now to data on parental care and activity of the HPA axis in human infants and children. These data, however, must be viewed in the context of developmental changes in HPA axis activity. We will describe these changes first.

Development of HPA activity during childhood

The HPA axis is highly responsive to stressors at birth; indeed, HPA stress responses are observed by as early as 18–20 weeks of gestation (Giannakoulopoulos et al., 1999). In adults, approximately 80% of circulating cortisol is usually bound to corticosteroid-binding globulin (CBG) and is thus rendered biologically inactive. However, CBG levels are low in the human newborn and do not achieve adult levels until about 6 months (Hadjian et al., 1972). As a result, plasma levels of total cortisol are low in the newborn and increase over the first months of life; while the free fraction of the hormone is as high or higher than levels observed in older infants and children (Gunnar et al., 1988). In addition, when salivary measures of cortisol are used which reflect only the free fraction of the hormone, newborns show marked elevations in cortisol to even very minor perturbations (e.g., undressing, weighing and measuring; Gunnar et al., 1992). Therefore, by 6 months of age, the HPA system of the human child is relatively mature.

Over the first year of life, the human HPA system becomes progressively less responsive to stressors (Gunnar et al., 1996). Indeed, by one year of age it becomes difficult to elevate cortisol to a variety of events, which nonetheless, elicit significant behavioral distress (Ramsay and Lewis, 1994; Gunnar and Donzella, 1999). As will be discussed below, this apparent hypo-responsivity of the HPA axis reflects the fact that during the first year of life the human HPA axis comes under strong social regulation or parental buffering (Gunnar and Donzella, 2002). During childhood, small increases in cortisol can be observed to the experimentally induced stressors (Blair et al., 2005), although in most cases only some children exhibit these elevations (Talge et al., 2007). Likewise, stressors such as the beginning kindergarten or a new school year and important school exams produce cortisol increases in some children (Boyce et al., 1995; Spangler, 1995; Davis et al., 1999; Bruce et al., 2002).

Epidemiological data indicate that onset of stress-related emotional disorders increases during adolescence (Angold and Rutter, 1992). It has been suggested that this heightened vulnerability to stressors may reflect increased reactivity of the HPA axis and other neurobiological changes over the transition to adolescence (Spear, 2000). There is now good evidence from both cross-sectional and longitudinal studies that basal cortisol levels increase as children progress into adolescence, perhaps related to pubertal changes (Gunnar and Vazquez, 2006). Indeed, a recent study found that adolescents at higher stages of pubertal development had more elevated overall diurnal cortisol curves reflected in cortisol levels sampled throughout the day (Adam, 2006). Several studies now suggest that over the pubertal transition the HPA system becomes more adult-like in its responses to stressors (Gunnar, in preparation; Stroud, in preparation). It thus seems likely that the impact of early care experiences on stress reactivity and regulation may not be fully evident until after the pubertal transition.

Normative variations in parental care

As in research on non-human primates, regulation of the human infant's HPA system has been viewed through the lens of attachment theory. By the end of the first year, infants develop strong attachment relationships with one or a few consistent caregivers (Bowlby, 1969). The attachment figure provides a source of comfort and stress modulation. Patterns of behavior in response to mildly stressful provocation (stranger approach and brief separations) have been described which presumably reflect differences in the child's sense of security in the attachment relationship (Ainsworth et al., 1978). Secure relationships are characterized by caregivers that are consistently responsive and sensitive to their children's needs for both exploration and reassurance (Ainsworth et al., 1978). Thus, both parental sensitivity/responsiveness and patterns of infant attachment behavior have been examined as a means of examining the parental buffering hypothesis in humans.

In general, both parental sensitivity/responsiveness and secure attachment relationships buffer the child's HPA axis in response to stressors. For example, Nachmias et al. (1996) exposed 18-month-olds to a series of events that elicited wary or fearful reactions in approximately half of the participants. Elevations in cortisol, however, were only observed in fearful toddlers who were insecurely attached to the parent who was with them during testing. Ahnert et al. (2004) examined cortisol responses of toddlers when they visited a new child care center in the presence of their mothers. In their mother's presence, but not during the child's initial days without mother at child care, securely attached infants exhibited markedly lower cortisol elevations compared to insecurely attached toddlers. In several studies, Spangler has demonstrated elevations in cortisol only among insecurely but not securely attached infants (for review, see Spangler and Grossmann, 1999).

Insensitive patterns of parental care early in the first year of life also predict larger cortisol responses to stressors (Gunnar et al., 1996). Indeed, when mothers were asked to play with their infants, infants with more sensitive mothers exhibited decreases in cortisol over the play period, while those with less sensitive mothers exhibited small increases in cortisol (Spangler et al., 1994). Harsh parenting (spanking) has also been shown to predict larger cortisol responses to mildly stressful events in infants (Bugental et al., 2003). Furthermore, when provided (randomly) with a sensitive and responsive surrogate caregiver, even 9-month-old infants with negative emotional temperaments did not show a cortisol elevation to 30 min of maternal separation; while those given an insensitive, unresponsive surrogate caregiver did (Gunnar et al., 1992). Similar findings have been obtained for toddlers and preschoolers in full-day out-of-home childcare (for review, see Gunnar and Donzella, 2002). These data support the hypothesis that children who experience less supportive and responsive care may frequently experience stressor-induced elevations in cortisol that might, over time, impact neurobehavioral development (see also De Kloet et al., 1996).

Studies of "at risk" children also support the argument that early care may have long-term impacts on vulnerability to stressful or traumatic events. A number of studies have been conducted on children whose mothers experience postpartum and/or major depression in the child's first years of life. During bouts of clinical depression, mothers are typically less responsive and more rejecting with their infants and young children (Ashman et al., 2002). Because bouts of clinical depression are associated with disturbances in parenting, associations between when the mother was depressed during the child's development and child outcomes provide a window into the timing of disturbances in parental care and neurobehavioral development.

Preschoolers whose mothers have a history of clinical depression or who report high number of depressive symptoms during the child's first 2 years, exhibit elevated home cortisol levels, particularly in the context of on-going stress in the family (Dawson and Ashman, 2000; Essex et al., 2002). School-aged children exposed to more prolonged periods of maternal depression in infancy and early childhood have been noted to have higher and more variable early morning cortisol levels even after controlling for maternal depression later in the child's life (Halligan et al., 2004; Lupien et al., 2000). It is not clear, however, whether maternal depression during a child's early development increases reactivity of the HPA axis, as this has been noted only for girls and then only for girls with concurrent internalizing behavior problems (Ashman et al., 2002). Several of these studies implicate elevated cortisol levels in the etiology of internalizing problems in childhood (Smider et al., 2002) and the emergence of depression in response to stressful events during adolescence (Halligan et al., 2007) for the offspring of depressed mothers. Unfortunately, although maternal depression has often been shown to be correlated with less sensitive and responsive care (Dawson and Ashman, 2000), in none of these studies was parental care examined as a mediator of the impact of maternal depressive symptoms on HPA axis activity. Furthermore, they do not allow us to disentangle genetic vulnerabilities in children of depressed mothers from the impact of postnatal care. That is, we do not know whether bouts of disturbed parental care would have the same impact on children who did not have a high genetic load for depression.

To summarize, studies of variations in parental care within the normal range in humans strongly indicate that less sensitive and responsive caregiving is associated with poorer concurrent regulation of the HPA axis in children. Sensitive and responsive care and associated secure attachment relationships appear to provide a powerful buffer for the HPA axis during early development. Children exposed to less sensitive/responsive care and those of mothers who use more harsh (though not necessarily abusive) discipline practices exhibit higher basal cortisol levels and their mother's presence does not prevent elevations in cortisol to the type of everyday stressors experienced by young children. Studies of the offspring of depressed mothers appear to indicate that these early care experiences may produce permanent alteration in HPA axis functioning. However, because the children of depressed mothers may have inherited genes that increase their vulnerability to variations in maternal care, it is difficult to know whether we can extrapolate these findings to children with different genetic heritages.

Severe deprivation, neglect and abuse

Thus far we have reviewed human studies examining care experiences, which though not perhaps optimal, would not reach criteria for maltreatment. Unfortunately for comparisons with animal studies, HPA reactivity to stressors has not yet been studied in children exposed to severe deprivation, neglect and abuse (see Heim et al., 2004, for such studies of adults exposed to abuse during childhood). Instead, the focus in the child research has been on basal activity of the axis. Evidence for both elevated and extremely low cortisol levels have been obtained. When extremely low levels have been noted, these have most often been seen in early morning, near the peak of the circadian rhythm (Gunnar and Vazquez, 2001). Because the diurnal slope of cortisol is largely determined by the height of the early morning peak, children with the extremely low morning levels have a flat daytime cortisol rhythm. Whether suppressed or elevated basal cortisol levels are noted, however, may depend on the age of the child, the type of maltreatment, whether the maltreatment is on-going and how long it has been since the child was placed in more supportive care arrangements.

There are now several studies of preschool-aged children in foster care that reveal quite different patterns of HPA axis functioning than those noted in studies of children of depressed mothers. Dozier et al. (2006) found that approximately 40% of their sample of foster preschoolers exhibited extremely *low* early morning cortisol levels, while about 20% exhibited markedly high levels relative to children living in non-maltreating families. Bruce et al. (under review) noted similar findings and also found that prior neglect, rather than abuse, predicted low early morning levels. In a study following preschoolers in foster care, children in regular foster care increasingly exhibited low early morning levels over time (Fisher et al., 2007). These children also were more likely than children of the same social class living with their non-maltreating parents to develop insecure patterns of attachment over this time period (Fisher et al., in press). These results are consistent with data on toddlers living in institutions (i.e., orphanages). Orphanage-reared children also have very low early morning cortisol levels and a relatively flat pattern of cortisol production over the day (Carlson and Earls, 1997). Institutional care tends to be characterized by little individualized care, low levels of adult–child interaction, but especially among infants and toddlers, little frank physical or sexual abuse (Zeanah et al., 2006). Benign neglect is a term used to typify institutional care of infants and young children, and hence these data appear remarkably similar to the results reported for preschool-aged children in foster care.

The findings for children exposed to significant neglect and deprivation are also similar to reports of hypo-cortisolism among adults who experience chronic stress (Heim et al., 2000). One mechanism invoked to explain the development of hypocortisolism is down-regulation of the HPA axis at the level of the pituitary in response to chronic CRH drive (Fries et al., 2005). Accordingly, one might expect that prior to appearance of very low early morning cortisol levels, there might be a period of exceptionally high cortisol (Fries et al., 2005). Unfortunately, most of the children examined in both the foster care and institutional care studies experienced adverse care from birth. Examining such children earlier in their care histories also means examining them when they are infants. We have already noted the marked changes in HPA axis activity over the course of the first year or two of life. Developmental change and duration of exposure to maltreatment are difficult to disentangle. Notably, however, the one foster care study examining infants, as opposed to preschoolers, reported exceptionally high cortisol levels for infants in regular foster care relative to infants of the same social class living with their non-maltreating parents (Dozier et al., in press).

There is little reason to believe that this hypocortisolism in basal activity of the HPA axis observed early in life in the context of neglect and deprivation is permanent. At least in work with adults, low cortisol output can be reversed by alleviating chronic stress conditions (for review, see Fries et al., 2005). Among infants in foster care, training foster parents to be more sensitive to the needs of previously neglected and abused infants results in normalization of basal cortisol levels with only 10 weeks of parent training (Dozier et al., in press). What is not clear is whether alterations in HPA axis activity persist once children have time to recover in more supportive care arrangements. In addressing this question we will not cover studies of children with chronic PTSD (Carrion et al., 2001, 2002; De Bellis et al., 1999a, b), although we will consider whether current psychiatric morbidity might explain associations between early care experiences and later HPA activity.

Two studies have now examined basal cortisol levels in children adopted from institutions. For children adopted from the type of global deprivation that characterized Romanian institutions in the early 1990s, elevated rather than suppressed cortisol levels were noted 6–7 years post-adoption (Gunnar et al., 2001). Most of these children were severely growth delayed at adoption due to poor health, nutrition and social stimulation in the orphanage (Rutter, 1998). In a subsequent study, growth delay at adoption rather than duration of institutional care or growth parameters at assessment predicted elevated early morning cortisol levels (Kertes et al., 2007). In this latter study, very few of the children who were growth delayed at adoption met criteria for internalizing problems at assessment; thus, it was unlikely that concurrent psychopathology moderated the relations between early care, growth delay and basal cortisol. Studies of maltreated children suggest that only those who experience the most severe and prolonged maltreatment may have elevated cortisol levels (Cicchetti and Rogosch, 2001a; De Bellis et al., 1999a). However, these factors also increase the risk of both affective and conduct disorder, and thus cortisol levels and psychopathology are difficult to disentangle (Cicchetti and Rogosch, 2001b). Even after their maltreatment has been exposed and child protective services are involved, on-going stress in the family, rather than maltreatment history, may explain increased HPA activity in the children (Kaufman et al., 1997).

Taken together, the studies of post-institutionalized children and children with maltreatment histories do not suggest that for most children, periods of adverse early care increases the basal set point of the HPA axis. Lack of a persistent increase in basal cortisol levels pursuant to early life stress would be consistent with findings in the animal literature where HPA reactivity but not basal levels appear to be affected. When early life stress results in long-term increases in basal HPA levels even after the conditions producing the stress have been improved, the child either appears to have experienced growth failure due to the stressful conditions or, as in the data on children of mothers with depression, may carry a high genetic load for depression or other affective disorders. However, because there are no studies on HPA stress reactivity in children exposed as infants to adverse care, we cannot conclude that their early experiences failed to impact how the axis responds to stressors later in development. This, rather than long-term changes in basal activity is what has been noted in the early experience animal literature and this is what remains to be studied in children exposed to significantly adverse care early in life. Obviously, studies of HPA axis stress reactivity pursuant to early deprivation and maltreatment are crucial to fully understanding how early parental care may impact vulnerability to PTSD. However, because a fully adult-like patterns of HPA responses to stressors may not emerge until sometime in adolescence, studies of the impact of adverse care on subsequent HPA responses to stressors will need to take into account the child's developmental level.

Conclusions

Pre-clinical human studies, animal models and recent studies with children deprived of adequate parenting provide evidence that the HPA axis is under strong social regulation early in life. Accordingly, variations in parental care have potent effects on concurrent regulation of the axis. Unlike in studies of rats, however, in neither non-human primates nor human children do early adverse care experiences appear to produce permanent alterations activity of the HPA axis. The exceptions to this conclusion can be found in studies of children of depressed mothers, children whose early care produced severe physical growth delays, and in children who develop chronic PTSD. Few studies of human children exposed to adverse early care have examined HPA responses to laboratory stressors or to real-life traumatic events. These studies are needed. Failure to observe long-term alterations in HPA axis activity in humans does not mean that failure of the caregiving system to provide regulation of this axis early in development is of no consequence. Rather, as in non-human primates, it may suggest that the HPA axis, being relatively mature at birth, is less susceptible to the long-term alterations noted in studies of infant rats. Nonetheless, failures in social regulation of the HPA axis in young children because of inadequate and/or abusive care may play a role in shaping cortico-limbic circuits involved in processing traumatic experiences. Evidence for this hypothesis, however, will require finding disturbances in HPA activity in early childhood that are reversible by the provision of more supportive care, but that nevertheless predict the development of individual differences in neurobiological systems associated with risk for PTSD. Longitudinal studies of this sort are just beginning to be conducted (Dozier et al., in press; Fisher et al., under review).

Abbreviations

CBG	corticosteroid-binding globulin
CRH	corticotropin-releasing hormone
GR	glucocorticoid receptor
HPA	hypothalamic-pituitary-adreno-cortical
PTSD	post-traumatic stress disorder

Acknowledgments

Preparation of this manuscript was supported by a National Institute of Mental Health pre-doctoral fellowship (T32 MH15755) to the second author, and a National Institute of Mental Health Senior Scientist Award (K05 MH66208) to the first author.

References

Adam, E.K. (2006) Transactions among adolescent trait and state emotion and diurnal and momentary cortisol activity in naturalistic settings. Psychoneuroendocrinology, 31(5): 664–679.

Ahnert, L., Gunnar, M.R., Lamb, M.E. and Barthel, M. (2004) Transition to child care: associations with infant–mother attachment, infant negative emotion, and cortisol elevations. Child Dev., 75(3): 639–650.

Ainsworth, M.S., Blehar, M.C., Waters, E. and Wall, S. (1978) Patterns of Attachment: A Psychological Study of the Strange Situation. Lawrence Erlbaum, Oxford, England, pp. 391.

Angold, A. and Rutter, M. (1992) Effects of age and pubertal status on depression in a large clinical sample. Dev. Psychopathol., 4(1): 5–28.

Ashman, S.B., Dawson, G., Panagiotides, H., Yamada, E. and Wilkinson, C.W. (2002) Stress hormone levels of children of depressed mothers. Dev. Psychopathol., 14(2): 333–349.

Blair, C., Granger, D. and Peters, R. (2005) Cortisol reactivity is positively related to executive function in preschool children attending head start. Child Dev., 76(3): 554–567.

Bowlby, J. (1969) Attachment. Volume 1 of Attachment and Loss. 2nd ed. Basic Books, New York, NY, pp. 425.

Boyce, W.T., Adams, S., Tschann, J.M., Cohen, F., Wara, D. and Gunnar, M.R. (1995) Adrenocortical and behavioral predictors of immune responses to starting school. Pediatr. Res., 38(6): 1009–1017.

Bruce, J., Davis, E.P. and Gunnar, M.R. (2002) Individual differences in children's cortisol response to the beginning of a new school year. Psychoneuroendocrinology, 27(6): 635–650.

Bruce, J., Pears, K.C., Levine, S. and Fisher, P.A. (under review) Morning cortisol levels in preschool-aged foster children: differential effects of maltreatment type. Oregon Social Learning Center, Eugene, Oregon.

Bugental, D.B., Martorell, G.A. and Barraza, V. (2003) The hormonal costs of subtle forms of infant maltreatment. Horm. Behav., 43(1): 237–244.

Caldji, C., Tannenbaum, B., Sharma, S., Francis, D., Plotsky, P.M. and Meaney, M.J. (1998) Maternal care during infancy regulates the development of neural systems mediating the expression of fearfulness in the rat. Proc. Natl. Acad. Sci. U.S.A., 95(9): 5335–5340.

Carlson, M. and Earls, F. (1997) Psychological and neuroendocrinological sequelae of early social deprivation in institutionalized children in Romania. Ann. N.Y. Acad. Sci., 807: 419–428.

Carrion, V.G., Weems, C.F., Eliez, S., Patwardhan, A., Brown, W., Ray, R.D., et al. (2001) Attenuation of frontal asymmetry in pediatric posttraumatic stress disorder. Biol. Psychiatry, 50(12): 943–951.

Carrion, V.G., Weems, C.F., Ray, R.D., Glaser, B., Hessl, D. and Reiss, A.L. (2002) Diurnal salivary cortisol in pediatric posttraumatic stress disorder. Biol. Psychiatry, 51(7): 575–582.

Cicchetti, D. and Rogosch, F.A. (2001a) Diverse patterns of neuroendocrine activity in maltreated children. Dev. Psychopathol., 13(3): 677–693.

Cicchetti, D. and Rogosch, F.A. (2001b) The impact of child maltreatment and psychopathology on neuroendocrine functioning. Dev. Psychopathol., 13(4): 783–804.

Cirulli, F., Alleva, E., Antonelli, A. and Aloe, L. (2000) NGF expression in the developing rat brain: effects of maternal separation. Dev. Brain Res., 123: 129–134.

Cirulli, F., Berry, A. and Alleva, E. (2003) Early disruption of the mother–infant relationship: effects on brain plasticity and implications for psychopathology. Neurosci. Biobehav. Rev., 27(1–2): 73–82.

Davis, E.P., Donzella, B., Krueger, W.K. and Gunnar, M.R. (1999) The start of a new school year: Individual differences in salivary cortisol response in relation to child temperament. Dev. Psychobiol., 35(3): 188–196.

Dawson, G. and Ashman, S. (2000) On the origins of a vulnerability to depression: the influence of early social environment on the development of psychobiological systems related to risk for affective disorder. In: Nelson C.A. (Ed.), The Effects of Adversity on Neurobehavioral Development. Minnesota Symposia on Child Psychology, Vol. 31. Erlbaum Associates, New York, pp. 245–278.

De Bellis, M.D., Baum, A.S., Birmaher, B., Keshavan, M.S., Eccard, C.H., Boring, A.M., et al. (1999a) Developmental traumatology. Part I: biological stress systems. Biol. Psychiatry, 45(10): 1259–1270.

De Bellis, M.D., Keshavan, M.S., Clark, D.B., Casey, B.J., Giedd, J.N., Boring, A.M., et al. (1999b) Developmental traumatology. Part II: brain development. Biol. Psychiatry, 45(10): 1271–1284.

Delahanty, D.L. and Nugent, N.R. (2006) Predicting PTSD prospectively based on prior trauma history and immediate biological responses. Ann. N.Y. Acad. Sci., 1071: 27–40.

Denenberg, V.H. (1999) Commentary: is maternal stimulation the mediator of the handling effects in infancy? Dev. Psychobiol., 34(1): 1–3.

Dettling, A.C., Feldom, J. and Pryce, C.R. (2002) Repeated parental deprivation in the infant common marmoset (*Callithrix jacchus*, primates) and analysis of its effects on early development. Biol. Psychiatry, 52(11): 1037–1046.

Dozier, M., Manni, M., Gordon, M.K., Peloso, E., Gunnar, M.R., Stovall-McClough, K.C., et al. (2006) Foster children's diurnal production of cortisol: an exploratory study. Child Maltreat., 11(2): 189–197.

Dozier, M., Peloso, E., Sepulveda, S., Manni, M., Lindhiem, O., Ackerman, J., et al. (in press) Preliminary evidence from a randomized clinical trial: intervention effects on behavioral and biobehavioral regulation of young foster children. J. Soc. Sci.

Essex, M.J., Klein, M.H., Cho, E. and Kalin, N.H. (2002) Maternal stress beginning in infancy may sensitize children to later stress exposure: effects on cortisol and behavior. Biol. Psychiatry, 52(8): 776–784.

Fisher, P., Gunnar, M., Dozier, M., Bruce, J. and Pears, K. (in press) Effects of therapeutic interventions for foster children on behavioral problems, caregiver attachment and stress regulatory neural systems. Ann. N.Y. Acad. Sci.

Fisher, P.A., Stoolmiller, M. and Gunnar, M.R. (2007) Effects of a therapeutic intervention for foster preschoolers on diurnal cortisol activity. Psychoneuroendocrinology, Jul 24 [Epub ahead of print].

Francis, D.D., Diorio, J., Plotsky, P.M. and Meaney, M.J. (2002) Environmental enrichment reverses the effects of maternal separation on stress reactivity. J. Neurosci., 22(18): 7840–7843.

Francis, D.D. and Meaney, M.J. (1999) Maternal care and the development of stress responses. Curr. Opin. Neurobiol., 9(1): 128–134.

Fries, E., Hesse, J., Hellhammer, J. and Hellhammer, D.H. (2005) A new view on hypocortisolism. Psychoneuroendocrinology, 30(10): 1010–1016.

Giannakoulopoulos, X., Teixeira, J., Fisk, N. and Glover, V. (1999) Human fetal and maternal noradrenaline responses to invasive procedures. Pediatr. Res., 45(4): 494–499.

Gunnar, M.R. and Fisher, P. (2006) Bringing basic research on early experience and stress neurobiology to bear on preventive interventions for neglected and maltreated children. Dev. Psychopathol., 18: 651–677.

Gunnar, M., Marvinney, D., Isensee, J. and Fisch, R.O. (1988) Coping with uncertainty: new models of the relations between hormonal behavioral and cognitive processes. In: Palermo D. (Ed.), Copying with Uncertainty: Biological, Behavioral, and Development Perspectives. Erlbaum, Hillsdale, NJ, pp. 101–130.

Gunnar, M.R., Brodersen, L., Krueger, K. and Rigatuso, J. (1996) Dampening of adrenocortical responses during infancy: normative changes and individual differences. Child Dev., 67(3): 877–889.

Gunnar, M.R. and Donzella, B. (1999) Looking for the Rosetta Stone: an essay on crying, soothing, and stress. In: Lewis M. and Ramsay D. (Eds.), Soothing and Stress. Lawrence Erlbaum Associates, Mahwah, NJ, pp. 39–56.

Gunnar, M.R. and Donzella, B. (2002) Social regulation of the cortisol levels in early human development. Psychoneuroendocrinology, 27(1–2): 199–220.

Gunnar, M.R., Larson, M.C., Hertsgaard, L., Harris, M.L., et al. (1992) The stressfulness of separation among nine-month-old infants: effects of social context variables and infant temperament. Child Dev., 63(2): 290–303.

Gunnar, M.R., Morison, S.J., Chisholm, K. and Schuder, M. (2001) Salivary cortisol levels in children adopted from Romanian orphanages. Dev. Psychopathol., 13(3): 611–628.

Gunnar, M.R. and Vazquez, D. (2006) Stress neurobiology and developmental psychopathology. In: Cicchetti, D. and Cohen, D. (Eds.), Developmental Psychopathology: Deve lopmental Neuroscience. 2nd ed., Vol. 2. Wiley, New York, pp. 533–577.

Gunnar, M.R., Wewerka, S., Frenn, K., Long, J.D. and Griggs, C. (in preparation) Developmental Changes in HPA Activity over the Transition to Adolescence: Normative Changes and Associations with Puberty. University of Minnesota.

Gunnar, M.R. and Vazquez, D.M. (2001) Low cortisol and a flattening of expected daytime rhythm: potential indices of risk in human development. Dev. Psychopathol., 13(3): 515–538.

Hadjian, A.J., Chedin, M. and Chambaz, E.M. (1972) Characteristics of cortisol binding to plasma proteins: peculiar aspects of the perinatal period. Arch. Fr. Pediatr., 29(6): 669–670.

Halligan, S.L., Herbert, J., Goodyer, L. and Murray, L. (2007) Disturbances in morning cortisol secretion in association with maternal postnatal depression predict subsequent depressive symptomatology in adolescents. Biol. Psychiatry, 62: 40–46.

Halligan, S.L., Herbert, J., Goodyer, I.M. and Murray, L. (2004) Exposure to postnatal depression predicts elevated cortisol in adolescent offspring. Biol. Psychiatry, 55: 376–381.

Heim, C., Ehlert, U. and Hellhammer, D.H. (2000) The potential role of hypocortisolism in the pathophysiology of stress-related bodily disorders. Psychoneuroendocrinology, 25(1): 1–35.

Heim, C. and Nemeroff, C.B. (2001) The role of childhood trauma in the neurobiology of mood and anxiety disorders: preclinical and clinical studies. Biol. Psychiatry, 49: 1023–1039.

Heim, C., Owens, M.J., Plotsky, P.M. and Nemeroff, C.B. (1997) The role of early adverse life events in the etiology of depression and posttraumatic stress disorder. Focus on corticotropin-releasing factor. In: Yehuda R.M. and Alexander C. (Eds.), Psychobiology of Posttraumatic Stress Disorder, Vol. 821. New York Academy of Sciences, New York, pp. 194–207.

Heim, C., Plotsky, P. and Nemeroff, C.B. (2004) The importance of studying the contributions of early adverse experiences to the neurobiological findings in depression. Neuropsychopharmacology, 29: 641–648.

Herman, J.P., Ostrander, M.M., Mueller, N.K. and Figueiredo, H. (2005) Limbic system mechanisms of stress regulation: hypothalamo-pituitary-adrenocortical axis. Prog. Neuropsychopharmacol. Biol. Psychiatry, 29(8): 1201–1213.

Kaufman, J., Birmaher, B., Perel, J., Dahl, R.E., Moreci, P., Nelson, B., et al. (1997) The corticotropin-releasing hormone challenge in depressed abused, depressed nonabused, and normal control children. Biol. Psychiatry, 42(8): 669–679.

Kertes, D.A., Gunnar, M.R., Madsen, N.J. and Long, J. (2007) Early deprivation and home basal cortisol levels: a study of internationally-adopted children. Dev. Psychopathol. (in press).

de Kloet, E.R. (1991) Brain corticosteroid receptor balance and homeostatic control. Front. Neuroendocrinol., 12(2): 95–164.

de Kloet, E.R., Rots, N.Y. and Cools, A.R. (1996) Brain-corticosteroid hormone dialogue: slow and persistent. Cell. Mol. Neurobiol., 16(3): 345–356.

Kraemer, G.W., Ebert, M.H., Schmidt, D.E. and McKinney, W.T. (1989) A longitudinal study of the effect of different social rearing conditions on cerebrospinal fluid norepinephrine and biogenic amine metabolites in rhesus monkeys. Neuropsychopharmacology, 2(3): 175–189.

Levine, S., Lyons, D.M. and Schatzberg, A.F. (1997) Psychobiological consequences of social relationships. Ann. N.Y. Acad. Sci., 807: 210–218.

Levine, S. and Wiener, S.G. (1988) Psychoendocrine aspects of mother–infant relationships in nonhuman primates. Psychoneuroendocrinology, 13(1–2): 143–154.

Lupien, S., King, S., Meaney, M.J. and McEwen, B. (2000) Child's stress hormone levels correlate with mother's socioeconomic status and depressive state. Biol. Psychiatry, 48: 976–980.

Meaney, M.J. and Szyf, M. (2005) Environmental programming of stress responses through DNA methylation: life at the interface between a dynamic environment and a fixed genome. Dialogues Clin. Neurosci., 7(2): 103–123.

Nachmias, M., Gunnar, M., Mangelsdorf, S., Parritz, R.H. and Buss, K. (1996) Behavioral inhibition and stress reactivity: the moderating role of attachment security. Child Dev., 67(2): 508–522.

Ramsay, D.S. and Lewis, M. (1994) Developmental change in infant cortisol and behavioral response to inoculation. Child Dev., 65(5): 1491–1502.

Roceri, M., Cirulli, F., Pessina, C., Peretto, P., Racagni, G. and Riva, M.A. (2004) Postnatal repeated maternal deprivation produces age-dependent changes of brain-derived neurotrophic factor expression in selected rat brain regions. Biol. Psychiatry, 55(7): 708–714.

Rosenblum, L.A. and Andrews, M.W. (1994) Influences of environmental demand on maternal behavior and infant development. Acta Paediatr. Suppl., 397: 57–63.

Rosenfeld, P., Suchecki, D. and Levine, S. (1992) Multifactorial regulation of the hypothalamic-pituitary-adrenal axis during development. Neurosci. Biobehav. Rev., 16(4): 553–568.

Rutter, M. (1998) Developmental catch-up, and deficit, following adoption after severe global early privation, English and Romanian Adoptees (ERA) Study Team. J. Child Psychol. Psychiatry, 39(4): 465–476.

Sanchez, M.M., Ladd, C.O. and Plotsky, P.M. (2001) Early adverse experience as a developmental risk factor for later psychopathology: evidence from rodent and primate models. Dev. Psychopathol., 13(3): 419–449.

Sanchez, M.M., Noble, P.M., Lyon, C.K., Plotsky, P.M., Davis, M., Nemeroff, C.B., et al. (2005) Alterations in diurnal cortisol rhythm and acoustic startle response in nonhuman primates with adverse rearing. Biol. Psychiatry, 57(4): 373–381.

Sanchez, M.M., Young, L.J., Plotsky, P. and Insel, T.R. (1999) Different rearing conditions affect the development of corticotropin-releasing hormone (CRF) and vasopressin (AVP) systems in the non-human primate. Paper presented at the Society for Neuroscience Abstracts. October 23–28. Miami Beach, Florida.

Siegel, S.J., Ginsberg, S.D., Hof, P.R., Foote, S.L., Young, W.G., Kraemer, G.W., et al. (1993) Effects of social deprivation in prepubescent rhesus monkeys: immunohistochemical analysis of the neurofilament protein triplet in the hippocampal formation. Brain Res., 619(1–2): 299–305.

Smider, N.A., Essex, M.J., Kalin, N.H., Buss, K.A., Klein, M.H., Davidson, R.J., et al. (2002) Salivary cortisol as a predictor of socioemotional adjustment during kindergarten: a prospective study. Child Dev., 73(1): 75–92.

Spangler, G. (1995) School performance, type A behavior and adrenocortical activity in primary school children. Anxiety, Stress Coping Int. J., 8(4): 299–310.

Spangler, G. and Grossmann, K. (1999) Individual and physiological correlates of attachment disorganization in infancy. In: Salomon J. and George C. (Eds.), Attachment Disorganization. Guilford, New York, NY, pp. 95–124.

Spangler, G., Schieche, M., Ilg, U., Maier, U. and Ackermann, C. (1994) Maternal sensitivity as an external organizer for biobehavioral regulation in infancy. Dev. Psychobiol., 27(7): 425–437.

Spear, L.P. (2000) The adolescent brain and age-related behavioral manifestations. Neurosci. Biobehav. Rev., 24(4): 417–463.

Stroud, L.R., Foster, E., Handwerger, K., Papandonatos, G.D., Granger, D.A., Kivlighan, K.T. & Niaura, R. (in preparation) Stress Response and the Adolescent Transition: Performance versus Peer Rejection Stress. Brown University Medical School.

Suchecki, D., Mazzafarian, D., Gross, G., Rosenfeld, P. and Levine, S. (1993a) Effects of maternal deprivation on the ACTH stress response in the infant rat. Neuroendocrinology, 57(2): 204–212.

Suchecki, D., Nelson, D.Y., Van Oers, H. and Levine, S. (1995) Activation and inhibition of the hypothalamic-pituitary-adrenal axis of the neonatal rat: effects of maternal deprivation. Psychoneuroendocrinology, 20(2): 169–182.

Suchecki, D., Rosenfeld, P. and Levine, S. (1993b) Maternal regulation of the hypothalamic-pituitary-adrenal axis in the infant rat: the roles of feeding and stroking. Dev. Brain Res., 75(2): 185–192.

Suomi, S.J. (1995) Influence of attachment theory on ethological studies of biobehavioral development in nonhuman primates. In: Goldberg M.S. and Kerr R.J. (Eds.), Attachment Theory: Social, Developmental, and Clinical Perspectives. The Analytic Press, Hillsdale, NJ, pp. 185–201.

Talge, N.M., Donzella, B. and Gunnar, M.R. (2007) Fearful temperament and stress reactivity among preschool-aged children. Infant and Child Development (in press).

Tang, A.C., Akers, K., Reeb, B., Romeo, R.D. and McEwen, B. (2006) Programming social, cognitive, and neuroendocrine development by early exposure to novelty. Proc. Nat. Acad. Sci., 103: 15716–15721.

Teicher, M.H., Andersen, S.L., Polcarri, A., Anderson, C.M. and Navalta, C.P. (2002) Developmental neurobiology of childhood stress and trauma. Psychiatric Clinics of North America, 25: 397–426.

Weaver, I.C., Champagne, F.A., Brown, S.E., Dymov, S., Sharma, S., Meaney, M.J., et al. (2005) Reversal of maternal programming of stress responses in adult offspring through methyl supplementation: altering epigenetic marking later in life. J. Neurosci. Biology of posttraumatic stress disorder, 25(47): 11045–11054.

Yehuda, R. (2001) Biology of posttraumatic stress disorder. J. Clin. Psychiatry, 62: 41–46.

Yehuda, R. (2002) Current status of cortisol findings in posttraumatic stress disorder. Psychiatr. Clin. North Am., 25(2): 341–368.

Zeanah, C.H., Smyke, A.T. and Settles, L. (2006) Children in orphanages. In: McCartney K. and Phillips D.A. (Eds.), Handbook of Early Childhood Development. Blackwell Publishing, Malden, MA, pp. 224–254.

Discussion: Chapter 10

JOËLS: I have two questions: one is about the Trier Social Stress Test. Did you see any differences between girls and boys?

GUNNAR: Yes, but the sex difference appears to be due to puberty. At 9 and 11 we did not observe sex differences, at 13 we did and at 15 we did not. When we examined the 13 year olds, it appeared that the sex difference was due to the girls being more advanced in pubertal development.

JOËLS: Okay, at 15 years they both respond.

GUNNAR: Yes, both boys and girls were responding at the age of 15.

JOËLS: I seem to remember that in adults there is a difference.

GUNNAR: It depends on what stage of the menstrual cycle; in one stage of the cycle women respond less than men, in the other there is no difference in response.

JOËLS: The second question is: do you have any indication or is it possible to test whether or not inhibition of the axis is a central effect like is known for GnRH-FSH/LH axis?

GUNNAR: You help me figure out what would be appropriate and ethical to do in healthy normal children.

JOËLS: Maybe with dexamethasone, but that is not ethical.

GUNNAR: Yes, it is difficult with healthy and normally developing children to propose any kind of drug manipulation. Nonetheless, getting at the question of changes in central regulation of the HPA axis around the pubertal transition is important.

YEHUDA: We have incubated live lymphocytes with dexamethasone (and potentially other compounds) and examined inhibition of lysozyme, an enzyme under the control of glucocorticoids. This allows an index of glucocorticoid responsiveness without administering any compounds to the subject. These kinds of methods can be very helpful in the study of children since all that is involved from the human subject point of view is a blood donation — the rest of the challenge occurs in the wet laboratory.

GUNNAR: Yes, we might be able to do that.

SECKL: Glucocorticoid action in the lymphocytes may not reflect the action of this hormone in the brain, particularly as in lymphocytes glucocorticoids predominantly use a GR promoter that is very different from the promoters in other tissues.

GUNNAR: Agreed. The need for specificity is one reason why non-human primate models are very helpful to understanding human development.

ZITMAN: You did your studies in a Romanian orphanage, isn't it? Can you tell us something about the circumstances in which these children were living? Can you explain a little bit of what you have found with the disturbance of the diurnal rhythm on those children?

GUNNAR: Circumstances in orphanages do vary but generally speaking children do not experience individualized care. Rooms are set up with similar-aged children so that care can be routinized. We observe the lack of a strong early morning peak in cortisol production. This might reflect a phase shift in the diurnal rhythm, although this has not been convincingly demonstrated. It is unlikely to reflect chaotic "zeitgebers", as the children experience highly consistent lights on/lights off times from day-to-day and feedings are at a consistent time. Indeed, variable routines are more likely in family-reared children. I should note that we have also observed low early AM cortisol levels in rhesus infants reared on cloth surrogates. So, I don't think we have a phenomenon that is restricted to infants living in orphanages, but rather to primates deprived of a consistent, responsive caregiver. I doubt, however, that it is a permanent effect. We note the presence of a normal early morning peak in cortisol among children adopted from institutions after they have been with their new families for a while.

E.R. de Kloet, M.S. Oitzl & E. Vermetten (Eds.)
Progress in Brain Research, Vol. 167
ISSN 0079-6123

CHAPTER 11

The functional neuroanatomy of PTSD: a critical review

Israel Liberzon* and Chandra Sekhar Sripada

Department of Psychiatry, University of Michigan, 1500 E. Medical Center Dr., MCHC, F6135, Ann Arbor, MI 48109, USA

Abstract: Neuroimaging provides an opportunity to understand core processes that mediate the experience of emotions in healthy individuals as well as dysregulation of these processes in conditions such as post-traumatic stress disorder (PTSD). The first decade of neuroimaging research produced symptom provocation, cognitive activation, and functional connectivity studies that highlighted the role of the medial prefrontal cortex (mPFC), amygdala, sublenticular extended amygdala (SLEA), and hippocampus, in mediating symptom formation in PTSD. There is a growing realization that a number of other psychological processes are relevant to PTSD, and they are emerging as a new focus of neuroimaging research. These include fear conditioning, habituation, and extinction; cognitive–emotional interactions; and self-related and social emotional processing. Neuroimaging findings are reviewed that suggest that the mPFC is implicated in a number of these processes. It is proposed that the mPFC plays a role in the "contextualization" of stimuli, and dysregulation of contextualization processes might play a key role in the generation of PTSD symptoms.

Keywords: PTSD; emotion regulation; neural circuitry; functional neuroimaging; medial prefrontal cortex; anterior cingulate cortex; amygdala; hippocampus

Introduction

Posttraumatic stress disorder (PTSD) is a debilitating illness characterized by exposure to a traumatic event followed by development of a constellation of symptoms. These symptoms typically include re-experiencing phenomena (e.g., flashbacks or nightmares), hyperarousal (e.g., vigilance or exaggerated startle response), and avoidance behavior (e.g., avoidance of persons or situations that are reminiscent of the traumatic episode). Over the past decade, interdisciplinary work in neurobiology and neuroimaging has led to formulation of neurocircuit models of PTSD (Rauch and Shin, 1997; Pitman et al., 2001; Liberzon and Phan, 2003). According to these models, PTSD can be conceptualized as a state of heightened responsivity to threatening stimuli and/or a state of insufficient inhibitory control over exaggerated threat-sensitivity. Accordingly, these models emphasize the centrality of threat-related processing in the pathophysiology of PTSD and extensive work has been conducted attempting to locate the specific derangements in threat-related circuitry associated with the disorder.

While studies that focus on threat-related processing have provided useful information and

*Corresponding author. Tel.: +1 734 615 0199; Fax: +1 734 647 8514; E-mail: liberzon@umich.edu

DOI: 10.1016/S0079-6123(07)67011-3

have guided the beginning of functional neuroanatomical and neurophysiological research in PTSD, it is becoming increasingly clear that the scope of these studies does not fully capture the complexity of changes occurring in trauma exposure and PTSD development. For instance, while the predominant notion of PTSD being a state of abnormal responsivity to threat explains some aspects of PTSD (such as hypervigilance and hyperarousal), it provides less clarification of the basis of other aspects of the disorder. These aspects include intrusive thoughts and memories, emotional numbing, vulnerability and resilience factors, and generalization of vigilance and avoidance from the initial traumatic event to other less closely related events.

Outside the direct study of PTSD, there have been significant efforts by investigators in cognitive neuroscience to clarify the psychological and neurobiological basis of a number of processes that appear relevant to development, maintenance, and/or recovery PTSD. These processes include the phenomena of conditioning, habituation, stimulus generalization, and (resistance to) extinction; cognitive–emotional modulation (involving appraisal and reappraisal); and social and self-related emotional processing. One important goal of this manuscript is to elaborate on some of these mechanisms as they may relate to the state of trauma exposure/PTSD and to initiate a discussion that may lead to more nuanced and integrative investigations in this area.

The first part of this chapter briefly reviews what is currently known on the basis of functional neuroimaging studies of subjects with PTSD. These consist mainly of symptom provocation studies and symptom correlation, and cognitive activation studies as well as functional connectivity analyses. The second part of the chapter focuses on specific psychological processes that have been implicated in PTSD symptom generation or pathophysiology. These include neuroimaging studies of fear conditioning, habituation, and extinction; cognitive–emotional interactions; and self-related and social emotional processing. The chapter concludes with a brief discussion of the threat-related processing model of PTSD and a modified and updated model is proposed in which a core process of PTSD involves impairments in the circuits that mediate the "contextualization" of stimuli. Finally, potential future directions for PTSD research are proposed.

Emotional responses in PTSD patients: functional neuroimaging studies

Functional neuroimaging studies in PTSD encompass a number of different imaging modalities including single-photon emission tomography (SPECT), positron emission tomography (PET), and functional magnetic resonance imaging (fMRI). In addition, a number of different types of imaging paradigms have been pursued, including symptom provocation studies, cognitive activation studies as well as functional connectivity analysis. The discussion that follows is a selective review of such studies in PTSD.

Symptom provocation studies

Symptom provocation studies were the first ones to provide replicable findings and still are the most common in the neuroimaging literature on PTSD. They involve provoking symptoms while attempting to capture the underlying neural substrates [as gleaned from blood flow and/or blood oxygen level dependent (BOLD) effects] and employ trauma-related stimuli of an autobiographical nature (e.g., narrative scripts of personal trauma) or general nature (e.g., nonpersonalized pictures and sounds).

In an early study, Rauch et al. (1996) used H_2O PET to examine regional cerebral blood flow (rCBF) changes in response to individualized trauma scripts in a group of eight PTSD subjects. They demonstrated increases in anterior paralimbic [right posterior medial orbitofrontal cortex (OFC), insular, anterior temporal polar, and medial temporal cortex] and limbic structures (amygdala) in the provoked versus control contrast. The same group investigated the specificity of emotional processing in combat veterans with and without PTSD using combat-related, emotionally negative, and neutral pictures paired with verbal descriptions (imagery). Combat veterans with PTSD had increased rCBF in ventral anterior cingulate cortex (ACC) and right amygdala when generating mental images of combat-related pictures but had decreased rCBF in the ACC in the combat image viewing versus neutral image viewing contrast (Shin et al., 1999). Though these early studies had methodological limitations, such as small and heterogeneous sample size and the lack of adequate control groups that limited the generalization of their findings, they set the stage for more detailed studies into the neural substrate of the symptomatic PTSD state.

In an ensuing study, our group used 99mTc-hexamethylpropyleneamineoxime (HMPAO) SPECT to examine the neural response to combat sounds versus white noise in three groups of subjects (14 combat PTSD subjects, 11 combat-exposed subjects without PTSD, and 11 combat-unexposed healthy subjects). Only the PTSD group showed increased rCBF in the left amygdaloid region (for the main contrast of combat sounds-white noise) (Liberzon et al., 1999b). Bremner and colleagues used combat-related pictures and sounds and PET in a group of combat veterans (10 with and 10 without PTSD). They found decreased blood flow in medial prefrontal cortex (mPFC) (area 25) and other areas in response to traumatic pictures and sounds in PTSD patients (PP), while non-PTSD control subjects activated the anterior cingulate (area 24) to a greater degree than PP (Bremner et al., 1999b). The same group also studied a different cohort of subjects [22 women with histories of childhood sexual abuse (CSA); 10 of whom had PTSD] with exposure to traumatic and neutral scripts and PET. The PTSD group showed rCBF increases in posterior cingulate (area 31) and anterolateral prefrontal cortex (PFC) (superior and middle frontal gyri bilaterally, areas 9 and 10). The PTSD group also showed deactivation in the subcallosal gyrus region of anterior cingulate (area 25) with a failure of activation in an adjacent portion of anterior cingulate (area 32) (Bremner et al., 1999a). In a study that used a similar group of subjects (16 subjects with CSA; 8 with PTSD), Shin et al. (1999) using script-driven imagery and PET, reported greater increases in rCBF in the OFC and temporal poles and deactivation of the medial prefrontal and left inferior frontal (Broca's) areas in the PTSD group versus the non-PTSD group in the traumatic versus neutral imagery contrast. A more recent script-driven imagery and PET study by Shin and colleagues of 17 Vietnam veterans with PTSD and 19 without PTSD showed rCBF decreases in the medial frontal gyrus for the traumatic versus neutral comparison in the PTSD group. This activity was inversely correlated with rCBF changes in the left amygdala and the right amygdala/periamygdaloid cortex. Only the male combat veteran subgroup (and not the female nurse veteran subgroup) showed increased rCBF in left amygdala (Shin et al., 2004a).

Several investigators have focused on correlational analysis in neuroimaging data to identify potential relationships between activations of neural regions and measures of symptom severity. These studies in general have pointed toward the same subset of limbic and cortical regions that have been implicated in earlier studies. Osuch et al. (2001) correlated rCBF response with flashback intensity in a personalized, script-driven imagery PET paradigm in eight chronic PTSD subjects. rCBF correlated directly with flashback intensity in the brain stem, insula, and hippocampus, and inversely in the prefrontal, right fusiform, and medial temporal cortices. In a script-driven imagery and PET study reported above, Shin and colleagues found that in the PTSD group, for the traumatic condition, symptom severity [as measured by the total score on the Clinician-Administered PTSD Scale (CAPS)] was positively related to rCBF in the right amygdala and negatively related to rCBF in medial frontal gyrus after controlling for depression severity score (Shin et al., 2004a). Some studies using correlational analysis however failed to replicate relationships between neural activations and symptom severity. For example, in an fMRI study, Lanius et al. (2002) reported that 7 CSA subjects with PTSD and concomitant dissociative responses to symptom provocation by scripts had increased activation in the ACC, mPFC, and several other cortical areas compared to 10 control subjects. However, none of these activations correlated with either dissociative or flashback intensity. It is possible that small sample size and significant comorbidity in this cohort played a role in this negative finding.

Symptom provocation studies have also examined the time course of neural responses with repeated presentation of trauma-related stimuli. These studies have tested whether abnormalities in brain activation in PTSD might stem from abnormal temporal response to traumatic stimuli, rather than from difference in the magnitude of response. For example, Hendler and colleagues used fMRI with combat veterans with and without PTSD to examine differences in neural responses to pictures with and without combat content in repeated versus novel presentations. Repeated presentations of the same combat visual stimuli resulted in less BOLD signal decrease in the lateral occipital cortex in PTSD subjects (vs. non-PTSD), suggestive of impaired habituation of the response to trauma-related stimuli (Hendler et al., 2001). In a block design fMRI study, Protopopescu et al. (2005) examined the time course of amygdala responses to trauma-relevant negative words, panic-relevant negative words (negative control condition), positive/safety words, and neutral words, in 9 predominantly sexual assault PP and 14 healthy controls. The PTSD group showed an increased left amygdala response to trauma-relevant negative versus neutral stimuli compared to controls in the first two (but not last two) runs, and this response correlated with the symptom severity (CAPS total score). Healthy controls showed the opposite pattern. In a recent block design fMRI study, 13 PP were compared with matched non-traumatized healthy subjects in a task that consisted of passively viewing 15 blocks of fearful face stimuli alternating pseudorandomly with 15 blocks of neutral faces (Williams et al., 2006). Time series data were used to examine amygdala–mPFC associations and changes across the first (Early) versus second (Late) phases of the experiment. Relative to healthy subjects, PTSD subjects showed a marked bilateral reduction in mPFC activity, especially in right ACC, which showed a different Early–Late pattern relative to non-traumatized subjects. Decreases in mPFC activity were also found to be correlated with measures of the degree of trauma impact and symptomatology. PTSD subjects also showed a small but significant enhancement in left amygdala activity, most apparent during the Late phase.

An important issue that emerged in symptom provocation studies of PTSD concerns whether the pattern of neural activations identified in response to trauma-related stimuli are specific to trauma-processing in PTSD or whether they are related to a broader category of emotion-related processing. Lanius and colleagues reported two studies where they used a script-driven symptom provocation paradigm and fMRI. The second study also included comparison of non-traumatic negative states — sad and anxious. They reported significantly decreased BOLD signal in the ACC [Brodmann area (BA) 32] and the thalamus in the PTSD group to both the traumatic and non-traumatic emotional states conditions, suggesting that some neuroimaging findings in PTSD may not be specific to traumatic stimuli, but might reflect more generalized abnormal emotional processing (Lanius et al., 2001, 2003). We recently reported a $[^{15}O]$ H_2O PET, script-driven imagery study of emotionally evocative and neutral autobiographic events in 16 combat veterans with PTSD (PP), 15 combat veterans without PTSD [combat controls (CC)], and 14 healthy, age-matched, noncombat control subjects [noncombat controls (NC)] giving us the ability to study changes that are trauma-related (PP vs. NC and CC vs. NC) and PTSD-specific (PTSD vs. CC). For the traumatic/stressful > neutral scripts contrast, all subjects deactivated the mPFC and activated the insula, the PP deactivated the rostral anterior cingulate cortex (rACC) more than both control groups (CC and NC), and these control groups also showed ventromedial prefrontal cortex (vmPFC) deactivation not found in PTSD. Trauma exposure per se (i.e., PP and CC groups) was associated with decreased amygdala activity regardless of presence or absence of PTSD diagnosis. On the other hand, deactivation of the rACC that was observed only in the PTSD group may reflect neural substrates specific to PTSD, whereas trauma-exposure related patterns (decreased amygdala activity) may represent compensatory changes (Britton et al., 2005).

In summary, symptom provocation studies have implicated several anterior paralimbic and limbic structures in the symptomatic state of PTSD including the posterior medial OFC, the insula, and the medial temporal cortex. The majority of studies, but not all, have demonstrated decreased activation in subregions of the mPFC and ACC. Increased responsivity of the amygdala has been observed in some studies, but has not been an "across the board" consistent finding. Several issues of design and/or methodology may contribute to these divergent findings, including the nature of symptom provocation method (trauma imagery vs. external stimuli), experimental tasks (passive viewing vs. active recall), scanning methods, and relatively small sample sizes, all of which may effect the ability to activate and/or detect amygdala response. Correlation analyses used to investigate the relationship between symptom-provoked activation and cross-sectional symptoms severity, rendered findings have been inconsistent for meaningful interpretation at this time.

Cognitive activation studies

Cognitive activation studies utilize a neurocognitive task (a "probe") that is expected to selectively activate neural circuits implicated in task-related processing. Selectively activating a region without eliciting symptoms has a substantial advantage in that this overcomes the confound that arises from simultaneously eliciting a large number of more general or nonspecific trauma-related responses. Investigators have used cognitive activation strategies to further examine a number of regions implicated in PTSD by symptom provocation studies, such as the amygdala, ACC, and hippocampus.

The amygdala is a region implicated in rapidly assessing the salience of emotion-related and especially threat-related stimuli (Davis and Whalen, 2001). Rauch et al. (2000) compared amygdala responses in nine PTSD subjects versus eight combat-exposed, non-PTSD subjects using a previously validated masked emotional faces paradigm. Contrasting fearful versus happy masked faces revealed exaggerated amygdala responses in the PTSD subjects. Furthermore, the magnitude of these responses distinguished PTSD subjects with 75% sensitivity and 100% specificity. These findings suggest that PTSD is associated with increased amygdala responsivity to threat-related (but not necessarily trauma-related) stimuli. In addition, this study suggested that nonconscious threat-related stimuli were able to elicit threat-related emotion responses in PTSD. Another group used a similar masked emotional faces paradigm to examine 13 subjects with acute, rather than chronic PTSD (Armony et al., 2005). There was a positive correlation between the severity of PTSD and the difference in amygdala responses between masked fearful and happy faces. These findings suggest that functional abnormalities in brain responses to emotional stimuli observed in chronic PTSD might be apparent already in the acute phase.

Shin et al. (2005) used overtly presented emotional facial expressions and fMRI to compare BOLD responses in 13 men with PTSD and 13 trauma-exposed men without PTSD. The PTSD group showed increased amygdala responses and decreased mPFC responses to overt fearful (vs. happy) facial expressions. BOLD signal changes in the amygdala in the PTSD subjects were negatively correlated with signal changes in the mPFC. Additionally, BOLD signal changes in the mPFC were inversely correlated with symptom severity (CAPS).

Another region that has been repeatedly implicated in PTSD is the ACC, albeit rather as hypofunctioning or failing to activate as compared to unaffected controls. The ACC is a region that has been activated by many functional neuroimaging studies and has been implicated in cognitive emotion interactions. A variety of evidence supports the existence of functional subdivisions in the ACC, with dorsal ACC supporting cognitive control and error-related processing, while rACC is involved in the assessment of salience of emotional information and the regulation of emotional responses (Bush et al., 2000). Bremner et al. (2004) used the Stroop task (color Stroop, emotional Stroop, and control task) and [^{15}O] H_2O PET to probe ACC function in 12 women with early CSA-related PTSD and 9 abused women without PTSD. The PTSD group had a relative decrease in ACC blood flow during the emotional but not the color Stroop task, which elicited increased rCBF in the ACC (BA 24 and 32) in both groups. Shin et al. (2001) also investigated ACC functioning in 16 Vietnam combat veterans (8 with PTSD) using fMRI and an emotional counting Stroop paradigm. Subjects were asked to count the number of combat-related, generally negative and neutral words while being scanned. In the comparison of combat related to generally negative words the non-PTSD group showed significant BOLD signal increases in rACC but the PTSD group did not.

A third region implicated in PTSD is the hippocampus, which plays a role in explicit memory processes as well as contextual learning (Eichenbaum, 2000; Corcoran and Maren, 2001). Shin et al. (2004b) used PET to investigate hippocampal function in 16 firefighters (8 with PTSD) using a word stem completion task. Subjects completed a three-letter word stem with deeply encoded/high recall and shallow encoded/low recall-words learned during a preceding training session. The PTSD group demonstrated greater rCBF in the hippocampi (bilateral) across conditions. In the main contrast of interest (high vs. low recall) the PTSD group (vs. control group) showed significantly smaller rCBF increases in the left hippocampus, which was primarily driven by relatively elevated rCBF in the low recall condition. Another group investigated mechanisms of updating working memory in PTSD using [^{15}O] H_2O PET and a variant of the n-back task (detection of trauma-unrelated target words under fixed and variable conditions; only the variable condition required target updating) in 10 patients with PTSD (mostly civilian trauma) compared to 10 healthy subjects (Shaw et al., 2002). Functional connectivity analysis during the working memory task revealed increased activation in bilateral inferior parietal lobules and left precentral gyrus, and reduced activation in inferior medial frontal lobe, bilateral middle frontal gyri, and right inferior temporal gyrus, in the PTSD group relative to the control subjects.

Functional connectivity analyses

Functional connectivity analysis refers to the application of specific statistical methods to functional neuroimaging data sets to identify correlated brain activity across various regions (Friston et al., 1993, 1996). This technique is particularly useful in light of a growing appreciation of the fact that complicated cognitive and emotional processes rely on the orchestrated interactions of distributed brain networks, rather than, or at least in addition to, activation of individual brain regions. Several recent studies have applied functional connectivity analysis to neuroimaging studies of PTSD. Gilboa et al. (2004) studied 20 individuals with a history of civilian trauma (10 with PTSD), using symptom provocation (autobiographical trauma-related and neutral scripts) and [^{15}O] H_2O PET. A multivariate analysis technique (partial least squares) was used to identify brain regions whose activity co-varied with two reference ("seed") voxels, one in right PFC (BA 10) and the other in right amygdala (both derived from a preliminary task). Amygdala activity was found to significantly influence activity in the visual cortex, subcallosal gyrus, and anterior cingulate in the PTSD subjects but not in the trauma-exposed controls. Correlational analysis, however, did not lend support for the failure of inhibition of the ACC over the amygdala.

Lanius et al. (2004) used functional connectivity analyses on data gathered during fMRI script-driven symptom provocation experiments in 11 subjects with PTSD from sexual abuse/assault or motor vehicle accident (MVA), and 13 trauma-exposed subjects without PTSD. Comparison of connectivity maps at a right ACC coordinate showed greater correlations in PTSD subjects (vs. controls) in the right posterior cingulate cortex (PCC) (BA 29), right caudate, right parietal lobe (BA 7 and 40), and right occipital lobe (BA 19). Subjects without PTSD had greater correlations (vs. PTSD subjects) in the left superior frontal gyrus (BA 9), left anterior ACC (BA 32), left striatum (caudate), left parietal lobe (BA 40 and 43), and left insula (BA 13). These findings are indeed intriguing; however our understanding of functional neural networks both in health and disease is still very limited. As methods for the analysis of functional connectivity continue to develop, and the knowledge base regarding coordinated activation of brain regions grows, these approaches will likely play an increasingly important role in delineating functional relationships between regions implicated in the pathophysiology of PTSD.

An important category of neuroimaging studies not reviewed in this chapter consists of studies using methods such as PET coupled with receptor-selective radiotracers and magnetic resonance spectroscopy, which are particularly well suited for illuminating molecular and receptor-level mechanisms in disorders including PTSD. For example, in a recent study, we used PET and the mu-opioid selective radiotracer [^{11}C] carfentanil in 16 male patients with PTSD contrasted with two non-PTSD male control groups, one with ($n = 14$) and the other without combat exposure ($n = 15$) (Liberzon et al., 2007). We found differences in mu-opioid receptor binding potential in a number of regions associated with emotion processing including amygdala and sublenticular extended amygdala (SLEA), as well regions associated with emotion regulation including OFC and subgenual ACC. Our results are consistent with other results that suggest abnormalities in the opioid system and pain processing in PTSD (see Geuze et al., 2007). Taken together, these findings lend additional support to the idea that PTSD involves heightened amygdala responsivity to stress and trauma cues, and suggest that abnormalities in the opioid system might play a role in this process. Overall, neuroimaging investigations aimed at clarifying the molecular and receptor-level mechanisms of PTSD represent important avenues for future research, and may be fruitfully combined with other kinds of neuroimaging results to provide converging lines of evidence about the pathophysiology of the disorder.

Summary of functional neuroimaging studies in PTSD

The studies reviewed above involve different cohorts (combat and CSA-related PTSD), different paradigms (symptom provocation vs. cognitive activation), and different modalities (fMRI, PET, and SPECT). Taken together, they lend tentative support to a neurocircuitry model that emphasizes the role of dysregulation in threat-related processing in PTSD. According to this model, trauma exposure sets off a cascade of neural changes that culminates in a state of amygdala hyperresponsivity to trauma-reminiscent and other threat-related stimuli. Amygdala hyperresponsivity is proposed to mediate symptoms of hyperarousal and vigilance associated with PTSD. The model also proposes associated inadequate top-down control by the mPFC, which helps maintain and perpetuate the state of amygdala hyperresponsivity, and also helps mediate the failure to suppress attention to trauma-related stimuli. Consistent with this model, several studies have demonstrated reduced activation of the mPFC (BA 10 and 11) and ACC (BA 32) in PTSD subjects compared to traumatized controls (Bremner et al., 1999b; Shin et al., 1999, 2001; Lanius et al., 2001, 2003). Other studies have reported increased responsivity of the amygdaloid region (Rauch et al., 1996, 2000; Liberzon et al., 1999b), though some have not (Bremner et al., 1999b; Shin et al., 1999; Lanius et al., 2001).

While the conceptualization of PTSD-related pathophysiology that emphasizes the role of threat-related processing has some initial empirical support, there is clearly a need for a broader conceptualization of the core processes implicated in the disorder. This is because deficits in threat-related processing explain only some aspects of PTSD, and other significant manifestations of PTSD remain unexplained by this model. These include intrusive thoughts and memories, emotional numbing, vulnerability and resilience factors, and generalization of vigilance and avoidance from the initial traumatic event to other less closely related events. Thus to understand these complex phenomena, additional relevant mechanisms that may assist in understanding the complex phenomenology of PTSD need to be explored. The following section includes a selective review of emerging neuroimaging research that focuses on a number of mechanisms that are potentially relevant to the pathophysiology of PTSD, including fear conditioning, cognitive–emotion interactions, and self-related and social emotional processing.

Part II: the medial prefrontal cortex and psychoneural processes relevant to PTSD

Neuroimaging studies of threat-related processing (fear conditioning, habituation, and extinction)

Studies using the fear conditioning paradigm in rats over the past two decades have helped outline a specialized threat-related neural network that involves several functionally connected regions including the subregions of the PFC, the amygdala, and the hippocampus. A key finding from these studies is that there appear to be two broad pathways in the processing of threat-related emotion — a subcortical "fast" pathway that transmits features of the stimulus rapidly, but with poor specificity and a cortical "slow" pathway that involves more integrated, and detailed cognitive processing of stimulus characteristics (Ledoux, 2000). Animal studies have identified the amygdaloid complex (specifically the central nucleus and lateral and basolateral nuclei) as a crucial substrate in the formation of stimulus response associations involved in the fear conditioning response as well as aversive learning. A number of investigators have used PET and fMRI to extend the study of fear-conditioning to humans (Buchel and Dolan, 2000).

In an interesting [^{15}O] H_2O PET study, Morris and colleagues used overtly presented and masked faces as the conditioned stimulus (CS) and a 1-s 100-dB burst of white noise as the unconditioned stimulus (US) in 10 healthy right-handed male subjects and demonstrated amygdala activation during CS+ trials (CS coupled with aversive US) but not during CS− trials (CS not associated with US). They also noted that the subconscious presentation of the CS (the masked CS+ minus masked CS− contrast) activated the right amygdala whereas conscious presentation of the CS (the unmasked CS+ minus unmasked CS− contrast) activated the left amygdala suggesting a differential lateralized response to automatic or implicit versus conscious or explicit processing of these stimuli (Morris et al., 1998). These findings provide direct evidence for the role of the amygdala in emotional learning (of behaviorally significant stimuli) suggesting that this learning can occur even in the absence of conscious perception of such stimuli.

Other neuroimaging studies have examined classical conditioning of aversive stimuli, typically aversive tones, or mild electrical shocks. For example, Buchel and colleagues used event-related fMRI to study the classical conditioning of faces paired with aversive tones in nine healthy right-handed subjects (LaBar et al., 1998). Comparison of the CS+ condition to the CS− condition revealed greater activation of the ACC. In addition, this comparison also revealed greater activation in the amygdala, though only in early trials suggesting a rapid habituation of the amygdala response. Other studies using related but slightly distinct fear conditioning paradigms also found activation in the ACC and amygdala, with a decrement in amygdala response suggesting rapid habituation (Buchel et al., 1998, 1999). These neuroimaging studies of fear conditioning provide evidence for the involvement of the ACC as well as the amygdala in the acquisition of fear conditioning in humans. As noted, both fear conditioning and the involved brain regions have been linked to cued hyperarousal symptoms in PTSD.

Potential differences in neural response over time in specific regions have been hypothesized in PTSD, and as we outlined above some empirical evidence supports this interpretation. Two relevant neural processes that involve a time component that might be relevant to PTSD are habituation and extinction. Habituation refers to the process by which repeated presentation of the same CS–US pairing leads to a decreasing conditioned response (CR), while extinction refers to a reduction and disappearance of a CR on account of learning about a new stimulus–response association (i.e., the CS is no longer associated with the US). These are considered adaptive processes for organisms as they provide the organism with flexibility to reallocate critical resources to threat-related stimuli in a constantly changing environment. The failure of habituation to trauma-related stimuli and/or the failure of extinction have been hypothesized to contribute to the development or maintenance of PTSD following exposure to trauma (i.e., trauma plays the role of the conditioning event). These phenomena have been extensively studied in animals and behaviorally in humans but these processes have only recently been the subject of neuroimaging investigations.

A number of neuroimaging studies have found that presentation of emotionally expressive faces, presented both overtly as well as in a masked manner (fearful or happy faces masked with a neutral face such that subjects consciously perceive only the neutral face) activates the amygdala, and this response rapidly habituates with repeated presentation regardless of the mode of presentation (overt or masked) (Breiter et al., 1996; Whalen et al., 1998). Several studies reviewed above in which a CS is repeatedly paired with an aversive US also found rapid habituation of the amygdala response (Buchel et al., 1998, 1999; LaBar et al., 1998).

One recent fMRI study suggests that repeated presentation of emotionally expressive faces may generate habituation in a regionally specific manner based on the valence of the facial stimulus. In this study fearful and happy faces were repeatedly presented in two 2-min runs to eight right-handed healthy male subjects. Significant fMRI signal decrement was observed in the left dorsolateral prefrontal cortex (dlPFC) and premotor cortex, and the right amygdala. The left dlPFC showed increased habituation to happy more than fearful faces, which possibly reflects differential responses of prefrontal versus subcortical structures to threat-related stimuli. Additionally, the right amygdala exhibited greater habituation to emotionally valenced stimuli (than the left) while the left amygdala responded significantly more to negatively versus positively valenced stimuli (relative to the right) (Wright et al., 2001). Our laboratory has also demonstrated rACC habituation with repeated emotional picture (aversive minus neutral/blank) presentation (Phan et al., 2003).

These studies provide evidence for habituation in the dlPFC, ACC, and the amygdala with some evidence for differential habituation in prefrontal versus subcortical regions to threat-related stimuli, as well as lateralized specialization of rapid versus sustained response to threat stimuli (Wright et al., 2001). Interestingly, the only study that has specifically addressed the time course of amygdala responses to trauma cues (trauma-relevant words) in PP and healthy controls found an increased left amygdala response to trauma-relevant negative versus neutral stimuli in the first two but not last two runs. This response correlated with the symptom severity (CAPS total score). However, while sensitization to non-trauma negative words was seen in the PTSD group, failure of habituation to trauma-related words was not seen (Protopopescu et al., 2005).

The process of extinction has also been the subject of recent neuroimaging studies. Phelps et al. (2004) used a simple discrimination, partial reinforcement fear conditioning paradigm with an event-related fMRI design. Colored squares were used for CS+ and CS− (blue and yellow) and US was a mild wrist shock. The study was conducted in three phases, an acquisition phase in which subjects were exposed to reinforced presentations of the CS, followed by day 1 extinction and day 2 extinction, in which subjects were exposed to unreinforced presentations of the CS. The authors reported that right amygdala activation predicted the CR in the early acquisition (positive correlation) and day 1 extinction phase (negative correlation). The vmPFC (the subgenual anterior cingulate region of interest) response positively correlated with the CR magnitude during day 2 extinction. These findings appear to be consistent with those of animal research that implicate the amygdala in acquisition and extinction and the vmPFC in the retention of the extinction learning process (Falls et al., 1992; Morgan et al., 1993, 2003; Morgan and Ledoux, 1995; Phelps et al., 2004). They are also intriguing in light of evidence reviewed earlier from human neuroimaging studies of altered connectivity between medial frontal regions and amygdala in PTSD.

Thus, the evidence from human neuroimaging studies discussed above implicates subregions of the mPFC and OFC, subdivisions of the ACC, the extended amygdala, the hippocampus, and nuclei of the thalamus in the processes of fear conditioning, habituation, and extinction. The fear conditioning paradigm and the associated phenomena of habituation and extinction (or failure of these processes) are relevant to aspects of psychopathological states, such as PTSD and phobic states. The neuroimaging studies of these processes in healthy humans provide a background for future work in extending these studies to patients with PTSD and other anxiety disorders.

Cognitive–emotional interactions: appraisal, reappraisal, and emotional regulation

Emotion regulation refers to the set of mental processes by which people amplify, attenuate, or otherwise modulate emotion states (Gross, 1998). Key features of PTSD, including emotional numbing and heightened and prolonged experience of fear, anxiety, and other negative affective states, suggest that poor emotion regulation plays a key role in this disorder and contributes significantly to behavioral dysfunction. For the purposes of this discussion, emotion regulation is understood in terms of a number of component processes that operate over different time scales. Appraisal refers to the cognitive interpretation of emotion-relevant stimuli by higher cortical centers. An increasing number of neuroimaging studies are providing evidence that cognitive appraisal can modulate emotional responses, which is reflected in changes in the activity of emotion processing areas. Cognitive reappraisal is a form of emotion regulation that involves volitionally reinterpreting the meaning of a stimulus to change one's emotional response to it.

Cognitive appraisal of emotions

A number of studies have manipulated the extent to which subjects cognitively attend to aspects of emotion-relevant stimuli. These studies suggest that even the simple process of labeling or rating an emotion can reduce the activity in structures that are responsive when the emotional stimulus is passively viewed or experienced.

Hariri and his colleagues examined the cognitive modulation of emotions by comparing the BOLD response in healthy subjects as they performed three different tasks (match, label, and control). In the match task, subjects were asked to match the affect of one of two faces to that of a simultaneously presented target face (angry or fearful) whereas in the label task, they were asked to assign one of two simultaneously presented linguistic labels (angry or afraid) to a target face. The control task involved matching a target shape. Matching was associated with increased activation in both the right and left amygdala whereas linguistically labeling the expression was associated with a decreased activation in the amygdala. Additionally, right PFC activity was inversely correlated with left amygdala activity. The authors interpreted these findings as evidence in support of prefrontal cortical structures being the neural substrate for the cognitive modulation of emotion via the process of interpretation and labeling (Hariri et al., 2000). This finding has been replicated using threatening and fearful pictures as well (Hariri et al., 2003). In our laboratory, we examined rCBF response in healthy subjects comparing a rating to a passive viewing condition (Taylor et al., 2003). Subjects saw aversive and neutral pictures from the international affective pure system (IAPS) (Lang et al., 1997), while they performed a passive viewing (PSVW) and rating (RTNG) task. During PSVW, for aversive minus neutral pictures, subjects activated foci in the area of the right insula/amygdala and left insula. RTNG was associated with increased activation of the dorsomedial prefrontal cortex (dmPFC) and the ACC. RTNG resulted in reduction in the intensity of sadness and reduced activation of the right insula/amygdala and left insula compared to PSVW. These findings demonstrate the involvement of the dmPFC and ACC in the cognitive rating task and suggest modulating effects of these structures on emotion-related structures, such as the amygdala and insula. These findings extend findings from animal studies that have demonstrated the inhibitory influence of the mPFC over the amygdala (Rosenkranz and Grace, 2002).

Cognitive modulation of emotions: reappraisal

Cognitive reappraisal refers to volitional reinterpretion of the meaning of a stimulus to modify one's emotional response. Recently, several groups have investigated the neural effects of reappraisal using functional neuroimaging methods. This line of work is likely very relevant to PTSD, where emotional dysregulation is a predominant feature. It is also of much interest in the investigation of brain mechanisms of cognitive behavioral therapy, an effective treatment for some patients with PTSD. Ochsner and colleagues used an event-related fMRI design and aversive IAPS pictures to study cognitive reappraisal in healthy female subjects. Subjects were asked to Attend (be aware of feelings elicited by the picture) or to Reappraise (reinterpret the picture so that it no longer elicits a negative emotional response) while being scanned (Ochsner et al., 2002). Reappraisal of highly negative scenes was successful. Reappraising (vs. attending) was associated with increased activation of the dorsal and ventral left lateral prefronal cortex, dmPFC, left temporal pole, right supramarginal gyrus (SMG), and left lateral occipital cortex. Greater activation in the right ACC and SMG correlated with greater decreases in negative affect (greater reappraisal success); left ventral PFC activation during reappraisal was inversely correlated with activity in the amygdala. Effective reappraisal resulted in increased activation in lateral PFC and mPFC regions implicated in working memory, cognitive control, and self-monitoring, and in decreased activation of medial OFC and amygdala, regions implicated in emotion processing.

Using a similar paradigm, Phan et al. (2005) showed highly aversive and arousing pictures from the IAPS to healthy subjects, who were instructed to either "maintain" (feel naturally) or "suppress" (by positive reframing or rationalizing) negative affect. Successful reduction of negative affect was associated with increasing activation of dmPFC, dorsal ACC, dlPFC, lateral OFC, and ventrolateral PFC/inferior frontal gyrus, and with decreasing activity in the left nucleus accumbens, left lateral PFC, and left extended amygdala. Additionally, right dorsal ACC, right anterior insula, bilateral dlPFC, and bilateral ventrolateral PFC activity inversely correlated with the intensity of negative affect. These studies provide additional evidence for the existence of emotion regulatory (including lateral PFC, dmPFC, SMG, and ACC) and emotionally responsive regions (including amygdala, insula, medial OFC) in the human brain. The observed difficulty among patients with PTSD to cognitively regulate their emotions can be hypothesized to be a result of dysfunctional cognitive–emotional processes (such as cognitive appraisal and reappraisal) subserved by some of these regions. The therapeutic mechanisms of cognitive behavioral therapy in PTSD may also be related to these processes and structures. There is therefore a need to extend these innovative paradigms to the study of PTSD.

Self-relatedness and social emotional processing

The tendency to interpret or perceive stimuli as self-relevant is of specific interest in PTSD given that core manifestations of PTSD include feelings of threat and guilt, difficulties in interpersonal and social functioning, and the observation that interpersonal trauma results in the highest rates of PTSD (Kessler et al., 1995). Healthy social functioning is pivotal to the survival of humans and their progeny and serves a protective function with regard to stressors and disease. Thus it makes intuitive sense that this important function be subserved by dedicated neural resources for the processing of social stimuli. Primate and human lesion studies have implicated the mPFC, OFC, superior temporal sulcus, amygdala, and other regions in processing social and related stimuli.

We investigated the concept of self-relatedness in the context of emotional processing in a series of studies. In the first study, we used aversive, positive, and neutral IAPS pictures in a trial-related fMRI design to compare the neural substrates underlying the assessment of the emotional intensity of the pictures versus the self-relatedness of their content, in healthy, right-handed volunteers (Phan et al., 2004). Individualized subjective ratings over these two dimensions (obtained postscan) were correlated with brain activity in a parametric factorial analysis. The appraisal of self-relatedness specifically engaged the mPFC and recruited the dmPFC and insula as the extent of self-relatedness increased. On the other hand, the amygdala activation was specific to affective judgment of emotional intensity. Both increasing emotional intensity and self-relatedness activated the nucleus accumbens. These findings suggest that appraisal of self-relatedness specifically recruits the mPFC, a region relevant to the symptomatology and possibly pathophysiology of PTSD.

In a recent study we extended our investigation of the neural substrates of emotion to the processing of social versus nonsocial stimuli in 12 healthy, right-handed volunteers (Britton et al., 2006). In a novel paradigm, subjects' viewed short video segments that evoked positive or negative emotions that were categorized as "social" (humor, sadness) or "nonsocial" (appetite, physical disgust). Following the video, static frames extracted from the video were viewed for 30 s to help subjects maintain the emotions evoked by the video clip; during this period fMRI images were acquired. Nonsocial and social–emotional experiences resulted in partially overlapping but somewhat separate neural patterns. Social positive and social negative conditions activated amygdala/SLEA, superior temporal gyrus, hippocampus, and posterior cingulate, whereas nonsocial positive and nonsocial negative conditions activated insula and visual cortex. Additional activations depended on both social context and valence: amygdala (nonsocial negative); ACC (nonsocial positive and social negative); and OFC and nucleus accumbens (social positive). In another recent study (Chua et al., unpublished), we explored the neural correlates of viewing messages that are more or less tailored to the characteristics of the viewer, and thus are more or less "self-relevant". Active smokers were recruited and completed a baseline survey of their smoking habits. This survey was used to construct two sets of smoking cessation health messages; "High-tailored" messages incorporated information about the subject's personal smoking habits into the smoking cessation message, while "Low-tailored" messages were linguistically comparable but lacked subject-specific information. In a mixed block and event-related fMRI design, subjects were presented with both High-tailored and Low-tailored smoking cessation messages while fMRI images were acquired. Consistent with our hypothesis, preliminary data with nine subjects revealed greater activation in the vmPFC region when contrasting High-tailored events to Low-tailored events. These studies independently replicate using different paradigms involvement of mPFC and ACC regions in both self-relatedness and social cognition.

An important role for the processes involving self-relatedness and social–emotion and cognition is in establishing a context for other kinds of learning processes. Animal studies demonstrate that the same stimulus can acquire different incentive values depending on the context in which the stimulus is presented. One kind of context is the social context, i.e., whether the stimulus is presented in the context of the self or another individual. We explored this possibility in an fMRI study in which subjects kept track of both the value of a stimulus and the social context in which the stimulus is presented (Self vs. Other) (Ho et al., unpublished). Consistent with the animal literature, we found that distinct neural circuits are associated with processing of contextual information. Context-related processing was associated with the vmPFC, anterior insula, inferior frontal gyrus, lateral PFC, frontal pole, anterior superior temporal gyrus, temporal parietal junction, superior parietal lobule, parahippocampal gyrus, hippocampus, caudate, putamen, and midbrain. The anterior and posterior cingulate cortices were associated with both valuational and contextual processing.

Overall, these preceding studies and findings highlight the roles of self-relatedness and of sociality linking these processes to activation of ventral regions of mPFC and of ACC, as well as amygdala and other regions. The demonstrated link between ACC and mPFC function and PTSD in this context also raises an interesting possibility that abnormal functioning of these regions in PTSD might explain the well-documented deficit in social functioning in this condition.

Functional neuroanatomy of neuroendocrine stress regulation

Neurobiological research over the past few decades has consistently suggested abnormalities in stress response systems, such as limbic hypothalamic pituitary adrenal (LHPA) axis systems in PTSD (Liberzon et al., 1999a; Phan et al., 2004). Furthermore, the role of stress hormones such as cortisol in mediating cognitive and emotional processes is also gradually being elucidated (Erickson et al., 2003). However, cortical regulation of these stress systems and their abnormalities in PTSD remain unknown. This has led to intriguing recent efforts to integrate functional neuroanatomical findings with those from neurochemical, neurophysiological, and neuroendocrinological studies in PTSD. Ottowitz et al. (2004) investigated the neural correlates of adrenocorticotropic hormone (ACTH) and cortisol regulation in a SPECT study in healthy controls. Sadness was induced in eight healthy women, and brain activation patterns were correlated with ACTH and cortisol levels during mood induction linking ACTH with rCBF in the left anterior cingulate and right insular cortices and cortisol with rCBF in the left insula. In a resting state study in PTSD, Bonne et al. (2003) compared 11 subjects with PTSD to 17 trauma-exposed subjects without PTSD and 17 non-traumatized healthy controls using HMPAO SPECT 6 months after the trauma. They found that cortisol level in PTSD was negatively correlated with medial temporal lobe perfusion. Anterior cingulate perfusion and cortisol levels were positively correlated in PTSD and negatively correlated in trauma survivors without PTSD.

Recently, we conducted a [^{15}O] H_2O PET study of a series of emotional challenges (aversive pictures and autobiographic narratives) in 16 combat PP, 15 CC, and 14 NC. Voxel-wise analyses showed ACTH responses covaried with rCBF in rACC and right insula in PP and rostral anterior cingulate and dmPFC in CC. These findings suggest involvement of insula, dmPFC, and rACC in HPA axis responses to trauma-related stimuli. Interestingly pre-stimulus cortisol level covaried with rCBF in subgenual ACC in PP and rACC in CC. This suggests that rACC may be a site of modulation by circulating cortisol in trauma-exposed subjects. Differential patterns of covariation between combat veterans with and without PTSD implicate dmPFC and rACC as areas of dysregulation of HPA axis responses in PTSD (Liberzon et al., 2007).

Taken together, these findings suggest that specific prefrontal cortical regions, the ACC, the insula, and the amygdala are intimately involved in the activation and cortical modulation of neuroendocrine stress responses. The same regions are implicated in emotional processing in general and social emotions in particular and in the symptomatology of PTSD. Activation of neuroendocrine stress responses in turn appears to have a modulatory effect in some of these areas. This raises interesting possibility that abnormal activation of these regions reported in PTSD might reflect aberrant neuroendocrine stress responses.

The failure of stimulus contextualization: a core process in PTSD?

The preceding discussion covered a number of seemingly heterogeneous processes potentially involved in PTSD, including habituation, extinction, cognition–emotion interactions, social and self-related processing, and cortical modulation of the HPA axis. Intriguingly, all of these processes have been linked to activation of various regions in the medial wall of the prefrontal lobe, described as dorsal and ventral regions of mPFC and ACC. While it is possible that all these functions engage the medial wall of the prefrontal lobe independently, it might be worth engaging in a bit of speculation about possible underlying themes that bring together these disparate processes in the hopes of producing further insight into the fundamental pathophysiology of PTSD.

One way of bringing together the preceding processes is to understand them as aspects of a larger process that might be labeled "contextualization". In order to survive and reproduce, an organism must be able to select the appropriate response to a particular stimulus from a large range of available alternatives. The same stimulus might represent different "values" that necessitates different responses depending on the context in which it appears. This task requires that the organism attend to the contextual cues in the environment, which includes cues present in the external environment as well as cues present in the organism's internal metabolic/physiological environment, that disambiguate which response is most appropriate given the organism's particular situation. Contextualization refers to the process by which key dimensions of the situational context are appraised, represented, and used to guide the selection of action.

The studies reviewed in the preceding part of the chapter suggest that mPFC might play a key role in contextualization. Extinction, cognitive–emotional interactions, reappraisal, social cognition, and self-relatedness are all processes that are based on contextualization. These processes all reassess the stimulus value based on new learning, cognitive information, context, social environment, or self-relatedness. Furthermore one can see cortical modulation of neuroendocrine secretion as modifying the physiological "context" in which stimuli are processed. Interestingly a number of recent animal studies have linked mPFC regions to contextual modulation of cue value in rats (Haddon and Killcross, 2006). Other findings that associate mPFC with reversal learning (Fellows and Farah, 2003) are also highly consistent with the "contextualization" perspective on mPFC function, since reversal learning can be understood as learning a new and different meaning in a new context.

Thus, there appear to be at least three categories of contextual variables represented in mPFC, cognitive context, social context and internal context, with each being represented along a caudal-rostral gradient in the mPFC. Cognitive context is established by processes that judge the relevance of stimuli to the organism's memories and strategic goals. Studies of cognitive control of emotion and reappraisal reviewed above suggest that cognitive context is primarily represented in dmPFC as well as dlPFC. Social context is established by processes that judge the extent to which stimuli are self-related, i.e., whether they are like the self or are not like the self, and to what degree. Studies of self-relatedness and social emotions suggest that vmPFC and other regions of mPFC represent information about the social context of stimuli. Internal context refers the overall homeostatic state of the internal milieu, including the state of drives, metabolic states, and overall physiological load. Evidence indicates that a network of inter-related neocortical regions, especially the rostral anterior insula and OFC, are involved in interoceptive attention and assessing and representing the internal context (Craig, 2004).

If indeed ACC and mPFC are involved in context-setting or contextualization function, and ACC/mPFC deficits are present in PTSD, this might shed new light on the specific deficits and processes of symptom generation in this disorder. The prevalent model of PTSD emphasizes dysfunction in threat-assessing neural circuits as the core pathophysiological process of the disorder. A key feature of the model is that it proposes dysfunction in mPFC circuitry leading to inadequate top-down control over the amygdala. Here we propose an expansion of this model based on the notion that mPFC circuitry has a much broader range of functions than merely providing inhibitory control over the amygdala. Indeed some recent animal (Sierra-Mercado et al., 2006) and neuroimaging data (Milad et al., 2006) demonstrate that mPFC can have an activating or facilitating role with respect to amygdala. Here we propose that mPFC plays a more complex role in contextualizing stimuli in terms of cognitive, social and internal contexts, thus helping guide the selection of appropriate responses suited to particular features of the environment.

It follows that dysfunction in mPFC circuitry in this case would be predicted to produce a number of disparate problems, and these problems do in fact appear to be characteristic of PTSD. Failures in discriminating contextual cues might lead to the inappropriate expression of trauma-related memories and emotions, thus contributing to reexperiencing phenomena. Cues resembling trauma are not perceived in the current context but rather independent of it — as if trauma is the actual context. Poor contextual discrimination might also contribute to emotional numbing, understood as failure to experience emotions that are normally and appropriately experienced in a particular context. Difficulties in emotion regulation might manifest as the inability to cognitively reappraise stimuli as safe or not threatening. Abnormal or dysregulated signals of internal body context arising from the rostral anterior insula and associated mPFC regions may give rise to anxiety, rumination, and avoidance behaviors (see Paulus and Stein, 2006). Other problems might include difficulties in social interactions due to exaggerated or inappropriate judgments that stimuli are self-related. Overall, failure to contextualize might be considered a core process of PTSD, explaining a large number of features of the disorder that have heretofore eluded adequate explanation under the currently dominant model of PTSD that places principal emphasis on dysfunction in threat-related processing.

Summary and future directions

Neuroimaging studies of PTSD over the past decade have been based on a model that conceptualizes the disorder as a state of heightened responsivity to threatening stimuli and/or a state of insufficient inhibitory control over exaggerated threat-sensitivity. Consistent with this model, several studies have demonstrated reduced activation of the mPFC (BA 10 and 11) and ACC (BA 32) in PTSD subjects compared to traumatized controls. Other studies have reported increased responsivity of the amygdaloid region though findings are not always consistent (see Fig. 1 for a summary of neural regions implicated in PTSD). Findings may be influenced by several methodological issues, such as small sample sizes, heterogeneous populations, and varying imaging methods that limit broad generalization. Despite the progress made, existing models and findings are unable to fully capture the complexity of PTSD. Innovative paradigms being developed in cognitive and social neuroscience suggest novel directions for future work that can broaden our understanding of a range of pathophysiological processes in PTSD. Future directions of research include neuroimaging studies of fear conditioning, habituation, and extinction as well as studies investigating emotion

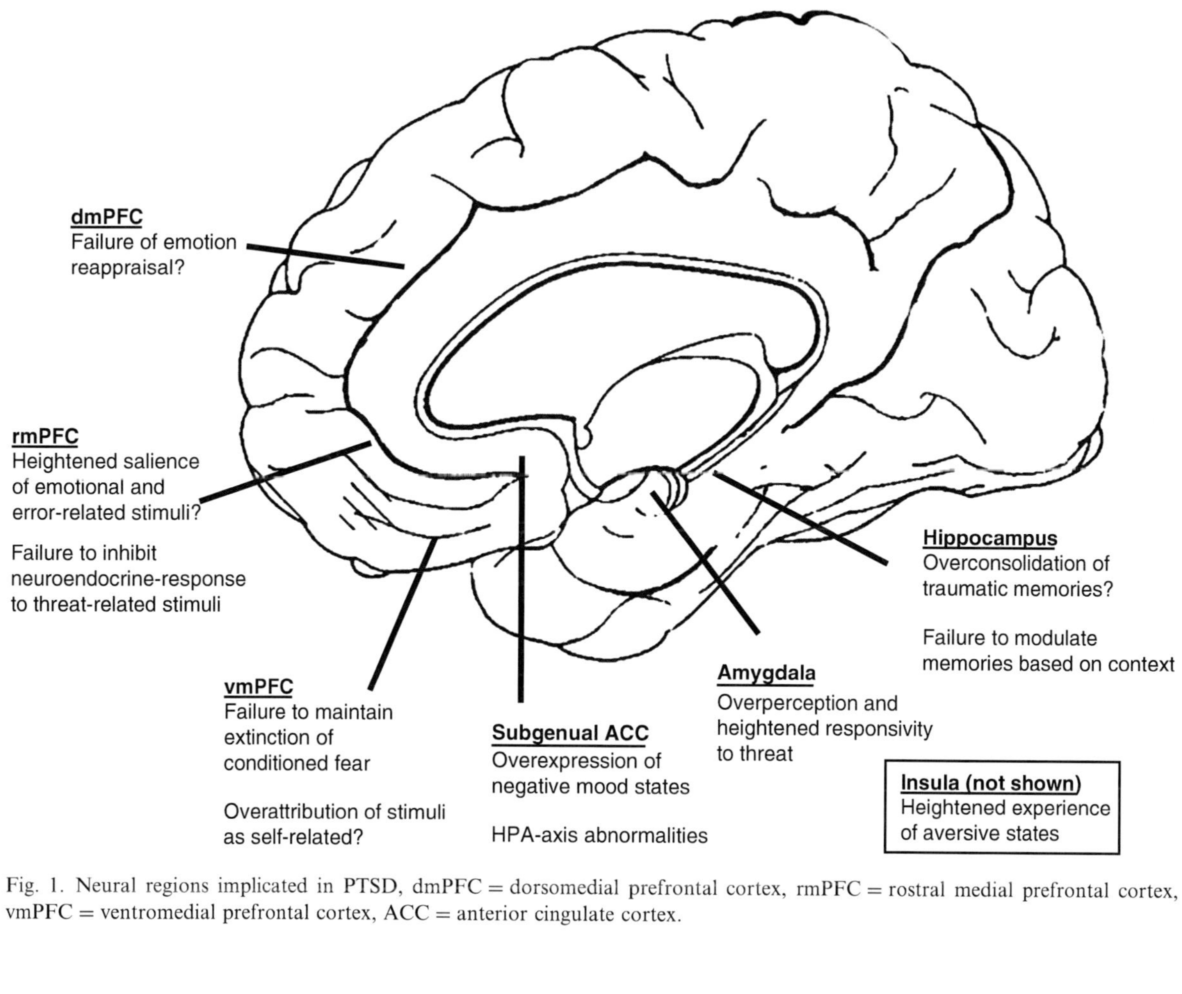

Fig. 1. Neural regions implicated in PTSD, dmPFC = dorsomedial prefrontal cortex, rmPFC = rostral medial prefrontal cortex, vmPFC = ventromedial prefrontal cortex, ACC = anterior cingulate cortex.

regulation processes, and self-related and social emotional processing. There is also a need for integrating different lines of inquiry, including genetic, neurochemical/receptor, HPA axis, and blood flow parameters in PTSD. This research holds the exciting promise of helping to identify neurobiological factors that may confer vulnerability or resilience to PTSD and offer meaningful clues to the pathophysiology of PTSD. This progress will be essential for the future development of effective prevention and treatment strategies for this disorder.

Abbreviations

ACC	anterior cingulate cortex
ACTH	adrenocorticotropic hormone
BA	Brodmann area
BOLD	blood oxygen level dependent
CAPS	Clinician-Administered PTSD Scale
CC	combat controls
CR	conditioned response
CS	conditioned stimulus
CSA	childhood sexual abuse
dlPFC	dorsolateral prefrontal cortex
dmPFC	dorsomedial prefrontal cortex
fMRI	functional magnetic resonance imaging
HMPAO	99mTc-hexamethylpropyleneamineoxime
IAPS	international affective pure system
LHPA	limbic hypothalamic pituitary adrenal
mPFC	medial prefrontal cortex
MVA	motor vehicle accident
NC	noncombat controls
OFC	orbitofrontal cortex
PCC	posterior cingulate cortex
PET	positron emission tomography
PFC	prefrontal cortex
PP	PTSD patients
PTSD	posttraumatic stress disorder
rACC	rostral anterior cingulate cortex
rCBF	regional cerebral blood flow
SLEA	sublenticular extended amygdala
SMG	supramarginal gyrus
SPECT	single-photon emission tomography
US	unconditioned stimulus
vmPFC	ventromedial prefrontal cortex

References

Armony, J.L., Corbo, V., Clement, M.H. and Brunet, A. (2005) Amygdala response in patients with acute PTSD to masked and unmasked emotional facial expressions. Am. J. Psychiatry, 162: 1961–1963.

Bonne, O., Gilboa, A., Louzoun, Y., Brandes, D., Yona, I., Lester, H., Barkai, G., Freedman, N., Chisin, R. and Shalev, A.Y. (2003) Resting regional cerebral perfusion in recent posttraumatic stress disorder. Biol. Psychiatry, 54: 1077–1086.

Breiter, H.C., Etcoff, N.L., Whalen, P.J., Kennedy, W.A., Rauch, S.L., Buckner, R.L., Strauss, M.M., Hyman, S.E. and Rosen, B.R. (1996) Response and habituation of the human amygdala during visual processing of facial expression. Neuron, 17: 875–887.

Bremner, J.D., Narayan, M., Staib, L.H., Southwick, S.M., Mcglashan, T. and Charney, D.S. (1999a) Neural correlates of memories of childhood sexual abuse in women with and without posttraumatic stress disorder. Am. J. Psychiatry, 156: 1787–1795.

Bremner, J.D., Staib, L.H., Kaloupek, D., Southwick, S.M., Soufer, R. and Charney, D.S. (1999b) Neural correlates of exposure to traumatic pictures and sound in Vietnam combat veterans with and without posttraumatic stress disorder: a positron emission tomography study. Biol. Psychiatry, 45: 806–816.

Bremner, J.D., Vermetten, E., Vythilingam, M., Afzal, N., Schmahl, C., Elzinga, B. and Charney, D.S. (2004) Neural correlates of the classic color and emotional stroop in women with abuse-related posttraumatic stress disorder. Biol. Psychiatry, 55: 612–620.

Britton, J.C., Phan, K.L., Taylor, S.F., Fig, L.M. and Liberzon, I. (2005) Corticolimbic blood flow in posttraumatic stress disorder during script-driven imagery. Biol. Psychiatry, 57: 832–840.

Britton, J.C., Phan, K.L., Taylor, S.F., Welsh, R.C., Berridge, K.C. and Liberzon, I. (2006) Neural correlates of social and nonsocial emotions: an fMRI study. Neuroimage, 31: 397–409.

Buchel, C. and Dolan, R.J. (2000) Classical fear conditioning in functional neuroimaging. Curr. Opin. Neurobiol., 10: 219–223.

Buchel, C., Dolan, R.J., Armony, J.L. and Friston, K.J. (1999) Amygdala-hippocampal involvement in human aversive trace conditioning revealed through event-related functional magnetic resonance imaging. J. Neurosci., 19: 10869–10876.

Buchel, C., Morris, J., Dolan, R.J. and Friston, K.J. (1998) Brain systems mediating aversive conditioning: an event-related fMRI study. Neuron, 20: 947–957.

Bush, G., Luu, P. and Posner, M.I. (2000) Cognitive and emotional influences in anterior cingulate cortex. Trends Cogn. Sci., 4: 215–222.

Corcoran, K.A. and Maren, S. (2001) Hippocampal inactivation disrupts contextual retrieval of fear memory after extinction. J. Neurosci., 21: 1720–1726.

Craig, A.D. (2004) Human feelings: why are some more aware than others? Trends Cogn. Sci., 8: 239–241.

Davis, M. and Whalen, P.J. (2001) The amygdala: vigilance and emotion. Mol. Psychiatry, 6: 13–34.

Eichenbaum, H. (2000) A cortical-hippocampal system for declarative memory. Nat. Rev. Neurosci., 1: 41–50.

Erickson, K., Drevets, W. and Schulkin, J. (2003) Glucocorticoid regulation of diverse cognitive functions in normal and pathological emotional states. Neurosci. Biobehav. Rev., 27: 233–246.

Falls, W.A., Miserendino, M.J. and Davis, M. (1992) Extinction of fear-potentiated startle: blockade by infusion of an NMDA antagonist into the amygdala. J. Neurosci., 12: 854–863.

Fellows, L.K. and Farah, M.J. (2003) Ventromedial frontal cortex mediates affective shifting in humans: evidence from a reversal learning paradigm. Brain, 126: 1830–1837.

Friston, K.J., Frith, C.D., Fletcher, P., Liddle, P.F. and Frackowiak, R.S. (1996) Functional topography: multidimensional scaling and functional connectivity in the brain. Cereb. Cortex, 6: 156–164.

Friston, K.J., Frith, C.D., Liddle, P.F. and Frackowiak, R.S. (1993) Functional connectivity: the principal-component analysis of large (PET) data sets. J. Cereb. Blood Flow Metab., 13: 5–14.

Geuze, E., Westenberg, H.G., Jochims, A., De Kloet, C.S., Bohus, M., Vermetten, E. and Schmahl, C. (2007) Altered pain processing in veterans with posttraumatic stress disorder. Arch. Gen. Psychiatry, 64: 76–85.

Gilboa, A., Shalev, A.Y., Laor, L., Lester, H., Louzoun, Y., Chisin, R. and Bonne, O. (2004) Functional connectivity of the prefrontal cortex and the amygdala in posttraumatic stress disorder. Biol. Psychiatry, 55: 263–272.

Gross, J.J. (1998) The emerging field of emotion regulation: an integrative review. Rev. Gen. Psychol., 2: 271–299.

Haddon, J.E. and Killcross, S. (2006) Prefrontal cortex lesions disrupt the contextual control of response conflict. J. Neurosci., 26: 2933–2940.

Hariri, A.R., Bookheimer, S.Y. and Mazziotta, J.C. (2000) Modulating emotional responses: effects of a neocortical network on the limbic system. Neuroreport, 11: 43–48.

Hariri, A.R., Mattay, V.S., Tessitore, A., Fera, F. and Weinberger, D.R. (2003) Neocortical modulation of the amygdala response to fearful stimuli. Biol. Psychiatry, 53: 494–501.

Hendler, T., Rotshtein, P. and Hadar, U. (2001) Emotion-perception interplay in the visual cortex: the eyes follow the heart. Cell. Mol. Neurobiol., 21: 733–752.

Kessler, R.C., Sonnega, A., Bromet, E., Hughes, M. and Nelson, C.B. (1995) Posttraumatic stress disorder in the National Comorbidity Survey. Arch. Gen. Psychiatry, 52: 1048–1060.

LaBar, K.S., Gatenby, J.C., Gore, J.C., Ledoux, J.E. and Phelps, E.A. (1998) Human amygdala activation during conditioned fear acquisition and extinction: a mixed-trial fMRI study. Neuron, 20: 937–945.

Lang, P.J., Bradley, M.M. and Cuthbert, B.N. (1997) International Affective Picture System (IAPS): Technical Manual and Affective Ratings. NIMH Center for the Study of Emotion and Attention, University of Florida, Gainesville, FL.

Lanius, R.A., Williamson, P.C., Boksman, K., Densmore, M., Gupta, M., Neufeld, R.W., Gati, J.S. and Menon, R.S. (2002) Brain activation during script-driven imagery induced dissociative responses in PTSD: a functional magnetic resonance imaging investigation. Biol. Psychiatry, 52: 305–311.

Lanius, R.A., Williamson, P.C., Densmore, M., Boksman, K., Gupta, M.A., Neufeld, R.W., Gati, J.S. and Menon, R.S. (2001) Neural correlates of traumatic memories in posttraumatic stress disorder: a functional MRI investigation. Am. J. Psychiatry, 158: 1920–1922.

Lanius, R.A., Williamson, P.C., Densmore, M., Boksman, K., Neufeld, R.W., Gati, J.S. and Menon, R.S. (2004) The nature of traumatic memories: a 4-T FMRI functional connectivity analysis. Am. J. Psychiatry, 161: 36–44.

Lanius, R.A., Williamson, P.C., Hopper, J., Densmore, M., Boksman, K., Gupta, M.A., Neufeld, R.W., Gati, J.S. and Menon, R.S. (2003) Recall of emotional states in posttraumatic stress disorder: an fMRI investigation. Biol. Psychiatry, 53: 204–210.

Ledoux, J.E. (2000) Emotion circuits in the brain. Annu. Rev. Neurosci., 23: 155–184.

Liberzon, I., Abelson, J.L., Flagel, S.B., Raz, J. and Young, E.A. (1999a) Neuroendocrine and psychophysiologic responses in PTSD: a symptom provocation study. Neuropsychopharmacology, 21: 40–50.

Liberzon, I., King, A.P., Britton, J.C., Phan, K.-L., Abelson, J.L. and Taylor, S.F. (2007) Paralimbic and medial prefrontal cortical involvement in neuroendocrine responses to traumatic stimuli. Am. J. Psychiatry, 164: 1250–1258.

Liberzon, I. and Phan, K.L. (2003) Brain-imaging studies of posttraumatic stress disorder. CNS Spectr., 8: 641–650.

Liberzon, I., Taylor, S.F., Amdur, R., Jung, T.D., Chamberlain, K.R., Minoshima, S., Koeppe, R.A. and Fig, L.M. (1999b) Brain activation in PTSD in response to trauma-related stimuli. Biol. Psychiatry, 45: 817–826.

Liberzon, I., Taylor, S.F., Phan, K.L., Britton, J.C., Fig, L.M., Bueller, J.A., Koeppe, R.A. and Zubieta, J.K. (2007) Altered central mu-opioid receptor binding after psychological trauma. Biol. Psychiatry, 61: 1030–1038.

Milad, M.R., Rauch, S.L., Pitman, R.K. and Quirk, G.J. (2006) Fear extinction in rats: implications for human brain imaging and anxiety disorders. Biol Psychol, 73: 61–71.

Morgan, M.A. and Ledoux, J.E. (1995) Differential contribution of dorsal and ventral medial prefrontal cortex to the acquisition and extinction of conditioned fear in rats. Behav. Neurosci., 109: 681–688.

Morgan, M.A., Romanski, L.M. and Ledoux, J.E. (1993) Extinction of emotional learning: contribution of medial prefrontal cortex. Neurosci. Lett., 163: 109–113.

Morgan, M.A., Schulkin, J. and Ledoux, J.E. (2003) Ventral medial prefrontal cortex and emotional perseveration: the memory for prior extinction training. Behav. Brain. Res., 146: 121–130.

Morris, J.S., Ohman, A. and Dolan, R.J. (1998) Conscious and unconscious emotional learning in the human amygdala. Nature, 393: 467–470.

Ochsner, K.N., Bunge, S.A., Gross, J.J. and Gabrieli, J.D. (2002) Rethinking feelings: an FMRI study of the cognitive regulation of emotion. J. Cogn. Neurosci., 14: 1215–1229.

Osuch, E.A., Benson, B., Geraci, M., Podell, D., Herscovitch, P., Mccann, U.D. and Post, R.M. (2001) Regional cerebral blood flow correlated with flashback intensity in patients with posttraumatic stress disorder. Biol. Psychiatry, 50: 246–253.

Ottowitz, W.E., Dougherty, D.D., Sirota, A., Niaura, R., Rauch, S.L. and Brown, W.A. (2004) Neural and endocrine correlates of sadness in women: implications for neural network regulation of HPA activity. J. Neuropsychiatry Clin. Neurosci., 16: 446–455.

Paulus, M.P. and Stein, M.B. (2006) An insular view of anxiety. Biol. Psychiatry, 60: 383–387.

Phan, K.L., Fitzgerald, D.A., Nathan, P.J., Moore, G.J., Uhde, T.W. and Tancer, M.E. (2005) Neural substrates for voluntary suppression of negative affect: a functional magnetic resonance imaging study. Biol. Psychiatry, 57: 210–219.

Phan, K.L., Liberzon, I., Welsh, R.C., Britton, J.C. and Taylor, S.F. (2003) Habituation of rostral anterior cingulate cortex to repeated emotionally salient pictures. Neuropsychopharmacology, 28: 1344–1350.

Phan, K.L., Taylor, S.F., Welsh, R.C., Ho, S.H., Britton, J.C. and Liberzon, I. (2004) Neural correlates of individual ratings of emotional salience: a trial-related fMRI study. Neuroimage, 21: 768–780.

Phelps, E.A., Delgado, M.R., Nearing, K.I. and Ledoux, J.E. (2004) Extinction learning in humans: role of the amygdala and vmPFC. Neuron, 43: 897–905.

Pitman, R.K., Shin, L.M. and Rauch, S.L. (2001) Investigating the pathogenesis of posttraumatic stress disorder with neuroimaging. J. Clin. Psychiatry, 62(Suppl 17): 47–54.

Protopopescu, X., Pan, H., Tuescher, O., Cloitre, M., Goldstein, M., Engelien, W., Epstein, J., Yang, Y., Gorman, J., Ledoux, J., Silbersweig, D. and Stern, E. (2005) Differential time courses and specificity of amygdala activity in posttraumatic stress disorder subjects and normal control subjects. Biol. Psychiatry, 57: 464–473.

Rauch, S.L. and Shin, L.M. (1997) Functional neuroimaging studies in posttraumatic stress disorder. Ann. N.Y. Acad. Sci., 821: 83–98.

Rauch, S.L., Van Der Kolk, B.A., Fisler, R.E., Alpert, N.M., Orr, S.P., Savage, C.R., Fischman, A.J., Jenike, M.A. and Pitman, R.K. (1996) A symptom provocation study of posttraumatic stress disorder using positron emission tomography and script-driven imagery. Arch. Gen. Psychiatry, 53: 380–387.

Rauch, S.L., Whalen, P.J., Shin, L.M., Mcinerney, S.C., Macklin, M.L., Lasko, N.B., Orr, S.P. and Pitman, R.K. (2000) Exaggerated amygdala response to masked facial stimuli in posttraumatic stress disorder: a functional MRI study. Biol. Psychiatry, 47: 769–776.

Rosenkranz, J.A. and Grace, A.A. (2002) Cellular mechanisms of infralimbic and prelimbic prefrontal cortical inhibition and dopaminergic modulation of basolateral amygdala neurons in vivo. J. Neurosci., 22: 324–337.

Shaw, M.E., Strother, S.C., Mcfarlane, A.C., Morris, P., Anderson, J., Clark, C.R. and Egan, G.F. (2002) Abnormal functional connectivity in posttraumatic stress disorder. Neuroimage, 15: 661–674.

Shin, L.M., Mcnally, R.J., Kosslyn, S.M., Thompson, W.L., Rauch, S.L., Alpert, N.M., Metzger, L.J., Lasko, N.B., Orr, S.P. and Pitman, R.K. (1999) Regional cerebral blood flow during script-driven imagery in childhood sexual abuse-related PTSD: a PET investigation. Am. J. Psychiatry, 156: 575–584.

Shin, L.M., Orr, S.P., Carson, M.A., Rauch, S.L., Macklin, M.L., Lasko, N.B., Peters, P.M., Metzger, L.J., Dougherty, D.D., Cannistraro, P.A., Alpert, N.M., Fischman, A.J. and Pitman, R.K. (2004a) Regional cerebral blood flow in the amygdala and medial prefrontal cortex during traumatic imagery in male and female Vietnam veterans with PTSD. Arch. Gen. Psychiatry, 61: 168–176.

Shin, L.M., Shin, P.S., Heckers, S., Krangel, T.S., Macklin, M.L., Orr, S.P., Lasko, N., Segal, E., Makris, N., Richert, K., Levering, J., Schacter, D.L., Alpert, N.M., Fischman, A.J., Pitman, R.K. and Rauch, S.L. (2004b) Hippocampal function in posttraumatic stress disorder. Hippocampus, 14: 292–300.

Shin, L.M., Whalen, P.J., Pitman, R.K., Bush, G., Macklin, M.L., Lasko, N.B., Orr, S.P., Mcinerney, S.C. and Rauch, S.L. (2001) An fMRI study of anterior cingulate function in posttraumatic stress disorder. Biol. Psychiatry, 50: 932–942.

Shin, L.M., Wright, C.I., Cannistraro, P.A., Wedig, M.M., Mcmullin, K., Martis, B., Macklin, M.L., Lasko, N.B., Cavanagh, S.R., Krangel, T.S., Orr, S.P., Pitman, R.K., Whalen, P.J. and Rauch, S.L. (2005) A functional magnetic resonance imaging study of amygdala and medial prefrontal cortex responses to overtly presented fearful faces in posttraumatic stress disorder. Arch. Gen. Psychiatry, 62: 273–281.

Sierra-Mercado, Jr., D., Corcoran, K.A., Lebron-Milad, K. and Quirk, G.J. (2006) Inactivation of the ventromedial prefrontal cortex reduces expression of conditioned fear and impairs subsequent recall of extinction. Eur J Neurosci, 24: 1751–1758.

Taylor, S.F., Phan, K.L., Decker, L.R. and Liberzon, I. (2003) Subjective rating of emotionally salient stimuli modulates neural activity. Neuroimage, 18: 650–659.

Whalen, P.J., Rauch, S.L., Etcoff, N.L., Mcinerney, S.C., Lee, M.B. and Jenike, M.A. (1998) Masked presentations of emotional facial expressions modulate amygdala activity without explicit knowledge. J. Neurosci., 18: 411–418.

Williams, L.M., Kemp, A.H., Felmingham, K., Barton, M., Olivieri, G., Peduto, A., Gordon, E. and Bryant, R.A. (2006) Trauma modulates amygdala and medial prefrontal responses to consciously attended fear. Neuroimage, 29: 347–357.

Wright, C.I., Fischer, H., Whalen, P.J., Mcinerney, S.C., Shin, L.M. and Rauch, S.L. (2001) Differential prefrontal cortex and amygdala habituation to repeatedly presented emotional stimuli. Neuroreport, 12: 379–383.

Discussion: Chapter 11

PITMAN: Very interesting talk. It is starting to raise an alternative view of what might be going on in the brain areas during PTSD besides fear conditioning. But at the end I think I heard you say something about contextualization as an alternative to mPFC inhibiting the amygdala. That did not sound quite right to me because inhibition of the amygdala is determined by a lot of neural inputs. So perhaps not being able to recognize the safe context in people with PTSD results in a failure to inhibit the amygdala.

LIBERZON: It is not that the process cannot include inhibition of amygdala. I think the contextualization is a larger process and you can have a failure to contextualize even in absence of amygdala activation ... So I am not saying that inhibition of amygdala can not be part of the PTSD picture. I am just saying that there is much more to the mPFC failure than inability to inhibit enhanced amygdala responses. It might be an inability to encode different information completely or to have appropriate communication between the hippocampus and mPFC, which has little to do with amygdala or fear responses. So, I guess I am just arguing that even in rats, mPFC which is really small, is doing more than simply inhibiting amygdala. It may, as you know, actually enhance amygdala responses in some cases, from the prelimbic region. I am suggesting that in human we have developed a very large mPFC because we have to deal with many contingencies, more than rats. We have to deal with social environment which is much more complex and requires new definitions of the stimuli within the social context, which is often much more important to humans than spatial context. mPFC is the region that is likely to be involved in performance of this function, and in PTSD it might not be functioning so well.

YEHUDA: Persons with PTSD are usually frightened when there is no need to be, but that this experience of being frightened without reason distresses them greatly. Does your finding speak to this? That is, the amygdala would be activated, but the person isn't sure why they are afraid, and then the mPFC wouldn't necessarily provide inhibitory feedback for this because it hasn't in a sense received an appropriate 'answer' from the hippocampus or other structures that provide context.

LIBERZON: So in a way it is describing the emotional experience "out of the context". It is the same way that stimulus is perceived out of context. Thus, there is a mismatch, an inability to match well the emotional experience and a particular stimulus.

YEHUDA: What type of brain abnormality would you see under these circumstances and how would you begin to differentiate that, from fear conditioning under situations where it is absolutely legitimate and appropriate to have fear right now?

LIBERZON: Right. So one example is the use of animal models. We have an animal model whereby, we compare contextual fear with conditioned fear, and examine acoustic startle. In our model conditioned freezing is pretty stable, however there is evidence for potentially abnormal contextual conditioning. So the effect we see is expressed in contextual processing but not in fear conditioning. Also in a manuscript that just came out, I believe in the Journal of Neuroscience, authors demonstrated that if you make lesions to medial prefrontal cortex, the rat is not able to discriminate between the two conditioned cues. The animal cannot use contextual information to know whether a particular cue will be rewarding, or will be threatening. I guess I am suggesting to use the same type of paradigm, to try and focus on the question of which component of the context is the most meaningful, how can we isolate it, and how to separate between the contextual cues and the first order conditioning cues.

E.R. de Kloet, M.S. Oitzl & E. Vermetten (Eds.)
Progress in Brain Research, Vol. 167
ISSN 0079-6123

CHAPTER 12

Structural and functional plasticity of the human brain in posttraumatic stress disorder

J. Douglas Bremner[1,*], Bernet Elzinga[2], Christian Schmahl[3] and Eric Vermetten[4]

[1]*Departments of Psychiatry and Behavioral Sciences and Radiology, Emory University School of Medicine, Atlanta, GA, USA and the Atlanta VAMC, Atlanta, GA, USA*
[2]*University of Leiden, Section of Clinical and Health Psychology, Leiden, The Netherlands*
[3]*Department of Psychosomatic Medicine and Psychotherapy, Central Institute of Mental Health, Mannheim, Germany*
[4]*Dutch Military Hospital and University of Utrecht, Utrecht, The Netherlands*

Abstract: Posttraumatic stress disorder (PTSD) is associated with long-term changes in neurobiology. Brain areas involved in the stress response include the medial prefrontal cortex, hippocampus, and amygdala. Neurohormonal systems that act on the brain areas to modulate PTSD symptoms and memory include glucocorticoids and norepinephrine. Dysfunction of these brain areas is responsible for the symptoms of PTSD. Brain imaging studies show that PTSD patients have increased amygdala reactivity during fear acquisition. Other studies show smaller hippocampal volume. A failure of medial prefrontal/anterior cingulate activation with re-experiencing of the trauma is hypothesized to represent a neural correlate of the failure of extinction seen in PTSD. The brain has the capacity for plasticity in the aftermath of traumatic stress. Antidepressant treatments and changes in environment can reverse the effects of stress on hippocampal neurogenesis, and humans with PTSD showed increased hippocampal volume with both paroxetine and phenytoin.

Keywords: PET; depression; cortisol; glucocorticoids; stress; PTSD

Introduction

Childhood abuse is a pervasive problem that is often associated with lasting psychopathology. For instance, 16% of women have a history of childhood sexual abuse (rape or fondling) based on nationwide surveys (McCauley et al., 1997). Ten percent of women (13 million) suffer from posttraumatic stress disorder (PTSD) at some time in their lives (Kessler et al., 1995), and PTSD is twice as common in women as in men. Childhood sexual abuse is the most common cause of PTSD in women (Kessler et al., 1995). This paper reviews the long-term effects of childhood abuse on brain and neurobiology, as well as the functional plasticity of the brain in the aftermath of trauma. Findings are reviewed in PTSD and other mental disorders related to early abuse, including borderline personality disorder (BPD) and dissociative identity disorder (DID).

Psychological effects of trauma

Trauma results in a range of mental symptoms, including PTSD, BPD, DID, substance abuse,

*Corresponding author. Tel.: +1 (404) 712 9569; Fax: +1 (404) 712-8442; E-mail: jdbremn@emory.edu

DOI: 10.1016/S0079-6123(07)67012-5

anxiety, and depression. Most of the research has been done in PTSD patients, however these patients frequently have co-morbid symptoms with these other disorders, which led to the use of the term "trauma-spectrum disorders" (Bremner, 2002). Risk factors for PTSD include prior history of stress, low years of education, prior psychiatric history, young age, and lack of social support (Bremner et al., 1995c). In one study Vietnam combat veterans with a history of childhood abuse had fourfold increased relative risk of PTSD (Bremner et al., 1993b). Childhood abuse was the factor most strongly associated with risk for PTSD, even after controlling for level of combat exposure, months in Vietnam, and participation in atrocities. Twin studies also show that there is a genetic contribution to PTSD risk (Goldberg et al., 1990).

Effects of stress on memory and the hippocampus

Studies in animals show that stress impacts adversely on the brain, especially on the hippocampus. Stress, acting through increased excitatory amino acids, decreased brain-derived neurotrophic factor (BDNF), and/or increased glucocorticoids, is associated with a loss of branching of neurons in the hippocampus and an inhibition of hippocampal neurogenesis (Uno et al., 1989; Sapolsky et al., 1990; Nibuya et al., 1995; Smith et al., 1995; Sapolsky, 1996; Duman et al., 1997). These effects are reversed by a variety of antidepressant treatments (Malberg et al., 2000; Duman et al., 2001; Santarelli et al., 2003; Duman, 2004). In addition, an enriched environment has been shown to promote hippocampal neurogenesis (Kempermann et al., 1997, 1998).

Consistent with the effects of stress on brain structures that mediate memory, including the hippocampus, prefrontal cortex, and amygdala, PTSD is associated with a wide range of memory deficits (Bremner, 2003). Memory can be categorized as declarative (memory for facts or lists, mediated in part by the hippocampus) or nondeclarative (memory for things like riding a bike, or conditioned responses) (Schacter, 1996). PTSD patients show deficits in declarative memory, enhanced responses to conditioning, and perseverative errors (possibly related to frontal lobe dysfunction) (Elzinga and Bremner, 2002).

Studies in PTSD showed deficits in hippocampal function as measured with neuropsychological tests of declarative memory function (Bremner et al., 1993a, 1995a, 2004b; Uddo et al., 1993; Yehuda et al., 1995; Vasterling et al., 2002, 2006; Vasterling and Bremner, 2006). One recent study showed a decline in verbal declarative memory function from before to after Iraq deployment, showing that combat exposure resulted in changes in cognitive function (Vasterling et al., 2006). Several studies have also shown smaller hippocampal volume and/or *N*-acetyl aspartate (NAA, a marker of neuronal integrity) measured with magnetic resonance imaging (MRI) in PTSD (Bremner et al., 1995b, 1997b, 2003c; Stein et al., 1997; Freeman et al., 1998; Schuff et al., 2001; Villarreal et al., 2002; Lindauer et al., 2004; Shin et al., 2004; Kitayama et al., 2005; Vythilingam et al., 2005; Jatzko et al., 2006). Two recent meta-analyses showed that this effect was seen for both left and right hippocampus, and was seen equally in men and women (Kitayama et al., 2005; Smith, 2005; Jatzko et al., 2006). However effects were only seen in adults (including those with early life stress) and not in children (De Bellis et al., 1999, 2001; Carrion et al., 2001). Findings from animal studies in fact show that early life stress may not have an immediate effect on the hippocampus, but may only manifest during the adult phase of development (Brunson et al., 2001).

Bremner has outlined a model of trauma-spectrum disorders (Bremner, 2002). These psychiatric disorders, ranging from depression to BPD, DID and PTSD, are all linked to stress and share (at least in part) common bases in the brain. Studies in these disorders in fact show that exposure to early childhood abuse is associated with smaller hippocampal volume, including depression (Vythilingam et al., 2002), PTSD (Bremner et al., 1997b, 2003c), BPD (Driessen et al., 2000; Schmahl et al., 2003), and DID (Vermetten et al., 2006a). In addition, these disorders are associated with increased cortisol response to symptom provoking stressors for PTSD (Bremner et al., 2003a; Elzinga et al., 2003) and BPD (Elzinga et al., unpublished data, 12/12/06).

BPD (Driessen et al., 2000; Schmahl et al., 2003) and DID (Vermetten et al., 2006a) (but not PTSD) are also associated with smaller amygdala volume.

Stress and neurohormonal systems

Alterations in the hypothalamic-pituitary-adrenal (HPA) axis have also been associated with stress-related psychiatric disorders. Corticotropin releasing factor (CRF) plays an important role in the stress response. Chronic stress exposure is associated with increases in CRF in animal studies (Arborelius et al., 1999). Central CRF administration is associated with fear-related behaviors (decreased exploration, increased startle, decreased grooming). Stress-induced lesions of the hippocampus result in a removal of inhibition of CRF release from the hypothalamus. Other findings from animal studies include a blunted adrenocorticotropin hormone (ACTH) response to CRF challenge, increased cortisol in the periphery, and resistance to negative feedback of dexamethasone (Arborelius et al., 1999). Two studies have shown increased concentrations of CRF in PTSD (Bremner et al., 1997a; Baker et al., 1999). Some studies (Yehuda et al., 1991b, 1994, 1996), but not others (Young and Breslau, 2004a, b) found decreased cortisol in 24 h urines or in diurnal salivary samples. Two studies using comprehensive measurement of plasma cortisol at multiple time points found lower cortisol concentrations in the afternoon (Yehuda et al., 1996; Bremner et al., 2007). Women with early childhood sexual abuse and PTSD were found to have lower afternoon cortisol and an increase in cortisol pulsatility compared to controls (Bremner et al., 2007). Other studies found increased lymphocyte glucocorticoid receptors (Yehuda et al., 1991a), super-suppression of cortisol with low-dose (0.5 mg) dexamethasone (Yehuda et al., 1993), blunted ACTH response to CRF, increased cortisol response to stressors (Bremner et al., 2003a) and to traumatic reminders of early trauma (Elzinga et al., 2003). Women with depression and early trauma also had increased cortisol response to public speaking (Heim et al., 2000).

There has been considerable interest in the relationship between stress, aging, and dehydroepiandosterone (DHEA). DHEA declines with aging (Orentreich et al., 1992; Barrett-Connor and Edelstein, 1994; Flynn, 1999; Johnson et al., 2002) and there has been considerable interest in the ability of DHEA supplements to block the normal effects of aging, although there is no convincing data that DHEA has such effects. DHEA also is important in the stress response. Chronic stress increases DHEA and DHEA-S (Fuller et al., 1984). DHEA also has antistress effects, blocking the effects of glucocorticoids on peripheral tissues as well as the hippocampus (Kimonides et al., 1998; Kaminska et al., 2000) and decreasing anxiety (Prasad et al., 1997). Studies of DHEA in patients with stress-related psychiatric disorders are contradictory, while studies in adult depressed patients showed both increases (Heuser et al., 1998) and decreases (Goodyer et al., 1996; Herbert et al., 1996) as well as no change (Michael et al., 2000; Young et al., 2002) in levels. Studies of DHEA in PTSD have been equally contradictory, with one study citing lower concentrations relative to controls (Kanter et al., 2001) while the other showed elevations (Spivak et al., 2000). We recently measured DHEA and DHEA-S at multiple time points over a 24 h period in women with early abuse and PTSD, and found elevations in both DHEA and DHEA-S (Bremner et al., 2007).

We performed a comprehensive assessment of memory, cortisol, DHEA, and the hippocampus in women with sexual abuse before 13, with and without PTSD, and healthy nonabused women. All subjects underwent assessment of hippocampal structure with MRI, assessment of hippocampal function with PET in conjunction with a paragraph encoding declarative memory task, assessment of HPA axis function at baseline and with a stressful challenge, and neuropsychological testing of declarative memory function. Early childhood sexual abuse before the age of 13 was defined as rape or molestation as assessed with the Early Trauma Inventory (Bremner et al., 2000). All subjects were free of psychotropic medication for 4 weeks before study.

Women with a history of early childhood sexual abuse and the diagnosis of PTSD ($N = 10$) were compared to abused non-PTSD women ($N = 12$) for hippocampal function using PET. All subjects were scanned during encoding of a paragraph and control task in conjunction with injection of 0–15 water and PET imaging of the brain. MR images were obtained for measurement of hippocampal volume, with an additional group of nonabused normal women (total $N = 33$). Subjects ($N = 56$) were also admitted to the GCRC for a 24 h period, for measurement of plasma cortisol, DHEA, and estradiol measured at 15 min intervals for 24 h. Salivary cortisol was measured after reading of a traumatic script related to personalized childhood abuse experiences.

In addition, salivary cortisol was measured before and after a 20 min cognitive challenge (arithmetic, color-word naming, problem solving under time pressure, and negative feedback). Women with abuse and PTSD had smaller hippocampal volume (Bremner et al., 2003b), a failure of hippocampal activation with declarative memory tasks (Bremner et al., 2003b), lower plasma cortisol concentrations in the afternoon (Bremner et al., 2007), increased cortisol pulsatility (Bremner et al., 2007), increased plasma DHEA concentrations (Bremner et al., 2007), increased cortisol response to stress (Bremner et al., 2003a), increased cortisol response to traumatic reminders (Elzinga et al., 2003), and impaired declarative memory measured with neuropsychological testing (Bremner et al., 2004b).

Neurohormonal modulation of memory

Glucocorticoids affect learning and memory. Elevations of glucocorticoids within the physiological range result in reversible deficits in memory function in animals (Oitzl and de Kloet, 1992; Bodnoff et al., 1995) as well as human subjects (Newcomer et al., 1994, 1999; Kirschbaum et al., 1996; Lupien et al., 1997, 1999, 2002; de Quervain et al., 2000; Wolf et al., 2001). Glucocorticoids released during stress, possibly acting through the hippocampus, may explain in part the acutely reversible as well as chronic effects that stress has on declarative memory (Kirschbaum et al., 1996; Porter and Landfield, 1998; de Kloet et al., 1999; Wolf, 2003). Greater deficits are seen in younger subjects in comparison to older subjects, hypothesized to be secondary to age-related decreases in glucocorticoid receptor density (Newcomer et al., 1995). Impairment of working memory by glucocorticoids may require noradrenergic stimulation to have its effect (Elzinga and Roelofs, 2005). We used a protocol of 1 mg of dexamethasone, followed by 2 mg one day later, and found an impairment in declarative memory function (percent retention of a paragraph after a delay) in healthy subjects, but not patients with depression (Bremner et al., 2004d) or PTSD (Bremner et al., 2005c). We hypothesized that this might be due to disease-related decreases in glucocorticoid receptor function. This is consistent with the idea of PTSD as an "accelerated aging" (Bremner and Narayan, 1998) related to common theories of progressive hippocampal atrophy and dysfunction in both processes. We have also shown that endogenous cortisol release stimulated by a cognitive stress challenge in healthy subjects impaired delayed recall of words and a spatial memory task at 24 h (Elzinga et al., 2005).

In women (with and without PTSD), with a history of early abuse, memory functioning was also affected after exposure to personalized scripts (Elzinga et al., 2003). For neutral paragraphs encoded after exposure to the trauma scripts there was an impairment in delayed recall relative to paragraphs encoded in a no-stress condition (Fig. 1). Recall 24 h later of an emotional paragraph presented immediately after the trauma scripts was positively correlated with cortisol response to the stressful challenge, meaning that cortisol enhanced consolidation of emotional memories. Another study in male healthy subjects has shown that endogenous cortisol levels in healthy subjects who became upset during a social speech task were positively correlated with enhanced delayed memory recall of pictures, which was especially prominent for recall of unpleasant pictures (Abercrombie et al., 2005). Taken together, these findings are consistent with animal models suggesting that glucocorticoid effects on learning require emotional arousal (Roozendaal, 2000).

Catecholamines released during stress also modulate the encoding and retrieval of memory (McGaugh, 2000). Administration of epinephrine

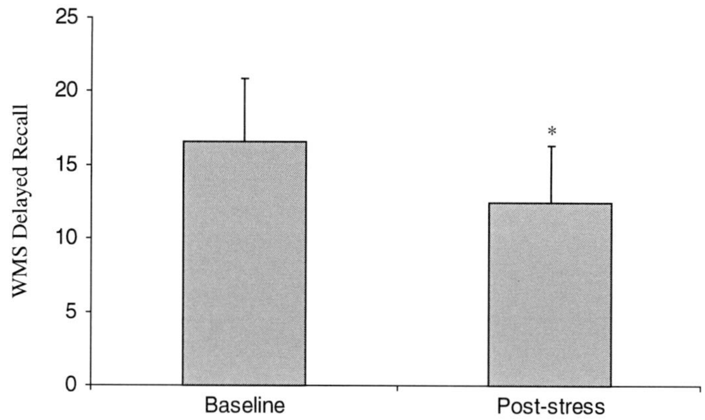

Fig. 1. Effects of a traumatic script on memory recall. There was a significant difference in delayed paragraph recall for paragraphs encoded after exposure to traumatic scripts compared to paragraphs encoded at a pre-stress baseline ($t(22) = -3.39$, $p<0.01$). This showed that stress impaired the ability to consolidate declarative memory.

(which is released from the adrenal) affects memory consolidation with an inverted U-shaped curve. Memory improves up to a point and decreases with high doses (Gold and van Buskirk, 1975; Liang et al., 1986). Lower doses of norepinephrine injected into the amygdala promote memory for an inhibitory avoidance task while higher doses inhibit memory (Liang et al., 1990). In humans, noradrenergic beta-blocker medications blocked the formation of emotional memories (Cahill et al., 1994), while enhanced norepinephrine release was associated with enhanced encoding of emotional memories (Southwick et al., 2002). Vasopressin and oxytocin have been shown to modulate memory formation in both animals (McGaugh, 2000) and human subjects (including those with PTSD) (Pitman et al., 1993).

Fear conditioning and extinction

One of the most classic laboratory paradigms that has been used as a model for PTSD is conditioned fear. In animal models, the pairing of light and shock leads to fear responses to the light alone. With exposure to light alone there is a gradual decrease in fear responding (called "extinction to fear") (Davis, 1992). Re-exposure to the light-shock at a later time point results in a rapid return of fear responding (Quirk, 2002). Medial prefrontal cortical inhibition of the amygdala (which plays a critical role in fear responses) is felt to represent the neural mechanism of extinction to fear responding (Quirk et al., 2006). This brain area is known to mediate emotion, as represented by the famous case of Phineas Gage (Damasio et al., 1994). Phineas Gage was a 19th century railroad worker who was injured by a spike that entered through his eye socket and lesioned his medial prefrontal cortex (mPFC). Areas involved included the orbitofrontal, anterior cingulate (25/24/32), and mesofrontal cortex (9). Speech and cognition remained intact. He had marked deficits in his ability to judge social contexts, behave appropriately in social contexts, and assess emotional nonverbal signals from others. Based on these findings and others, the mPFC has been judged to play a critical role in the emotion and social function.

This medial prefrontal area also plays an important role in the modulation of the neurohormonal response to stress. This area mediates peripheral cortisol and sympathetic responses to stress (Diorio et al., 1993). Dysfunction of this area could explain altered neurohormonal responses to stress in PTSD patients.

Studies in PTSD have shown dysfunction in the medial prefrontal cortical response to stress and traumatic reminders. We previously found decreased medial prefrontal function in combat veterans with PTSD exposed to combat-related PTSD

(Bremner et al., 1999b). In a second study we showed that women with PTSD related to early childhood sexual abuse had a decrease in medial prefrontal function in response to scripts of early childhood sexual abuse (Bremner et al., 1999a). A second study of exposure to emotional word pairs (e.g., rape-mutilate) showed decreased medial prefrontal and hippocampal function in abused women with PTSD (Bremner et al., 2003d). Another study used "Stroop" words (say the color of a color word, e.g., green, which leads to slowing of response time, due to inhibition of response) with an emotional Stroop component (name the color of a word like "rape"). The Stroop paradigm is associated with activation of anterior cingulate. Studies of the emotional Stroop (e.g., say the color of the word rape) has been associated with a slower response time in abuse-related (Foa et al., 1991) or combat-related PTSD (McNally et al., 1990). We studied neural correlates of the emotional Stroop in women with a history of early abuse with and without PTSD. We found that performance of the emotional Stroop was associated with decreased function in the mPFC in the PTSD patients (Bremner et al., 2004c).

We also have assessed neural correlates of conditioned fear in PTSD. Pairing of light and shock leads to increased fear responding and increased startle to light alone (conditioned fear). Conditioned fear and startle response are mediated by the central nucleus of the amygdala. Failure of extinction occurs with lesions of the mPFC (which inhibits the amygdala).

We studied fear conditioning with PET in women with a history of early abuse and PTSD and healthy nonabused women (Bremner et al., 2005b). Subjects were exposed to repeated and intermittent exposure to a blue square on a screen in the absence of shock (habituation), exposure to a blue square with a shock (fear acquisition), and then exposure to the blue square in the absence of shock (extinction). On a separate control day, they received random shocks instead of paired exposures; otherwise the protocol was the same. PTSD subjects experienced increased anxiety with fear acquisition and extinction. PTSD subjects also had increased amygdala blood flow during fear acquisition and decreased medial prefrontal blood flow during extinction. Increased amygdala blood flow during fear acquisition in the PTSD patients was correlated with increased PTSD symptoms, anxiety, and dissociation during fear acquisition. Increased amygdala blood flow during fear acquisition was correlated with decreased medial prefrontal blood flow during fear extinction in all of the subjects. There was a highly significant negative correlation between increased anxiety and decreased medial prefrontal blood flow during extinction in the PTSD patients ($r = -0.90$; $p = 0.006$).

Effects of treatment on the brain in PTSD

We have also assessed the effects of the selective serotonin reuptake inhibitor paroxetine on brain and cognition in PTSD. Previous multisite randomized placebo-controlled trials have shown efficacy for paroxetine over placebo in PTSD (Marshall et al., 2001; Tucker et al., 2001). Antidepressants have also been shown to promote neurogenesis in the hippocampus, a brain area involved in learning and memory (Duman et al., 1997). In an open-label study we showed a 5% increase in hippocampal volume after 9 months of treatment with paroxetine, as well as a 30% improvement in verbal declarative memory function measured with neuropsychological testing (Vermetten et al., 2003). Paroxetine treatment was also associated with a decreased cortisol and heart rate response to a stressful task (Vermetten et al., 2006b).

Glutamate, dissociation, and PTSD

Alterations in glutamatergic function has also been implicated in PTSD as well as dissociation (Krystal et al., 1996; Chambers et al., 1999). Symptoms of dissociation are an important part of the psychopathological response to stress. Symptoms of dissociation measured with the Clinician Administered Dissociative States Scale (CADSS) (Bremner et al., 1998) are:

> Do things seem to be moving in slow motion?
> Do things seem to be unreal to you, as if you are in a dream?

Do you feel as if you are watching the situation as an observer or spectator?
Do you feel disconnected from your own body?
Do you see things as if you were in a tunnel, or looking through a wide-angle photographic lens?
Does this experience seem to take much longer than you would have expected?

Increased dissociative symptoms at the time of trauma predict long-term PTSD (Bremner et al., 1992; Marmar et al., 1994). Although symptoms of dissociation are not part of the DSM criteria for PTSD, they are part of the criteria for acute stress disorder, and symptoms of dissociation are frequently seen in PTSD patients. PTSD patients are observed clinically to have an increased dissociative response to the original trauma, and then have chronic increased susceptibility to dissociative responses to minor stressors and traumatic reminders.

Although the neurobiology of dissociation has been studied less than PTSD, alterations in stress hormones likely play a role in these symptoms. One particular neurotransmitter system that has been hypothesized to play a role in dissociative symptoms is the excitatory amino acid glutamate (Krystal et al., 1994, 1996; Chambers et al., 1999). Glutamate is released during stress (Moghaddam et al., 1997), and high levels of glutamate are associated with toxicity to the hippocampus. Glutamate acts at the *N*-methyl-D-aspartic acid (NMDA) receptor, and is highly concentrated in the hippocampus. Glutamate is involved in memory at the molecular level. Excessive levels of glutamate can cause cytotoxicity as seen in patients with epilepsy. Stress inhibits glucose utilization, and thereby impairs reuptake of glutamate in glia with associated cytotoxicity.

Several lines of evidence support alterations of glutamatergic function in dissociation. The NMDA antagonist, Ketamine, when administered to normal subjects, results in an increased dissociative symptoms as measured with the CADSS (Krystal et al., 1994). In addition, increased

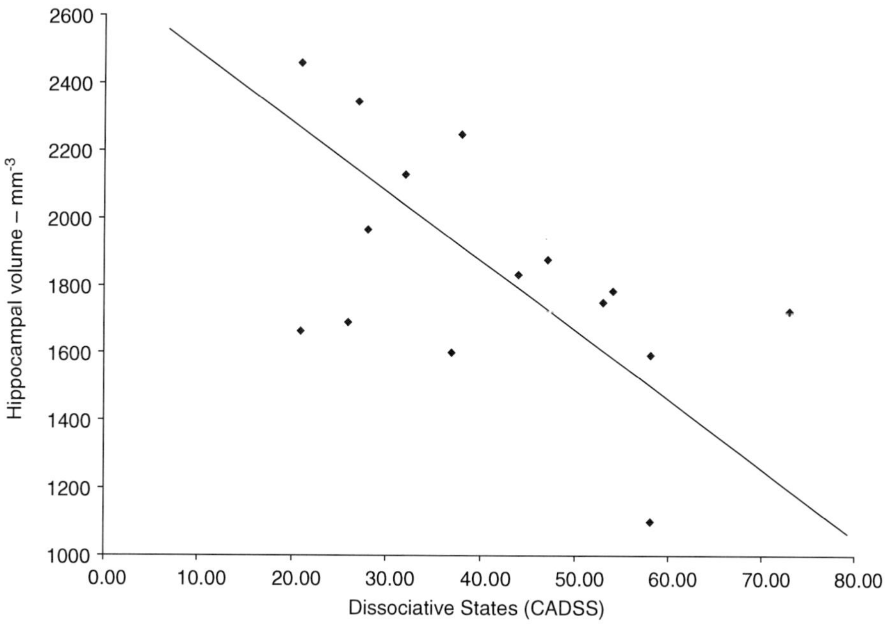

Fig. 2. Relationship between hippocampal volume measured with MRI and dissociative states measured with the CADSS in women with early abuse and the diagnosis of dissociative identity disorder. There was a significant negative correlation between hippocampal volume and dissociative states ($r = -0.54$; df $= 14$; $p < 0.05$), suggesting that increased levels of dissociation were related to smaller hippocampal volume. This correlation was not shown for amygdala volume. In addition there was not an association between level of PTSD symptoms and hippocampal volume in these patients.

dissociative states correlate with smaller volume of the hippocampus (which as noted above has a high concentration of NMDA receptors) in women with early abuse and PTSD (Stein et al., 1997; Bremner et al., 2003b). A correlation between dissociative states as measured with the CADSS and smaller hippocampal volume was seen in women with early abuse and DID (Fig. 2). Phenytoin (dilantin) is an antiepileptic drug that is efficacious in the treatment of epilepsy. Phenytoin modulates glutamatergic function and blocks the effects of stress on the hippocampus in animal studies (Watanabe et al., 1992). We conducted a pilot project in nine PTSD subjects of the effect of phenytoin on symptoms of PTSD and the brain. Phenytoin resulted in a decrease in PTSD symptoms (Bremner et al., 2004a) as well as a 5% increase in right hippocampal and right cerebral volume (Bremner et al., 2005a).

Conclusions

We have presented evidence for long-term alterations in brain and neurobiology in PTSD. Brain areas involved in the stress response include the mPFC, hippocampus, and amygdala. Neurohormonal systems that act on the brain areas to modulate PTSD symptoms and memory include glucocorticoids and norepinephrine. Dysfunction of these brain areas is responsible for symptoms of PTSD.

The related symptom area of dissociation is felt to be related to alterations in glutamatergic function; however, more research is needed in this area.

Brain imaging studies show that PTSD patients have increased amygdala reactivity during fear acquisition. Other studies show smaller hippocampal volume. A failure of medial prefrontal/anterior cingulate activation with re-experiencing of the trauma is hypothesized to represent a neural correlate of the failure of extinction seen in PTSD.

The brain has the capacity for plasticity in the aftermath of traumatic stress. Antidepressant treatments and changes in environment can reverse the effects of stress on hippocampal neurogenesis. In humans with PTSD, paroxetine increases hippocampal volume and improves memory function in conjunction with improving PTSD symptoms. Phenytoin, which blocks the effects of stress on the hippocampus in animal studies, also increases hippocampal volume in PTSD patients.

Future studies should use brain imaging and neurobiology to assess plasticity in PTSD. These can include both functional neuroimaging and neuroreceptor imaging to track the course of change during treatment, or to predict which traumatized individuals will develop chronic PTSD. The information from such studies will provide valuable information that will guide the development of new treatments.

Abbreviations

ACTH	adrenocorticotropin
BDNF	brain-derived neurotrophic factor
BPD	borderline personality disorder
CRF	corticotropin releasing factor
DHEA	dehydroepiandosterone
DID	dissociative identity disorder
mPFC	medial prefrontal cortex
MRI	magnetic resonance imaging
NAA	*N*-acetyl aspartic acid
NMDA	*N*-methyl-D-aspartic acid
PET	positron emission tomography
PTSD	posttraumatic stress disorder

References

Abercrombie, H.C., Speck, N.S. and Monticelli, R.M. (2005) Endogenous cortisol elevations are related to memory facilitation only in individuals who are emotionally aroused. Psychoneuroendocrinology, 31: 187–196.

Arborelius, L., Owens, M.J., Plotsky, P.M. and Nemeroff, C.B. (1999) The role of corticotropin-releasing factor in depression and anxiety disorders. J. Endocrinol., 160: 1–12.

Baker, D.G., West, S.A., Nicholson, W.E., Ekhator, N.N., Kasckow, J.W., Hill, K.K., Bruce, A.B., Orth, D.N. and Geracioti, T.D. (1999) Serial CSF corticotropin-releasing hormone levels and adrenocortical activity in combat veterans with posttraumatic stress disorder. Am. J. Psychiatry, 156: 585–588.

Barrett-Connor, E. and Edelstein, S.L. (1994) A prospective study of dehydroepiandrosterone sulfate and cognitive function in an older population: the Rancho Bernardo study. J. Am. Geriatr. Soc., 42: 420–423.

Bodnoff, S.R., Humphreys, A.G., Lehman, J.C., Diamond, D.M., Rose, G.M. and Meaney, M.J. (1995) Enduring effects of chronic corticosterone treatment on spatial learning, synaptic plasticity, and hippocampal neuropathology in young and mid-aged rats. J. Neurosci., 15: 61–69.

Bremner, J.D. (2002) Does Stress Damage the Brain? Understanding Trauma-related Disorders from a Mind-Body Perspective. W.W. Norton, New York.

Bremner, J.D. (2003) Functional neuroanatomical correlates of traumatic stress revisited 7 years later, this time with data. Psychopharmacol. Bull., 37: 6–25.

Bremner, J.D., Krystal, J.H., Putnam, F., Marmar, C., Southwick, S.M., Lubin, H. and Charney, D.S. (1998) Measurement of dissociative states with the Clinician Administered Dissociative States Scale (CADSS). J. Trauma. Stress, 11: 125–136.

Bremner, J.D., Licinio, J., Darnell, A., Krystal, J.H., Owens, M., Southwick, S.M., Nemeroff, C.B. and Charney, D.S. (1997a) Elevated CSF corticotropin-releasing factor concentrations in posttraumatic stress disorder. Am. J. Psychiatry, 154: 624–629.

Bremner, J.D., Mletzko, T., Welter, S., Siddiq, S., Reed, L., Williams, C., Heim, C.M. and Nemeroff, C.B. (2004a) Treatment of posttraumatic stress disorder with phenytoin: an open label pilot study. J. Clin. Psychiatry, 65: 1559–1564.

Bremner, J.D., Mletzko, T., Welter, S., Quinn, S., Williams, C., Brummer, M., Siddiq, S., Reed, L., Heim, C.M. and Nemeroff, C.B. (2005a) Effects of phenytoin on memory, cognition and brain structure in posttraumatic stress disorder: a pilot study. J. Psychopharmacol., 19: 159–165.

Bremner, J.D. and Narayan, M. (1998) The effects of stress on memory and the hippocampus throughout the life cycle: implications for childhood development and aging. Dev. Psychopathol., 10: 871–886.

Bremner, J.D., Narayan, M., Staib, L.H., Southwick, S.M., McGlashan, T. and Charney, D.S. (1999a) Neural correlates of memories of childhood sexual abuse in women with and without posttraumatic stress disorder. Am. J. Psychiatry, 156: 1787–1795.

Bremner, J.D., Randall, P.R., Capelli, S., Scott, T.M., McCarthy, G. and Charney, D.S. (1995a) Deficits in short-term memory in adult survivors of childhood abuse. Psychiatry Res., 59: 97–107.

Bremner, J.D., Randall, P.R., Scott, T.M., Bronen, R.A., Delaney, R.C., Seibyl, J.P., Southwick, S.M., McCarthy, G., Charney, D.S. and Innis, R.B. (1995b) MRI-based measurement of hippocampal volume in patients with combat-related posttraumatic stress disorder. Am. J. Psychiatry, 152: 973–981.

Bremner, J.D., Randall, P.R., Vermetten, E., Staib, L., Bronen, R.A., Mazure, C.M., Capelli, S., McCarthy, G., Innis, R.B. and Charney, D.S. (1997b) MRI-based measurement of hippocampal volume in posttraumatic stress disorder related to childhood physical and sexual abuse: a preliminary report. Biol. Psychiatry, 41: 23–32.

Bremner, J.D., Scott, T.M., Delaney, R.C., Southwick, S.M., Mason, J.W., Johnson, D.R., Innis, R.B., McCarthy, G. and Charney, D.S. (1993a) Deficits in short-term memory in posttraumatic stress disorder. Am. J. Psychiatry, 150: 1015–1019.

Bremner, J.D., Southwick, S.M., Brett, E., Fontana, A., Rosenheck, A. and Charney, D.S. (1992) Dissociation and posttraumatic stress disorder in Vietnam combat veterans. Am. J. Psychiatry, 149: 328–332.

Bremner, J.D., Southwick, S.M. and Charney, D.S. (1995c) Etiological factors in the development of posttraumatic stress disorder. In: Mazure C.M. (Ed.), Does Stress Cause Psychiatric Illness? Vol. 46. American Psychiatric Press, Washington, DC, pp. 149–186.

Bremner, J.D., Southwick, S.M., Johnson, D.R., Yehuda, R. and Charney, D.S. (1993b) Childhood physical abuse and combat-related posttraumatic stress disorder in Vietnam veterans. Am. J. Psychiatry, 150: 235–239.

Bremner, J.D., Staib, L., Kaloupek, D., Southwick, S.M., Soufer, R. and Charney, D.S. (1999b) Neural correlates of exposure to traumatic pictures and sound in Vietnam combat veterans with and without posttraumatic stress disorder: a positron emission tomography study. Biol. Psychiatry, 45: 806–816.

Bremner, J.D., Vermetten, E. and Kelley, M.E. (2007) Cortisol, dehydroepiandrosterone, and estradiol measured over 24 h in women with childhood sexual abuse-related posttraumatic stress disorder. J. Nerv. Ment. Dis. (in press).

Bremner, J.D., Vermetten, E. and Mazure, C.M. (2000) Development and preliminary psychometric properties of an instrument for the measurement of childhood trauma: the Early Trauma Inventory. Depress. Anxiety, 12: 1–12.

Bremner, J.D., Vermetten, E., Nafzal, N. and Vythilingam, M. (2004b) Deficits in verbal declarative memory function in women with childhood sexual abuse-related posttraumatic stress disorder (PTSD). J. Nerv. Ment. Dis., 192: 643–649.

Bremner, J.D., Vermetten, E., Schmahl, C., Vaccarino, V., Vythilingam, M., Afzal, N., Grillon, C. and Charney, D.S. (2005b) Positron emission tomographic imaging of neural correlates of a fear acquisition and extinction paradigm in women with childhood sexual abuse-related posttraumatic stress disorder. Psychol. Med., 35: 791–806.

Bremner, J.D., Vermetten, E., Vythilingam, M., Afzal, N., Schmahl, C., Elzinga, B. and Charney, D.S. (2004c) Neural correlates of the classic color and emotional stroop in women with abuse-related posttraumatic stress disorder. Biol. Psychiatry, 55: 612–620.

Bremner, J.D., Vythilingam, M., Vermetten, E., Adil, J., Khan, S., Nazeer, A., Afzal, N., McGlashan, T., Anderson, G., Heninger, G.R., Southwick, S.M. and Charney, D.S. (2003a) Cortisol response to a cognitive stress challenge in posttraumatic stress disorder (PTSD) related to childhood abuse. Psychoneuroendocrinology, 28: 733–750.

Bremner, J.D., Vythilingam, M., Vermetten, E., Newcomer, J.W. and Charney, D.S. (2004d) Effects of glucocorticoids on declarative memory function in major depression. Biol. Psychiatry, 55: 811–815.

Bremner, J.D., Vythilingam, M., Vermetten, E., Newcomer, J.W. and Charney, D.S. (2005c) Effects of dexamethasone on declarative memory function in posttraumatic stress disorder (PTSD). Psychiatry Res., 129: 1–10.

Bremner, J.D., Vythilingam, M., Vermetten, E., Southwick, S.M., McGlashan, T., Nazeer, A., Khan, S., Vaccarino, L.V., Soufer, R., Garg, P., Ng, C.K., Staib, L.H., Duncan, J.S. and Charney, D.S. (2003b) MRI and PET study of deficits in hippocampal structure and function in women with childhood sexual abuse and posttraumatic stress disorder (PTSD). Am. J. Psychiatry, 160: 924–932.

Bremner, J.D., Vythilingam, M., Vermetten, E., Southwick, S.M., McGlashan, T., Nazeer, A., Khan, S., Vaccarino, L.V., Soufer, R., Garg, P.K., Ng, C.K., Staib, L.H., Duncan, J.S. and Charney, D.S. (2003c) MRI and PET study of deficits in hippocampal structure and function in women with childhood sexual abuse and posttraumatic stress disorder. Am. J. Psychiatry, 160: 924–932.

Bremner, J.D., Vythilingam, M., Vermetten, E., Southwick, S.M., McGlashan, T., Staib, L., Soufer, R. and Charney, D.S. (2003d) Neural correlates of declarative memory for emotionally valenced words in women with posttraumatic stress disorder (PTSD) related to early childhood sexual abuse. Biol. Psychiatry, 53: 289–299.

Brunson, K.L., Eghbal-Ahmadi, M., Bender, R., Chen, Y. and Baram, T.Z. (2001) Long-term, progressive hippocampal cell loss and dysfunction induced by early-life administration of corticotropin-releasing hormone reproduce the effects of early-life stress. Proc. Natl. Acad. Sci. U.S.A., 98: 8856–8861.

Cahill, L., Prins, B., Weber, M. and McGaugh, J.L. (1994) Beta-adrenergic activation and memory for emotional events. Nature, 371: 702–704.

Carrion, V.G., Weems, C.F., Eliez, S., Patwardhan, A., Brown, W., Ray, R.D. and Reiss, A.L. (2001) Attenuation of frontal asymmetry in pediatric posttraumatic stress disorder. Biol. Psychiatry, 50: 943–951.

Chambers, R.A., Bremner, J.D., Moghaddam, B. and Krystal, J.H. (1999) Glutamate and posttraumatic stress disorder: toward a psychobiology of dissociation. Semin. Clin. Neuropsychiatry, 4: 274–281.

Damasio, H., Grabowski, T., Frank, R., Galaburda, A.M. and Damasio, A.R. (1994) The return of Phineas Gage: clues about the brain from the skull of a famous patient. Science, 264: 1102–1105.

Davis, M. (1992) The role of the amygdala in fear and anxiety. Annu. Rev. Neurosci., 15: 353–375.

De Bellis, M.D., Hall, J., Boring, A.M., Frustaci, K. and Moritz, G. (2001) A pilot longitudinal study of hippocampal volumes in pediatric maltreatment-related posttraumatic stress disorder. Biol. Psychiatry, 50: 305–309.

De Bellis, M.D., Keshavan, M.S., Clark, D.B., Casey, B.J., Giedd, J.N., Boring, A.M., Frustaci, K. and Ryan, N.D. (1999) A.E. Bennett Research Award: developmental traumatology: Part II. Brain development. Biol. Psychiatry, 45: 1271–1284.

Diorio, D., Viau, V. and Meaney, M.J. (1993) The role of the medial prefrontal cortex (cingulate gyrus) in the regulation of hypothalamic-pituitary-adrenal responses to stress. J. Neurosci., 13: 3839–3847.

Driessen, M., Herrmann, J., Stahl, K., Zwaan, M., Meier, S., Hill, A., Osterheider, M. and Petersen, D. (2000) Magnetic resonance imaging volumes of the hippocampus and the amygdala in women with borderline personality disorder and early traumatization. Arch. Gen. Psychiatry, 57: 1115–1122.

Duman, R.S. (2004) Depression: a case of neuronal life and death? Biol. Psychiatry, 56: 140–145.

Duman, R.S., Heninger, G.R. and Nestler, E.J. (1997) A molecular and cellular theory of depression. Arch. Gen. Psychiatry, 54: 597–606.

Duman, R.S., Malberg, J.E. and Nakagawa, S. (2001) Regulation of adult neurogenesis by psychotropic drugs and stress. J. Pharmacol. Exp. Ther., 299: 401–407.

Elzinga, B.M., Bakker, A. and Bremner, J.D. (2005) Stress-induced cortisol elevations are associated with impaired delayed, but not immediate recall. Psychiatry Res., 134: 211–223.

Elzinga, B.M. and Bremner, J.D. (2002) Are the neural substrates of memory the final common pathway in PTSD? J. Affect. Disord., 70: 1–17.

Elzinga, B.M. and Roelofs, K. (2005) Cortisol induced impairments of working memory require acute sympathetic activation. Behav. Neurosci., 119: 98–103.

Elzinga, B.M., Schmahl, C.S., Vermetten, E., van Dyck, R. and Bremner, J.D. (2003) Higher cortisol levels following exposure to traumatic reminders in abuse-related PTSD. Neuropsychopharmacology, 28: 1656–1665.

Flynn, M.A. (1999) Dehydroepiandrosterone replacement in aging humans. J. Clin. Endocrinol. Metab., 84: 1527–1533.

Foa, E.B., Feske, U., Murdock, T.B., Kozak, M.J. and McCarthy, P.R. (1991) Processing of threat related information in rape victims. J. Abnorm. Psychology, 100: 156–162.

Freeman, T.W., Cardwell, D., Karson, C.N. and Komoroski, R.A. (1998) In vivo proton magnetic resonance spectroscopy of the medial temporal lobes of subjects with combat-related posttraumatic stress disorder. Magn. Reson. Med., 40: 66–71.

Fuller, G.B., Hobson, W.C., Reyes, F.I., Winter, J.S. and Faiman, C. (1984) Influence of restraint and ketamine anesthesia on adrenal steroids, progesterone, and gonadotropins in rhesus monkeys. Proc. Soc. Exp. Biol. Med., 175: 487–490.

Gold, P.E. and van Buskirk, R. (1975) Facilitation of time-dependent memory processes with posttrial epinephrine injections. Behav. Biol., 13: 145–153.

Goldberg, J., True, W.R., Eisen, S.A. and Henderson, W.G. (1990) A twin study of the effects of the Vietnam War on posttraumatic stress disorder. JAMA, 263: 1227–1232.

Goodyer, I.M., Herbert, J., Altharn, P.M., Pearson, J., Secher, S.M. and Shiers, H.M. (1996) Adrenal steroids during major depression in 8- to 16-year-olds, I. Altered diurnal rhythms in salivary cortisol and dehydroepiandrosterone (DHEA) presentation. Psychol. Med., 26: 245–256.

Heim, C., Newport, D.J., Heit, S., Graham, Y.P., Wilcox, M., Bonsall, R., Miller, A.H. and Nemeroff, C.B. (2000) Pituitary-adrenal and autonomic responses to stress in women after sexual and physical abuse in childhood. JAMA, 284: 592–597.

Herbert, J., Goodyer, I.M., Altham, P.M., EPearson, J., Secher, S.M. and Shiers, H.M. (1996) Adrenal secretion and major depression in 8- to 16-year-olds, II: influence of co-morbidity at presentation. Psychol. Med., 26: 257–263.

Heuser, I., Deuschle, M., Luppa, P., Schweiger, U., Standhardt, H. and Weber, B. (1998) Increased diurnal plasma concentrations of dehydroepiandrosterone in depressed patients. J. Clin. Endocrinol. Metab., 83: 3130–3133.

Jatzko, A., Rothenhofer, S., Schmitt, A., Gaser, C., Demirakca, T., Weber-Fahr, W., Wessa, M., Magnotta, V. and Braus, D.F. (2006) Hippocampal volume in chronic posttraumatic stress disorder (PTSD): MRI study using two different evaluation methods. J. Affect. Disord., 94: 121–126.

Johnson, M.D., Bebb, R.A. and Sirrs, S.M. (2002) Uses of DHEA in aging and other disease states. Ageing Res. Rev., 1: 29–41.

Kaminska, M., Harris, J., Gijsbers, K. and Dubrovsky, B. (2000) Dehydroepiandrosterone sulfate (DHEAS) counteracts decremental effects of corticosterone on dentate gyrus LTP: implications for depression. Brain Res. Bull., 52: 229–234.

Kanter, E.D., Wilkinson, C.W., Radant, A.D., Petrie, E.C., Dobie, D.J., McFall, M.E., Peskind, E.R. and Raskind, M.A. (2001) Glucocorticoid feedback sensitivity and adrenocortical responsiveness in posttraumatic stress disorder. Biol. Psychiatry, 50: 238–245.

Kempermann, G., Kuhn, H.G. and Gage, F.H. (1997) More hippocampal neurons in adult mice living in an enriched environment. Nature, 386: 493–495.

Kempermann, G., Kuhn, H.G. and Gage, F.H. (1998) Experience-induced neurogenesis in the senescent dentate gyrus. J. Neurosci., 18: 3206–3212.

Kessler, R.C., Sonnega, A., Bromet, E., Hughes, M. and Nelson, C.B. (1995) Posttraumatic stress disorder in the national comorbidity survey. Arch. Gen. Psychiatry, 52: 1048–1060.

Kimonides, V.G., Khatibi, N.H., Svendsen, C.N., Sofroniew, M.V. and Herbert, J. (1998) Dehydroepiandrosterone (DHEA) and DHEA-sulfate (DHEAS) protect hippocampal neurons against excitatory amino acid-induced neurotoxicity. Proc. Natl. Acad. Sci. U.S.A., 95: 1852–1857.

Kirschbaum, C., Wolf, O.T., May, M., Wippich, W. and Hellhammer, D.H. (1996) Stress- and-treatment-induced elevations of cortisol levels associated with impaired declarative memory in healthy adults. Life Sci., 58: 1475–1483.

Kitayama, N., Vaccarino, V., Kutner, M., Weiss, P. and Bremner, J.D. (2005) Magnetic resonance imaging (MRI) measurement of hippocampal volume in posttraumatic stress disorder: a meta-analysis. J. Affect. Disord., 88: 79–86.

de Kloet, E.R., Oitzl, M.S. and Joels, M. (1999) Stress and cognition: are corticosteroids good or bad guys? Trends Neurosci., 22: 422–426.

Krystal, J.H., Bennett, A., Bremner, J.D., Southwick, S.M. and Charney, D.S. (1996) Recent developments in the neurobiology of dissociation: implications for posttraumatic stress disorder. In: Michelson L.K. and Ray W.J. (Eds.), Handbook of Dissociation: Theoretical, Empirical, and Clinical Perspectives. Plenum Press, New York, pp. 163–190.

Krystal, J.H., Karper, L.P., Seibyl, J.P., Freeman, G.K., Delaney, R., Bremner, J.D., Heninger, G.R., Bowers, M.B. and Charney, D.S. (1994) Subanesthetic effects of the non-competitive NMDA antagonist, ketamine, in humans: psychotomimetic, perceptual, cognitive, and neuroendocrine responses. Arch. Gen. Psychiatry, 51: 199–214.

Liang, K.C., Juler, R.G. and McGaugh, J.L. (1986) Modulating effects of post-training epinephrine on memory: involvement of the amygdala noradrenergic system. Brain Res., 368: 125–133.

Liang, K.C., McGaugh, J.L. and Yao, H.Y. (1990) Involvement of amygdala pathways in the influence of post-training intra-amygdala norepinephrine and peripheral epinephrine on memory storage. Brain Res., 508: 225–233.

Lindauer, R.J., Vlieger, E.J., Jalink, M., Olff, M., Carlier, I.V., Majoie, C.B., den Heeten, G.J. and Gersons, B.P. (2004) Smaller hippocampal volume in Dutch police officers with posttraumatic stress disorder. Biol. Psychiatry, 56: 356–363.

Lupien, S.J., Gaudreau, S., Tchiteya, B.M., Maheu, F., Sharma, S., Nair, N.P., Hauger, R.L., McEwen, B.S. and Meaney, M.J. (1997) Stress-induced declarative memory impairment in healthy elderly subjects: relationship to cortisol reactivity. J. Clin. Endocrinol. Metab., 82: 2070–2075.

Lupien, S.J., Gillin, C.J. and Hauger, R.L. (1999) Working memory is more sensitive than declarative memory to the acute effects of corticosteroids: a dose-response study in humans. Behav. Neurosci., 113: 420–430.

Lupien, S.J., Wilkinson, C.W., Briere, S., Ng Ying Kin, N.M.K., Meaney, M.J. and Nair, N.P.V. (2002) Acute modulation of aged human memory by pharmacological manipulation of glucocorticoids. J. Clin. Endocrinol. Metab., 87: 3798–3807.

Malberg, J.E., Eisch, A.J., Nestler, E.J. and Duman, R.S. (2000) Chronic antidepressant treatment increases neurogenesis in adult rat hippocampus. J. Neurosci., 20: 9104–9110.

Marmar, C.R., Weiss, D.S., Schlenger, D.S., Fairbank, J.A., Jordan, B.K., Kulka, R.A. and Hough, R.L. (1994) Peritraumatic dissociation and posttraumatic stress in male Vietnam theater veterans. Am. J. Psychiatry, 151: 902–907.

Marshall, R.D., Beebe, K.L., Oldham, M. and Zaninelli, R. (2001) Efficacy and safety of paroxetine treatment for chronic PTSD: a fixed-dose, placebo-controlled study. Am. J. Psychiatry, 158: 1982–1988.

McCauley, J., Kern, D.E., Kolodner, K., Dill, L., Schroeder, A.F., DeChant, H.K., Ryden, J., Derogatis, L.R. and Bass, E.G. (1997) Clinical characteristics of women with a history of childhood abuse: unhealed wounds. JAMA, 277: 1362–1368.

McGaugh, J.L. (2000) Memory: a century of consolidation. Science, 287: 248–251.

McNally, R.J., Kaspi, R.J., Riemann, B.C. and Zeitlin, S.B. (1990) Selective processing of threat cues in posttraumatic stress disorder. J. Abnorm. Psychol., 99: 398–402.

Michael, A., Jenaway, A., Paykel, E.S. and Herbert, J. (2000) Altered salivary dehydroepiandrosterone levels in major depression in adults. Biol. Psychiatry, 48: 989–995.

Moghaddam, B., Adams, B., Verma, A. and Daly, D. (1997) Activation of glutamatergic neurotransmission by ketamine: a novel step in the pathway from NMDA receptor blockade to dopaminergic and cognitive disruptions associated with the prefrontal cortex. J. Neurosci., 17: 2127–2912.

Newcomer, J.W., Craft, S., Hershey, T., Askins, K. and Bardgett, M.E. (1994) Glucocorticoid-induced impairment in declarative memory performance in adult humans. J. Neurosci., 14: 2047–2053.

Newcomer, J.W., Selke, G., Kelly, A.K., Paras, L. and Craft, S. (1995) Age-related differences in glucocorticoid effect on memory in human subjects. Soc. Neurosci. Abstr., 21: 161.

Newcomer, J.W., Selke, G., Melson, A.K., Hershey, T., Craft, S., Richards, K. and Alderson, A.L. (1999) Decreased memory performance in healthy humans induced by stress-level cortisol treatment. Arch. Gen. Psychiatry, 56: 527–533.

Nibuya, M., Morinobu, S. and Duman, R.S. (1995) Regulation of BDNF and trkB mRNA in rat brain by chronic electroconvulsive seizure and antidepressant drug treatments. J. Neurosci., 15: 7539–7547.

Oitzl, M.S. and de Kloet, E.R. (1992) Selective corticosteroid antagonists modulate specific aspects of spatial orientation learning. Behav. Neurosci., 106: 62–71.

Orentreich, N., Brind, J.L., Vogelman, J.H., Andres, R. and Baldwin, H. (1992) Long-term longitudinal measurements of plasma dehydroepiandrosterone sulfate in normal men. J. Clin. Endocrinol. Metab., 75: 1002–1004.

Pitman, R.K., Orr, S.P. and Lasko, N.B. (1993) Effects of intranasal vasopressin and oxytocin on physiologic responding during personal combat imagery in Vietnam veterans with posttraumatic stress disorder. Psychiatry Res., 48: 107–117.

Porter, N.M. and Landfield, P.W. (1998) Stress hormones and brain aging. Nat. Neurosci., 1: 3–4.

Prasad, A., Imamura, M. and Prasad, C. (1997) Dehydroepiandrosterone decreases behavioral despair in high- but not low-anxiety rats. Physiol. Behav., 62: 1053–1057.

de Quervain, D.J.-F., Roozendaal, B., Nitsch, R.M., McGaugh, J.L. and Hock, C. (2000) Acute cortisone administration impairs retrieval of long-term declarative memory in humans. Nat. Neurosci., 3: 313–314.

Quirk, G.J. (2002) Memory for extinction of conditioned fear is long-lasting and persists following spontaneous recovery. Learn. Mem., 9: 402–407.

Quirk, G.J., Garcia, R. and Gonzalez-Lima, F. (2006) Prefrontal mechanisms in extinction of conditioned fear. Biol. Psychiatry, 60: 337–343.

Roozendaal, B. (2000) Glucocorticoids and the regulation of memory consolidation. Psychoneuroendocrinology, 25: 213–238.

Santarelli, L., Saxe, M., Gross, C., Surget, A., Battaglia, F., Dulawa, S., Weisstaub, N., Lee, J., Duman, R., Arancio, O., Belzung, C. and Hen, R. (2003) Requirement of hippocampal neurogenesis for the behavioral effects of antidepressants. Science, 301: 805–809.

Sapolsky, R.M. (1996) Why stress is bad for your brain. Science, 273: 749–750.

Sapolsky, R.M., Uno, H., Rebert, C.S. and Finch, C.E. (1990) Hippocampal damage associated with prolonged glucocorticoid exposure in primates. J. Neurosci., 10: 2897–2902.

Schacter, D.L. (1996) Searching for Memory: The Brain, the Mind, and the Past. Basic Books, New York.

Schmahl, C.G., Vermetten, E., Elzinga, B.M. and Bremner, J.D. (2003) Magnetic resonance imaging of hippocampal and amygdala volume in women with childhood abuse and borderline personality disorder. Psych. Res.: Neuroimaging, 122: 193–198.

Schuff, N., Neylan, T.C., Lenoci, M.A., Du, A.T., Weiss, D.S., Marmar, C.R. and Weiner, M.W. (2001) Decreased hippocampal *N*-acetylaspartate in the absence of atrophy in posttraumatic stress disorder. Biol. Psychiatry, 50: 952–959.

Shin, L.M., Shin, P.S., Heckers, S., Krangel, T.S., Macklin, M.L., Orr, S.P., Lasko, N., Segal, E., Makris, N., Richert, K., Levering, J., Schacter, D.L., Alpert, N.M., Fischman, A.J., Pitman, R.K. and Rauch, S.L. (2004) Hippocampal function in posttraumatic stress disorder. Hippocampus, 14: 292–300.

Smith, M.A., Makino, S., Kvetnansky, R. and Post, R.M. (1995) Stress and glucocorticoids affect the expression of brain-derived neurotrophic factor and neurotrophin-3 mRNA in the hippocampus. J. Neurosci., 15: 1768–1777.

Smith, M.E. (2005) Bilateral hippocampal volume reduction in adults with post-traumatic stress disorder: a meta-analysis of structural MRI studies. Hippocampus, 15: 798–807.

Southwick, S.M., Horner, B., Morgan, C.A., Bremner, J.D., Davis, M., Cahill, L., Gold, P.B. and Charney, D.S. (2002) Relationship of enhanced norepinephrine activity during memory consolidation to enhanced long-term memory in humans. Am. J. Psychiatry, 159: 1420–1422.

Spivak, B., Maayan, R., Kotler, M., Mester, R., Gil-Ad, I., Shtaif, B. and Weizman, A. (2000) Elevated circulatory level of GABA-A antagonistic neurosteroids in patients with combat-related posttraumatic stress disorder. Psychol. Med., 30: 1227–1231.

Stein, M.B., Koverola, C., Hanna, C., Torchia, M.G. and McClarty, B. (1997) Hippocampal volume in women victimized by childhood sexual abuse. Psychol. Med., 27: 951–959.

Tucker, P., Zaninelli, R., Yehuda, R., Ruggiero, L., Dillingham, K. and Pitts, C.D. (2001) Paroxetine in the treatment of chronic posttraumatic stress disorder: results of a placebo-controlled flexible-dosage trial. J. Clin. Psychiatry, 62: 860–868.

Uddo, M., Vasterling, J.J., Braily, K. and Sutker, P.B. (1993) Memory and attention in posttraumatic stress disorder. J. Psychopathol. Behav. Assess., 15: 43–52.

Uno, H., Tarara, R., Else, J.G., Suleman, M.A. and Sapolsky, R.M. (1989) Hippocampal damage associated with prolonged and fatal stress in primates. J. Neurosci., 9: 1705–1711.

Vasterling, J.J. and Bremner, J.D. (2006) The impact of the 1991 Gulf War on the mind and brain: findings from neuropsychological and neuroimaging research. Philos. Trans. R Soc. Lond. Biol. Sci., 361: 593–604.

Vasterling, J.J., Duke, L.M., Brailey, K., Constans, J.I., Allain, A.N. and Sutker, P.H. (2002) Attention, learning, and memory performance and intellectual resources in Vietnam veterans: PTSD and no disorder comparisons. Neuropsychologia, 16: 5–14.

Vasterling, J.J., Proctor, S.P., Amoroso, P., Kane, R., Heeren, T. and White, R.F. (2006) Neuropsychological outcomes of army personnel following deployment to the Iraq War. JAMA, 296: 519–529.

Vermetten, E., Schmahl, C., Lindner, S., Loewenstein, R.J. and Bremner, J.D. (2006a) Hippocampal and amygdala volumes in dissociative identity disorder. Am. J. Psychiatry, 163: 1–8.

Vermetten, E., Vythilingam, M., Schmahl, C., De Kloet, C., Southwick, W.M., Charney, D.S. and Bremner, J.D. (2006b) Alterations in stress reactivity after long-term treatment with paroxetine in women with posttraumatic stress disorder. Ann. N.Y. Acad. Sci., 1071: 184–202.

Vermetten, E., Vythilingam, M., Southwick, S.M., Charney, D.S. and Bremner, J.D. (2003) Long-term treatment with paroxetine increases verbal declarative memory and hippocampal volume in posttraumatic stress disorder. Biol. Psychiatry, 54: 693–702.

Villarreal, G., Hamilton, D.A., Petropoulos, H., Driscoll, I., Rowland, L.M., Griego, J.A., Kodituwakku, P.W., Hart, B.L., Escalona, R. and Brooks, W.M. (2002) Reduced hippocampal volume and total white matter in posttraumatic stress disorder. Biol. Psychiatry, 52: 119–125.

Vythilingam, M., Heim, C., Newport, C.D., Miller, A.H., Vermetten, E., Anderson, E., Bronen, R., Staib, L., Charney, D.S., Nemeroff, C.B. and Bremner, J.D. (2002) Childhood trauma associated with smaller hippocampal volume in women with major depression. Am. J. Psychiatry, 159: 2072–2080.

Vythilingam, M., Luckenbaugh, D.A., Lam, T., Morgan III, C.A., Lipschitz, D., Charney, D.S., Bremner, J.D. and Southwick, S.M. (2005) Smaller head of the hippocampus in Gulf War-related posttraumatic stress disorder. Psychiatry Res., 139: 89–99.

Watanabe, Y.E., Gould, H., Cameron, D., Daniels, D. and McEwen, B.S. (1992) Phenytoin prevents stress and corticosterone induced atrophy of CA3 pyramidal neurons. Hippocampus, 2: 431–436.

Wolf, O.T. (2003) HPA axis and memory. Best Pract. Res. Clin. Endocrinol. Metab., 17: 287–299.

Wolf, O.T., Schommer, N.C., Hellhammer, D.H., McEwen, B.S. and Kirschbaum, C. (2001) The relationship between stress induced cortisol levels and memory differs between men and women. Psychoneuroendocrinology, 26: 711–720.

Yehuda, R., Keefe, R.S., Harvey, P.D., Levengood, R.A., Gerber, D.K., Geni, J. and Siever, L.J. (1995) Learning and memory in combat veterans with posttraumatic stress disorder. Am. J. Psychiatry, 152: 137–139.

Yehuda, R., Lowry, M.T., Southwick, S.M., Mason, J.W. and Giller, E.L. (1991a) Increased number of glucocorticoid receptors in posttraumatic stress disorder. Am. J. Psychiatry, 148: 499–504.

Yehuda, R., Southwick, S.M., Krystal, J.H., Bremner, J.D., Charney, D.S. and Mason, J. (1993) Enhanced suppression of cortisol with low dose dexamethasone in posttraumatic stress disorder. Am. J. Psychiatry, 150: 83–86.

Yehuda, R., Southwick, S.M., Nussbaum, E.L., Giller, E.L. and Mason, J.W. (1991b) Low urinary cortisol in PTSD. J. Nerv. Ment. Dis., 178: 366–369.

Yehuda, R., Teicher, M.H., Levengood, R.A., Trestman, R.L. and Siever, L.J. (1994) Circadian regulation of basal cortisol levels in posttraumatic stress disorder. Ann. N.Y. Acad. Sci.: 378–380.

Yehuda, R., Teicher, M.H., Trestman, R.L., Levengood, R.A. and Siever, L.J. (1996) Cortisol regulation in posttraumatic stress disorder and major depression: a chronobiological analysis. Biol. Psychiatry, 40: 79–88.

Young, A.H., Gallagher, P. and Porter, R.J. (2002) Elevation of the cortisol-dehydroepiandrosterone ratio in drug-free depressed patients. Am. J. Psychiatry, 159: 1237–1239.

Young, E.A. and Breslau, N. (2004a) Cortisol and catecholamines in posttraumatic stress disorder: an epidemiologic community study. Arch. Gen. Psychiatry, 61: 394–401.

Young, E.A. and Breslau, N. (2004b) Saliva cortisol in posttraumatic stress disorder: a community epidemiologic study. Biol. Psychiatry, 56: 205–209.

Discussion: Chapter 12

OITZL: Did I miss something about the temporal aspect of PTSD? That is, are there differences between individuals who may have had PTSD for a limited duration like 6 months and those with chronic PTSD who have had PTSD for many years?

BREMNER: We're starting to study this in people that are coming from Iraq. The argument that we make is that people coming from Iraq in the first 6 months to a year after their return have a different type of PTSD than some of the older PTSD veterans from Vietnam. Until now the subjects we have studied have chronic forms of PTSD, such as the PTSD related to abuse in early childhood or Vietnam combat veterans with PTSD. We are looking at adolescents with abuse-related PTSD in which the PTSD is more acute, but that study is not completed. And then we do have a current study that is looking at returning Iraqi veterans trying to get people in the first 6 months after they were discharged from service. We are doing a brain imaging and intervention that involves mindfulness-based stress reduction.

JOËLS: The number of newborn cells in adult brain is extremely low. It is not well known how many newborn cells there are in children, but from the little that is known it is not a lot. Even if these numbers are doubled and you look over a couple of weeks, the increase in hippocampal volume or 5% you see in your studies after paroxetine treatment cannot easily be explained by these newborn cells. So where do you think this increase of volume comes from? Is it maybe a change in blood flow or something else?

BREMNER: It could be that in addition to new neurons developing with paroxetine treatment, there are other contributions to increased volume such as an increase in dendritic branching. It could also be that treatment results in an increase in water content. However, we have preliminary data showing an increase in *N*-acetyl aspartate (NAA) measured with magnetic resonance spectroscopy (MRS) in the hippocampus of PTSD patients. NAA is a marker of neuronal integrity, so this suggests that paroxetine has effects on neuronal structure.

SECKL: I would like to ask a slightly different type of question and also about your findings with DHEA, which among us as endocrinologists has been a graveyard for many people's careers. But it is a pretty unique association: Do you think there is a fundamental underlying diathesis in this condition with this altered 17 hydroxylase? What is your thinking about this finding?

BREMNER: Our study showed that DHEA is elevated in women with abuse-related PTSD, and also that there was an elevation in DHEA to cortisol ratio where there was even more of a difference between the patients and the controls. From the standpoint of PTSD pathology, you can argue that it is a paradoxical finding since DHEA supposedly should have antiglucocorticoid-protective effect on the hippocampus. Here we are getting elevated levels, and smaller hippocampal volume and lower cortisol in the same subjects. In terms of the pathophysiology of why it is elevated, I do not know if I have a good explanation. If you look at the depression literature it is a mixed literature. In PTSD there are two studies that have a single sample, one significantly decreased and one significantly increased. Do you want to comment? I am not an endocrinologist.

YEHUDA: You have a lot of data and it is really impressive that you can get so many measures on the same subjects, and that is really what helps you make progress in the field because we tend to make these observations usually in different subjects; to me that is terrific. However, I am wondering I think there is a point in your presentation where you present like this is the theory and here are the data, the data don't fit the theory, and then we come back then about the theory, e.g., you have one slide about traumatic stress spectrum and all that, and depression and PTSD should be alike and elevated cortisol and cortisol damages the brain. Ok all that's fine and we know that is in the literature out there. But now you have collected your data, you are one of the few people on the planet that have multiple measures in the same cohort. So you can say, "Hey wait a minute, I have low cortisol at baseline and small hippocampuses, ha!" So what does that do to those slides in the beginning of the talk that summarize the literature and form the basis of your model? How do you feel about extending that?

BREMNER: There's some pattern that you know, e.g., that we can see correlations between memory performance measured at the time of the 24 h cortisol measure. In normal subjects there is an inverse correlation, so the higher the cortisol the lower the memory function which makes sense in terms of what is known about how administration of cortisol impairs declarative memory function in normal subjects. Because of time I did not show the trial in which we have given dexamethasone to healthy subjects in a 3-day protocol. After 3 days memory function becomes impaired in healthy subjects but not in the PTSD patients. We can see low cortisol and small hippocampal volume in the same subjects. And then we put them through a trauma-specific task that gives them anxiety and cortisol release is increased. So there could be some pattern cortisol release that causes hippocampal damage and memory impairment. But...

YEHUDA: How can it do that when it is low? How does low basal cortisol cause hippocampal damage?

BREMNER: Well low basal cortisol does not cause hippocampal damage. And we do not find correlations between hippocampal volume and cortisol in our patients with PTSD from traumas 20 years ago.

YEHUDA: You did not show cortisol and hippocampal volume correlation?

BREMNER: No we don't have this correlation; there is not a correlation in PTSD. So you know whether cortisol at one point in time resulted at the time of trauma in a kind of damage that would be a speculation. I do not know if we could ever really look at that. But what it looks like more is that there is a pattern of findings in subjects at baseline that, maybe, points at some central pathological process that leads to lower cortisol and smaller hippocampal volume that may not even be related in terms of the pathogenesis.

YEHUDA: It is at the heart of the matter of when we decide that maybe cortisol does not damage the hippocampus. I mean how much data do you need to look at the literature and say "Ha, this does not apply here." But it does not sound that there ever is going to be evidence of such a correlation for some of us.

DE KLOET: There is a literature out there, in which it is stated explicitly that high cortisol concentrations are damaging to the hippocampus, but in my opinion that cannot be generalized because there many situations like exercise that also produce high cortisol, but that are paradoxical not damaging at all. Only if levels of cortisol are chronically elevated for prolonged periods of time under conditions of distress one may see deterioration of the immune system, metabolic changes, and impairment of brain function. The question then is: What is special in the pattern of cortisol secretion that it damages the brain?

YEHUDA: The point is that at some time the data that we get should force us to have new models and abandon models that don't fit the data. Because then we are just trying to make that data fit models that are no good and you can have low cortisol and small hippocampal volume and somehow it still becomes important to say at some time there was no high cortisol, why, maybe, maybe not, maybe that does not fit, I am just wondering at what point we do that.

DE KLOET: I think the jury is out. There is a need for good controlled studies to measure the pulsatile patterns of cortisol to demonstrate what the actual significance of cortisol is, whether it is a predisposing factor or a consequence of the PTSD condition.

E.R. de Kloet, M.S. Oitzl & E. Vermetten (Eds.)
Progress in Brain Research, Vol. 167
ISSN 0079-6123

CHAPTER 13

Commentary: biological findings in PTSD — too much or too little?

Arieh Y. Shalev* and Ronen H. Segman

Department of Psychiatry, Hadassah University Hospital, P.O. Box 12000, Jerusalem 90815, Israel

Abstract: Summarizing the contributions in this section of the book, this chapter addresses questions regarding the complex etiology of PTSD, and the relative strength of discernable biological indicators of the disorder. It outlines two major approaches to exploring the biology of the disorder and discusses the reason for the many non-replications of findings. It defines the constructs of multicausality, equifinality, and multifinality, and evaluates their main implication for studies of PTSD, namely that no biological signal can be properly appraised without taking into account its context. Such context, in PTSD, includes both concurring biological systems and regulatory mechanisms, and environmental–psychosocial input. Studies of gene expression of PTSD exemplify one way of studying the context of putative biological signals. The role of biological alterations as templates for responding to psychosocial challenges is discussed.

Keywords: stress disorder post traumatic; biological markers; HPA axis; norepinephrine

Introduction

This conference brought together clinical- and basic-scientists, around issues related to stress exposure and its aftermath. For scholars in both areas, such a dialogue offers a unique opportunity to evaluate similarities and differences in their approaches to supposedly analogous problems. It also brings an often-sobering realization of one's own boundaries — and of generic limitations of this field of research.

Coming to discuss the clinical and conceptual section of the conference, three such limitations emerge. (a) The inherent complexity of the etiology and the pathophysiology of human stress disorders; (b) the complementary contribution to the occurrence of stress disorders, of both biological and psychosocial factors; and (c) the resulting difficulties to identify productive heuristics and valid translations of laboratory findings to clinical operations.

This comment addresses these limitations. It uses the session's papers as a point of origin. It then illustrates the above-mentioned difficulties by assessing the practical yield of two major biological approaches to post-traumatic stress disorder (PTSD). It also evaluates the potential promise of high-throughput methods, such as DNA microarrays, to advance the field beyond its current boundaries.

Complexity is too often used as an excuse for lack of progress. Rather than doing that, this chapter emphasizes the challenges of complex causation and symptom-maintenance in PTSD, and outlines ways to reduce their potentially paralyzing effect.

*Corresponding author. Tel.: +972 2 6777184; Fax: +972 2 6413642; E-mail: ashalev@cc.huji.ac.il

DOI: 10.1016/S0079-6123(07)67013-7

Why PTSD?

PTSD is one of the better-studied psychopathological effect of major stressors in humans. However, PTSD is far from being the only, or even the most frequent consequence of stressful exposure, the effects of which extend from positive learning to the reactivation of virtually any preexisting mental disorder (O'Donnell et al., 2004; Zoellner and Maercker, 2006).

In a way, therefore, PTSD is already a reduction: it is but one of many consequences of traumatic exposure and it is clearly delineated by a set of typical symptoms and typical time course. PTSD is also a robust phenotype, found across traumatic experiences and time-lags from the triggering event (Asmundson et al., 2000; Davidson et al., 2004).

Because of its phenotypical robustness and because of the apparent analogy between PTSD and stress-induced behavioral changes in animals (Yehuda et al., 2006), PTSD could be expected to also have prominent biological characteristics. Some of the best studies of the biology of PTSD were summarized in this session of the conference. These can clearly illustrate where we are in our attempts to uncover biological consistency behind the disorder's robust phenotypical façade.

Dr. Yehuda, who elsewhere summarized the biological findings in PTSD (Yehuda, 2002a, b), reported new findings linking parental PTSD and PTSD-related biological abnormalities in an offspring. Specifically, offsprings of Holocaust survivors with PTSD showed significantly lower 24 h mean urinary cortisol excretion and salivary cortisol levels, and enhanced plasma cortisol suppression by low doses of dexamethasone. This demonstration of trans-generational similarity of biological traits raises questions about the mechanism of transmission and about the actual link between putative HPA axis sensitivity and the occurrence of PTSD, i.e., whether a "hypersensitive" HPA axis is a vulnerability trait for PTSD, the transmission of which is independent from that of the disorder?

The observed association between endocrine abnormalities in offspring and *maternal* PTSD points to one possible transmission mechanism: maternal behavior following specific stressors (e.g., early maternal separation; Meaney, 2001; de Kloet et al., 2005) significantly mitigates their long-term endocrine effects. Could Holocaust surviving mothers have been irresponsive or inappropriately responsive (e.g., lacking a sense of security) during critical developmental stages in their offspring's lives?

As to the link between putative endocrine "vulnerability factors" and PTSD, these as well as other biological "markers" of PTSD do not properly separate trauma survivors with and without the disorder: they often lack predictive power, and have very little specificity and sensitivity as indicators of the disorder (Shalev et al., in press). They, so to speak, "drown" in the sea of other biological, environmental, and behavioral occurrences that underlay the occurrence of the disorder. We will expand on this point later in this paper.

Clearly related to the above, Dr. Gunner's report of significant effect of parental neglect on the regulation of the HPA axis in children illustrates another way in which the "environment" may affect human's bodily responsiveness to stressors. It extends previous findings of an early postnatal "epigenetic" effect of maternal separation (and subsequent grooming) on the HPA axis (Meaney, 2001) to a much later period in childhood maturation, during which parental *neglect* — rather than separation — disturbs the relatively dampened responsiveness of the HPA axis to stressors.

By extension, can one conceive of other circumstances, in which the HPA axis regains its sensitivity to "reprogramming?" Could the first days, or weeks, following traumatic experiences of "fear, helplessness, and horror" induce a reopening of a "source code" of gene-expression to rewriting, which then can last for years and decades? The progressive development, in PTSD, of an exaggerated heart rate (HR) response to starting tones, during the few months that follow a traumatic exposure, illustrates such a progressive effect, firstly shown in a prospective study (Shalev et al., 2000) and subsequently corroborated by a cross-section study of Vietnam veterans and their monozygotic twin brothers (Orr et al., 2003).

Summarizing years of brain-imaging studies in PTSD, Dr. Bremner provides evidence of both reduced hippocampal volume in PTSD and a reversal of this effect with SSRIs or phenytoin. A parsimonious reading of these findings suggests that these allegedly "structural" changes might be more reversible than permanent.

Finally, Dr. Liberzon provides a critical review of functional brain-imaging studies of PTSD, arguing that the hypothetical negative effect of a dysfunctional medial prefrontal cortex (mPFC) on the extinction of fear conditioned responses (Milad et al., 2006) is short of explaining the pathogenesis of the disorder. Dr. Liberzon extends the discussion of the potential role of mPFC in PTSD to include this brain structure's role in self-relatedness, reappraisal, social emotions, and contextualization processes.

Indeed, a focus on fear response and its extinction in PTSD (Pitman and Delahanty, 2005) should not let us forget that this disorder also, and essentially, involves a biased interpretation of contextual cues (Davis, 2006), as well as the loss of a sense of "safe territory" (Fanselow, 2000).

Undoubtedly a stress-initiated condition, PTSD is probably more than the sum of its parts. In longitudinal studies, our group has shown that *PTSD does not "develop" with time* but rather "does not recover" with time (King et al., 2003; Peleg and Shalev, 2006). Furthermore, Freedman et al. (1999) and King et al. (in press) have shown that *early depressive symptoms are powerful predictors of chronic PTSD* among recent survivors. Other prospective studies found that early *hyperarousal has a major contribution to PTSD symptoms' trajectories* (Schell et al., 2004). Ehlers and Clark (2000) showed a *significant effect of cognitive appraisal of one's own reaction on recovery from early PTSD* symptoms. Recent studies of injured Iraq war returnees show a *twofold increase in the incidence of PTSD in the year that followed homecoming* (Grieger et al., 2006), suggesting that the social context within which one reappraises one's past experiences has major effect on, at least, the overt expression of the disorder.

This array of putative risk factors illustrate an important aspect of the complexity of PTSD, namely its underlying *multicausality* and potential *equifinality*, i.e., the fact that PTSD is always the product of many causes, and the related fact that PTSD might be the common outcome of several sets of concurring etiologies. The fact that a potentially traumatic event leads to diverse consequences illustrates the inherent *multifinality* of major stressors, namely their ability to trigger multiple outcomes.

This truly sets the context within which I wish to discuss the relative role of biological factors in PTSD, and the boundaries to discovering such a role. Specifically, the following text will critically appraise the likelihood of a single biological system to provide a good-enough explanation of PTSD. Looking at the constant effect of environmental input on the CNS, it will refine the distinction between neuronal "bottom-up" and "top-down" processes, as applied to PTSD, and argue that PTSD might be a specific result of a mutual amplification of these two elements, whereas recovery from trauma involves a self-regulatory interaction between the two.

Biological studies of PTSD

Three areas of biological studies yielded replicable findings in PTSD: neuroendocrine studies, brain-imaging studies, and psychophysiological explorations. Because brain-imaging studies are reviewed elsewhere in this volume, the following discussion concerns neuroendocrine and psychophysiological studies.

Studies of the HPA axis (Yehuda, 2001; Yehuda et al., 2006) have explored the idea that PTSD is linked with a hypersensitive or hyperresponsive HPA axis. Supportive evidences for that hypothesis were sought in measures of peripheral hormones, evaluations of the HPA axis' diurnal variation, and studies of the axis' responses to challenge tests (de Kloet et al., 2006).

Often converging, the ensemble of these studies did not yield a robust signal, i.e., a signal strong enough to be consistently captured, and characterize PTSD patients across traumatic conditions, age and gender differences, time lags from the triggering trauma, and other "real world" contingencies. Remote response modifiers (such as lifetime exposure to violence) had major effect on endocrine responses in the recently traumatized (Resnick et al., 1995). However, the chain of causation from sensitizing life events, via trauma exposure to PTSD has not been demonstrated.

Provocation tests axes were somewhat more consistent (see for review, de Kloet et al., 2006) showing both enhanced hormonal response to *behavioral provocation* and enhanced cortisol suppression by *low doses of dexamethasone*. Recent studies, however, have shown a main effect of trauma exposure (rather than PTSD) on both dexamethasone suppression test (DST) and lymphocytes glucocorticoid receptors density levels (de Kloet et al., 2007).

One of the major postulates of the HPA axis hypersensitivity hypothesis concerned contributions of an HPA axis dysfunction, namely a failure to mount sufficient levels of circulating cortisol at the time of the traumatic event, to the occurrence of PTSD. To test this proposition, our group prospectively measured plasma and saliva cortisol, urine excretion of cortisol, plasma ACTH, and glucocorticoid receptors' density upon admission to an emergency room (ER), following trauma, and 10 days, 1 month, and 5 months later, in 155 civilian survivors of traumatic events (Shalev et al., in press). Survivors who developed PTSD at 5 months ($N = 31$) did not differ from those who did not develop PTSD in any hormone measures, at any time (plasma cortisol levels are reported in Table 1). Cortisol levels in the ER did not predict PTSD symptom severity (see also, Delahanty et al., 2000, 2005).

A most intriguing aspect of our study is that, while hormone levels were very similar, survivors who developed PTSD showed major differences (at times more than threefold) in both PTSD and depression symptoms (Table 1). Clinically as well, PTSD subjects were very different from those without PTSD: they were much more distressed and showed extensive avoidance and intense hyperarousal.

Looking at the larger picture, the co-occurrence of remarkable phenotypical differences *and* little variation in endocrine measures is the rule, rather than the exception, in endocrine studies of PTSD. Hormonal measures in both PTSD and control subjects are rarely out of the normal physiological range. Indeed, this is a generic finding, seen across other mental disorders — significant behavioral deviance (e.g., in chronic schizophrenia) and minor biological alterations.

Table 1. Group differences in symptom levels (PTSD and depression) and in plasma cortisol levels in a longitudinal study of PTSD

	PTSD ($n = 31$)	No PTSD ($n = 124$)
Age	31.2±11.6	31.2±10.9
Gender (M(%)//F(%))	16(51%)// 15(48%)	75 (60%)// 49(40%)
BMI	40.2±6.7	40.8±6.8
PTSD symptoms (IES-R)		
Ten days	64.1±21.8***	34.4±22.7
One month	59.2±21.4***	25.3±19.8
Five months	56.2±21.7***	17.5±15.5
Depression symptoms (BDI)		
Ten days	19.7±9.4***	8.9±7.0
One month	16.4±10.7***	5.9±6.2
Five months	17.9±12.3***	4.5±5.6
Plasma cortisol levels[a]		
Emergency room	13.0±6.7	13.6±5.9
Ten days	13.0±5.9	12.6±4.9
One month	12.7±6.3	11.8±5.9
Five months	10.5±3.6	11.7±4.9

Note: BMI = body mass index; IES-R = impact of events scale — revised; BDI = Beck depression inventory.
***$p < 0.001$.
[a]Subject numbers may differ due to missing biological samples.

Studies of adrenergic responses to traumatic events were fueled by similarly tempting hypothesis: because blood levels of catecholamines enhance the acquisition of fear conditioned responses (McGaugh and Roozendaal, 2002), it was hoped that PTSD would be associated with higher hormone levels following trauma exposure. Studies of initial hormone levels in PTSD, however, were inconsistent (Delahanty et al., 2000, 2005). In the above-mentioned prospective neuroendocrine study of PTSD (Shalev et al., in press), ER plasma norepinephrine (NE) levels did not differentiate survivors with PTSD at 5 months from those without PTSD (281±132 and 320±146 pg/ml, respectively, in PTSD and non-PTSD subjects). Urinary NE excretion in the ER was 1289±928 ng/h in PTSD and 1981±1519 ng/h in survivors who did not develop PTSD — again a non-significant difference.

Studies of chronic PTSD similarly yielded inconsistent findings, with some showing higher levels of catecholamines in PTSD (Pitman and Orr, 1990; Yehuda et al., 1998; Liberzon et al., 1999; Young and Breslau, 2004) and others showing either lower levels of catecholamines in PTSD or no difference (Murburg et al., 1995; Marshall et al., 2002).

Thus, one of the clearest narratives concerning the etiology of PTSD is not supported by endocrine studies of the disorder. It is not that catecholamines are not involved in the etiology of the disorder, rather their putative involvement does not translate into measurable endocrine signature.

Preventive endocrine treatments for PTSD included increasing cortisol levels following trauma (Schelling et al., 2004a, b, 2006) and dampening the early adrenergic response. Among the latter, Pitman et al. (2002) showed that administering propranolol to survivors of traumatic events, shortly after exposure, reduces the magnitude of physiological responses to trauma reminders 6 months later (see also, Vaiva et al., 2003). However, Orr et al. (2006) and van Stegeren et al. (2002) failed to show an effect of beta-adrenergic blockade on the acquisition and the retention of conditioned responses to aversive stimuli in humans.

Psychophysiological exploration of PTSD (e.g., autonomic responses to reminders of the traumatic event, auditory startle responses) yielded somewhat more consistent findings (reviewed in Orr et al., 2002). PTSD patients regularly show elevated HR, skin conductance and facial muscles responses to reminders of the traumatic event, and elevated autonomic responses to startling tones (Shalev et al., 2000). An initial finding of elevated HR responses to traumatic events in survivors who develop PTSD (Shalev et al., 1998a) has been replicated by others (Blanchard et al., 2002; Bryant et al., 2003; Zatzick et al., 2005; Bryant, 2006; Kraemer et al., 2007) and there are, obviously, non-replications (Blanchard et al., 2002).

Two non-replications by our group can illustrate the extent to which the context of measuring HR as well as minor modifications in sample characteristics and phenotype definition can affect this otherwise salient finding. In 1998, we reported a significant relationship between HR in the ER and PTSD in 86 non-injured subjects diagnosed according to DSM III-R criteria (Shalev et al., 1998b). We have subsequently replicated this finding in a large group ($n = 354$) of road traffic accident victims, *but not in survivors of concurrent terror attacks* ($n = 39$; Shalev and Freedman, 2005). Finally, in the above-mentioned prospective study of stress hormones and PTSD we found non-significant ER HR difference between PTSD patients and non-PTSD controls (86.9 ± 14.0 BPM in PTSD vs. 83.2 ± 13.0 BPM in non-PTSD; Shalev et al., in press).

Looking at possible reasons for these non-replications, we firstly found that all our terror survivors, regardless of subsequent PTSD, had elevated HR in the ER (specifically, 93.4 ± 18.2 BPM in PTSD and 94.9 ± 20.3 BPM in non-PTSD). In the previous (1998a) study, this HR level was seen in the PTSD group alone.

For everyone who experienced the extremely stressful environment of an ER following a terrorist attack, the source of the elevated HR among all subjects is all too clear. Here, therefore, the noisy *context* of measuring HR could have confounded a possible group difference.

The non-replication in the more recent work is probably due to the use of *DSM-IV* criterion of a "traumatic event" (i.e., exposure *and* strong reaction of fear or horror) as entry criterion to the study. A study sample selected by having had an event and having *a strong immediate reaction* differs from those previously recruited using DSM III-R definition of a traumatic event. The latter concerned exposure alone, regardless of early responses.

However, the general lesson, here, is that despite its saliency and converging replications, this HR risk indicator of PTSD is easily confounded by small contextual variations. Similar "context" effects have existed in studies of NE levels in PTSD, many of which consisted of measuring "baseline" hormonal levels in subjects who were expecting a stressful test. Also, a finding of elevated eyeblink responses to startling tones (Morgan et al., 1996) took place in an environment that was later interpreted as being stressful for the examinees.

Genetic studies of PTSD

The path of discovery of the genetics of PTSD resembles the one taken by other biological markers. Familial transmission studies firstly showed preliminary evidence of family clustering of PTSD patients and other anxiety disorders (Davidson et al., 1985, 1998). Twin studies then showed robust but complex contributions of inheritance to both the likelihood of trauma (combat) exposure and, separately, to each of the three symptom clusters of PTSD (Goldberg et al., 1990; True et al., 1993). Twin studies of putative biomarkers for PTSD (Gilbertson et al., 2002, p. 1959; Orr et al., 2003; Pitman et al., 2006) have helped separate inherited from acquired biological traits (small hippocampus = inherited; abnormal startle = acquired).

Explorations of genotypic variation, however (Segman et al., 2002), have not risen to the challenge of major discovery: none has been replicated so far, and attempts to identify predictive polymorphic variation in a single locus have been abandoned in an area of total genome scans.

With time, our understanding of the genetic contribution has moved from an essentially deterministic views (either simple "Mandelian," or complex "small effect genes") to perceiving the genetics of mental disorder as mainly transmitting templates for environment-responsive gene-expression effects (Segman and Shalev, 2003). Recently, high throughput techniques enabled a simultaneous evaluation of tens of thousands of genes, and, temporarily, challenged the idea of hypothesis-driven research (since these techniques could test a very large number of previously stated and novel hypotheses).

In a study of peripheral gene expression (Segman et al., 2005), we showed a good enough prediction of PTSD and PTSD symptoms, at 4 months, from differential gene expression in the ER, and a cluster of differentiating genes 4 months after the event. The former can be seen as signaling a "pathogenic" event, and the latter disease-maintaining mechanisms.

Peripheral gene expression, however, is not the signal. It is, at best, the noise that accompanies those CNS alterations that lead to PTSD — their peripheral "signature." Like the cloud of dust that accompanies rushing convoys in the desert, they indicate the presence of a movement, but are not the vehicles themselves.

Will this expensive approach generate new biological hypothesis for PTSD? Looking at the differentiating genes in our study, one does find some that have biological "relevance" (i.e., are in accord with current views of the disorder), such as genes that are also expressed in the amygdala, genes that modulate the immune response (Fig. 1). Other families of differentiating genes include those that mediate apoptosis, neural plasticity, etc. There might be place for discovery, if these findings are replicated, and the rather large cluster of differentiating genes is reduced into a manageable dimension.

But the path from peripheral genes to putative CNS markers is long, and any single differentiating might be lost when studied separately. Arguably, the main advantage of gene expression profiles is that they do address large clusters of contributing factors, thereby capturing an extremely complex, and possibly unstable signal that links trauma exposure to PTSD.

Is the "context" a noise — or a signal?

Ultimately, seeing how brittle are biological measures in "real-world" studies should lead us to reconsider the role of experimental noise in studies of PTSD. The previously mentioned ideas of multicausality, equifinality, and multifinality suggest that the effect of any etiological factor can only be studied "in context," i.e., within a dynamic relationship with other co-occurring factors.

Importantly for biological studies of PTSD, the relevant "context" is not only biological (e.g., the concurrent activity of other bodily systems, such as the concurrent effect of endocrine and immune systems) but also psychosocial; indeed an array of concurring psychosocial effects. For example, the psychological effects of combat injury, a major risk factor for PTSD (Schnyder, 2001, p. 2684; Koren et al., 2005, p. 166) may only be revealed *if* adversity following homecoming is encountered. This late psychosocial context may determine

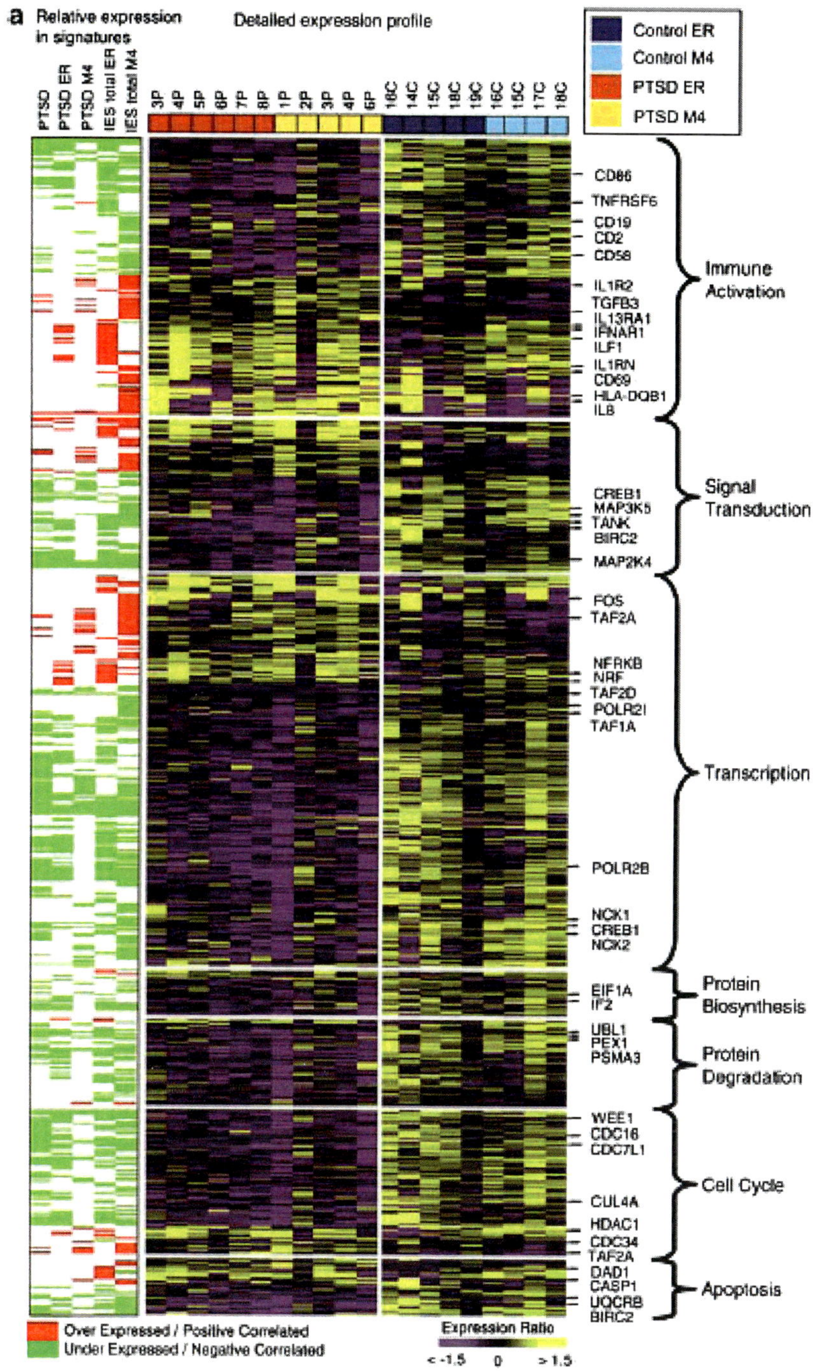

Fig. 1. Functional attributes of differentially expressed genes in recent trauma survivors with and without PTSD (Segman et al., 2005). (Adapted with permission from Nature Publication Group.)

which of the many possible psychological outcomes of combat injury will be expressed. Extending this example, the biology of combat-related PTSD might consist of creating vulnerability to further assault.

These "top-down" processes, i.e., those social cognitions that belatedly determine where will a potentially shattering CNS effect of stressors lead, might be as strong if not more potent than the biological alterations set forward by a trauma. Indeed, we are back to the licking and grooming mothers of Meaney (2001) pups' handling experiments, whose "top-down" soothing effects on offspring CNS determined the long-term endocrine effect of that early misery.

Embedding the noise in biological studies of PTSD is a major methodological challenge. Without it, we might be repeatedly, and rather unsuccessfully, exploring a few of the abundant risk factors — a never-ending story.

Acknowledgment

Study supported by an NIMH grant R34-MH71651 to Dr. Shalev.

References

Asmundson, G.J., Frombach, I., Mcquaid, J., Pedrelli, P., Lenox, R. and Stein, M.B. (2000) Dimensionality of posttraumatic stress symptoms: a confirmatory factor analysis of DSM-IV symptom clusters and other symptom models. Behav. Res. Ther., 38: 203–214.

Blanchard, E.B., Hickling, E.J., Galovski, T. and Veazey, C. (2002) Emergency room vital signs and PTSD in a treatment seeking sample of motor vehicle accident survivors. J. Trauma. Stress, 15: 199–204.

Bryant, R.A. (2006) Longitudinal psychophysiological studies of heart rate: mediating effects and implications for treatment. Ann. N.Y. Acad. Sci., 1071: 19–26.

Bryant, R.A., Harvey, A.G., Guthrie, R.M. and Moulds, M.L. (2003) Acute psychophysiological arousal and posttraumatic stress disorder: a two-year prospective study. J. Trauma. Stress, 16: 439–443.

Davidson, J.R., Tupler, L.A., Wilson, W.H. and Connor, K.M. (1998) A family study of chronic post-traumatic stress disorder following rape trauma. J. Psychiatr. Res., 32: 301–309.

Davidson, J.R.T., Stein, D.J., Shalev, A.Y. and Yehuda, R. (2004) Posttraumatic stress disorder: acquisition, recognition, course, and treatment. J. Neuropsychiatry Clin. Neurosci., 16: 135–147.

Davidson, J., Swartz, M., Storck, M., Krishnan, R.R. and Hammett, E. (1985) A diagnostic and family study of posttraumatic stress disorder. Am. J. Psychiatry, 142: 90–93.

Davis, M. (2006) Neural systems involved in fear and anxiety measured with fear-potentiated startle. Am. Psychol., 61: 741–756.

Delahanty, D.L., Nugent, N.R., Christopher, N.C. and Walsh, M. (2005) Initial urinary epinephrine and cortisol levels predict acute PTSD symptoms in child trauma victims. Psychoneuroendocrinology, 30: 121–128.

Delahanty, D.L., Raimonde, A.J. and Spoonster, E. (2000) Initial posttraumatic urinary cortisol levels predict subsequent PTSD symptoms in motor vehicle accident victims. Biol. Psychiatry, 48: 940–947.

Ehlers, A. and Clark, D.M. (2000) A cognitive model of posttraumatic stress disorder. Behav. Res. Ther., 38: 319–345.

Fanselow, M.S. (2000) Contextual fear, gestalt memories, and the hippocampus. Behav. Brain Res., 110: 73–81.

Freedman, S.A., Brandes, D., Peri, T. and Shalev, A. (1999) Predictors of chronic post-traumatic stress disorder: a prospective study. Br. J. Psychiatry, 174: 353–359.

Gilbertson, M.W., Shenton, M.E., Ciszewski, A., Kasai, K., Lasko, N.B., Orr, S.P. and Pitman, R.K. (2002) Smaller hippocampal volume predicts pathologic vulnerability to psychological trauma. Nat. Neurosci., 5: 1242–1247.

Goldberg, J., True, W.R., Eisen, S.A. and Henderson, W.G. (1990) A twin study of the effects of the Vietnam War on posttraumatic stress disorder. JAMA, 263: 1227–1232.

Grieger, T.A., Cozza, S.J., Ursano, R.J., Hoge, C., Martinez, P.E., Engel, C.C. and Wain, H.J. (2006) Posttraumatic stress disorder and depression in battle-injured soldiers. Am. J. Psychiatry, 163: 1777–1783 quiz 1860.

Koren, D., Norman, D., Cohen, A., Berman, J. and Klein, E.M. (2005) Increased PTSD risk with combat-related injury: a matched comparison study of injured and uninjured soldiers experiencing the same combat events. Am. J. Psychiatry, 162: 276–282.

King, D.W., King, L.A., Shalev, A.Y., Doron-LaMarca, S. and McArdle, J. (in press) Sequential temporal dependencies in the comorbidity of depression and posttraumatic stress disorder.

King, L.A., King, D.W., Salgado, D.M. and Shalev, A.Y. (2003) Contemporary longitudinal methods for the study of trauma and posttraumatic stress disorder. CNS Spectr., 8: 686–692.

de Kloet, C.S., Vermetten, E., Geuze, E., Kavelaars, A., Heijnen, C.J. and Westenberg, H.G. (2006) Assessment of HPA-axis function in posttraumatic stress disorder: pharmacological and non-pharmacological challenge tests, a review. J. Psychiatr. Res., 40: 550–567.

de Kloet, C.S., Vermetten, E., Heijnen, C.J., Geuze, E., Lentjes, E.G. and Westenberg, H.G. (2007) Enhanced cortisol suppression in response to dexamethasone administration in traumatized veterans with and without posttraumatic stress disorder. Psychoneuroendocrinology, 32: 215–226.

de Kloet, E.R., Sibug, R.M., Helmerhorst, F.M. and Schmidt, M. (2005) Stress, genes and the mechanism of programming the brain for later life. Neurosci. Biobehav. Rev., 29: 271–281.

Kraemer, B., Moergeli, H., Roth, H., Hepp, U. and Schnyder, U.J. (2007) Contribution of initial heart rate to the prediction of posttraumatic stress symptom level in accident victims. J. Psychiatr. Res. [Epub ahead of print]

Liberzon, I., Abelson, J.L., Flagel, S.B., Raz, J. and Young, E.A. (1999) Neuroendocrine and psychophysiologic responses in PTSD: a symptom provocation study. Neuropsychopharmacology, 21: 40–50.

Marshall, R.D., Blanco, C., Printz, D., Liebowitz, M.R., Klein, D.F. and Coplan, J. (2002) A pilot study of noradrenergic and HPA axis functioning in PTSD vs. panic disorder. Psychiatry Res., 110: 219–230.

Mcgaugh, J.L. and Roozendaal, B. (2002) Role of adrenal stress hormones in forming lasting memories in the brain. Curr. Opin. Neurobiol., 12: 205–210.

Meaney, M.J. (2001) Maternal care, gene expression, and the transmission of individual differences in stress reactivity across generations. Annu. Rev. Neurosci., 24: 1161–1192.

Milad, M.R., Rauch, S.L., Pitman, R.K. and Quirk, G.J. (2006) Fear extinction in rats: implications for human brain imaging and anxiety disorders. Biol. Psychol., 73: 61–71.

Morgan, C.A., Grillon, C., Southwick, S.M., Davis, M. and Charney, D.S. (1996) Exaggerated acoustic startle reflex in Gulf War veterans with posttraumatic stress disorder. Am. J. Psychiatry, 153: 64–68.

Murburg, M.M., McFall, M.E., Lewis, N. and Veith, R.C. (1995) Plasma norepinephrine kinetics in patients with posttraumatic stress disorder. Biol. Psychiatry, 38: 819–825.

O'Donnell, M.L., Creamer, M. and Pattison, P. (2004) Posttraumatic stress disorder and depression following trauma: understanding comorbidity. Am. J. Psychiatry, 161: 1390–1396.

Orr, S.P., Metzger, L.J., Lasko, N.B., Macklin, M.L., HU, F.B., Shalev, A.Y. and Pitman, R.K. (2003) Physiologic responses to sudden, loud tones in monozygotic twins discordant for combat exposure: association with posttraumatic stress disorder. Arch. Gen. Psychiatry, 60: 283–288.

Orr, S.P., Metzger, L.J. and Pitman, R.K. (2002) Psychophysiology of post-traumatic stress disorder. Psychiatr. Clin. North Am., 25: 271–293.

Orr, S.P., Milad, M.R., Metzger, L.J., Lasko, N.B., Gilbertson, M.W. and Pitman, R.K. (2006) Effects of beta blockade, PTSD diagnosis, and explicit threat on the extinction and retention of an aversively conditioned response. Biol. Psychol., 73: 262–271.

Peleg, T. and Shalev, A.Y. (2006) Longitudinal studies of PTSD: overview of findings and methods. CNS Spectr., 11: 589–602.

Pitman, R.K. and Delahanty, D.L. (2005) Conceptually driven pharmacologic approaches to acute trauma. CNS Spectr., 10: 99–106.

Pitman, R.K., Gilbertson, M.W., Gurvits, T.V., MAY, F.S., Lasko, N.B., Metzger, L.J., Shenton, M.E., Yehuda, R. and Orr, S.P. (2006) Clarifying the origin of biological abnormalities in PTSD through the study of identical twins discordant for combat exposure. Ann. N.Y. Acad. Sci., 1071: 242–254.

Pitman, R.K. and Orr, S.P. (1990) 24-Hour urinary cortisol and catecholamine excretion in combat-related posttraumatic stress disorder. Biol. Psychiatry, 27: 245–247.

Pitman, R.K., Sanders, K.M., Zusman, R.M., Healy, A.R., Cheema, F., Lasko, N.B., Cahill, L. and Orr, S.P. (2002) Pilot study of secondary prevention of posttraumatic stress disorder with propranolol. Biol. Psychiatry, 51: 189–192.

Resnick, H.S., Yehuda, R., Pitman, R.K. and Foy, D.W. (1995) Effect of previous trauma on acute plasma cortisol level following rape. Am. J. Psychiatry, 152: 1675–1677.

Schell, T.L., Marshall, G.N. and Jaycox, L.H. (2004) All symptoms are not created equal: the prominent role of hyperarousal in the natural course of posttraumatic psychological distress. J. Abnorm. Psychol., 113: 189–197.

Schelling, G., Kilger, E., Roozendaal, B., de Quervain, D.J., Briegel, J., Dagge, A., Rothenhausler, H.B., Krauseneck, T., Nollert, G. and Kapfhammer, H.P. (2004a) Stress doses of hydrocortisone, traumatic memories, and symptoms of posttraumatic stress disorder in patients after cardiac surgery: a randomized study. Biol. Psychiatry, 55: 627–633.

Schelling, G., Roozendaal, B. and De, Q.D.J.F. (2004b) Can posttraumatic stress disorder be prevented with glucocorticoids? Ann. N.Y. Acad. Sci., 1032: 158–166.

Schelling, G., Roozendaal, B., Krauseneck, T., Schmoelz, M., de Quervain, D. and Briegel, J. (2006) Efficacy of hydrocortisone in preventing posttraumatic stress disorder following critical illness and major surgery. Ann. N.Y. Acad. Sci., 1071: 46–53.

Segman, R.H., Kooper-Kazaz, R., Macciardi, F., Gulcer, T., Chalfon, Y., Dubroborski, T. and Shalev, A.Y. (2002) Association between the dopamine transporter gene and posttraumatic stress disorder. Mol. Psychiatry, 7: 903–907.

Segman, R.H. and Shalev, A.Y. (2003) Genetics of posttraumatic stress disorder. CNS Spectr., 8: 693–698.

Segman, R.H., Shefi, N., Goltser-Dubner, T., Friedman, N., Kaminski, N. and Shalev, A.Y. (2005) Peripheral blood mononuclear cells gene expression profiles identify persisting posttraumatic stress disorder among trauma survivors. Mol. Psychiatry, 10: 500–514.

Shalev, A.Y. and Freedman, S. (2005) PTSD following terrorist attacks: a prospective evaluation. Am. J. Psychiatry, 162: 1188–1191.

Shalev, A.Y., Freedman, S., Peri, T., Brandes, D., Sahar, T., Orr, S.P. and Pitman, R.K. (1998a) Prospective study of posttraumatic stress disorder and depression following trauma. Am. J. Psychiatry, 155: 630–637.

Shalev, A.Y., Peri, T., Brandes, D., Freedman, S., Orr, S.P. and Pitman, R.K. (2000) Auditory startle response in trauma survivors with posttraumatic stress disorder: a prospective study. Am. J. Psychiatry, 157: 255–261.

Shalev, A.Y., Sahar, T., Freedman, S., Peri, T., Glick, N., Brandes, D., Orr, S.P. and Pitman, R.K. (1998b) A prospective study of heart rate response following trauma and the subsequent development of posttraumatic stress disorder. Arch. Gen. Psychiatry, 55: 553–559.

Shalev, A.Y., Videlock, E.J., Peleg, T., Segman, R. and Yehuda, R. (in press) Stress hormones and post-traumatic stress disorder in civilian trauma victims: a longitudinal study. Part I: HPA-axis responses. Int. J. Neuropsychopharm.

van Stegeren, A.H., Everaerd, W. and Gooren, L.J. (2002) The effect of beta-adrenergic blockade after encoding on memory of an emotional event. Psychopharmacology (Berl.), 163(2): 202–212.

True, W.R., Rice, J., Eisen, S.A., Heath, A.C., Goldberg, J., Lyons, M.J. and Nowak, J. (1993) A twin study of genetic and environmental contributions to liability for posttraumatic stress symptoms. Arch. Gen. Psychiatry, 50: 257–264.

Vaiva, G., Ducrocq, F., Jezequel, K., Averland, B., Lestavel, P., Brunet, A. and Marmar, C.R. (2003) Immediate treatment with propranolol decreases posttraumatic stress disorder two months after trauma. Biol. Psychiatry, 54: 947–949.

Yehuda, R. (2001) Biology of posttraumatic stress disorder. J. Clin. Psychiatry, 62(Suppl 17): 41–46.

Yehuda, R. (2002a) Clinical relevance of biologic findings in PTSD. Psychiatr. Q, 73: 123–133.

Yehuda, R. (2002b) Post-traumatic stress disorder. N. Engl. J. Med., 346: 108–114.

Yehuda, R., Flory, J.D., Southwick, S. and Charney, D.S. (2006) Developing an agenda for translational studies of resilience and vulnerability following trauma exposure. Ann. N.Y. Acad. Sci., 1071: 379–396.

Yehuda, R., Siever, L.J., Teicher, M.H., Levengood, R.A., Gerber, D.K., Schmeidler, J. and Yang, R.-K. (1998) Plasma norepinephrine and 3-methoxy-4-hydroxyphenylglycol concentrations and severity of depression in combat posttraumatic stress disorder and major depressive disorder. Biol. Psychiatry, 44: 56–63.

Young, E.A. and Breslau, N. (2004) Cortisol and catecholamines in posttraumatic stress disorder: an epidemiologic community study. Arch. Gen. Psychiatry, 61: 394–401.

Zatzick, D.F., Russo, J., Pitman, R.K., Rivara, F., Jurkovich, G. and Roy-Byrne, P. (2005) Reevaluating the association between emergency department heart rate and the development of posttraumatic stress disorder: a public health approach. Biol. Psychiatry, 57: 91–95.

Zoellner, T. and Maercker, A. (2006) Posttraumatic growth in clinical psychology: a critical review and introduction of a two component model. Clin. Psychol. Rev., 26: 626–653.

General Discussion: Section III

RICHTER: I would like to come back to something that came up before: is there a syndrome called PTSD, or do we have a family of syndromes that trauma can lead to?

LIBERZON: I think somatically it is not such a difficult thing, but we need more than one syndrome, because syndrome per definition does not require any common pathophysiology, it requires a set of symptoms. Yes, it is a single syndrome; yes, because different manifestations can be represented in the same syndrome. You can have congestive heart failure due to pulmonary fibrosis, and you can have congestive heart failure due to basic cardiomyopathy, or vascular or due to anemia, eventually. So I guess I am trying to take a metaphor from the basic treatment of congestive heart failure saying that the syndromes are useful concepts that allow heterogeneity, different etiology, and pathpophysiological processes that might be treated with the same therapeutic approach in the beginning. So I think there are stages for treating the syndrome. But we are not in the stage of claiming that this is a single disorder of any kind.

BREMNER: I would just like to add what I call trauma spectrum disorders and the concept that you can have the same exposure leading to multiple outcomes. Arik Shalev had on his slides the concepts of equifinality and multifinality meaning that you can have different causes and the same outcome. So if women have been abused in early childhood some develop borderline personality disorder, some develop PTSD. You know there has been attention in the field to focus on PTSD with the exclusion of other things like borderline personality disorder, or dissociative disorders, which is unfortunate. Eric Vermetten has a paper on hippocampal volume in dissociative disorders and the magnitude of the changes is even greater than in PTSD. From my perspective I say these things so people understand, but you may not be able to have an animal model that differentiates dissociative disorder from PTSD.

YEHUDA: I think they move in and out of the same symptoms. That is what our longitudinal data suggest. A simple answer to your question is that PTSD is a singular syndrome. But what you seem to really be asking also is whether there are subtypes or different syndromes based on what happens to you. The question then is to get more precision around PTSD.

LIBERZON: If I can just add, just a touch. This is not just a problem of PTSD, it is a question of psychiatric nosology. We have created the nosology without the phenotypes and the genotypes. Kraepelin made very important observations 100 years ago suggesting that there is schizophrenia and bipolar disorder, or in fact a psychosis. But as we go to genetic predispositions and look that the risk factors for segregating are on the same chromosomes and suddenly you find out that if you look at it carefully it appears to be a bipolar disorder with psychotic features and so on, then it becomes complicated. So I think, it is a problem.

RICHTER: I agree, I asked this question because trauma is not a sufficient explanation for the syndrome. I am referring to the siblings of a Holocaust patient — siblings that never have experienced a trauma, but still show symptoms of PTSD. As a basic scientist, if I want to come back to my model I have to ask if there is a special diagnosis of something related to a specific trauma or not?

YEHUDA: Not all children of holocaust survivors have PTSD. They are three times more likely to meet, to develop PTSD to a trauma that they may experience, particularly one that involve interpersonal violence or loss. You can interview many children of holocaust survivors, even parental PTSD, and they may not come up with an event that is traumatic or they can come up with an event that is traumatic but is not on the list. It is just that they are more likely to react with PTSD than demographically similar people.

JOELS: I would like to raise the issue of definition, and I would like to raise it to this panel. I do not know if there is an exception, but I, sort of, have an uncomfortable feeling that there is a huge gap between basic research and the clinic, and that maybe the clinicians after hearing all these complicated stories yesterday thought, well, I cannot work with these confusing data. And also this morning from the clinical work there are a lot of examples to make this translational step difficult. There are some aspects of PTSD that are extremely difficult to model in rodents; for instance anything that has to do with the linguistic aspect of the disease, how are you going to model that in rodents. I also realized that animal studies allow to control conditions; for instance, if corticosteroids are released within the context of a learning situation they help to remember that situation. If out of context, the corticosteroids impair. That's basically different from giving high dosages of corticosteroids exogenously and then to compare it with what needs to be remembered. The other example is neurogenesis; we showed clearly that there are very profound functional changes in the dentate gyrus that have nothing to do with neurogenesis. So neurogenesis is a window to look at plasticity, but it may not explain functional changes. I believe because of this there is still a big gap between basic and clinical studies of PTSD.

SHALEV: Basic studies are really important for me as a clinician in order to understand the links between brain-imaging finding and neuroendocrinological findings and behavioral studies.

BREMNER: In humans you do neuropsychology testing, that is accepted as a measure of hippocampal function; you can measure volume; you can do a memory task. But some of the studies are cross-sectional and some are longitudinal. Then we formulate a hypothesis based on the animal findings and interpret our clinical data relative to the animal findings. I don't do the research on neurogenesis; I have read the literature, and I offer it as a possible explanation, what the findings could mean.

I think the point that Rachel is making is correct that where, in our data, can I point to evidence that cortisol is associated with hippocampal damage and memory deficits? Nowhere. The basic research on memory may not be directly applicable to what we see in patients but does that mean that it is irrelevant? No. It is relevant, and in this domain it is specifically relevant. And there are some things that we want to understand better, like, you say, that corticosteriods can facilitate memory and specifically under arousal conditions; I never thought about it before. But the question is whether we can test that in humans and what would be the best way to do that.

LIBERZON: Basic science tests specific models, specific mechanisms. Clinicians try to learn from it and try understanding the principle to generate some kind of pathophysiological hypothesis and models. However, no animal model would address any type of complete disorder, but it is very useful.

YEHUDA: The gap can be smaller if we make the agenda more specific and the questions more focussed. The gap will remain large if we have a very big agenda like an animal model for PTSD. If we have a specific question about either what can or cannot happen in a brain region or what system or different behaviour or not. The iterative process that we just talked about is important because, that is to say, let's put the third piece of information that resolved the discrepancy between our two pieces of data. When I started in the field and was talking to Robert Sapolsky about the low cortisol in PTSD, he said low couldn't be. That is what he said to me. I was a graduate postdoc. I know I am wrong, couldn't be me, must be wrong. Because low cortisol is incompatible with stress. Now a person can say, well if you are telling me I have low cortisol I am telling you I am stressed, so low cortisol isn't that incompatible with stress. So in other words part of the iterative process is we say: you mean to those Holocaust survivors that it wasn't stressful? And basic scientists, let me see if we can broaden it. So I think we can make the gap smaller.

GUNNAR: A very quick question from the developmental standpoint: how does the development of the prefrontal cortex and that of the hippocampus play a role in PTSD? In children in the third year we see huge effects that I have not even mentioned here, i.e., the development of white matter tracks, which are profoundly affected by early experience. I have not heard how that might play into the development and function of the brain.

LEVINE: I wanted to remain quiet. I have been listening to this discourse and I ask myself how many times have I heard it before? It is a discourse that has been heard. I look at the models you put out there, the amygdala, the hippocampus, all of that information came from basic science. Animal observations, all of the tracking of circuitry came from those basic studies. The first Klüver-Bücy study, which implicated the whole neuroanatomical basis of affect, the Papez limbic circuit, all of this was basic research. How can one possibly make the argument that basic research does not in some way contribute. The argument troubles me, it is just too redundant… . One other thing: I just want to add that the world is complex.

SECTION IV

The Clinical Perspective: Targets for Therapy in PTSD

E.R. de Kloet, M.S. Oitzl & E. Vermetten (Eds.)
Progress in Brain Research, Vol. 167
ISSN 0079-6123

CHAPTER 14

Models of PTSD and traumatic stress: the importance of research "from bedside to bench to bedside"

Robert J. Ursano*, He Li, Lei Zhang, Chris J. Hough, Carol S. Fullerton, David M. Benedek, Thomas A. Grieger and Harry C. Holloway

Department of Psychiatry and Center for the Study of Traumatic Stress, Uniformed Services University, Bethesda, MD, USA

Abstract: The epidemiology and psychology of PTSD noted above is not often considered in neurobiological models of PTSD. Neurobiological models tend to focus on symptoms. This is an important perspective but it does not capture the brains total response to traumatic events. Similarly, neurobiologists have rarely used the extensive knowledge of animal behavioral responses to stress as a means to define the human stress phenomenology, looking for the human equivalent (rather than the other way around). The development of animal models for PTSD and other traumatic stress-related brain changes is an important part of advancing our neurobiological understanding of the disease process as well as recovery, resilience, and possible therapeutic targets. Animal models should address symptoms but also other aspects of PTSD that are seen in clinical care including the waxing and waning of symptoms, Understanding "forgetting", toxic exposure, failure to recover and how the neural systems fail rather than function are important perspectives on developing animal models. The cognitive process of identification is another important animal model to develop. Using these perspectives recent work has shown new avenues for understanding the trauma response in animal models and human brain tissue of individuals with PTSD. The 5-HT2A receptor and p11 protein and associated regulators are avenues of new investigation that warrant study and consideration in models of PTSD.

Keywords: animal models; PTSD; posttraumatic stress disorder; neurobiology

The majority of people exposed to traumatic events and disasters do well; however, some individuals develop psychiatric disorders, distress, or health risk behaviors such as an increased in alcohol or tobacco use. At times, traumatic events may even have unexpected beneficial effects by serving as organizing events and providing a sense of purpose and an opportunity for positive growth experiences (Ursano, 1987; Tedeschi et al., 1998).

Exposure to a traumatic event, the essential element for the development of acute stress disorder (ASD) or posttraumatic stress disorder (PTSD), is a relatively common experience. Approximately 50–70% of the US population is exposed to a traumatic event sometime during their lifetime. However, only 5–12% develop PTSD. In a nationally representative study of 5877 people

*Corresponding author. Tel.: +1 301 295-3293; Fax: +1 301 295-2536; E-mail: rursano@usuhs.mil

DOI: 10.1016/S0079-6123(07)67014-9

aged 15–45 in the US, the National Comorbidity Study (NCS) (Kessler et al., 1995) found the lifetime prevalence of exposure to trauma to be 60.7% in men and 51.2% in women. In a nationally representative sample of women in the US, the National Women's Study (NWS) (Resnick et al., 1993) found that 69% of women were exposed to a traumatic event at sometime in their lives. Over a lifetime, any given individual is very likely to be exposed to a traumatic event.

Posttraumatic stress disorder is not uncommon following many traumatic events, from terrorism to motor vehicle accidents to industrial explosions (Breslau et al., 1991, 2005). In its acute form, PTSD may be more like the common cold, experienced at some time in ones life by nearly all. Some colds of course progress to pneumonia — and may create substantial illness, impairment of function, and be debilitating. Similarly, PTSD when it becomes chronic requires psychotherapeutic and/or pharmacological intervention.

Importantly, PTSD is not the only trauma-related disorder, nor perhaps the most common (Fullerton and Ursano, 1997; North et al., 1999; Norris et al., 2002, 2005). People exposed to traumatic events and disasters are at increased risk for depression (e.g., Miguel-Tobal et al., 2006), generalized anxiety disorder, panic disorder, and increased substance use, an important health risk behavior different from addiction or abuse (Breslau et al., 1991; Kessler et al., 1995; North et al., 1999, 2002; Vlahov et al., 2002). Nearly 40.5% of disaster workers following a plane crash met criteria for at least one diagnosis (i.e., ASD, PTSD, or depression) in a 13-month longitudinal study (Fullerton et al., 2004). Exposed disaster workers with ASD were 7.33 times more likely to meet PTSD criteria at 13 months. Forty-five percent of survivors of the Oklahoma City bombing had a post disaster psychiatric disorder. Of these 34.3% had PTSD and 22.5% had major depression (North et al., 1999). Nearly 40% of those with PTSD or depression had no previous history of psychiatric illness (North et al., 1999). Death is a common aspect of traumatic events and disasters in particular. Those who identify with the dead — that is they think "It could have been me" — are at increased risk of PTSD (Ursano et al., 1992, 1999).

Implications of epidemiology and psychology of PTSD for neurobiological models of PTSD

The epidemiology and psychology of PTSD noted above is not often considered in neurobiological models of PTSD. Neurobiological models tend to focus on symptoms. This is an important perspective but it does not capture the brains total response to traumatic events. The breadth of what we know about traumatic responses is important to our ability to understand the brains response to traumatic events. Similarly, neurobiologists have rarely used the extensive knowledge of animal behavioral responses to stress as a means to define the human stress phenomenology, looking for the human equivalent (rather than the other way around). The phenomenology of our traumatic stress-related disorders could be well informed by such a strategy. Rather than assume that only some human aspects of traumatic stress can be modeled in animals, perhaps the aspects of behavior seen in animal models should be sought more diligently in human responses. Alternatively their absence calls for understanding the genomic or proteomic differences that may explain this response pattern difference.

PTSD is a waxing and waning disorder — perhaps more like multiple sclerosis in this way than atherosclerosis. That is the disorder can come and go, be more or less present across a long period of time. Animal models rarely model such vacillation in symptoms. Most examine fixed symptomatology. There are animal models of the fading of behavioral changes after stress and their ease of return after or without re-exposure.

Animal models of PTSD have rarely considered that everyone does not develop the traumatic response after the event. Most animal studies examine the animals that show symptoms; behavioral changes after the stressor rather than the individual differences.

PTSD is also a "toxic exposure", not only metaphorically but also in the stimuli (i.e., threat to life, exposure to death) that result in PTSD. The "toxic model" brings a number of questions into focus, in addition to being understandable to many non scientists. Toxins have dose, duration, and susceptible organs or functions. Is the toxicity of threat to life the same as the toxicity of witnessing death of a loved one? Is the dose, duration, and response curve the same? What are the useful animal models of witnessing death?

PTSD is a disorder of forgetting perhaps even more than of remembering. It is the inability to forget that leads to the pathology and suffering in PTSD. Forgetting is often overlooked or even avoided in clinical practice. Yet it is a critical component of recovery. Of course if we could not "forget", our brains would rapidly be cluttered with information and observations and perhaps limit our cognitive control functions for other activities (Kuhl et al., 2007).

Extinction is one avenue for examining "forgetting" (Quirk et al., 2000; Myers and Davis, 2002; Sharot et al., 2004). Extinction — which may also be conceived of as new learning — is a potentially important mechanism for PTSD formation and also for its treatment (Ressler et al., 2004). Similarly, differences in brain regions in their response to reversal of cues (Morris and Dolan, 2004) indicating a threatening or painful state is no longer present may lead to new avenues of considering forgetting, extinction, and the inability to recognize safety after exposure to threat.

Importantly PTSD can also be considered a "breakdown" of our usual neural functions. Engineering models are often built around the concept of "failure mode analysis", that is, "How can we break the machine? How does it breakdown?" Breakdown is not the same as function. It draws our attention to the neurobiology that may be least redundant, most subject to interference or disruption. How our "forgetting system" may breakdown is intrinsically a different question from how does it function.

If PTSD is the failure to recover (rather than the onset of disease), what is the breakdown that has occurred? Perhaps some aspect of our "immunological" response to traumatic stress has not operated sufficiently or perhaps at all? The contribution of identification with the dead ("It could have been me. It could have been my child") as a cognitive risk factor for development of PTSD (Ursano et al., 1992, 1999) indicates a failure of a normally adaptive and health promoting cognitive mechanism, identification.

We most often think of identification as growth promoting. PTSD from this vantage is an autoimmune disorder — a mechanism usually growth enhancing and protective (e.g., identification) is causing disease. Identification with the aggressor and the Stockholm syndrome are additional examples of identification gone awry. Why in high stress settings is identification likely or possibly able to induce pathological neurobiological and psychological changes?

Identification appears to be a one of the neural systems that supports cognitive up-regulation of emotion — both negative and positive. Medial prefrontal regions (BA32) have been implicated in regulation of emotion related to focusing on personal relevance of negative emotions (Ochsner et al., 2004). In these studies the participants imagined themselves or their loved one as the central figure in the situation. This is in contrast to cognitions related to altered anticipation of future events which recruit lateral prefrontal regions. Thus different cognitive approaches to traumatic stimuli may involve different neurocircuitry and modulate amygdala response in different ways.

Animal models

The combination of the ability to examine human brain tissue and animal models holds great promise for furthering studies of traumatic stress responses and PTSD in particular. Historically, adequate animal models have enhanced our ability to develop effective therapeutic treatments. The development of animal models is critical for the study of therapeutic and prophylactic treatments of stress-associated psychiatric disorders, such as PTSD. The primary animal models of interest have included models of (a) predator stress, (b) social defeat, (c) shock, (d) restraint and shock, and (e) serial prolonged stress. Different strains as well as knock out and knock in strains offer additional sources of study.

In the predator stress model, rats are exposed for 5 min in a novel room where they are threatened but not attacked by a cat. In the social defeat model, rats are exposed to a more aggressive rat from the same or different strain once on 5 successive days for 10–60 min per day. After the first defeat the rat is usually protected from the more aggressive rat but is in a cage in sight and smell of the more aggressive rat. In the shock model over one or a few (2–5) sessions (days), one long (10 s–3 min) or a few short shocks (1–10 times, 1–6 s) are delivered through a grid floor or to the tail. Longer times cause behavioral inhibition that dissipates within 3 days. In the restraint and tail-shock model, the rat is restrained and shocked randomly on the tail for 2 h on 3 successive days. In serial prolonged stress rats are exposed sequentially to 2 h of restraint, 20 min of swimming, and exposed to ether until loss of consciousness. In the mouse foot shock model B6N strain is most susceptible to PTSD like symptoms. This strain also had the widest variance in responses. B6JOla strain was less susceptible (Hough et al., 2007).

The emotional and health-related consequences of aversive experiences are usually much worse when the organism has no control over the aversive event. Therefore chronic uncontrollable stress (learned helplessness) is an animal model of traumatic stress and depressive illness of particular interest (Minor and Hunter, 2002; Drevets, 2003; Maier and Watkins, 2005). Indeed, chronic, uncontrollable stress causes behavioral, endocrine, immune, and neurotransmitter disturbances in the animals that are similar to those observed in depressive illness and associated with anxiety symptoms (Leonard, 2001). Since PTSD, to some extent, overlaps with depressive illness in terms of etiology and certain features, the animal model of PTSD modified from learned helplessness paradigm has been shown to mimic to a substantial extent symptoms of PTSD in terms of pathophysiological measurements (Servatius et al., 1995; Garrick et al., 2001; Braga et al., 2004). Our lab (Li, H and Zhang, L) and several others have examined the inescapable tail-shock model of stress in rats and verified the behavioral and neurobiological alterations induced by this rat model (as shown in Table 1) are similar in many aspects to those found in PTSD subjects (Servatius et al., 1995; Garrick et al., 2001).

Using this animal model, we recently found that the 5-hydroxytryptophan (5-HT)2A receptor and messenger RNA are down-regulated after traumatic stress exposure (H. Li laboratory, unpublished observation) indicating 5-HT2A receptor-mediated functions are impaired after exposure to traumatic stress. Alterations in the 5-HT2A receptor function in the amygdala as a function of stress are being now further examined by electrophysiological, molecular, and behavioral approaches in our ongoing research.

5-HT2 receptor in amygdala: stress and traumatic memory

The ability to react to threatening events in the environment is one of the vital adaptive survival

Table 1. Comparison of symptoms of PTSD in humans to dysfunction related to stress in rats

PTSD in humans	Inescapable tail-shock model of stress in rats
Weight loss of comorbid depression (Sutker et al., 1990; Myers et al., 2005; Braga et al., 2004)	Suppressed feeding and body weight loss (Hu et al., 2000; Harris et al., 2002)
Difficulty falling or staying asleep, nightmares (Maher et al., 2006)	Altered sleep patterns (Adrien et al., 1991)
Psychomotor numbness (Epstein et al., 1998; Lopez-Ibor, 2002)	Persistent behavioral abnormalities, i.e., suppressed open-field activity, longer hanging wire latencies (Minor et al., 1984; Pare, 1994)
Poor concentration; memory deficits (Green, 2003; Bremner et al., 2004; Isaac et al., 2006; Jelinek et al., 2006)	Deficits in escape/avoidance learning and learning of an appetitive task (Maier, 2001)
Hypervigilance and/or exaggerated startle response (Pitman et al., 1999; Orr and Roth, 2000; Orr et al., 2002)	Exaggerated startle (Servatius et al., 1995; Garrick et al., 2001; Manion et al., 2007)
Hyperresponsiveness of the noradrenergic system (Maes et al., 1999; Orr and Roth, 2000)	Hyperresponsiveness of the noradrenergic system (Simson and Weiss, 1988)

traits of organisms. Through our ability to recognize "lions, tigers and bears" we have been able to survive as a species. Our brain is a well-developed organism for identifying reminders of past threats in order to mobilize our response and protect our lives. PTSD in many ways is a deficit in this process. The brain is identifying threats which are not present. Through over generalization or sensitivity to cues or hyper responsive neural circuitry, those with PTSD respond as if threat is present even when it is not. Pharmacological intervention targeting stress-induced over activation of neurotransmitter receptors serves as one model that leads to a key strategy to prevent the pathophysiological alteration and promote the resilience process of those with PTSD.

Serotonin, stress, and stress-related psychiatric disorders

Cumulative evidence indicates that serotonin dysregulation involves in the pathophysiology of trauma-related symptoms. In humans, dysfunction of the serotonergic system has been closely associated with depression (Brown and Linnoila, 1990), anxiety, aggression, impulsivity (Brown and Linnoila, 1990), and suicidal behaviors (Stanley and Stanley, 1990). Many of these symptoms are also associated with PTSD. Deficits in serotonin (5-HT) metabolism (Arora et al., 1993; Fichtner et al., 1995; Maes et al., 1999; van Praag, 2004b) and serotonin receptor disturbances (Davis et al., 1997; van Praag, 2004a, b) have been consistently observed in patients with PTSD and in certain subgroups of depressive patients. The alterations of serotonin 5-HT1A and 5-HT2 receptors in these disorders appear to be associated with a cognitive deficiency observed in PTSD and certain subgroups of depressive illnesses (van Praag, 2004a, b).

Alterations in the serotonin system in trauma-related disorders is seen in experimental animals after sustained stress (van Praag, 2004b). For instance, sustained stress leads to a decrease in 5-HT turnover in rat brain (van Praag, 2004b). Chronic stress also alters 5-HT1A and 5-HT2 receptor expression and signaling in the brain regions that participate in stress and emotion response in rodents, including the hippocampus (van Riel et al., 2003; Dwivedi et al., 2005; Matsumoto et al., 2005), prefrontal cortex (PFC, Harvey et al., 2003; Dwivedi et al., 2005), amygdala (Wu et al., 1999), and hypothalamus (Wu et al., 1999; Dwivedi et al., 2005).

Over the past two decades, numerous studies have used experimental animals, in an inescapable stress condition, to understand the role of the serotonergic system in pathogenesis of depression and anxiety-like symptoms (Minor and Hunter, 2002; Maier and Watkins, 2005; Amat et al., 2005; Robbins, 2005). These studies reveal that inescapable stress strongly activates the neuronal activity in dorsal raphe nucleus (DRN), which dramatically increases serotonin release in its projected brain regions, including the amygdala, prefrontal cortex, hippocampus, and hypothalamus (see Fig. 1). Since the amygdala is the brain region with the intense serotonergic innervations from the DRN (Tork, 1990; Jacobs and Azmitia, 1992), the over activation of DRN 5-HTergic neurons following inescapable stress, can induce heightened 5-HT concentrations in the amygdala complex for a prolonged period of time (Amat et al., 1998; Minor and Hunter, 2002; Maier and Watkins, 2005). Thus, such long-term exposure of the amygdala neuron to elevated levels of 5-HT, may contribute to the development of debilitative consequences of stress response (Minor and Hunter, 2002).

Genetic variations of the human serotonin transporter and serotonin synthesizing enzyme, tryptophan hydroxylase 2 (TPH2), which is a rate-limiting enzyme in the biosynthesis of serotonin from tryptophan in the brain, can affect amygdala neuronal activity in response to emotional stimuli and might predispose the individual to anxiety and stress disorders (Hariri et al., 2002, 2005; Brown et al., 2005; Canli et al., 2005).

Pharmacological blockade of 5-HT2 receptors has a specific influence on mood and amygdala-related physiological or pathological conditions such as fear conditioning, anxiety, and PTSD (Hertzberg et al., 1998). Prophylactic administration of selective 5-HT2A receptor antagonist, EMD 281014, could prevent predator stress-induced

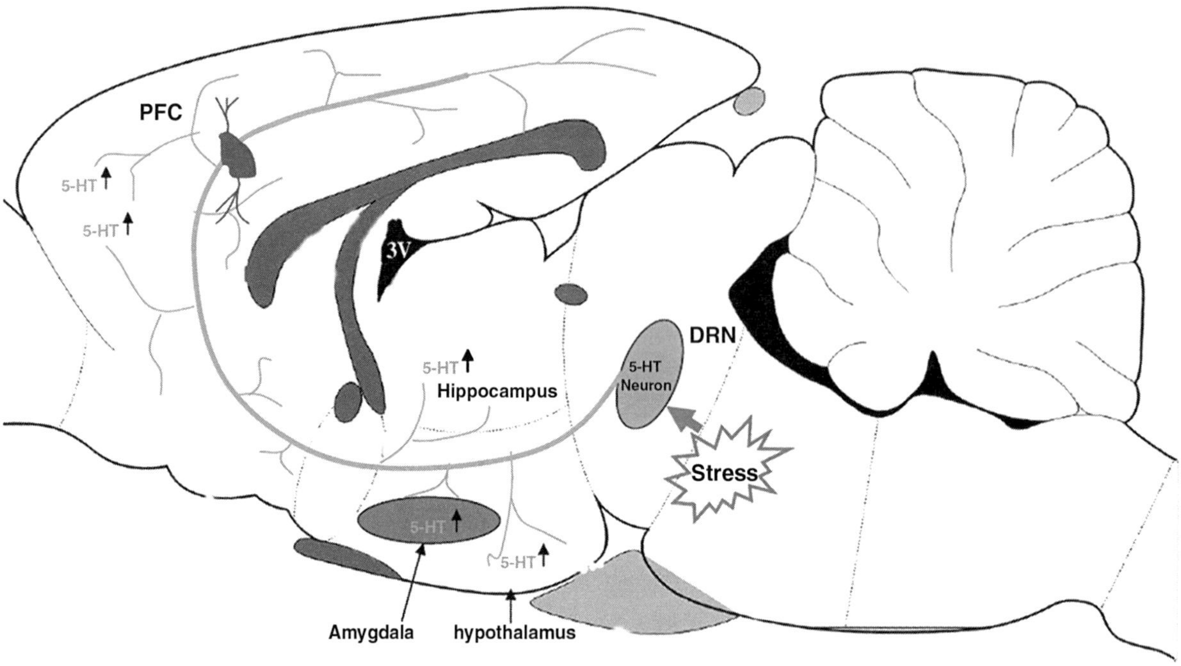

Fig. 1. Inescapable stress strongly activates the dorsal raphe nucleus (DRN) and dramatically enhances the 5-HT levels in its principal projecting regions, including the amygdala, prefrontal cortex, hippocampus, and hypothalamus.

enhanced acoustic startle and increased open arm avoidance test in a dose-dependent manner, indicating a potential clinical application of selective 5-HT2 receptor agents in prevention of traumatic stress-induced anxiety disorders (Adamec et al., 2004). 5-HT2-receptor activation has also been found to potentiate NMDA receptor-mediated currents and potentials (Blank et al., 1996; Chen et al., 2003b), NMDA-induced depolarization of neocortical neurons (Rahman and Neuman, 1993), to enhance NMDA receptor activation induced increase of intracellular calcium signaling and NMDA receptor-dependent long-term potentiation, a cellular model of learning and memory, in amygdala circuitry (Chen et al., 2003b). Thus, the efficacy of prophylactic administration of 5-HT2A receptor antagonist appears to be mediated by preventing traumatic memory formation or consolidation in the brain region associated with 5-HT2 receptor expression.

Several recent studies also reveal that 5-HT receptor agonists could modulate amygdala excitability by exciting interneurons and increasing GABAergic synaptic transmission in the basolateral amygdala (BLA), especially at relatively low concentrations (Stutzmann et al., 1998; Stutzmann and LeDoux, 1999; Rainnie, 1999b; Stein et al., 2000). The receptors involved in this effect appear to be mediated via the 5-HT2 receptor subtype (Rainnie, 1999a; Stein et al., 2000). The roles of 5-HT2 receptor in amygdala GABA transmission as well as in the orchestration and modulation of the organism's response to aversive, stressful events remains to be determined and is an important area for study of the "hyper responsive" fear circuitry.

Extinction and memory mechanisms in amygdala neurocircuitry

Traumatic memories associated with life threatening events are readily retrieved (manifested as flashbacks) in subjects suffering from PTSD, indicating that such memory traces are consolidated in the associated neuronal structure in the brain. The amygdala complex is activated during emotional experiences via cortical and thalamic afferents to the lateral amygdala (LA), basolateral (BLA), and central (CA) amygdaloid nuclei. Traumatic memory traces such as fear conditioning learning experiences have been demonstrated to enhance synaptic efficacy in the amygdala neuronal circuitry (Rumpel et al., 2005b). Such synaptic modification in the process of fear learning are similar to the induction of long-term synaptic potentiation (Aroniadou-Anderjaska et al., 2001; Li et al., 2001; Chen et al., 2003a). It requires activation of NMDA receptors as well as transfer of GluR1 AMPA receptors to the postsynaptic membrane (Rumpel et al., 2005a). In addition, interference with GluR1 receptor trafficking impairs amygdala LTP as well as fear conditioning, which indicates that GluR1 AMPA receptor may serve as a molecule that contributes to memory formation (Rumpel et al., 2005c). As reported in the literature, approximately 30% of those who encounter traumatic experiences develop PTSD after 3 months, indicating that there is a failed extinction of traumatic memory in these subjects. In studies of experimental extinction in a fear conditional paradigm in rats, successful reduction in fear response was achieved by unpairing the conditioned stimulus (CS) from the unconditioned stimulus (US). However, the gradually declining fear memory could be reinstated if the CS was recoupled with the US, indicating that extinction training does not erase initially acquired traumatic memory but instead forms a new inhibitory learning process that prevents the expression of the original aversive memory. In a recent study using an enhanced extinction paradigm based on bilateral infusion of D-cycloserine, a partial agonist of the glycine site on the NMDA receptor (Mao et al., 2006b), into the amygdala 30 min before extinction training, the extinction training response was augmented and elicited a reduction in startle and reversed the conditioning-induced increase in GluR2 AMPA receptor (Mao et al., 2006a). However, when D-cycloserine was administered 24 h after extinction training, extinction training reduced only the enhanced startle response without influencing increased GluR1 receptor expression. These results indicate that appropriate therapeutic approaches must be applied within a critical time window after exposure to traumatic stress in order to achieve the most efficacious treatment of PTSD.

Most recently studies of Depue et al. (2007) have provided evidence that brain has an active process for suppressing memories (e.g., forgetting). In an fMRI study in which subjects were asked to not think of a particular stimuli, two areas of the PFC were activated, the right inferior frontal gyrus and right medial frontal gyrus. The amygdala, hippocampus, and thalamus showed decreased activity. These findings support the active process of "forgetting". This circuitry is important to both resilience and recovery from PTSD — the ability to forget traumatic memories. Forgetting thus may be accompanied by decreased emotional response (thalamus, amydala) as well as reducing demand on cognitive control that involve the anterior cingulated, ventral, and dorsal lateral PFC freeing neural processing for other activities (Kuhl et al., 2007).

Human brain studies of traumatic stress and PTSD: potential biomarkers and novel therapeutic targets for PTSD

Biomarkers are increasingly used to diagnose disease conditions promptly and accurately and to identify individuals at high-risk for certain diseases (Schmidt, 2006). The diagnosis of PTSD is currently dependant on the assessment of clinical symptoms. Biomarkers may be related to the diagnosis of PTSD or risk factors to predict a vulnerability to the disorder. In addition, however, biomarkers may be able to aid in predicting treatment response or course of illness (Vieweg et al., 2006). Combinations of new technologies such as peripheral blood gene assays and neuroimaging may create unique opportunities for diagnostic determination, assessment of response to treatment, and/or prediction of course of illness/recovery. The development of such tools requires a strategy to identify biomarkers for PTSD, advanced research into the underlying molecular mechanisms of PTSD, and a detailed neuroimaging understanding of onset, course, and recovery from PTSD. Examination of gene expression profiling and the examination of gene products (proteins) are particularly important approaches to advance our ability to identify potential PTSD biomarkers.

High-throughput "omic" approaches including genomics, proteomics, and metabolomics have been used in the study of biomarker selection and identification. These high-throughput "omic" approaches have greatly facilitated the process of biomarker searching. However, even though we have increased our efforts in the search for PTSD biomarkers, progress has been slow and often frustrating, because of the undefined validation process, the lack of brain tissues from living PTSD patients, and the complexity of the molecular mechanism of the disease.

Recently we and the Traumatic Stress Brain Study Group have initiated a program to identify potential biomarkers for PTSD based on research utilizing an animal model of PTSD and postmortem brain tissue of PTSD patients. These approaches have allowed us to make initial observations on a potential biomarker in PTSD patients and to validate it in an animal model of traumatic stress. This potential biomarker is a depression-associated protein, called p11 (Svenningsson et al., 2006).

We (Laboratory of L. Zhang and H. Li) observed that p11 is over-expressed in PFC area 46 of PTSD postmortem brain, an area associated with PTSD (Fig. 2). These data were validated in a rat model of traumatic stress through the use of a DNA microarray and quantitative real time PCR. p11 was up-regulated in the PFC of rats exposed to inescapable shock. These rats showed behavioral disturbances similar to those seen in PTSD patients and p11 protein over-expression similar to that seen in the postmortem PFC of PTSD patients.

To determine whether p11 could serve as a biomarker for PTSD, we also measured p11 gene expression in peripheral blood mononuclear cells (PBMC) of PTSD patients ($n = 13$) and normal control subjects ($n = 15$) by real time PCR. We were surprised to find that p11 mRNA significantly decreased in PTSD patients compared to controls ($p < 0.05$). These findings provide evidence that p11 gene expression is down-regulated in peripheral blood, although it is unregulated in the CNC. Although these findings are early and preliminary they indicate a process for moving forward in identifying biomarkers for PTSD and traumatic stress responses.

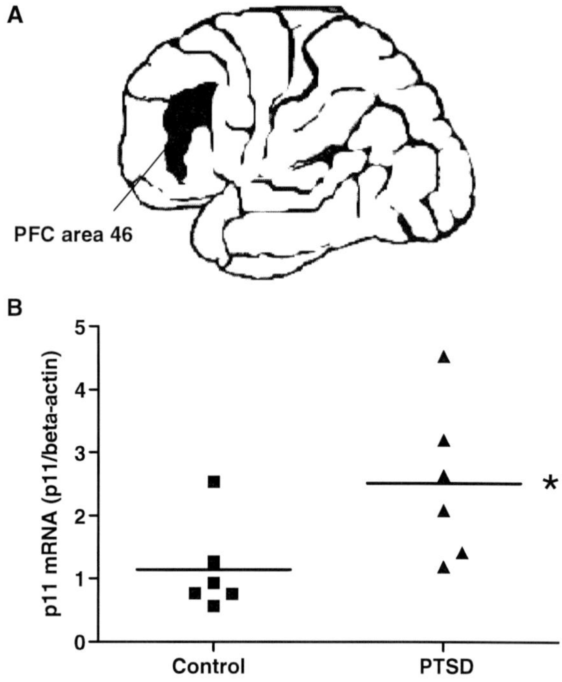

Fig. 2. p11 mRNA expression in postmortem PFC is significantly increased in patients with PTSD compared to age- and sex-matched controls. (A) Perspective of the human brain to show the prefrontal cortex (area 46). (B) p11 mRNA in PFC (area 46) is significantly greater in PTSD patients compared to age- and sex-matched controls ($n = 6$ per group). Data are shown as means $\pm$SEM, $^{*}p < 0.05$ (control vs. PTSD) and have been analyzed by student's t test (Laboratory of L. Zhang and the Traumatic Stress Brain Studies Group, CSTS).

p11 and potential therapeutic targets for PTSD

Recently, a study demonstrated that Dex, a synthetic glucocorticoid, can up-regulate p11, a S-100 calcium-binding protein (Yao et al., 1999). p11 has been shown to form a complex with 5-hydroxytryptamine-1B (5-HT1B) receptors (Svenningsson et al., 2006). It can then transport 5-HT1B receptors from the cytosol to the cell surface membrane, where it regulates 5-HT release. These observations have led to two hypotheses: first, p11 expression is mediated by glucocorticoid receptors in PTSD patients under traumatic stress. Second, up-regulated p11 affects 5-HT1B receptor translocation and function. Indeed, we (Laboratory of L. Zhang and H. Li) found traumatic stress-induced induction of p11 in the PFC and elevated levels of corticosterone in the plasma of rats that have experienced inescapable tail-shock. Dex has been shown to up-regulate p11 expression in SH-SY5Y cells through glucocorticoid response elements (GREs) within the p11 promoter. This response can be attenuated by either the glucocorticoid receptor antagonist, RU486, or by mutating two of the three GREs (GRE2 and GRE3) in the p11 promoter. p11 not only expresses in the neuron but also can be up-regulated by glucocorticoid treatment. These data support the idea that p11 can be a therapeutic target in PTSD. Direct blockade of glucocorticoid receptors appears to block traumatic stress-induced p11 over-expression, thus enhancing membrane expression of the 5-HT1B receptor. Alternatively, direct blockade of the formation of p11/5-HT1B complex will decrease 5-HT1B function on the membrane.

5-HT1B, formerly designated 5-HT 1D beta receptors are receptors that mediate serotonergic neurotransmission (Pauwels, 1997). In the brain, they are highly enriched in the globus pallidus and the substantia nigra. Presynaptic 5-HT1B receptors are involved in controlling the release of 5-HT, acetylcholine, glutamate, dopamine, noradrenalin, and gamma-aminobutyric acid. Since selective blockade of central 5-HT1B autoreceptors facilitates 5-HT neurotransmission, the use of a 5-HT1B antagonist is a novel approach to antidepressant therapy. Abnormalities of 5-HT1B receptors are associated with depression (Moret and Briley, 2000), a disease that is highly comorbid with PTSD. Chronic administration of selective serotonin reuptake inhibitors leads to the desensitisation of terminal 5-HT1B autoreceptors (Moret and Briley, 2000). Hyper locomotion is mediated by 5-HT1B receptors. Moreover, the 5-HT1B receptors also affect aggressive behavior. Thus, 5-HT1B receptors also represent important potential pharmacologic targets for the treatment of several trauma-related disorders. Intervention of glucocorticoid activation or p11/5-HT1B receptor complex formation therefore may be novel therapeutic targets for the treatment of PTSD.

Conclusion

The development of animal models for PTSD and other traumatic stress-related brain changes is an important part of advancing our neurobiological understanding of the disease process as well as recovery, resilience, and possible therapeutic targets. Animal models should address symptoms but also other aspects of PTSD that are seen in clinical care. Understanding "forgetting", identification and how the neural systems fail rather than function are important perspectives on developing animal models. Using these perspectives recent work has shown new avenues for understanding the trauma response in animal models and human brain tissue of individuals with PTSD. The 5-HT2A receptor and p11 protein and associated regulators are avenues of new investigation that warrant study and consideration in models of PTSD.

References

Adamec, R., Creamer, K., Bartoszyk, G.D. and Burton, P. (2004) Prophylactic and therapeutic effects of acute systemic injections of EMD 281014, a selective serotonin 2A receptor antagonist on anxiety induced by predator stress in rats. Eur. J. Pharmacol., 504: 79–96.

Adrien, J., Dugovic, C. and Martin, P. (1991) Sleep–wakefulness patterns in the helpless rat. Physiol. Behav., 49: 257–262.

Amat, J., Baratta, M.V., Paul, E., Bland, S.T., Watkins, L.R. and Maier, S.F. (2005) Medial prefrontal cortex determines how stressor controllability affects behavior and dorsal raphe nucleus. Nat. Neurosci., 8: 365–371.

Amat, J., Matus-Amat, P., Watkins, L.R. and Maier, S.F. (1998) Escapable and inescapable stress differentially alter extracellular levels of 5-HT in the basolateral amygdala of the rat. Brain Res., 812: 113–120.

Aroniadou-Anderjaska, V., Post, R.M., Rogawski, M.A. and Li, H. (2001) Input-specific LTP and depotentiation in the basolateral amygdala. Neuroreport, 12: 635–640.

Arora, R.C., Fichtner, C.G., O'Connor, F. and Crayton, J.W. (1993) Paroxetine binding in the blood platelets of post-traumatic stress disorder patients. Life Sci., 53: 919–928.

Blank, T., Zwart, R., Nijholt, I. and Spiess, J. (1996) Serotonin 5-HT2 receptor activation potentiates N-methyl-D-aspartate receptor-mediated ion currents by a protein kinase C-dependent mechanism. J. Neurosci. Res., 45: 153–160.

Braga, M.F., roniadou-Anderjaska, V., Manion, S.T., Hough, C.J. and Li, H. (2004) Stress impairs alpha(1A) adrenoceptor-mediated noradrenergic facilitation of GABAergic transmission in the basolateral amygdala. Neuropsychopharmacology, 29: 45–58.

Bremner, J.D., Vermetten, E., Afzal, N. and Vythilingam, M. (2004) Deficits in verbal declarative memory function in women with childhood sexual abuse-related posttraumatic stress disorder. J. Nerv. Ment. Dis., 192: 643–649.

Breslau, N., Davis, G.C., Andreski, P. and Peterson, E. (1991) Traumatic events and posttraumatic stress disorder in an urban population of young adults. Arch. Gen. Psychiatry, 48: 216–222.

Breslau, N., Reboussin, B.A., Anthony, J.C. and Storr, C.L. (2005) The structure of posttraumatic stress disorder. latent class analysis in 2 community samples. Arch. Gen. Psychiatry, 62: 1343–1351.

Brown, G.L. and Linnoila, M.I. (1990) CSF serotonin metabolite (5-HIAA) studies in depression, impulsivity, and violence. J. Clin. Psychiatry, 51(Suppl.): 31–41.

Brown, S.M., Peet, E., Manuck, S.B., Williamson, D.E., Dahl, R.E., Ferrell, R.E. and Hariri, A.R. (2005) A regulatory variant of the human tryptophan hydroxylase-2 gene biases amygdala reactivity. Mol. Psychiatry, 10: 884–888 805.

Canli, T., Congdon, E., Gutknecht, L., Constable, R.T. and Lesch, K.P. (2005) Amygdala responsiveness is modulated by tryptophan hydroxylase-2 gene variation. J. Neural Transm., 112: 1479–1485.

Chen, A., Hough, C.J. and Li, H. (2003a) Serotonin type II receptor activation facilitates synaptic plasticity via *n*-methyl-aspartate-mediated mechanism in the rat basolateral amygdala. Neuroscience, 119: 53–63.

Chen, A., Hough, C.J. and Li, H. (2003b) Serotonin type II receptor activation facilitates synaptic plasticity via *N*-methyl-D-aspartate-mediated mechanism in the rat basolateral amygdala. Neuroscience, 119: 53–63.

Davis, L.L., Suris, A., Lambert, M.T., Heimberg, C. and Petty, F. (1997) Post-traumatic stress disorder and serotonin: new directions for research and treatment. J. Psychiatry Neurosci., 22: 318–326.

Depue, B.E., Curran, T. and Banich, M.T. (2007) Prefrontal regions orchestrate suppression of emotional memories via a two-phase process. Science, 317(5835): 215–219.

Drevets, W.C. (2003) Neuroimaging abnormalities in the amygdala in mood disorders. Ann. N.Y. Acad. Sci., 985: 420–444.

Dwivedi, Y., Mondal, A.C., Payappagoudar, G.V. and Rizavi, H.S. (2005) Differential regulation of serotonin (5HT)2A receptor mRNA and protein levels after single and repeated stress in rat brain: role in learned helplessness behavior. Neuropharmacology, 48: 204–214.

Epstein, R.S., Fullerton, C.S. and Ursano, R.J. (1998) Posttraumatic stress disorder following an air disaster: a prospective study. Am. J. Psychiatry, 155: 934–938.

Fichtner, C.G., O'Connor, F.L., Yeoh, H.C., Arora, R.C. and Crayton, J.W. (1995) Hypodensity of platelet serotonin uptake sites in posttraumatic stress disorder: associated clinical features. Life Sci., 57: L37–L44.

Fullerton, C.S. and Ursano, R.J. (Eds.). (1997) Posttraumatic stress disorder: acute and long term responses to trauma and disaster. American Psychiatric Press, Washington, DC.

Fullerton, C.S., Ursano, R.J. and Wang, L. (2004) Acute stress disorder, posttraumatic stress disorder, and depression in disaster or rescue workers. Am. J. Psychiatry, 161: 1370–1376.

Garrick, T., Morrow, N., Shalev, A.Y. and Eth, S. (2001) Stress-induced enhancement of auditory startle: an animal model of posttraumatic stress disorder. Psychiatry, 64: 346–354.

Green, B. (2003) Post-traumatic stress disorder: symptom profiles in men and women. Curr. Med. Res. Opin., 19: 200–204.

Hariri, A.R., Drabant, E.M., Munoz, K.E., Kolachana, B.S., Mattay, V.S., Egan, M.F. and Weinberger, D.R. (2005) A susceptibility gene for affective disorders and the response of the human amygdala. Arch. Gen. Psychiatry, 62: 146–152.

Hariri, A.R., Mattay, V.S., Tessitore, A., Kolachana, B., Fera, F., Goldman, D., Egan, M.F. and Weinberger, D.R. (2002) Serotonin transporter genetic variation and the response of the human amygdala. Science, 297: 400–403.

Harris, R.B.S., Mitchell, T.D., Simpson, J., Redmann Jr., S.M., Youngblood, B.D. and Ryan, D.H. (2002) Weight loss in rats exposed to repeated acute restraint stress is independent of energy or leptin status. Am. J. Physiol. Regul. Integr. Comp. Physiol., 282: R77–R88.

Harvey, B.H., Naciti, C., Brand, L. and Stein, D.J. (2003) Endocrine, cognitive and hippocampal/cortical 5HT1A/2A receptor changes evoked by a time-dependent sensitisation (TDS) stress model in rats. Brain Res., 983: 97–107.

Hertzberg, M.A., Feldman, M.E., Beckham, J.C., Moore, S.D. and Davidson, J.R. (1998) Open trial of nefazodone for combat-related posttraumatic stress disorder. J. Clin. Psychiatry, 59: 460–464.

Hough, C.J., the Posttraumatic Stress Brain CSTS Group (2007) Animal models of PTSD, presented to Neurobiology of PTSD and Stress Seminar, Center for the Study of Traumatic Stress, Bethesda, MD, U.S.A., July 2007.

Hu, Y., Cardounel, A., Gursoy, E., Anderson, P. and Kalimi, M. (2000) Anti-stress effects of dehydroepiandrosterone: protection of rats against repeated immobilization stress-induced weight loss, glucocorticoid receptor production, and lipid peroxidation. Biochem. Pharmacol., 59: 753–762.

Isaac, C.L., Cushway, D. and Jones, G.V. (2006) Is posttraumatic stress disorder associated with specific deficits in episodic memory? Clin. Psychol. Rev., 26: 939–955.

Jacobs, B.L. and Azmitia, E.C. (1992) Structure and function of the brain serotonin system. Physiol. Rev., 72: 165–229.

Jelinek, L., Jacobsen, D., Kellner, M., Larbig, F., Biesold, K.H., Barre, K. and Moritz, S. (2006) Verbal and nonverbal memory functioning in posttraumatic stress disorder (PTSD). J. Clin. Exp. Neuropsychol., 28: 940–948.

Kessler, R.C., Sonnega, A., Bromet, E., Hughes, M. and Nelson, C.B. (1995) Posttraumatic stress disorder in the National Comorbidity Survey. Arch. Gen. Psychiatry, 52: 1048–1060.

Kuhl, B.A., Dudukovic, N.M., Kahn, I. and Wagner, A.D. (2007) Decreased demands on cognitive control reveal the neural processing benefits of forgetting. Nat. Neurosci., 10: 908–914.

Leonard, B.E. (2001) Stress, norepinephrine and depression. J. Psychiatry Neurosci., 26(Suppl): S11–S16.

Li, H., Chen, A., Xing, G., Wei, M.L. and Rogawski, M.A. (2001) Kainate receptor-mediated heterosynaptic facilitation in the amygdala. Nat. Neurosci., 4: 612–620.

Lopez-Ibor, J.J. (2002) The classification of stress-related disorders in ICD-10 and DSM-IV. Psychopathology, 35: 107–111.

Maes, M., Lin, A.H., Verkerk, R., Delmeire, L., Van, G.A., Van der, P.M. and Scharpe, S. (1999) Serotonergic and noradrenergic markers of post-traumatic stress disorder with and without major depression. Neuropsychopharmacology, 20: 188–197.

Maher, M.J., Rego, S.A. and Asnis, G.M. (2006) Sleep disturbances in patients with post-traumatic stress disorder: epidemiology, impact and approaches to management. CNS Drugs, 20: 567–590.

Maier, S.F. (2001) Exposure to the stressor environment prevents the temporal dissipation of behavioral depression/learned helplessness. Biol. Psychiatry, 49: 763–773.

Maier, S.F. and Watkins, L.R. (2005) Stressor controllability and learned helplessness: the roles of the dorsal raphe nucleus, serotonin, and corticotropin-releasing factor. Neurosci. Biobehav. Rev., 29: 829–841.

Manion, S.T., Gamble, E.H. and Li, H. (2007) Prazosin administered prior to inescapable stressor blocks subsequent exaggeration of acoustic startle response in rats. Pharmacol. Biochem. Behav., 86: 559–565.

Mao, S.C., Hsiao, Y.H. and Gean, P.W. (2006a) Extinction training in conjunction with a partial agonist of the glycine site on the NMDA receptor erases memory trace. J. Neurosci., 26: 8892–8899.

Mao, S.C., Hsiao, Y.H. and Gean, P.W. (2006b) Extinction training in conjunction with a partial agonist of the glycine site on the NMDA receptor erases memory trace. J. Neurosci., 26: 8892–8899.

Matsumoto, M., Higuchi, K., Togashi, H., Koseki, H., Yamaguchi, T., Kanno, M. and Yoshioka, M. (2005) Early postnatal stress alters the 5-HTergic modulation to emotional stress at postadolescent periods of rats. Hippocampus, 15: 775–781.

Miguel-Tobal, J.J., Cano-Vindel, A., Gonzalez-Ordi, H., Iruarrizaga, I., Rudenstine, S., Vlahov, D. and Galea, S. (2006) PTSD and depression after the Madrid March 11 train bombings. J. Trauma. Stress, 19: 69–80.

Minor, T.R. and Hunter, A.M. (2002) Stressor controllability and learned helplessness research in the United States: sensitization and fatigue processes. Integr. Physiol. Behav. Sci., 37: 44–58.

Minor, T.R., Jackson, R.L. and Maier, S.F. (1984) Effects of task-irrelevant cues and reinforcement delay on choice-escape learning following inescapable shock: evidence for a deficit in selective attention. J. Exp. Psychol. Anim. Behav. Process, 10: 543–556.

Moret, C. and Briley, M. (2000) The possible role of 5-HT(1B/D) receptors in psychiatric disorders and their potential as a target for therapy. Eur. J. Pharmacol., 404: 1–12.

Morris, J.S. and Dolan, R.J. (2004) Dissociable amygdale and orbitofrontal responses during reversal fear conditioning. Neuroimage, 22: 372–380.

Myers, K.M. and Davis, M. (2002) Behavioral and neural analysis of extinction. Neuron, 36: 567–584.

Myers, M.W., Kimbrell, T.A., Booe, L.Q. and Freeman, T.W. (2005) Weight loss and PTSD symptom severity in former POWs. J. Nerv. Ment. Dis., 193: 278–280.

Norris, F.H., Baker, C.K., Murphy, A.D. and Kaniasty, K. (2005) Social support mobilization and deterioration after Mexico's 1999 flood: effects of context, gender, and time. Am. J. Community Psychol., 36: 15–28.

Norris, F.H., Friedman, M.J., Watson, P.J., Byrne, C.M., Diaz, E. and Kaniasty, K. (2002) 60,000 disaster victims speak, Part I. An empirical review of the empirical literature: 1981–2001. Psychiatry, 65: 207–239.

North, C.S., Nixon, S.J., Shariat, S., Mallonee, S., McMillen, J.C., Spitznagel, E.L. and Smith, E.M. (1999) Psychiatric disorders among survivors of the Oklahoma City bombing. J. Am. Med. Assoc., 282: 755–762.

Ochsner, K.N., Ray, R.D., Cooper, J.C., Robertson, E.R., Chopra, S., Gabrieli, J.D.E. and Gross, J.J. (2004) For better or for worse: neural systems supporting the cognitive down- and up-regulation of negative emotion. Neuroimage, 23: 483–499.

Orr, S.P., Metzger, L.J. and Pitman, R.K. (2002) Psychophysiology of post-traumatic stress disorder. Psychiatr. Clin. North Am., 25: 271–293.

Orr, S.P. and Roth, W.T. (2000) Psychophysiological assessment: clinical applications for PTSD. J. Affect. Disord., 61: 225–240.

Pare, W.P. (1994) Open field, learned helplessness, conditioned defensive burying, and forced-swim tests in WKY rats. Physiol. Behav., 55: 433–439.

Pauwels, P.J. (1997) 5-HT 1B/D receptor antagonists. Gen. Pharmacol., 29: 293–303.

Pitman, R.K., Orr, S.P., Shalev, A.Y., Metzger, L.J. and Mellman, T.A. (1999) Psychophysiological alterations in post-traumatic stress disorder. Semin. Clin. Neuropsychiatry, 4: 234–241.

van Praag, H.M. (2004b) Can stress cause depression? Prog. Neuropsychopharmacol. Biol. Psychiatry, 28: 891–907.

van Praag, H.M. (2004a) The cognitive paradox in posttraumatic stress disorder: a hypothesis. Prog. Neuropsychopharmacol. Biol. Psychiatry, 28: 923–935.

Quirk, G.J., Russo, G.K., Barron, J.L. and Lebron, K. (2000) The role of ventromedial prefrontal cortex in recovery of extinguished fear. J. Neursosci., 20: 6225–6231.

Rahman, S. and Neuman, R.S. (1993) Activation of 5-HT2 receptors facilitates depolarization of neocortical neurons by N-methyl-D-aspartate. Eur. J. Pharmacol., 231: 347–354.

Rainnie, D.G. (1999a) Serotonergic modulation of neurotransmission in the rat basolateral amygdala. J. Neurophysiol., 82: 69–85.

Rainnie, D.G. (1999b) Serotonergic modulation of neurotransmission in the rat basolateral amygdala. J. Neurophysiol., 82: 69–85.

Resnick, H.S., Kilpatrick, D.G., Dansky, B.S., Saunders, B.E. and Best, C.L. (1993) Prevalence of civilian trauma and posttraumatic stress disorder in a representative national sample of women. J. Consult. Clin. Psychol., 61: 984–991.

Ressler, K.J., Rothbaum, B.O., Tannenbaum, L., Anderson, P., Graap, K., Zimand, E., Hodges, L. and Davis, M. (2004) Arch. Gen. Psychiatry, 61: 1136–1144.

van Riel, E., Meijer, O.C., Steenbergen, P.J. and Joels, M. (2003) Chronic unpredictable stress causes attenuation of serotonin responses in cornu ammonis 1 pyramidal neurons. Neuroscience, 120: 649–658.

Robbins, T.W. (2005) Controlling stress: how the brain protects itself from depression. Nat. Neurosci., 8: 261–262.

Rumpel, S., LeDoux, J., Zador, A. and Malinow, R. (2005a) Postsynaptic receptor trafficking underlying a form of associative learning. Science, 308: 83–88.

Rumpel, S., LeDoux, J., Zador, A. and Malinow, R. (2005b) Postsynaptic receptor trafficking underlying a form of associative learning. Science, 308: 83–88.

Rumpel, S., LeDoux, J., Zador, A. and Malinow, R. (2005c) Postsynaptic receptor trafficking underlying a form of associative learning. Science, 308: 83–88.

Schmidt, C.W. (2006) Signs of the times: biomarkers in perspective. Environ. Health Perspect., 114: A700–A705.

Servatius, R.J., Ottenweller, J.E. and Natelson, B.H. (1995) Delayed startle sensitization distinguishes rats exposed to one or three stress sessions: further evidence toward an animal model of PTSD. Biol. Psychiatry, 38: 539–546.

Sharot, T., Delgado, M.R. and Phelps, E.A. (2004) How emotion enhances the feeling of remembering. Nat. Neurosci., 7: 1376–1380.

Simson, P.E. and Weiss, J.M. (1988) Altered activity of the locus coeruleus in an animal model of depression. Neuropsychopharmacology, 1: 287–295.

Stanley, M. and Stanley, B. (1990) Postmortem evidence for serotonin's role in suicide. J. Clin. Psychiatry, 51(Suppl): 22–28.

Stein, C., Davidowa, H. and Albrecht, D. (2000) 5-HT(1A) receptor-mediated inhibition and 5-HT(2) as well as 5-HT(3) receptor-mediated excitation in different subdivisions of the rat amygdala. Synapse, 38: 328–337.

Stutzmann, G.E. and LeDoux, J.E. (1999) GABAergic antagonists block the inhibitory effects of serotonin in the lateral amygdala: a mechanism for modulation of sensory inputs related to fear conditioning. J. Neurosci., 19: 8RC.

Stutzmann, G.E., McEwen, B.S. and LeDoux, J.E. (1998) Serotonin modulation of sensory inputs to the lateral amygdala: dependency on corticosterone. J. Neurosci., 18: 9529–9538.

Sutker, P.B., Galina, Z.H., West, J.A. and Allain, A.N. (1990) Trauma-induced weight loss and cognitive deficits among former prisoners of war. J. Consult Clin. Psychol., 58: 323–328.

Svenningsson, P., Chergui, K., Rachleff, I., Flajolet, M., Zhang, X., El Yacoubi, M., Vaugeois, J.M., Nomikos, G.G. and Greengard, P. (2006) Alterations in 5-HT1B receptor function by p11 in depression-like states. Science, 311: 77–80.

Tedeschi, R.G., Park, C.L. and Calhoun, R.G. (1998) Posttraumatic Growth: Positive Changes in the Aftermath of Crises. Lawrence Erlbaum, New Jersey.

Tork, I. (1990) Anatomy of the serotonergic system. Ann. N.Y. Acad. Sci., 600: 9–34.

Ursano, R.J. (1987) Commentary. Posttraumatic stress disorder: the stressor criterion. J. Nerv. Ment. Dis., 175: 273–275.

Ursano, R.J., Fullerton, C.S., Vance, K. and Kao, T.C. (1999) Posttraumatic stress disorder and identification in disaster workers. Am. J. Psychiatry, 156: 353–359.

Ursano, R.J., Kao, T.C. and Fullerton, C.S. (1992) Posttraumatic stress disorder and meaning: structuring human chaos. J. Nerv. Ment. Dis., 180: 756–759.

Vieweg, W.V., Julius, D.A., Fernandez, A., Beatty-Brooks, M., Hettema, J.M. and Pandurangi, A.K. (2006) Posttraumatic stress disorder: Clinical features, pathophysiology, and treatment. Am. J. Med., 119: 383–390.

Vlahov, D., Galea, S., Resnick, H., Boscarino, J.A., Bucuvalas, M., Gold, J. and Kilpatrick, D. (2002) Increased use of cigarettes, alcohol, and marijuana among Manhattan, New York, residents after the September 11 terrorist attacks. Am. J. Epidemiol., 155: 988–996.

Wu, J., Kramer, G.L., Kram, M., Steciuk, M., Crawford, I.L. and Petty, F. (1999) Serotonin and learned helplessness: a regional study of 5-HT1A, 5-HT2A receptors and the serotonin transport site in rat brain. J. Psychiatr. Res., 33: 17–22.

Yao, X.L., Cowan, M.J., Gladwin, M.T., Lawrence, M.M., Angus, C.W. and Shelhamer, J.H. (1999) Dexamethasone alters arachidonate release from human epithelial cells by induction of p11 protein synthesis and inhibition of phospholipase A2 activity. J. Biol. Chem., 274: 17202–17208.

Discussion: Chapter 14

SHALEV: There is continuity between levels in responses to traumatic events. Would it be advisable to formulate a threshold level for treatment rather than to optimize treatment to those who are more severely affected?

URSANO: I absolutely agree of course. It highlights one of the dimensions of PTSD that we don't often focus upon but has become a target of great concern at the Veterans Administration in the USA which is the level of disability. Disability and altered function have a relationship to symptoms and symptom levels are not the same. One can have very high levels of symptoms and very low disability or very low levels of symptoms and very high levels of disability. Impaired function is perhaps the best indicators in our experience of when intervention is needed. Richard Bryant has reported repeatedly that when you intervene too early for ASD you may do more harm than good. Even though very high rates of ASD may be present early on after a severe traumatic event, most people who get PTSD did not have ASD and most people who have ASD don't get PTSD. So looking at the question of functional impairment is a critical aspect of considering when to intervene.

DE KLOET: How do you decide that something breaks down if you do not know how it works. Can you give a specific example on what criteria you would decide there is a breakdown?

URSANO: For example, we can understand that the important mechanisms involved in a bicycle are the chain and the wheels. If we are looking at how it works, we may put most of our attention in these areas. However, the breakdown of a bicycle may actually be due to a flat tire more often than a broken chain. The breakdown requires consideration of the process in context and other demands present. Recent studies on "forgetting" indicate that cognitive control of other brain processes may be the cost of not forgetting. For example, we are not freeing neuro processes for other work. Thus the breakdown is a loss of ability to process in areas very different from the impairment. Certainly we cannot push the breakdown versus how it works analogy too far — both are important. But they are not the same.

VERMETTEN: What are the most important targets of clinical care?

At the individual and the population level targeting interpersonal withdrawal (increased perhaps by comorbid depression) may be much more important than targeting PTSD symptoms. At the population level this may mean it is more important for us to screen for depression than for PTSD, because of the risk with depression and because we have excellent screening tools. In addition, screening for depression is already a part of many primary care settings. By screening for depression we pick up those people who have perhaps the highest degree of impairment from PTSD as well. Kessler has shown that in general the greater the number of comorbidities the greater the illness impairment.

E.R. de Kloet, M.S. Oitzl & E. Vermetten (Eds.)
Progress in Brain Research, Vol. 167
ISSN 0079-6123

CHAPTER 15

What is it that a neurobiological model of PTSD must explain?

Chris R. Brewin*

Subdepartment of Clinical Health Psychology, University College London, Gower Street, London WC1E 6BT, UK

Abstract: PTSD is a complex disorder that involves far more than a fear response, and cannot be explained by a simple conditioning model. Both individual vulnerability and specific reactions during and after the trauma are involved in maintaining the disorder. A consideration of risk factors implicates the experience of being "overwhelmed" at the time of the trauma, accompanied by possible downregulation of the prefrontal cortex. Also important are reactions to symptoms post-trauma and specific strategies adopted to manage symptoms, such that there is a continuing inability to process trauma memories. An analysis of the characteristic forms of autobiographical memory in PTSD implicates two memory systems, one predominantly image-based and one predominantly verbal. These systems are likely to be differentially impacted by hormonal responses to extreme stress, leading to an imbalance in the representation of trauma in the two systems. Exposure to trauma reminders leads to retrieval competition between the two sets of memories, with retrieval of verbal memories able to inhibit inappropriate amygdala responses. Evidence to support this analysis is described, drawing on experimental studies of memory for trauma and a meta-analysis of memory for emotionally neutral information in PTSD. The implications for neurobiological studies of PTSD are discussed.

Keywords: PTSD; stress; memory; amygdala; hippocampus

Posttraumatic stress disorder (PTSD) is a condition that may occur following an overwhelming event that typically involves actual or threatened death or injury. The symptoms consist of the re-experiencing of the event in the form of intrusive memories or nightmares, avoidance of reminders of the event, emotional numbing, and a permanent state of high arousal. In order to evaluate the potential contribution of neurobiological research to the understanding of PTSD it is necessary to summarize the current state of knowledge about clinical features of this disorder, its risk factors, and the psychological mechanisms that contribute to its onset and maintenance. This information will provide a useful benchmark with which to gauge the strengths and weaknesses of a neuroscience perspective, and to assess those areas in which it is likely to make an important contribution.

Clinical features of PTSD

It is clear that PTSD is associated with overwhelming stress, but is not defined by exposure to overwhelming stress. The majority of individuals exposed to extreme stress do not develop PTSD. In this respect PTSD is similar to other psychiatric

*Corresponding author. Tel.: +44 207 679 5927;
Fax: +44 207 916 1989; E-mail: c.brewin@ucl.ac.uk

DOI: 10.1016/S0079-6123(07)67015-0

disorders in that onset of a particular episode tends to follow a stressor but is more likely to do so in vulnerable individuals. PTSD requires a diathesis-stress model in which aspects of event exposure interact with individual characteristics and response patterns (Brewin, 2003). Similarly, the traumatic memories of individuals with PTSD are different to the traumatic memories of non-sufferers: once an initial period of adjustment has passed, memory characteristics are less a function of the event than of the individual who has experienced it (Brewin, 2007).

PTSD symptoms also have in common with the symptoms of other psychiatric disorders that they are not in themselves abnormal, but are commonly experienced in the immediate aftermath of a traumatic event. What defines the disorder is not the unique nature of the symptoms, but their frequency, longevity, and the associated impairment. Further, PTSD symptoms overlap to a large extent with those characteristic of other conditions associated with exposure to stress, such as bereavement and depression. In both these conditions there tends to be a spontaneous intrusion of memories that the individual tries to avoid, ruminative thoughts, social withdrawal and emotional numbing, and arousal symptoms including lack of concentration, irritability, and sleeplessness.

Although traumatic events are defined by the DSM-IV as generating extreme fear, helplessness, and horror, it is a mistake to think that PTSD is just a disorder of fear. Studies have shown that other emotions, such as anger, are as if not more common than fear (Reynolds and Brewin, 1999). The whole question of the relationship between PTSD and emotions needs to be considered separately in terms of onset and maintenance, and a variety of different causal pathways and psychological mechanisms may be relevant. For example, shame appears to be important in determining the longitudinal course of the disorder (Andrews et al., 2000). Thus there is a significant degree of overlap between PTSD and other disorders in emotions as well as symptoms.

Despite the fact that phobic symptoms are often part of PTSD, the disorder is unlike a simple phobia. In phobia, anxiety is only present when the individual is confronted with their feared situation. In PTSD, by contrast, there is a sense of continuing, current threat that leads to a permanent state of arousal (Ehlers and Clark, 2000). This sense of threat may be present even though the individual accepts that it is logically impossible for the trauma to happen again. PTSD is also associated with much more profound changes to sufferers' sense of identity than in phobia: It is not unusual for patients to feel strongly that they have been changed irrevocably by their experiences and that they can no longer see the world in the same way (Brewin, 2003). This often goes along with wide-ranging changes to their social relationships.

It has also been argued that PTSD is distinguished from other disorders by two main symptoms, flashbacks and traumatic nightmares. Both symptoms involve the repeated reliving of the traumatic event, often in a stereotyped way. The concept of reliving is as yet poorly understood, and it is commonplace for emotional memories, for example in depression, to contain an element of repetition and recognition of the original emotions. What appears to differentiate flashbacks from these common intrusive memories is that they involve the sense that events are happening again in the present. This distortion in the sense of time is an aspect of dissociation, a term that refers to a variety of ways in which the usually integrated functions of consciousness, memory, identity, or perception of the environment may be disrupted. For example, speeding up or slowing of subjective time, feelings of numbness and detachment, or out-of-body experiences, are common during and after traumatic events. In extreme cases traumatized individuals may become totally absorbed in their memories and unable to apprehend events in the environment. This kind of phenomenon has not been described in depressive disorders, where intrusive memories tend to be experienced as belonging in the past, like ordinary autobiographical memories.

Insights from risk factors

Risk factor research has been summarized in two recent meta-analyses (Brewin et al., 2000; Ozer et al., 2003). Background factors, such as greater socioeconomic disadvantage, prior trauma, adverse parenting, prior psychopathology, and a family history of psychopathology, are risk factors for PTSD just as they are for most psychiatric conditions. There have been several longitudinal studies indicating that lower intellectual ability confers some risk, or alternatively that greater intellectual ability confers a degree of protection. A recent twin study suggests that this effect is accounted for by verbal rather than non-verbal abilities (Gilbertson et al., 2006). These factors only account for a relatively small amount of the variance, however.

A second set of risk factors describes responses during the traumatic event (peri-traumatic factors). High stressor intensity, particularly subjectively experienced intensity as measured through a perceived threat to life or extreme fear, helplessness, or horror, is strongly related to risk of later PTSD. A sense of mental defeat, as having given up any attempt to control the outcome, is a related phenomenon that also acts as a risk factor (Ehlers et al., 2000). Greater dissociation is also well established as a risk factor. This is interesting, as dissociation might be expected to interfere with encoding of the traumatic events into memory. Indeed, some theories have assumed that PTSD involves enhanced encoding of traumatic material and have even predicted that greater dissociation should protect against PTSD for this reason (Tryon, 1999).

A third set of risk factors involves processes that occur post-trauma and that may interfere with normal adaptation. Ehlers and Clark's (2000) cognitive model of PTSD maintenance describes numerous factors that have now been shown to predict a worse outcome, including negative appraisals of the trauma, of people's actions, and of the patient's own symptoms, and inappropriate coping strategies such as avoidance, thought suppression, and adoption of safety behaviors (see Brewin and Holmes, 2003, for a review).

Consistent with the major psychological theories of PTSD, disruption and fragmentation to narrative memories of the trauma also appears to be a risk factor for the development of the disorder. This has been shown in two longitudinal studies, one measuring narratives as early as 1 week post-trauma (Halligan et al., 2003; Jones et al., 2007). Narratives characterized by repetition of utterances and the presence of non-consecutive chunks were associated with more severe PTSD at 3 months. Additionally, disorganization in narrative memories is associated with peri-traumatic dissociation. These findings are compelling, but it should be noted that they depend on having independent blind raters categorize the content of patients' narratives. Attempts to have patients rate disorganization in their own narratives, or to use computer programs to index this, have often failed to find a relationship between PTSD and memory disorganization.

This review has emphasized the variety of processes shown to be associated with PTSD, and it is important to be open to the possibility that there are alternative pathways leading to the disorder. For example, many PTSD patients exposed to trauma stimuli experience an increase in heart rate (Pitman et al., 2000), and there is evidence that increased heart rate immediately post-trauma predicts the likelihood of developing PTSD (Shalev et al., 1998; Bryant et al., 2000). Some individuals with PTSD, on the other hand, respond to trauma stimuli with dissociative reactions, a type of response that has on several occasions been linked to reductions in heart rate (Griffin et al., 1997; Koopman et al., 2004). Likewise, longitudinal studies of trauma survivors suggest that those who experience high levels of dissociative symptoms initially, and those who have high levels of reexperiencing and arousal, are both at risk for PTSD but that they form non-overlapping groups (Brewin et al., 1999). These contrasting patterns have been related to the experience of alternative behavioral responses to threat (fight/flight versus freezing) that are thought to possess different biological substrates (Nijenhuis et al., 1998).

Explaining flashbacks and nightmares: sensory versus verbal memory

In order to explain why traumatized individuals experience vivid involuntary memories while at the same time having difficulty in deliberately retrieving coherent narrative memories, clinicians have from the end of the 19th century onwards argued for the existence of parallel verbal and sensory memories of trauma (see Brewin, 2003, for a review). Verbal memories require high levels of cortical processing, resulting in them being contextualized within the autobiographical memory system, whereas sensory memories receive less in the way of higher processing and are stored as isolated images devoid of context. Metcalfe and Jacobs (1998) outlined evidence that high levels of stress may simultaneously impair the operation of brain structures supporting autobiographical memory (e.g., the prefrontal cortex and hippocampus), while simultaneously facilitating the operation of brain structures such as the amygdala that support image-based memories.

In the most recent incarnation of this approach, the dual representation theory of PTSD, Brewin (Brewin et al., 1996; Brewin, 2003) discusses in some detail how image-based ("situationally accessible") forms of memory can be retrieved automatically by trauma cues and how the lack of contextual coding results in the brain responding as thought the trauma were occurring again in the present. In this theory a prerequisite for the development of flashbacks as well as the poor quality of intentional recall is enhanced encoding of situationally accessible trauma memories coupled with impaired encoding of the same material into the autobiographical ("verbally accessible") memory system. This preferential encoding may be a product of peri-traumatic dissociation and the prefrontal cortex temporarily going "off-line" in response to a level of stress that exceeds the individual's coping abilities. Over the next days and weeks flashbacks provide an opportunity for this material to receive additional conscious processing and to be re-encoded into verbally accessible memory. The contextual information provided by this form of memory enables the brain to classify the trauma as having happened in the past and inhibits the retrieval of corresponding sensory memories. Thus imbalances in encoding into the two memory systems can be naturally corrected in a relatively short space of time, allowing defensive arousal to subside.

In PTSD, however, dual representation theory suggests that normal adaptation does not occur because this process of re-encoding never takes place. The high levels of behavioral and cognitive avoidance characteristic of the disorder result in intrusive sensory images not being consciously attended to. Corresponding verbally accessible memories remain impoverished and never contain sufficient information about critical retrieval cues to enable them to inhibit the intrusions. As a result exposure to reminders of the trauma continues to elicit the intrusion of unwanted sensory memories accompanied by high levels of emotion and perceived threat, shortly to be followed by secondary consequences such as depression and social withdrawal.

Evidence for dual representations of trauma

Experimental studies of intrusions

In order to test hypotheses about intrusive trauma memories being supported by a sensory memory system, Holmes et al. (2004) had healthy volunteers watch a traumatic film while carrying out a secondary, concurrent task. Separate measures were taken of recall, recognition, and number of intrusive film-related memories experienced during the following week. In both studies a concurrent visuospatial tapping task had the effect of reducing later intrusions relative to a control no-task condition. Holmes et al. ruled out a number of explanations of their findings. The effect could not be simply due to distraction since a concurrent verbal task had the opposite effect of increasing subsequent intrusions relative to a control condition. Measures of recall and recognition were unrelated to the number of intrusions, and demand characteristics could not account for the pattern of results. A subsequent study (Stuart et al., 2006) replicated the effects of a concurrent visuospatial task on intrusions using a within-subjects design and an alternative task.

Holmes et al. (2004, Expts. 1 and 2) found that the more participants reported dissociative experiences (derealization and depersonalization) while they watched the film, the more likely they were to have intrusive memories of the film over the next week. These results provide an experimental parallel to the clinical studies that, as noted above, show that reports of dissociation during an actual traumatic event are related to a greater risk of developing subsequent PTSD. Holmes et al. also showed that the lower participants' heart rate while they watched the film the more likely they were to report later intrusions. Moreover, the specific scenes that intruded for any individual during the following week were associated with a lower heart rate during those sections of the film. Importantly, the measures of involuntary memory (intrusions) were consistently unrelated to measures of voluntary memory for the film, such as recall and recognition.

Verbal and visual memory capacity in PTSD

The theory suggests that PTSD is likely to be associated with impaired verbal memory but a well-functional sensory memory system. In order to test whether there are general deficits in verbal memory that apply even to emotionally neutral material, Brewin et al. (2007) recently conducted a meta-analysis of verbal and visual memory functioning in PTSD. The analysis summarized investigations that compared a PTSD group diagnosed according to the DSM with healthy controls on their performance on standardized memory tests. They found 27 relevant studies that included a total of 812 controls and 660 individuals with PTSD arising from a variety of different trauma types. PTSD was associated with impaired memory functioning generally and, consistent with dual representation theory, patients' verbal memory was significantly worse than their visual memory. These differences could not be explained by potential confounding factors such as trauma exposure, general intellectual ability, head injury, or substance abuse.

The analysis does not tell us whether these neuropsychological differences predate the onset of PTSD, whether they are a result of PTSD, or both. Relevant evidence is provided in a study by Gilbertson et al. (2006) of combat veterans and their identical twins who were not exposed to combat. By comparing the performance of veterans with and without PTSD, and their non-exposed co-twins, it is possible to distinguish the consequence of combat exposure and PTSD from familial vulnerability factors that increase or decrease the risk of disorder. Gilbertson et al. reported that the identical co-twins of PTSD combat veterans, who had neither combat exposure nor PTSD themselves, showed a similar pattern to their PTSD brothers in having poorer verbal memory. They suggested that better neuropsychological functioning in these areas acted as a source of pre-existing resilience when veterans were later faced with traumatic events. Consistent with the results of the meta-analysis, they found no evidence that poorer visual memory was associated with PTSD symptoms or acted as a risk or resiliency factor.

Experimental studies of PTSD

Hellawell and Brewin (2004) described the difference between flashbacks, involving a marked sense of reliving events in the present, and ordinary memories to people with PTSD and then had them write a detailed narrative of their traumatic event. At the completion of the narrative participants retrospectively identified periods of writing during which they experienced each of the two types of memory. All the participants reported recognizing and being able to distinguish between the two types of memory as they wrote about their trauma, but there was great individual variation in how many reliving periods they identified, how long these lasted, and where in the narrative they occurred.

Consistent with prediction, during parts of the narrative involving reliving they used more words describing seeing, hearing, smelling, tasting, and bodily sensations, as well as more verbs and references to motion, than they did during ordinary memory sections. Fear, helplessness, horror, and thoughts of death, all reactions associated with the traumatic moments themselves, were prominent during the reliving sections. In contrast, emotions associated with later appraisals and interpretations of the events, such as sadness and anger, were more prominent during the ordinary memory sections. These data support the argument that reliving is not just a function of extreme emotion, but of specific emotions such as fear that occur at the moment of trauma (see also Reynolds and Brewin, 1999, for evidence supporting a specific relationship between fear and reliving).

As part of this study, Hellawell and Brewin (2002) investigated whether flashbacks were predominantly image-based, using visuospatial resources, and ordinary memories predominantly verbal. They reasoned that if flashbacks use visuospatial resources, then they should interfere with performance on other tasks that also made visuospatial demands but not interfere with unrelated tasks. So, while participants with PTSD were writing their narratives, they were stopped on two occasions, once when they were in a reliving phase and once when they were in an ordinary memory phase, and made to carry out two tasks. One task, trail-making, involved visuospatial abilities and the other, counting backwards in threes, involved more verbal abilities. The results showed that trail-making performance was much worse when participants had been halted during a reliving phase of their narrative than when they had been halted during an ordinary memory phase, whereas counting backwards in threes was adversely affected to an equal extent in both phases. This supports the idea that there is a qualitative difference between flashbacks and ordinary memories.

A possible neural substrate for trauma memories

Within the declarative memory system concerned with conscious knowledge of facts and events the hippocampus appears to be specialized for the learning of context (including temporal context; Kesner, 1998), and for learning relational properties among stimuli. It is thought to be crucial in binding together the separate features or elements of an episode to make a coherent and integrated ensemble. Eichenbaum (1997) proposed that the hippocampus encodes separate stimulus elements and the relations between them such that the representations can be utilized flexibly and accessed in a variety of ways. It has also been suggested that the hippocampal system is particularly associated with memories of conscious experience (Moscovitch, 1995).

From the perspective of dual representation theory it would appear that hippocampal processing is likely to be a critical aspect of verbally accessible (narrative) memories that form the basis of deliberate appraisal and communication concerning the trauma (Brewin, 2001, 2005). The hippocampus is highly sensitive to stress, being well supplied with receptors that are occupied by stress hormones. A wealth of animal studies confirm that severe stress impairs hippocampal function and memory performance, and there is a corresponding literature in humans. This demonstrates impaired explicit memory performance associated with raised levels of glucocorticoids, adrenal steroid hormones that are released after stressful experiences, in the context of naturally occurring conditions or experimental treatments (Alderson and Novack, 2002). Acutely elevated levels of glucocorticoids have been found to be associated with impaired memory and reduced blood flow in areas of the medial temporal lobe adjacent to the hippocampus (de Quervain et al., 2003). At present it is unclear whether high levels of stress hormones impair the consolidation of memories, their retrieval, or both.

Importantly, studies reveal that mild to moderate levels of stress have opposite effects to severe stress on memory function. This has been related to the levels of occupancy of different types of receptor in the hippocampus. At low levels of stress there is heavy occupancy of mineralocorticoid receptors and low-to-moderate occupancy of glucocorticoid receptors. Memory impairments appear to be associated with the high levels of occupancy of glucocorticoid receptors that are associated with severe stress (Alderson and Novack, 2002; Kim and Diamond, 2002; Sapolsky, 2003). At present it is unclear whether reduced hippocampal functioning in PTSD is primarily related to the effects of severe stress, to pre-existing vulnerabilities, or both.

Whereas explicit memory is associated with the hippocampus, the various forms of implicit memory (e.g., priming, fear conditioning) do not appear to have any particular locus in the brain. The involuntary memories that are characteristic of PTSD may be considered to be related to implicit memory, in that they share the characteristics of being automatically retrieved in a predictable, cue-driven manner, and being hard to control. They are unlike normal examples of implicit memory, however, in that they possess explicit trauma-related content that is immediately recognized by the person experiencing them. In the dual representation theory of PTSD, both flashbacks and fear conditioning are considered to be products of the situationally accessible memory system.

It is possible for fear-relevant information to reach the amygdala via a number of different routes, independently of the hippocampus. For example, the visual areas of the inferior temporal cortex, which are involved in the late stages of sensory processing, project strongly to the amygdala. The pathway from the thalamus to the amygdala has a less sophisticated processing capacity and would be capable of transmitting lower level sensory features of frightening situations. Memories formed in these ways would not be open to deliberate recall, but could be accessed automatically by reminders, particularly perceptual features, similar to those recorded in the fear memory.

As noted by several authors (Metcalfe and Jacobs, 1998; Elzinga and Bremner, 2002), high levels of stress appear to have very different effects on the hippocampus and the amygdala. For example, the same stress experience produces dendritic atrophy and debranching in the hippocampus at the same time as producing enhanced dendritic arborization in the amygdala (Vyas et al., 2002, 2003). Thus, sensory or situationally accessible memories (such as flashbacks) could be enhanced at the same time as narrative or verbally accessible memories were impaired.

Conclusions: implications for a neurobiological model of PTSD

The results of existing data from structural and functional imaging studies of PTSD show some consistency, for example in increased amygdala activation coupled with hypoactivation of the medial prefrontal cortex, but also a high degree of variability, for example with respect to hippocampal activation (Hull, 2002; Shin et al., 2005a; Francati et al., 2007) that has the potential to impede progress in this area. Research into the clinical features, risk factors, and psychological mechanisms associated with PTSD has a number of important implications for designing better and more tightly focused studies, and reducing this variability. Among the most obvious are the findings concerning the high incidence of comorbidity, the role of maintenance as well as onset processes, the different pathways to the development of PTSD, and the existence of different forms of memory.

Given the overlap in symptoms between PTSD and other disorders, the use of comparison groups needs to be reconsidered. It is already accepted as good practice to include a control group of individuals who have been exposed to similar traumatic events without ever developing PTSD, in order to rule out the possibility that observed group differences are simply due to trauma exposure. Most neuroimaging studies, however, do not include comparison groups suffering from other disorders such as depression (see Shin et al., 2005a, for a review). This creates very real problems of interpreting structural and functional differences between PTSD patients and controls. For example, the interpretation of studies showing reduced hippocampal volume in PTSD is complicated by studies with similar findings in patients selected for major depressive disorder (Neumeister et al., 2005; Saylam et al., 2006).

The fact that it is the persistence of PTSD symptoms that is pathological, rather than their form, also has important implications. It suggests that neurobiological differences between PTSD patients and controls are as likely to reflect differences in the way individuals have adapted or responded to having symptoms as they are to reflect differences in their initial response to trauma. This is particularly important given that the vast majority of neuroimaging studies, for example, have been conducted on samples with very chronic PTSD (Shin et al., 2005a). PTSD is associated with the use of well-marked behavioral and cognitive strategies including avoidance, thought suppression, and the adoption of safety behaviors. It is these habitual responses that may be detected by functional imaging studies of patients exposed to trauma stimuli, for example in the script-driven imagery paradigm.

Risk factor research has identified the importance of processes that operate before, during, and after traumatic events, with the latter two sets of processes accounting for most of the variance. This suggests that neurobiological studies conducted at encoding (in healthy volunteers exposed to experimental stress) or immediately post-trauma (e.g., on admission to an emergency room) are essential to disentangle the two sets of effects. Further, there will be value to studies that focus more intensively on the first 2–3 weeks post-trauma, when high levels of initial symptoms subside in most exposed individuals but not in the subgroup who go on to develop PTSD. Shalev (2003), for example, has described how during this period a process of increasing sensitization differentiates those who fail to recover. Sensitization refers to systematic changes in normal thresholds of stimulus discrimination and responding, illustrated for example by a slowly developing abnormal startle response. There are also specific emotional states that may only develop post-trauma and yet have been shown to predict a worse outcome. It will be important to correct the over-emphasis on emotions present at encoding (such as fear) in favor of emotions present during recovery (such as anger and shame).

Risk factor research has also drawn attention to the possibility that there is more than one pathway to the development of PTSD, involving acute reactions characterized either by excessive arousal and high heart rate, or by dissociation and lowered heart rate. Both routes have received independent support from experimental research either with PTSD patients or healthy volunteers. Apart from alerting researchers to the possible existence of subgroups of PTSD patients demonstrating different forms of pathology, these data underscore the importance of individual differences, for example in response to tasks in the laboratory such as being exposed to trauma reminders. These different reactions are likely to be associated with alternative patterns of neural as well as psychophysiological responses. Lanius and colleagues, for example, reported that individuals with a history of sexual abuse evidenced different kinds of responses to script-driven imagery, some experiencing reliving and increased heart rate while others experienced a dissociative response. They went on to analyze differential patterns of neural activation (Lanius et al., 2002) and interregional brain activity co-variations (Lanius et al., 2005) in these groups.

Equally, psychological studies have drawn attention to the existence of different forms of intrusive cognition that may underlie specific symptoms and that may interact to determine outcome. This implies that researchers need to develop strategies that enable them to focus more explicitly on individual symptoms rather than on PTSD as a whole. Osuch et al. (2001), for example, investigated the correlation between regional cerebral blood flow and the intensity of flashbacks. Researchers need to pay more attention to individual responses to trauma stimuli, script-driven imagery, etc. These may elicit a variety of responses including rumination, ordinary remembering, flashbacks, thought suppression, or dissociation, all of which may correspond to distinct patterns of neural activation. Also, neurobiological models may need to consider in more detail how to model interactions between different areas of the brain. A good example of this is the existing work showing an inverse association between activation of the amygdala and prefrontal cortex (Shin et al., 2005b; Francati et al., 2007).

Acknowledgments

The author's research is supported by the Wellcome Trust.

References

Alderson, A.L. and Novack, T.A. (2002) Neurophysiological and clinical aspects of glucocorticoids and memory: a review. J. Clin. Exp. Neurophys., 24: 335–355.

Andrews, B., Brewin, C.R., Rose, S. and Kirk, M. (2000) Predicting PTSD symptoms in victims of violent crime: the role of shame, anger, and childhood abuse. J. Abnormal Psychol., 109: 69–73.

Brewin, C.R. (2001) A cognitive neuroscience account of post-traumatic stress disorder and its treatment. Behav. Res. Ther., 39: 373–393.

Brewin, C.R. (2003) Post-traumatic stress disorder: malady or myth? Yale University Press, New Haven, CT.

Brewin, C.R. (2005) Encoding and retrieval of traumatic memories. In: Vasterling J.J. and Brewin C.R. (Eds.), Neuropsychology of PTSD: Biological, Cognitive, and Clinical Perspectives. Guilford Press, New York, pp. 131–150.

Brewin, C.R. (2007) Autobiographical memory for trauma: update on four controversies. Memory, 15: 227–248.

Brewin, C.R., Andrews, B., Rose, S. and Kirk, M. (1999) Acute stress disorder and posttraumatic stress disorder in victims of violent crime. Am. J. Psychiatry, 156: 360–366.

Brewin, C.R., Andrews, B. and Valentine, J.D. (2000) Meta-analysis of risk factors for posttraumatic stress disorder in trauma-exposed adults. J. Consult. Clin. Psychol., 68: 748–766.

Brewin, C.R., Dalgleish, T. and Joseph, S. (1996) A dual representation theory of post traumatic stress disorder. Psychol. Rev., 103: 670–686.

Brewin, C.R. and Holmes, E.A. (2003) Psychological theories of posttraumatic stress disorder. Clin. Psychol. Rev., 23: 339–376.

Brewin, C.R., Kleiner, J.S., Vasterling, J.J. and Field, A.P. (2007) Memory for emotionally neutral information in posttraumatic stress disorder: a meta-analytic investigation. J. Abnormal Psychol., 116: 448–463.

Bryant, R.A., Harvey, A.G., Guthrie, R.M. and Moulds, M. (2000) A prospective study of psychophysiological arousal, acute stress disorder, and posttraumatic stress disorder. J. Abnormal Psychol., 109: 341–344.

Ehlers, A. and Clark, D.M. (2000) A cognitive model of posttraumatic stress disorder. Behav. Res. Ther., 38: 319–345.

Ehlers, A., Maercker, A. and Boos, A. (2000) Posttraumatic stress disorder following political imprisonment: the role of mental defeat, alienation, and perceived permanent change. J. Abnormal Psychol., 109: 45–55.

Eichenbaum, H. (1997) Declarative memory: insights from cognitive neurobiology. Ann. Rev. Psychol., 48: 547–572.

Elzinga, B.M. and Bremner, J.D. (2002) Are the neural substrates of memory the final common pathway in posttraumatic stress disorder (PTSD)? J. Affect. Disord., 70: 1–17.

Francati, V., Vermetten, E. and Bremner, J.D. (2007) Functional neuroimaging studies in posttraumatic stress disorder: review of current methods and findings. Depress Anxiety, 24: 202–218.

Gilbertson, M.W., Paulus, L.A., Williston, S.K., Gurvits, T.V., Lasko, N.B., Pitman, R.K. and Orr, S.P. (2006) Neurocognitive function in monozygotic twins discordant for combat exposure: relationship to posttraumatic stress disorder. J. Abnormal Psychol., 115: 484–495.

Griffin, M.G., Resick, P.A. and Mechanic, M.B. (1997) Objective assessment of peritraumatic dissociation: psychophysiological indicators. Am. J. Psychiatry, 154: 1081–1088.

Halligan, S.L., Michael, T., Clark, D.M. and Ehlers, A. (2003) Posttraumatic stress disorder following assault: the role of cognitive processing, trauma memory, and appraisals. J. Consult. Clin. Psychol., 71: 419–431.

Hellawell, S.J. and Brewin, C.R. (2002) A comparison of flashbacks and ordinary autobiographical memories of trauma: cognitive resources and behavioural observations. Behav. Res. Ther., 40: 1139–1152.

Hellawell, S.J. and Brewin, C.R. (2004) A comparison of flashbacks and ordinary autobiographical memories of trauma: content and language. Behav. Res. Ther., 42: 1–12.

Holmes, E.A., Brewin, C.R. and Hennessy, R.D. (2004) Trauma films, information processing, and intrusive memory development. J. Exp. Psychol. Gen., 133: 3–22.

Hull, A.M. (2002) Neuroimaging findings in post-traumatic stress disorder: systematic review. Br. J. Psychiat., 181: 102–110.

Jones, C., Harvey, A.G. and Brewin, C.R. (2007) The organisation and content of trauma memories in survivors of road traffic accidents. Behav. Res. Ther., 45: 151–162.

Kesner, R.P. (1998) Neural mediation of memory for time: role of the hippocampus and medial prefrontal cortex. Psychonom. Bull. Rev., 5: 585–596.

Kim, J.J. and Diamond, D.M. (2002) The stressed hippocampus, synaptic plasticity, and lost memories. Nat. Rev. Neurosci., 3: 453–462.

Koopman, C., Carrion, V., Butler, L.D., Sudhakar, S., Palmer, L. and Steiner, H. (2004) Relationships of dissociation and childhood abuse and neglect with heart rate in delinquent adolescents. J. Trauma. Stress, 17: 47–54.

Lanius, R.A., Williamson, P.C., Bluhm, R.L., Densmore, M., Boksman, K., Neufeld, R.W.J., Gati, J.S. and Menon, R.S. (2005) Functional connectivity of dissociative responses in posttraumatic stress disorder: a functional magnetic resonance imaging investigation. Biol. Psychiatry, 57: 873–884.

Lanius, R.A., Williamson, P.C., Boksman, K., Densmore, M., Gupta, M., Neufeld, R.W.J., Gati, J.S. and Menon, R.S. (2002) Brain activation during script-driven imagery induced dissociative responses in PTSD: a functional magnetic resonance imaging investigation. Biol. Psychiatry, 52: 305–311.

Metcalfe, J. and Jacobs, W.J. (1998) Emotional memory: the effects of stress on 'cool' and 'hot' memory systems. In: Medin D.L. (Ed.), The Psychology of Learning and Motivation, Vol. 38. Academic Press, New York, pp. 187–222.

Moscovitch, M. (1995) Recovered consciousness: a hypothesis concerning modularity and episodic memory. J. Clin. Exp. Neuropsychol., 17: 276–290.

Neumeister, A., Wood, S., Bonne, O., Nugent, A.C., Luckenbaugh, D.A., Young, T., Bain, E.E., Charney, D.S. and Drevets, W.C. (2005) Reduced hippocampal volume in unmedicated, remitted patients with major depression versus control subjects. Biol. Psychiatry, 57: 935–937.

Nijenhuis, E.R.S., Vanderlinden, J. and Spinhoven, P. (1998) Animal defensive reactions as a model for trauma-induced dissociative reactions. J. Trauma. Stress, 11: 243–260.

Osuch, E.A., Benson, B., Geraci, M., Podell, D., Herscovitch, P., McCann, U.D. and Post, R.M. (2001) Regional cerebral blood flow correlated with flashback intensity in patients with posttraumatic stress disorder. Biol. Psychiatry, 50: 246–253.

Ozer, E.J., Best, S.R., Lipsey, T.L. and Weiss, D.S. (2003) Predictors of posttraumatic stress disorder and symptoms in adults: a meta-analysis. Psychol. Bull., 129: 52–73.

Pitman, R.K., Shalev, A.Y. and Orr, S.P. (2000) Posttraumatic stress disorder: emotion, conditioning, and memory. In: Gazzaniga M.S. (Ed.), The New Cognitive Neurosciences (2nd ed.). MIT Press, Cambridge, MA, pp. 1133–1147.

de Quervain, D.J.-F., Henke, K., Aerni, A., Treyer, V., McGaugh, J.L., Berthold, T., Nitsch, R.M., Buck, A., Roozendaal, B. and Hock, C. (2003) Glucocorticoid-induced impairment of declarative memory retrieval is associated with reduced blood flow in the medial temporal lobe. Eur. J. Neurosci., 17: 1296–1302.

Reynolds, M. and Brewin, C.R. (1999) Intrusive memories in depression and posttraumatic stress disorder. Behav. Res. Ther., 37: 201–215.

Sapolsky, R.M. (2003) Stress and plasticity in the limbic system. Neurochem. Res., 28: 1735–1742.

Saylam, C., Ucerler, H., Kitis, O., Ozand, E. and Gonul, A.S. (2006) Reduced hippocampal volume in drug-free depressed patients. Surg. Radiolog. Anat., 28: 82–87.

Shalev, A.Y. (2003) Psycho-biological perspectives on early reactions to traumatic events. In: Ørner R. and Schnyder U. (Eds.), Reconstructing Early Intervention After Trauma. Oxford University Press, Oxford, pp. 57–64.

Shalev, A.Y., Sahar, T., Freedman, S., Peri, T., Glick, N., Brandes, D., Orr, S.P. and Pitman, R.K. (1998) A prospective study of heart rate response following trauma and the subsequent development of posttraumatic stress disorder. Arch. Gen. Psychiatry, 55: 553–559.

Shin, L.M., Rauch, S.L. and Pitman, R.K. (2005a) Structural and functional anatomy of PTSD: findings from neuroimaging research. In: Vasterling J.J. and Brewin C.R. (Eds.), Neuropsychology of PTSD: Biological, Cognitive, and Clinical Perspectives. Guilford, New York, pp. 59–82.

Shin, L.M., Wright, C.I., Cannistraro, P.A., Wedig, M.M., McMullin, K., Martis, B., Macklin, M.L., Lasko, N.B., Cavanagh, S.R., Krangel, T.S., Orr, S.P., Pitman, R.K., Whalen, P.J. and Rauch, S.L. (2005b) A functional magnetic resonance imaging study of amygdala and medial prefrontal cortex responses to overtly presented fearful faces in posttraumatic stress disorder. Arch. Gen. Psychiatry, 62: 273–281.

Stuart, A.D.P., Holmes, E.A. and Brewin, C.R. (2006) The influence of a visuospatial grounding task on intrusive images of a traumatic film. Behav. Res. Ther., 44: 611–619.

Tryon, W.W. (1999) A bidirectional associative memory explanation of posttraumatic stress disorder. Clin. Psychol. Rev., 19: 789–818.

Vyas, A., Bernal, S. and Chattarji, S. (2003) Effects of chronic stress on dendritic arborization in the central and extended amygdala. Brain Res., 965: 290–294.

Vyas, A., Mitra, R., Rao, B.S.S. and Chattarji, S. (2002) Chronic stress induces contrasting patterns of dendritic remodeling in hippocampal and amygdaloid neurons. J. Neurosci., 22: 6810–6818.

Discussion: Chapter 15

SHALEV: Looking into the nature and origin of intrusive recollections is extremely important to the understanding of PTSD. For me they also occur in response to the incongruence and grotesqueness of the experience, which takes us beyond the fight/flight sequence paradigm. Relating this to work in basic science, there is evidence that exposure to species-specific 'grotesque' stimuli can evoke, in some animal species, a sequence of first attending to the aversive stimulus and second avoiding it (Hebb, D.O. (1946) On the nature of fear. Psychol. Rev., 53: 259–276; Humphrey, N.K. and Keeble, G.R. (1974) The reaction of monkeys to 'fearsome' pictures. Nature, 251: 502–502). This sequence of attending/avoiding might be analogous to the sequence of intrusive recollections and their avoidance in PTSD. I would argue that stressful events do not become traumatic unless they include an element of incongruity — or otherwise unacceptable, shocking or unimaginable novelty. Stern's book "The Buffalo Creek Disaster" (Random House, 1977, pp. 46–47), for example, describes survivors who were satisfied to have struggled and pulled themselves out of the sliding mud, but then became extremely distressed as they saw other survivors — women and children — being swept away in the water. For me, this marks the boundary between stressful (fight/flight) and potentially traumatic (incongruous novelty) experience. Animal models of dealing with similarly incongruous novelty might teach us about this part of PTSD.

BREWIN: I think this is a very important point. Not enough attention has been paid to the content of intrusive images in PTSD. There is a difference between horror and fear for example, and we really have no idea about how reactions involving these two emotions are similar or different. Sometimes in a life threatening event it is almost as though a curtain between life and death gets torn aside, even if only for a second, and the person gets a glimpse of some terrifying other world involving their own death which they may never have to have contemplated. Is that the same as the sort of grotesqueness that someone might experience at seeing a burned body in a car? I think there are some very interesting questions here that will take us further.

YEHUDA: It is a discussion question for later, but I think that this whole idea of the utility of studying a symptom in isolation of the syndrome is questionable. I also wonder if it is useful to study something that occurs in normals, and call it a symptom because it reminds us of a similar phenomenon that occurs in someone who is impaired. Somehow the occurrence of a faulty memory process in the context of an illness seems like something different completely because it is embedded within the larger syndrome, and therefore may reflect something quite different. I don't think we bridge the gap between basic and clinical neuroscience this way, I think we widen it.

BREWIN: I agree with you it is important to be cautious, and we have certainly thought about this issue a lot. What has impressed us about our data from the analogue studies of trauma is that the same risk factors that have been identified in clinical studies came out as predictive in the nonclinical samples as well. So it seems to us there are concrete grounds to argue that analogue studies may be useful in developing our knowledge.

SECKL: You started off with a fantastic premise, trying to bridge clinical science to basic science in order to formulate questions in a clinical model in a way that basic scientists might be able to address. One problem we've had in this meeting is that basic scientists have lovely models, but clinicians find they are poor facsimiles of the patients we see. And yet, those who see humans with disorders also tend poorly to analyze them in the context of the underpinning basic science of the brain. You describe what to a neuroscientist are 'black box' concepts: VAM and SAM. I wonder if you have thought how these constructs that are quite hypothetical notions, might be formulated so that basic neuroscientists could address them. How are you going to come from your interesting position to talk to the basic scientists and say "let's construct an experiment?" I find myself concerned that we are in danger of making things even more inaccessible to each other rather than trying to bridge the gap, trying to discuss back and forth from bench-to-bedside and from bedside-to-bench. In my view we need to be able to formulate questions and problems in ways that allow us to examine pathogenesis as a fundamental biological entity as well as very complex psychological constructions.

BREWIN: I don't think that anybody has suggested that psychological constructs are anything but that, hypothetical. I would not want to claim that I am the person who can bridge this gap and somehow make this translation into basic science. What I've been trying to do in this talk is to identify what we as clinicians have observed about PTSD, and that is likely to be relevant to other scientists who are trying to understand the condition. The sort of model that I put forward was not designed to be tested by basic scientists, but rather to be informed by basic science, and particularly by what we know about the brain. The model is testable within the framework of cognitive psychology, but extending it further may depend on whether imagery processes, for example, are of interest to scientists who work with animals. It seems to me that in creating their models basic scientists may benefit from understanding in greater detail the clinical features of PTSD.

SECKL: The point that I am making is that I think we are speaking two different languages. And there is difficulty in interaction, hampering translation which actually is the critical thing. We are in danger of developing a dualist view of PTSD, with the mind a 'soul', remote from genetics and cells and molecules that underlie disorders of body and brain. I find myself uncomfortable with this.

E.R. de Kloet, M.S. Oitzl & E. Vermetten (Eds.)
Progress in Brain Research, Vol. 167
ISSN 0079-6123

CHAPTER 16

Post-traumatic stress disorder in somatic disease: lessons from critically ill patients

Gustav Schelling*

Department of Anaesthesiology, Ludwig-Maximilians University, 81377 Munich, Germany

Abstract: Post-traumatic stress disorder (PTSD) is a well-recognized complication of severe illness. PTSD has been described in patients after multiple trauma, burns, or myocardial infarction with a particularly high incidence in survivors of acute pulmonary failure (Acute Respiratory Distress Syndrome) or septic shock. Many patients with evidence of PTSD after critical illness have been treated in intensive care units (ICUs). Studies in long-term survivors of ICU treatment demonstrated a clear and vivid recall of different categories of traumatic memory such as nightmares, anxiety, respiratory distress, or pain with little or no recall of factual events. A high number of these traumatic memories from the ICU has been shown to be a significant risk factor for the later development of PTSD in long-term survivors. In addition, patients in the ICU are often treated with stress hormones like epinephrine, norepinephrine, or cortisol. The number of the above-mentioned categories of traumatic memory increased with the totally administered dosages of catecholamines and cortisol, and the evaluation of these categories at different time points after discharge from the ICU showed better memory consolidation with higher dosages of stress hormones administered. Conversely, the prolonged administration of *β-adrenergic antagonists* during the recovery phase after cardiac surgery resulted in a lower number of traumatic memories and a lower incidence of stress symptoms at 6 months after surgery. Findings with regard to the administration of the stress hormone cortisol were more complex, however. Several studies from our group have demonstrated that the administration of stress doses of cortisol to critically ill patients resulted in a significant reduction of PTSD symptoms measured after recovery without influencing the number of categories of traumatic memory. This can possibly be explained by a cortisol-induced temporary impairment in traumatic memory retrieval that has previously been demonstrated in both rats and humans. ICU therapy of critically ill patients can serve as a stress model that allows the delineation of stress hormone effects on traumatic memory and PTSD development. This could also result in new approaches for prophylaxis and treatment of stress-related disorders.

Keywords: intensive care unit; critical illness; post-traumatic stress disorder; PTSD; catecholamines; glucocorticoids

Incidence of PTSD after critical illness

Post-traumatic stress disorder (PTSD) has been described as a consequence of serious medical conditions like myocardial infarction, cardiac surgery, hemorrhage and stroke, childbirth, miscarriage,

*Corresponding author. Tel.: +49897095-0 (operator); Fax: +49897095-8886;
E-mail: gustav.schelling@med.uni-muenchen.de

DOI: 10.1016/S0079-6123(07)67016-2

Table 1. Incidence of PTSD in survivors of major injury or critical illness

Patients	References	Number of patients	Follow-up period	Instruments	Incidence
Traffic accidents	Mayou et al. (2002)	1441	3/12 months	Questionnaires	23.1/16.5%
	Schnyder and Buddeberg (1996)	104	13 days/1 year	Interview, CAPS[a]	4.7/1.9%
Burns	Difede et al. (1997)	51	2/6/12 months	Interview	35.3/40/45.2%
	Madianos et al. (2001)	40	12 months	Interview	20%
Awareness during anesthesia	Osterman et al. (2001)	26	17.9 years post-op.	Interview	56.3%
Coronary heart disease/ myocardial infarction	Kutz et al. (1988)	100	Prevalence study	?	Chronic PTSD = 16%; acute PTSD = 9%
Cardiopulmonary resuscitation	Ladwig et al. (1999)	21	39 months (mean)	Interview	38.1%
Heart transplantation	Stukas et al. (1999)	145–191	7/12/36 months	Interview	9.7/15.6/17%
Heart surgery	Stoll et al. (2000)	80	5–6 months	Questionnaires	15%
	Schelling et al. (2003a)	148	Pre-op. and 6 months	Questionnaires	4.8/18.2%
Intensive care therapy	Capuzzo et al. (2005)	84	1 week/3 months	Questionnaires	5% (PTSD symptoms)
General intensive care	Schelling et al. (1998)	80	4.3 ± 3.0 years	Questionnaires	27.5%
Acute pulmonary failure (ARDS)	Kapfhammer et al. (2004)	46	8 years (median)	Interview	24%
Septic shock	Schelling et al. (1999)	27 cortisol/ 27 controls	4 years (median)	Questionnaires	18.5% with cortisol/ 59% without

[a]Clinician administered PTSD scale.

abortion and gynecological procedures, intensive care treatment, human immunodeficiency virus infection, or awareness under anesthesia. The highest prevalence rates are described in individuals treated in intensive care units (ICUs) where patients with acute respiratory failure or septic shock showed the highest risk (Tedstone and Tarrier, 2003; Krauseneck et al., 2005) (Table 1). Furthermore, PTSD symptoms may not only appear in patients after ICU treatment but also in close relatives of such individuals who were confronted with life-threatening illness and ICU treatment of their beloved ones (Azoulay et al., 2005).

Traumatic memories from the ICU and PTSD development after severe illness

PTSD in patients treated for life-threatening illness has consistently been linked to traumatic memories from ICU treatment (Schelling et al., 1998; Jones et al., 2001; Schelling et al., 2003a; Kapfhammer et al., 2004; Deja et al., 2006). These traumatic memories consist of a clear and vivid recall of emotionally relevant experiences such as hallucinations and nightmares (Schelling et al., 1998; Jones et al., 2000) but relatively little recall of factual events during their ICU stay (Jones et al., 1979, 2000). Particularly well remembered potentially traumatic events are respiratory distress, anxiety/panic (Deja et al., 2006), or fear of imminent death that were vividly present up to 10 years after discharge from the ICU (Schelling et al., 1998; Stoll et al., 1999). Several studies found a significant association between number and type of traumatic memories from the ICU and the incidence and intensity of PTSD symptoms (Schelling et al., 1998; Kapfhammer et al., 2004; Deja et al., 2006) (Fig. 1).

Use of stress hormones as therapeutic agents in the ICU and traumatic memories in patients after critical illness

In contrast to other patient populations at risk for PTSD, patients who are treated in an ICU often

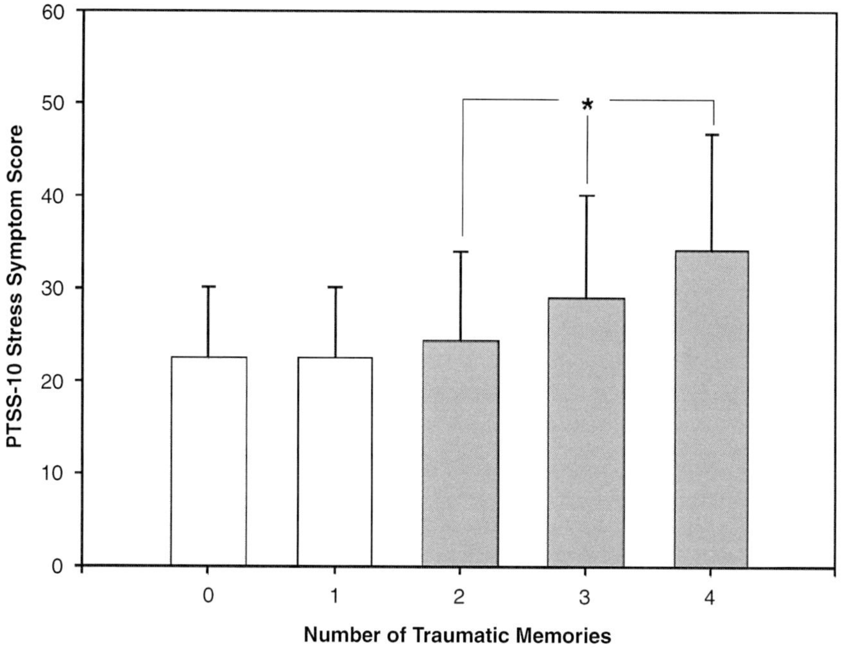

Fig. 1. Comparison of PTSD symptom scores between patients ($n = 148$) with 0–4 categories of traumatic memory from post-operative ICU treatment after cardiac surgery. PTSD scores were measured and traumatic memories were evaluated 6 months after discharge from the ICU (Schelling et al., 2003). The top of the bars indicate the mean, the "whiskers" show standard deviation. PTSD symptom scores were measured with the PTSS-10 instrument that has been validated in patients after ICU therapy (Stoll et al., 1999). Patients with multiple (>1, $n = 80$) categories of traumatic memory (gray bars) had significantly higher PTSD scores when compared to patients with 0 or only 1 category of traumatic memory ($*p<0.01$, t-test).

require exogenously administered stress hormones including epinephrine (for cardiac failure), norepinephrine (for vascular failure with hypotension), or glucocorticoids (to downregulate the systemic inflammatory reaction that is often seen in patients with critical illness like septic shock) (Briegel et al., 1999; Annane et al., 2002). There is extensive evidence from the literature that stress hormones may cause consolidation of emotional memory in both animals (McGaugh, 2002) and humans (Buchanan and Lovallo, 2001; Andreano and Cahill, 2006). Because of the aforementioned relationship between traumatic memories from the ICU and the development of PTSD, the exogenous administration of stress hormones could theoretically increase the incidence of traumatic memories or PTSD after critical illness. We tested this hypothesis in a recent prospective cohort study in cardiac surgical patients (Schelling et al., 2003a). The number of traumatic memories from the ICU in these patients did indeed correlate with the totally administered dosage of epinephrine whereas other variables such as the duration of ICU therapy or disease severity had no effect. On the other hand, patients who had received a *β-adrenergic antagonist* during the recovery phase after surgery had a significantly lower number of traumatic memories and lower PTSS-10 scores. Interestingly, this protective effect of β-adrenergic antagonists was only seen in female patients whereas the presumed memory consolidating effects of β-adrenergic stimulation with epinephrine was more pronounced in male individuals (Schelling et al., 2003b). These findings are corroborated by other studies in non-ICU patients at risk for PTSD. Increased heart rates after a traumatic event (indicative of increased catecholaminergic activity) was positively associated with the later development of PTSD symptoms (Shalev et al., 1998; Delahanty et al., 2003) while the use of the β-adrenergic antagonist propranolol in patients after trauma showed some protective effects regarding PTSD development (Pitman et al., 2002).

Whereas there is some evidence of a positive relationship between β-adrenergic activation with regard to traumatic memory formation and PTSD development in critically ill patients from our previous studies (Schelling et al., 2000, 2003a), our findings with regard to glucocorticoid effects on traumatic memories and PTSD appeared to be more complex.

Glucocorticoids and PTSD development in patients after ICU treatment

Glucocorticoids enhance memory consolidation for emotionally aversive experiences in both animals (McGaugh and Roozendaal, 2002) and healthy humans (Buchanan and Lovallo, 2001; Andreano and Cahill, 2006). In addition, we found that the intensity of PTSD symptoms increases with the number of categories of traumatic memory (Fig. 1). One could therefore assume that glucocorticoids might also enhance the incidence and intensity of PTSD in patients after critical illness. This possibility was tested in several controlled studies that systematically compared individuals who were treated with stress doses of hydrocortisone during critical illness with patients who received either standard therapy (Schelling et al., 1999, 2001, 2004) or placebo (Schelling et al., 2001; Weis et al., 2006). Patients from the hydrocortisone groups of these studies consistently had either lower stress symptom scores or a lower incidence of PTSD when compared to the respective control groups with no differences regarding the number and type of traumatic memories. Figures 2 and 3 give an example of glucocorticoid effects on traumatic memories from the ICU and PTSD symptoms in patients after cardiac surgery (Weis et al., 2006).

These data do not support the above-mentioned assumption that exogenously administered glucocorticoids increase consolidation of traumatic information and the likelihood for critically ill patients to develop PTSD after exposure to high

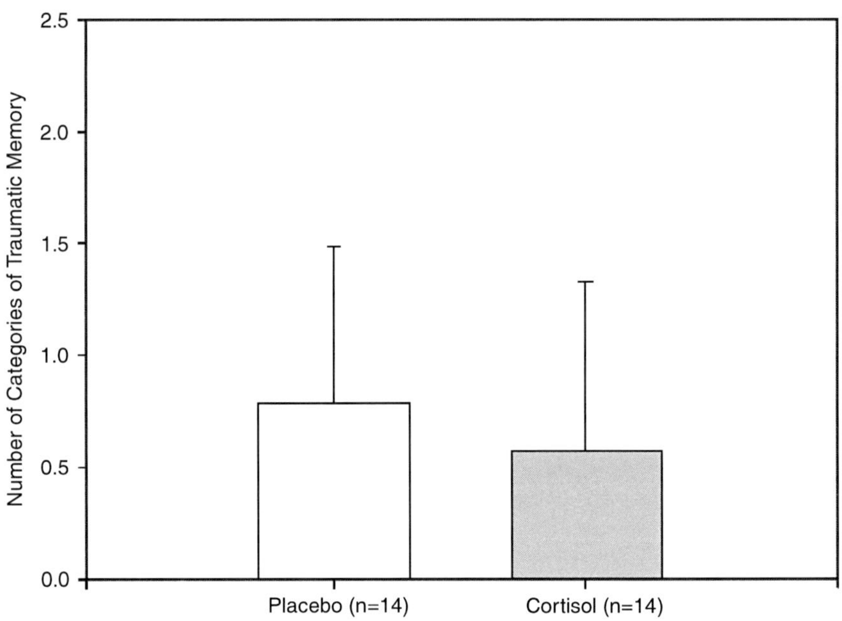

Fig. 2. Number of different categories of traumatic memory in patients who received stress doses of hydrocortisone during the perioperative period of cardiac surgery (the hydrocortisone group, $n = 14$, gray bars) and patients from the placebo group ($n = 14$, white bars) at 6 months after cardiac surgery (Weis et al., 2006). The top of the bars indicate the mean, the "whiskers" show standard deviation. The difference between both groups was not statistically significant ($p = 0.44$).

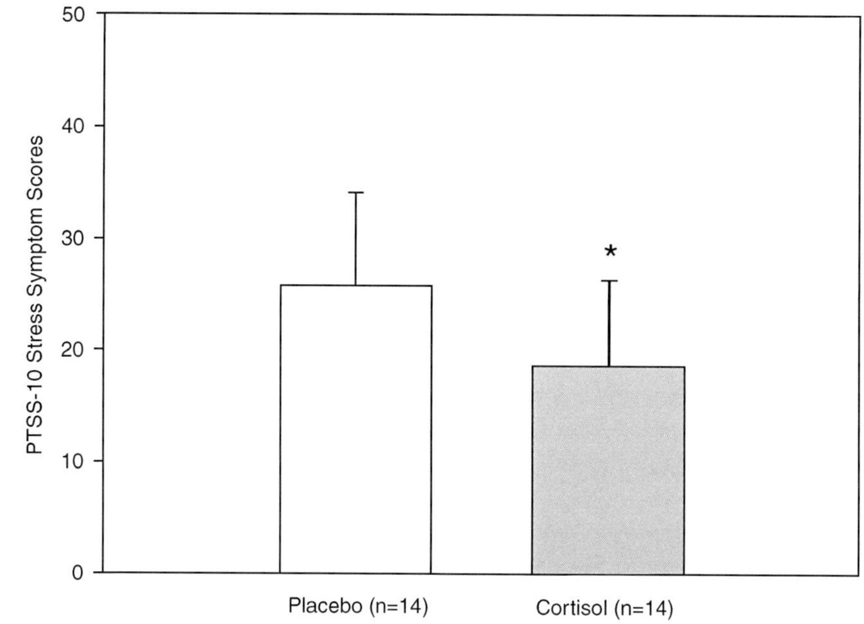

Fig. 3. Comparison of PTSD symptom scores between patients who received stress doses of hydrocortisone during the perioperative period of cardiac surgery (the hydrocortisone group, $n = 14$, gray bars) and patients from a control group which received placebo ($n = 14$, white bars) (Weis et al., 2006). * indicates $p = 0.03$ when compared to patients from the control group. The top of the bars indicate the mean, the "whiskers" show standard deviation. Stress symptom scores were measured using a modified version of the PTSS-10 questionnaire (Stoll et al., 1999).

stress but rather show the opposite, namely a protective effect of stress doses of hydrocortisone. These findings can possibly be explained by the fact that glucocorticoids are not only involved in memory consolidation but also interfere with memory recall (Roozendaal, 2002). Indeed, it has been shown in both rats (de Quervain et al., 1998; Roozendaal et al., 2003a, b) and humans (de Quervain et al., 2000; Wolf et al., 2001) that exposure to stress levels of glucocorticoids may result in a temporary impairment in memory retrieval (Domes et al., 2005; Kuhlmann et al., 2005). Low effective serum cortisol levels such as found during critical illness (Cooper and Stewart, 2003), in some patients after trauma at risk for PTSD (Delahanty et al., 2000) or in some patients with established PTSD (Yehuda et al., 1990; Boscarino, 1996; Yehuda, 2002), could thus lead to excessive retrieval of traumatic information. In addition, because it has been proposed that PTSD develops over time because of positive feedback mechanisms in which traumatic memories are constantly retrieved and restored (Pitman and Delahanty, 2005), increased cortisol levels after a massive stress exposure could inhibit the recall of traumatic information thus reducing the probability of PTSD. Of particular interest is the fact, that glucocorticoid effects on memory appears to be more pronounced with regard to emotionally relevant information as compared to emotionally neutral material. In rats, both memory-consolidating effects of glucocorticoids (Roozendaal et al., 2006b) and glucocorticoid effects on memory retrieval require concurrent noradrenergic activation (Roozendaal et al., 2004, 2006a). In addition, the prolonged glucocorticoid-induced impairment of memory retrieval could also facilitate extinction of traumatic memories. High circulating levels of glucocorticoids have been shown to facilitate extinction behaviors of traumatic experiences in rats whereas a blockade of glucocorticoid synthesis prevents extinction of conditioned fear (Bohus and De Wied, 1980; Barrett and Gonzalez-Lima, 2004; Yang et al., 2006).

Conclusions

Critically ill patients in the ICU are at risk for the formation of traumatic memories and the development of PTSD. Both of these adverse outcomes are facilitated by increased levels of adrenergic stimulation. Whereas the use of β-adrenergic antagonists during the recovery phase after critical illness may have a protective effect, the prolonged administration of stress doses of glucocorticoids does not influence traumatic memory formation but may result in a lower incidence of PTSD symptoms in survivors. Among other factors, this effect could be due to a more efficient downregulation and termination of the stress response, an interference with traumatic memory retrieval, a facilitated extinction of traumatic information, or a combination of these factors. In addition, these data strongly suggest that abnormalities in circulating stress hormones may play an important role in the pathogenesis of PTSD.

Abbreviations

PTSD	post-traumatic stress disorder
ICU	intensive care unit

References

Andreano, J.M. and Cahill, L. (2006) Glucocorticoid release and memory consolidation in men and women. Psychol. Sci., 17(6): 466–470.

Annane, D., Sebille, V., Charpentier, C., Bollaert, P.E., Francois, B., Korach, J.M., Capellier, G., Cohen, Y., Azoulay, E., Troche, G., Chaumet-Riffaut, P. and Bellissant, E. (2002) Effect of treatment with low doses of hydrocortisone and fludrocortisone on mortality in patients with septic shock. JAMA, 288(7): 862–871.

Azoulay, E., Pochard, F., Kentish-Barnes, N., Chevret, S., Aboab, J., Adrie, C., Annane, D., Bleichner, G., Bollaert, P.E., Darmon, M., Fassier, T., Galliot, R., Garrouste-Org, M., Goulenok, C., Goldgran-Toledano, D., Hayon, J., Jourdain, M., Kaidomar, M., Laplace, C., Larche, J., Liotier, J., Papazian, L., Poisson, C., Reignier, J., Saidi, F. and Schlemmer, B. (2005) Risk of post-traumatic stress symptoms in family members of intensive care unit patients. Am. J. Respir. Crit. Care Med., 171(9): 987–994.

Barrett, D. and Gonzalez-Lima, F. (2004) Behavioral effects of metyrapone on Pavlovian extinction. Neurosci. Lett., 371(2-3): 91–96.

Bohus, B. and De Wied, D. (1980) In: Chester-Jones I. and Henderson I.W. (Eds.), General, Comparative and Clinical Endocrinology of the Adrenal Cortex. Academic Press, London, pp. 265–347.

Boscarino, J.A. (1996) Posttraumatic stress disorder, exposure to combat, and lower plasma cortisol among Vietnam veterans: findings and clinical implications. J. Consult. Clin. Psychol.: 191–201.

Briegel, J., Forst, F., Haller, M., Schelling, G., Kilger, E., Kuprat, G., Hemmer, B., Hummel, T., Lenhart, A., Heyduck, M., Stoll, C. and Peter, K. (1999) Stress doses of hydrocortisone reverse hyperdynamic septic shock: a prospective, randomized, double-blind, single center study. Crit. Care Med., 27(4): 723–732.

Buchanan, T.W. and Lovallo, W.R. (2001) Enhanced memory for emotional material following stress-level cortisol treatment in humans. Psychoneuroendocrinology, 26(3): 307–317.

Capuzzo, M., Valpondi, V., Cingolani, E., Gianstefani, G., De Luca, S., Grassi, L. and Alvisi, R. (2005) Post-traumatic stress disorder-related symptoms after intensive care. Minerva Anestesiol., 71(4): 167–179.

Cooper, M.S. and Stewart, P.M. (2003) Corticosteroid insufficiency in acutely ill patients. N. Engl. J. Med., 348(8): 727–734.

Deja, M., Denke, C., Weber-Carstens, S., Schroder, J., Pille, C.E., Hokema, F., Falke, K.J. and Kaisers, U. (2006) Social support during intensive care unit stay might improve mental impairment and consequently health-related quality of life in survivors of severe acute respiratory distress syndrome. Crit. Care, 10(5): R147.

Delahanty, D.L., Raimonde, A.J. and Spoonster, E. (2000) Initial posttraumatic urinary cortisol levels predict subsequent PTSD symptoms in motor vehicle accident victims. Biol. Psychiatry, 48(9): 940–947.

Delahanty, D.L., Raimonde, A.J., Spoonster, E. and Cullado, M. (2003) Injury severity, prior trauma history, urinary cortisol levels, and acute PTSD in motor vehicle accident victims. J. Anxiety Disord., 17(2): 149–164.

Difede, J.A., Jaffe, A.B., Musngi, G., Perry, S. and Yurt, R. (1997) Determinants of pain expression in hospitalized burn patients. Pain, 72(1-2): 245–251.

Domes, G., Rothfischer, J., Reichwald, U. and Hautzinger, M. (2005) Inverted-U function between salivary cortisol and retrieval of verbal memory after hydrocortisone treatment. Behav. Neurosci., 119(2): 512–517.

Jones, C., Griffiths, R.D. and Humphris, G. (2000) Disturbed memory and amnesia related to intensive care. Memory, 8(2): 79–94.

Jones, C., Giffiths, R.D., Humphris, G. and Skirrow, P.M. (2001) Memory, delusions, and the development of acute posttraumatic stress disorder-related symptoms after intensive care. Crit. Care Med., 29(3): 573–580.

Jones, J., Hoggart, B., Withey, J., Donaghue, K. and Ellis, B.W. (1979) What the patients say: a study of reactions to an intensive care unit. Intensive Care Med., 5(2): 89–92.

Kapfhammer, H.P., Rothenhausler, H.B., Krauseneck, T., Stoll, C. and Schelling, G. (2004) Posttraumatic stress disorder and health-related quality of life in long-term survivors of acute respiratory distress syndrome. Am. J. Psychiatry, 161(1): 45–52.

Krauseneck, T., Rothenhausler, H.B., Schelling, G. and Kapfhammer, H.P. (2005) PTSD in somatic disease. Fortschr. Neurol. Psychiatr., 73(4): 206–217.

Kuhlmann, S., Kirschbaum, C. and Wolf, O.T. (2005) Effects of oral cortisol treatment in healthy young women on memory retrieval of negative and neutral words. Neurobiol. Learn. Mem., 83(2): 158–162.

Kutz, I., Garb, R. and David, D. (1988) Post-traumatic stress disorder following myocardial infarction. Gen. Hosp. Psychiatry, 10(3): 169–176.

Ladwig, K.H., Schoefinius, A., Dammann, G., Danner, R., Gurtler, R. and Herrmann, R. (1999) Long-acting psychotraumatic properties of a cardiac arrest experience. Am. J. Psychiatry, 156(6): 912–919.

Madianos, M.G., Papaghelis, M., Ioannovich, J. and Dafni, R. (2001) Psychiatric disorders in burn patients: a follow-up study. Psychother. Psychosom., 70(1): 30–37.

Mayou, R.A., Ehlers, A. and Bryant, B. (2002) Posttraumatic stress disorder after motor vehicle accidents: 3-year follow-up of a prospective longitudinal study. Behav. Res. Ther., 40(6): 665–675.

McGaugh, J.L. and Roozendaal, B. (2002) Role of adrenal stress hormones in forming lasting memories in the brain. Curr. Opin. Neurobiol., 12(2): 205–210.

Osterman, J.E., Hopper, J., Heran, W.J., Keane, T.M. and Van der Kolk, B.A. (2001) Awareness under anesthesia and the development of posttraumatic stress disorder. Gen. Hosp. Psychiatry, 23(4): 198–204.

Pitman, R.K. and Delahanty, D.L. (2005) Conceptually driven pharmacologic approaches to acute trauma. CNS Spectr., 10(2): 99–106.

Pitman, R.K., Sanders, K.M., Zusman, R.M., Healy, A.R., Cheema, F., Lasko, N.B., Cahill, L. and Orr, S.P. (2002) Pilot study of secondary prevention of posttraumatic stress disorder with propranolol. Biol. Psychiatry, 51(2): 189–192.

de Quervain, D.-F., Roozendaal, B. and McGaugh, J.L. (1998) Stress and glucocorticoids impair retrieval of long-term spatial memory. Nature, 394(6695): 787–790.

de Quervain, D.J., Roozendaal, B., Nitsch, R.M., McGaugh, J.L. and Hock, C. (2000) Acute cortisone administration impairs retrieval of long-term declarative memory in humans. Nat. Neurosci., 3(4): 313–314.

Roozendaal, B. (2002) Stress and memory: opposing effects of glucocorticoids on memory consolidation and memory retrieval. Neurobiol. Learn. Mem., 78: 578–595.

Roozendaal, B., Griffith, Q.K., Buranday, J., de Quervain, D.J.-F. and McGaugh, J.L. (2003a) The hippocampus mediates glucocorticoid-induced impairment of spatial memory retrieval: dependence on the basolateral amygdala. Proc. Natl. Acad. Sci. U.S.A., 100: 1328–1333.

Roozendaal, B., Hahn, E.L., Nathan, S.V., de Quervain, D.J. and McGaugh, J.L. (2004) Glucocorticoid effects on memory retrieval require concurrent noradrenergic activity in the hippocampus and basolateral amygdala. J. Neurosci., 24(37): 8161–8169.

Roozendaal, B., Okuda, S., de Quervain, D.J. and McGaugh, J.L. (2006a) Glucocorticoids interact with emotion-induced noradrenergic activation in influencing different memory functions. Neuroscience, 138(3): 901–910.

Roozendaal, B., Okuda, S., van der Zee, E.A. and McGaugh, J.L. (2006b) Glucocorticoid enhancement of memory requires arousal-induced noradrenergic activation in the basolateral amygdala. Proc. Natl. Acad. Sci. U.S.A., 103(17): 6741–6746.

Roozendaal, B., de Quervain, D.J.-F., Schelling, G. and McGaugh, J.L. (2003b) A systemically administered β-adrenoceptor antagonist blocks corticosterone-induced impairment of contextual memory retrieval in rats. Neurobiol. Learn. Mem., 81(2): 150–154.

Schelling, G., Briegel, J., Roozendaal, B., Stoll, C., Rothenhausler, H.B. and Kapfhammer, H.P. (2001) The effect of stress doses of hydrocortisone during septic shock on posttraumatic stress disorder in survivors. Biol. Psychiatry, 50(12): 978–985.

Schelling, G., Briegel, J., Stoll, C., Rothenhausler, H.B. and Kapfhammer, H.P. (2000) The effect of the norepinephrine dosage/serum cortisol ratio during septic shock on post-traumatic stress disorder in survivors. Anesthesiology, 94: A516 Ref Type: Abstract.

Schelling, G., Kilger, E., Roozendaal, B., Briegel, J., Dagge, A., Rothenhausler, H.B., Nollert, G. and Kapfhammer, H. (2004) Stress doses of hydrocortisone, traumatic memories and post-traumatic stress disorder in patients after cardiac surgery: a randomized study. Biol. Psychiatry, 55(6): 627–633.

Schelling, G., Richter, M., Roozendaal, B., Rothenhausler, H., Stoll, C., Nollert, G., Schmidt, M. and Kapfhammer, H. (2003a) Exposure to high stress in the ICU may have negative effects on health-related quality of life outcomes after cardiac surgery. Crit. Care Med., 31(7): 1971–1980.

Schelling, G., Rothenhausler, H., Krauseneck, T., Nollert, G. and Kapfhammer, H. (2003b) Sex-related effects of beta-adrenergic blockade on memory for traumatic events from the perioperative period of cardiac surgery. Anesthesiology, 99: A310.

Schelling, G., Stoll, C., Haller, M., Briegel, J., Manert, W., Hummel, T., Lenhart, A., Heyduck, M., Polasek, J., Meier, M., Preuss, U., Bullinger, M., Schüffel, W. and Peter, K. (1998) Health-related quality of life and post-traumatic stress disorder in survivors of the acute respiratory distress syndrom (ARDS). Crit. Care Med., 25(4): 651–659.

Schelling, G., Stoll, C., Kapfhammer, H.P., Rothenhausler, H.B., Krauseneck, T., Durst, K., Haller, M. and Briegel, J.

(1999) The effect of stress doses of hydrocortisone during septic shock on posttraumatic stress disorder and health-related quality of life in survivors. Crit. Care Med., 27(12): 2678–2683.

Schnyder, U. and Buddeberg, C. (1996) Psychosocial aspects of accidental injuries: an overview [see comments]. Langenbecks Arch. Chir., 381(3): 125–131.

Shalev, A.Y., Sahar, T., Freedman, S., Peri, T., Glick, N., Brandes, D., Orr, S.P. and Pitman, R.K. (1998) A prospective study of heart rate response following trauma and the subsequent development of posttraumatic stress disorder. Arch. Gen. Psychiatry, 55(6): 553–559.

Stoll, C., Kapfhammer, H.P., Haller, H., Briegel, J., Krauseneck, T., Durst, K. and Schelling, G. (1999) Sensitivity and specificity of a screening test to document traumatic experiences and to diagnose post-traumatic stress disorder in patients after intensive care treatment. Intensive Care Med., 25: 697–704.

Stoll, C., Schelling, G., Goetz, A.E., Kilger, E., Bayer, A., Kapfhammer, H.P., Rothenhausler, H.B., Kreuzer, E., Reichart, B. and Peter, K. (2000) Health-related quality of life and post-traumatic stress disorder in patients after cardiac surgery and intensive care treatment. J. Thorac. Cardiovasc. Surg., 120(3): 505–512.

Stukas Jr., A.A., Dew, M.A., Switzer, G.E., DiMartini, A., Kormos, R.L. and Griffith, B.P. (1999) PTSD in heart transplant recipients and their primary family caregivers. Psychosomatics, 40(3): 212–221.

Tedstone, J.E. and Tarrier, N. (2003) Posttraumatic stress disorder following medical illness and treatment. Clin. Psychol. Rev., 23(3): 409–448.

Weis, F., Kilger, E., Roozendaal, B., de Quervain, D.J.-F., Lamm, P., Schmidt, M., Schmoelz, M., Briegel, J. and Schelling, G. (2006) Stress doses of hydrocortisone reduce chronic stress symptoms and improve health-related quality of life in high risk patients after cardiac surgery: a randomized study. J. Thorac. Cardiovasc. Surg., 131(2): 277–282.

Wolf, O.T., Convit, A., McHugh, P.F., Kandil, E., Thorn, E.L., de Santi, S., McEwen, B.S. and de Leon, M.J. (2001) Cortisol differentially affects memory in young and elderly men. Behav. Neurosci., 115(5): 1002–1011.

Yang, Y.L., Chao, P.K. and Lu, K.T. (2006) Systemic and intra-amygdala administration of glucocorticoid agonist and antagonist modulate extinction of conditioned fear. Neuropsychopharmacology, 31: 912–924.

Yehuda, R. (2002) Current status of cortisol findings in post-traumatic stress disorder. Psychiatr. Clin. North Am., 25(2): 341–vii.

Yehuda, R., Southwick, S.M., Nussbaum, G., Wahby, V., Giller Jr., E.L. and Mason, J.W. (1990) Low urinary cortisol excretion in patients with posttraumatic stress disorder. J. Nerv. Ment. Dis., 178(6): 366–369.

Discussion: Chapter 16

LIBERZON: If I understand the norepi/cort ratio correctly, you measured norepinephrine in the periphery.

SCHELLING: We measured the average daily-administered dosage of norepinephrine; the dosage correlates with circulating norepinehprine levels, but we did not measure norepi plasma levels.

LIBERZON: So you did not measure norepi levels.

SCHELLING: Yes. We did another study where we correlated the norepinephrine dosages given to these individuals with biologic endpoints measured during follow-up. Norepinephrine dosages correlated with traumatic memories and stress symptom scores and this suggests that norepinephrine dosages may reflect the actual levels that are achieved in these patients.

LIBERZON: It is surprising to me because norepinephrine levels are mostly coming from a muscle, they don't really come from the adrenal source. So I am a little confused that we are talking from different biologies. The central norepinephrine versus the peripheral — how do you reconcile on that?

SCHELLING: If you give norepinephrine to animals you get a central activation of the brain noradrenergic system, stimulating β-adrenergic receptors on the vagal nerve and this projects to the amygdala. So if you activate the peripheral adrenergic system you get central activation in parallel. If epinephrine is given in high dosages to an individual you presumably get pronounced central activation and I think it is pretty safe to assume this relationship.

DE KLOET: In the critically ill patient glucocorticoid action is inadequate and that's why the replacement with glucocorticoids is considered advisable and you demonstrated that as a consequence later PTSD symptoms are suppressed. What criteria are being used to decide that glucocorticoid administration is required, and when do you administer?

SCHELLING: In our patients with septic shock, the administration of stress doses of glucocorticoids starts as soon as the diagnosis is established. In the cardiac surgery studies, glucocorticoids are given at induction of anesthesia, before the actual stress exposure of surgery. The time point of the intervention does not seem to be critical because under both circumstances, the protective effect of glucocorticoids could be shown.

GUNNAR: You had some influence on traumatic memory. Have you checked to see whether non-traumatic memory is affected?

SCHELLING: No, we did not do this. It is known from experiments on human volunteers that non-traumatic memory is also affected. Benno Roozendaal and Dominique de Quervain have, however, shown that glucocorticoid effects on memory are particularly pronounced with simultaneous activation of the noradrenergic system. We in some way may have mimicked this clinical situation. ...

LIBERZON: I am trying to imagine the patients with respect to the dosages. Who would get more norepinephrine and less cortisol? Who would get more cortisol and less norepinephrine? It seems to me that these will be different patients.

SCHELLING: What we see are patients from randomized studies where stress doses of hydrocortisone or placebo are given. Patients with stress doses of cortisol have higher cortisol levels and need lower norepinephrine dosages and patients who receive placebo have lower glucocorticoid levels and high noradrenergic stimulation. Because of randomization, both patient groups should be identical in many other aspects and should not be different from each other.

E.R. de Kloet, M.S. Oitzl & E. Vermetten (Eds.)
Progress in Brain Research, Vol. 167
ISSN 0079-6123

CHAPTER 17

Glucocorticoid-induced reduction of traumatic memories: implications for the treatment of PTSD

Dominique J.-F. de Quervain*

Division of Psychiatry Research, University of Zürich, Lenggstr. 31, 8032 Zürich, Switzerland

Abstract: Post-traumatic stress disorder (PTSD) is an anxiety disorder that can occur after a traumatic event such as military combat, terrorist attacks, or accidents. The disorder is characterized by traumatic memories that manifest as reexperiencing symptoms including daytime recollections, traumatic nightmares, or flashbacks in which components of the event are relived. These symptoms result from excessive retrieval of traumatic memories that often retain their vividness and power to evoke distress for decades or even a lifetime. We have reported previously that elevated glucocorticoid levels inhibit memory retrieval in animals and healthy human subjects. We therefore hypothesized that the administration of cortisol might also inhibit the retrieval of traumatic memories in patients with PTSD. In a recent pilot study we found the first evidence to support this hypothesis. During a 3-month observation period, low-dose cortisol (10 mg per day) was administered orally for 1 month to three patients with chronic PTSD using a double-blind, placebo-controlled, crossover design. In each patient investigated, there was a significant treatment effect with cortisol-related reductions in one of the daily-rated symptoms of traumatic memories without causing adverse side effects. Furthermore, we have reported evidence for a prolonged effect of the cortisol treatment. Persistent retrieval and reconsolidation of traumatic memories is a process that keeps these memories vivid and thereby the disorder alive. By inhibiting memory retrieval, cortisol may weaken the traumatic memory trace and thus reduce symptoms even beyond the treatment period. Future studies with more patients and longer treatment periods are required to evaluate the efficacy of cortisol treatment for PTSD.

Keywords: PTSD; glucocorticoids; cortisol; memory; retrieval; traumatic memory; treatment

Traumatic memories in post-traumatic stress disorder

Post-traumatic stress disorder (PTSD) is a response to a traumatic event and characterized by the following features: reexperiencing the traumatic event, avoidance of stimuli associated with the trauma, and hyperarousal. Reexperiencing symptoms include daytime recollections, traumatic nightmares, and flashbacks in which components of the event are relived (American Psychiatric Association, 1994; Yehuda, 2002b). These reexperiencing symptoms result from excessive retrieval of traumatic memories that often retain their vividness and power to evoke distress for decades or even a lifetime. Importantly, traumatic reexperiencing phenomena are again consolidated (reconsolidated) into memory that cements the traumatic memory trace. In fact, persistent retrieval, reexperiencing, and reconsolidation of traumatic memories is a process that keeps these memories vivid

*Corresponding author. Tel.: +41 44 384 2601;
Fax: +41 44 384 2686; E-mail: quervain@bli.unizh.ch

DOI: 10.1016/S0079-6123(07)67017-4

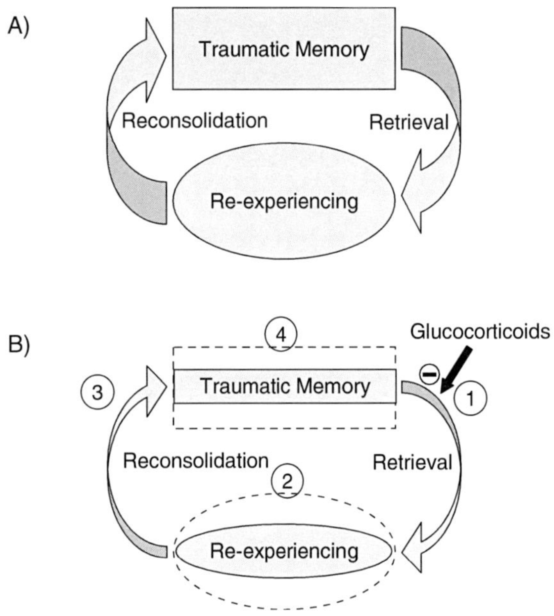

Fig. 1. Model on the role of glucocorticoids in the reduction of traumatic memory. (A) Persistent retrieval, reexperiencing, and reconsolidation of traumatic memories is a process that keeps these memories vivid and thereby the disorder alive. (B) Glucocorticoid-induced reduction of traumatic memory. By inhibiting memory retrieval, glucocorticoids partly interrupt the vicious cycle of retrieving (1), reexperiencing (2), and reconsolidating (3) traumatic memories and, thereby, promote forgetting and extinction processes (4) (see text for details).

and thereby the disorder alive (Fig. 1A). Therefore, it would be desirable to have a drug that reduces excessive retrieval of traumatic memories, as this would result in less reexperiencing phenomena and, consequently, in a weakening of the traumatic memory trace.

Glucocorticoids and memory retrieval

Glucocorticoids, stress hormones released from the adrenal cortex, are known to influence memory processes and growing evidence suggests that glucocorticoids have differential effects on discrete memory phases. In animal and human subjects, single administration of glucocorticoids enhances the consolidation of new memories (Kovacs et al., 1977; Flood et al., 1978; Roozendaal, 2000; Buchanan and Lovallo, 2001; Kuhlmann and Wolf, 2006). In contrast, we found that glucocorticoids impair memory retrieval processes (de Quervain et al., 1998). Specifically, we reported that 30 min after an electric footshock, rats have impaired retrieval of spatial memory acquired 24 h earlier. Interestingly, memory retrieval was not impaired 2 min or 4 h after the footshock. These time-dependent effects on retrieval performance corresponded to the circulating corticosterone levels at the time of testing, which suggested that the retrieval impairment is directly related to increased adrenocortical function. In support of this idea, we found that suppression of corticosterone synthesis with metyrapone blocks the stress-induced retention impairment. In addition, systemic corticosterone administered to non-stressed rats 30 min before retention testing induced dose-dependent retention impairment. Because corticosterone did not affect acquisition or immediate recall, the corticosterone-induced impairment in retention performance is attributable to a selective influence on long-term memory retrieval. In a next step we have translated these findings to healthy humans and found that a single administration of 25 mg cortisone impairs the recall of words learned 24 h earlier (de Quervain et al., 2000). Several further studies from different laboratories have indicated that impaired memory retrieval after the administration of glucocorticoids is a consistent finding in both animals and humans (Wolf et al., 2001; de Quervain et al., 2003; Roozendaal et al., 2003, 2004a; Buss et al., 2004; Het et al., 2005; Kuhlmann et al., 2005a; Sajadi et al., 2007). Moreover, there is recent evidence that emotionally arousing information is especially sensitive to the retrieval-impairing effects of glucocorticoids (Kuhlmann et al., 2005a, b; de Quervain et al., 2007).

Whereas elevated glucocorticoid levels are certainly detrimental when information should be retrieved (e.g., during exams), they may actually be beneficial in conditions when memory retrieval is distressing. As detailed above, PTSD is such a condition. We therefore hypothesized that by inhibiting the retrieval of traumatic memories, the administration of cortisol may be beneficial in patients with PTSD.

Glucocorticoids reduce traumatic memories in PTSD

Recently, we investigated the effects of cortisol treatment on the retrieval of traumatic memories in a small number of patients with chronic PTSD (Aerni et al., 2004). During a 3-month observation period, low-dose cortisol (10 mg per day) was administered orally for 1 month using a double-blind, placebo-controlled, crossover design. The administration of this low dose of cortisol for 1 month does not cause major side effects and does not suppress endogenous cortisol production (Cleare et al., 1999). To assess possible treatment effects on retrieval of traumatic memories, the patients daily rated the intensity and frequency of the feeling of reliving the traumatic event and the physiological distress felt in response to traumatic memories and nightmares (self-administered rating scales from the Clinician Administered PTSD Scale questions). Patient 1 was a 50-year-old man who survived a terrorist attack 4.5 years before inclusion into the study. There was a significant treatment effect for the intensity of the feeling of reliving the traumatic event (Fig. 2). Of interest, the intensity ratings during the last study month (with no medication) were significantly lower compared to those during the first month (placebo), suggesting a carryover effect of cortisol. There was also a significant treatment effect for the intensity of physiological distress. Patient 2 was a 40-year-old woman who experienced a life-threatening physical assault 1 year before inclusion in the study. There was a significant treatment effect for the frequency of nightmares (Fig. 2). Patient 3 was a 55-year-old man who had a severe car accident 8 years before inclusion in the study. To control for possible treatment order effects, this patient

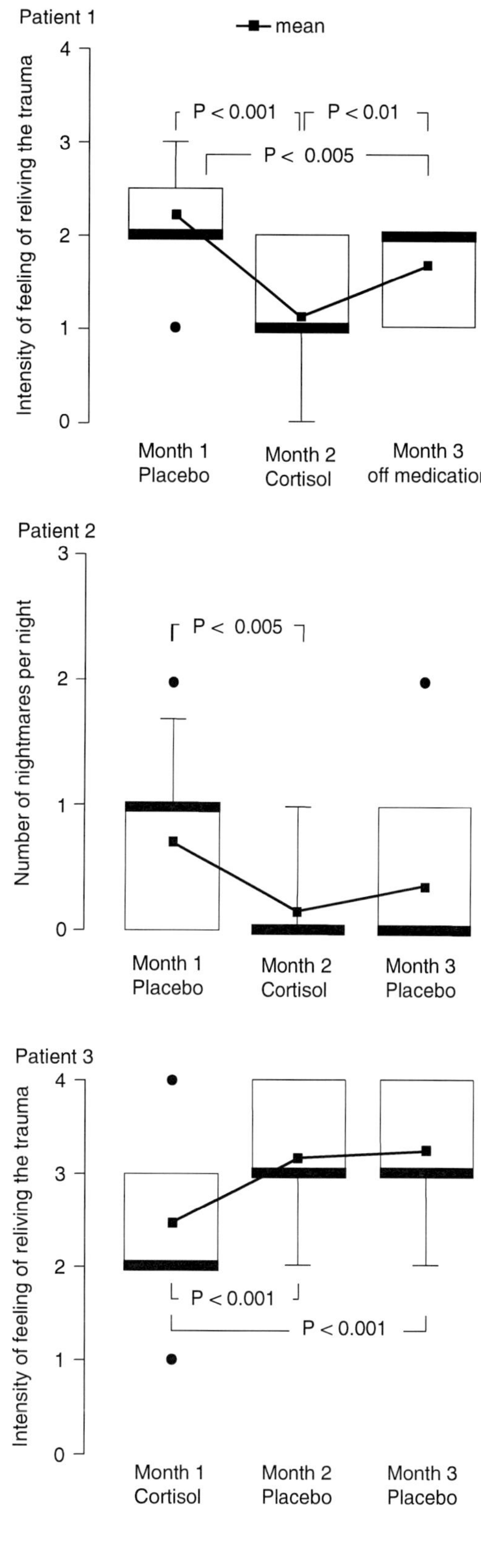

Fig. 2. Effects of cortisol on traumatic memories in PTSD. Most significant treatment-related change in frequency or intensity among the daily-rated symptoms of traumatic memories in each patient. The thick black line indicates the median. Black dots indicate outliers, whiskers which start from the boxes indicate the 10th and the 90th percentiles of the distribution, respectively; top and bottom of each box indicate the 75th and 25th percentiles, respectively. For further details, see Aerni et al. (2004).

received cortisol in the first month, followed by 2 months of placebo medication. Significant treatment effects were detected for the intensity of the feeling of reliving the traumatic event (Fig. 2), the physiological distress, and the frequency of nightmares. None of the patients complained about treatment-related disturbances of everyday memory upon questioning. Taken together, in all patients investigated, low-dose cortisol treatment had beneficial effects with significant reductions of at least 38% in one of the daily-rated symptoms of traumatic memories.

In recent experiments in patients with phobia we additionally found evidence that glucocorticoids may not only reduce retrieval of traumatic memory in patients with PTSD but also retrieval of fear memory in patients with phobia (Soravia et al., 2006). Phobic disorders are characterized by marked and persistent fear that is excessive or unreasonable, cued by the presence or anticipation of a specific object or situation (American Psychiatric Association, 1994; Barlow and Liebowitz, 1995). Exposure to a phobic stimulus almost invariably provokes retrieval of stimulus-associated fear memory (Cuthbert et al., 2003). In addition, phobic individuals tend to construct highly negative images of a phobic situation that substantially contributes to anticipatory anxiety as well as negative post-event processing. Such images are usually associated with explicit fearful memories of past phobic experiences that reinforce negative beliefss that are difficult to suppress and may strengthen the phobic response (Rapee and Heimberg, 1997; Fehm and Margraf, 2002). We hypothesized that glucocorticoids might inhibit retrieval of fear memory in phobia and thereby reduce stimulus-induced fear. We tested this hypothesis by administering glucocorticoids to 40 subjects with social phobia and 20 subjects with spider phobia in two double-blind, placebo-controlled studies (Soravia et al., 2006). In the social phobia study, cortisone (25 mg) administered orally 1 h before a socio-evaluative stressor significantly reduced self-reported fear during the anticipation-, exposure-, and recovery phase of the stressor. Moreover, the stress-induced release of cortisol in placebo-treated subjects correlated negatively with fear ratings, suggesting that endogenously released cortisol in the context of a phobic situation buffers fear symptoms. In the spider phobia study, repeated oral administration of cortisol (10 mg), but not placebo, 1 h before exposure to a spider photograph induced a progressive reduction of stimulus-induced fear. This effect was maintained when subjects were exposed to the stimulus again 2 days after the last cortisol administration, suggesting that cortisol may also have facilitated the extinction of phobic fear. Importantly, cortisol treatment did not reduce general, phobia-unrelated anxiety. These experiments indicate that by a common mechanism of reducing memory retrieval, glucocorticoids may be suited for the treatment of PTSD as well as phobias.

Possible mode of action of glucocorticoids in the reduction of traumatic memories

In the PTSD-study detailed above (Aerni et al., 2004) we found that the administration of cortisol reduces reexperiencing symptoms, which is a direct measure of traumatic memory retrieval. Extensive evidence from studies in amnesic patients, human-imaging studies, and lesion studies in animals indicates that the medial temporal lobe (MTL) is crucially involved in memory retrieval and that activation of the MTL is associated with successful memory retrieval (Squire, 1992; Moser and Moser, 1998; Cabeza and Nyberg, 2000). Moreover, a functional magnetic resonance imaging (fMRI) study in patients with PTSD showed that the MTL becomes activated by viewing masked traumatic images (Sakamoto et al., 2005). Using positron emission tomography (PET) imaging in healthy humans, we found that acutely administered cortisone reduces blood flow in the MTL during memory retrieval, an effect that correlated with the degree of memory retrieval impairment (de Quervain et al., 2003). Furthermore, systemic administration of glucocorticoids to rats shortly before retention testing induced memory retrieval impairments for contextual memory (Roozendaal et al., 2004b), a task that depends on the MTL (Squire, 1992), and local infusions of a glucocorticoid receptor agonist into the hippocampus of rats induced retrieval impairments comparable to those seen after systemic administration

(Roozendaal et al., 2003). Together, these findings suggest that elevated cortisol levels may have reduced the retrieval of traumatic memories by inhibiting MTL activity.

In the PTSD-study (Aerni et al., 2004), we additionally found evidence for a prolonged effect of the cortisol treatment. Persistent retrieval, reexperiencing, and reconsolidation of traumatic memories is a process that keeps these memories vivid and thereby the disorder alive (Fig. 1A). By inhibiting memory retrieval, cortisol may weaken the traumatic memory trace and thus reduce symptoms even beyond the treatment period. Specifically, by inhibiting memory retrieval, cortisol may partly interrupt the vicious cycle of spontaneous retrieving, reexperiencing, and reconsolidating traumatic memories and, thereby, promote forgetting (Fig. 1B), a spontaneous process that occurs when memory is not reactivated. Furthermore, and in line with findings in animals (Bohus and Lissak, 1968), cortisol may facilitate the extinction of conditioned responses to traumatic memory cues. Accordingly, because of the cortisol-induced reduction of memory retrieval, a traumatic memory cue would not be followed by the usual traumatic memory retrieval and reexperiencing but, instead, may become associated with a non-traumatic experience that would be stored as extinction memory.

In addition to the inhibitory effect on memory retrieval, elevated glucocorticoid levels are known to enhance the long-term consolidation of memories (Kovacs et al., 1977; Flood et al., 1978; Roozendaal, 2000; Buchanan and Lovallo, 2001; Kuhlmann and Wolf, 2006). It is therefore possible that glucocorticoids may have further promoted extinction of the traumatic memory by facilitating the storage of corrective experiences, as evidenced by recent findings indicating that glucocorticoids enhance the consolidation of fear extinction memory (Barrett and Gonzalez-Lima, 2004; Cai et al., 2006).

Role of endogenous cortisol in PTSD

Patients with PTSD often show low endogenous cortisol levels (Mason et al., 1986; Yehuda et al., 1995; Yehuda, 2002a). However, some studies also found normal (Young and Breslau, 2004) or higher (Pitman and Orr, 1990) cortisol levels. Furthermore, evidence indicates that a reduced cortisol excretion in response to a traumatic event may be associated with a higher risk of developing subsequent PTSD (McFarlane et al., 1997; Yehuda et al., 1998; Delahanty et al., 2000). The idea that higher cortisol levels may be protective with regard to the development of PTSD is strongly supported by the work of Schelling et al. (2001, 2004) who showed that the prolonged administration of stress doses of cortisol during intensive care treatment in critically ill patients reduces the risk for later PTSD. We propose that cortisol may influence both risk and symptoms of PTSD by controlling the amount of retrieved traumatic memories. Elevated cortisol levels may decrease risk and symptoms of PTSD by inhibiting excessive retrieval of traumatic memories, whereas low endogenous cortisol levels may promote development and symptomatology of PTSD by a disinhibition of traumatic memory retrieval. This notion is in line with the broader view that glucocorticoid release during acute stress represents an adaptive response that helps the organism to deal with a wide spectrum of internal and external demands (McEwen, 1998; de Kloet et al., 1999).

Conclusions

In a first small study we found evidence that the administration of low-dose cortisol reduces reexperiencing symptoms in patients with PTSD. This finding indicates that the inhibiting effect of glucocorticoids on memory retrieval is not restricted to episodic memory in healthy humans but also applies to traumatic memories in patients with PTSD. Furthermore, by inhibiting memory retrieval, glucocorticoids may weaken the traumatic memory trace and thus reduce symptoms even beyond the treatment period. Additional studies with more patients and longer treatment periods are needed to further evaluate the therapeutic efficacy and safety of low-dose cortisol for the treatment of PTSD.

Abbreviations

fMRI	functional magnetic resonance imaging
MTL	medial temporal lobe
PET	positron emission tomography
PTSD	post-traumatic stress disorder

Acknowledgment

Supported by a grant from the Swiss National Science Foundation to D.Q. (PP00B-106708).

References

Aerni, A., Traber, R., Hock, C., Roozendaal, B., Schelling, G., Papassotiropoulos, A., Nitsch, R.M., Schnyder, U. and de Quervain, D.J. (2004) Low-dose cortisol for symptoms of posttraumatic stress disorder. Am. J. Psychiatry, 161(8): 1488–1490.

American Psychiatric Association (1994) Diagnostic and Statistical Manual of Mental Disorders, fourth edition (DSM-IV). American Psychiatric Association, Washington, DC.

Barlow, D.H. and Liebowitz, M.R. (1995) Specific and social phobias. In: Kaplan H.I. and Sadock B.J. (Eds.), Comprehensive Textbook of Psychiatry, Vol. VI. Williams and Wilkins, New York, pp. 1204–1218.

Barrett, D. and Gonzalez-Lima, F. (2004) Behavioral effects of metyrapone on Pavlovian extinction. Neurosci. Lett., 371(2–3): 91–96.

Bohus, B. and Lissak, K. (1968) Adrenocortical hormones and avoidance behaviour of rats. Int. J. Neuropharmacol., 7(4): 301–306.

Buchanan, T.W. and Lovallo, W.R. (2001) Enhanced memory for emotional material following stress-level cortisol treatment in humans. Psychoneuroendocrinology, 26(3): 307–317.

Buss, C., Wolf, O.T., Witt, J. and Hellhammer, D.H. (2004) Autobiographic memory impairment following acute cortisol administration. Psychoneuroendocrinology, 29(8): 1093–1096.

Cabeza, R. and Nyberg, L. (2000) Imaging cognition II: an empirical review of 275 PET and fMRI studies. J. Cogn. Neurosci., 12(1): 1–47.

Cai, W.H., Blundell, J., Han, J., Greene, R.W. and Powell, C.M. (2006) Postreactivation glucocorticoids impair recall of established fear memory. J. Neurosci., 26(37): 9560–9566.

Cleare, A.J., Heap, E., Malhi, G.S., Wessely, S., O'Keane, V. and Miell, J. (1999) Low-dose hydrocortisone in chronic fatigue syndrome: a randomised crossover trial. Lancet, 353(9151): 455–458.

Cuthbert, B.N., Lang, P.J., Strauss, C., Drobes, D., Patrick, C.J. and Bradley, M.M. (2003) The psychophysiology of anxiety disorder: fear memory imagery. Psychophysiology, 40(3): 407–422.

Delahanty, D.L., Raimonde, A.J. and Spoonster, E. (2000) Initial posttraumatic urinary cortisol levels predict subsequent PTSD symptoms in motor vehicle accident victims. Biol. Psychiatry, 48(9): 940–947.

Fehm, L. and Margraf, J. (2002) Thought suppression: specificity in agoraphobia versus broad impairment in social phobia? Behav. Res. Ther., 40(1): 57–66.

Flood, J.F., Vidal, D., Bennett, E.L., Orme, A.E., Vasquez, S. and Jarvik, M.E. (1978) Memory facilitating and anti-amnesic effects of corticosteroids. Pharmacol. Biochem. Behav., 8(1): 81–87.

Het, S., Ramlow, G. and Wolf, O.T. (2005) A meta-analytic review of the effects of acute cortisol administration on human memory. Psychoneuroendocrinology, 30(8): 771–784.

de Kloet, E.R., Oitzl, M.S. and Joels, M. (1999) Stress and cognition: are corticosteroids good or bad guys? Trends Neurosci., 22(10): 422–426.

Kovacs, G.L., Telegdy, G. and Lissak, K. (1977) Dose-dependent action of corticosteroids on brain serotonin content and passive avoidance behavior. Horm. Behav., 8(2): 155–165.

Kuhlmann, S., Kirschbaum, C. and Wolf, O.T. (2005a) Effects of oral cortisol treatment in healthy young women on memory retrieval of negative and neutral words. Neurobiol. Learn. Mem., 83(2): 158–162.

Kuhlmann, S., Piel, M. and Wolf, O.T. (2005b) Impaired memory retrieval after psychosocial stress in healthy young men. J. Neurosci., 25(11): 2977–2982.

Kuhlmann, S. and Wolf, O.T. (2006) Arousal and cortisol interact in modulating memory consolidation in healthy young men. Behav. Neurosci., 120(1): 217–223.

Mason, J.W., Giller, E.L., Kosten, T.R., Ostroff, R.B. and Podd, L. (1986) Urinary free-cortisol levels in posttraumatic stress disorder patients. J. Nerv. Ment. Dis., 174(3): 145–149.

McEwen, B.S. (1998) Protective and damaging effects of stress mediators. N. Engl. J. Med., 338(3): 171–179.

McFarlane, A.C., Atchison, M. and Yehuda, R. (1997) The acute stress response following motor vehicle accidents and its relation to PTSD. Ann. N.Y. Acad. Sci., 821: 437–441.

Moser, M.B. and Moser, E.I. (1998) Distributed encoding and retrieval of spatial memory in the hippocampus. J. Neurosci., 18(18): 7535–7542.

Pitman, R.K. and Orr, S.P. (1990) Twenty-four hour urinary cortisol and catecholamine excretion in combat- related posttraumatic stress disorder. Biol. Psychiatry, 27(2): 245–247.

de Quervain, D.J., Aerni, A. and Roozendaal, B. (2007) Preventive effect of {beta}-adrenoceptor blockade on glucocorticoid-induced memory retrieval deficits. Am. J. Psychiatry, 164(6): 967–969.

de Quervain, D.J., Henke, K., Aerni, A., Treyer, V., McGaugh, J.L., Berthold, T., Nitsch, R.M., Buck, A., Roozendaal, B. and Hock, C. (2003) Glucocorticoid-induced impairment of declarative memory retrieval is associated with reduced blood flow in the medial temporal lobe. Eur. J. Neurosci., 17(6): 1296–1302.

de Quervain, D.J., Roozendaal, B. and McGaugh, J.L. (1998) Stress and glucocorticoids impair retrieval of long-term spatial memory. Nature, 394(6695): 787–790.

de Quervain, D.J., Roozendaal, B., Nitsch, R.M., McGaugh, J.L. and Hock, C. (2000) Acute cortisone administration impairs retrieval of long-term declarative memory in humans. Nat. Neurosci., 3(4): 313–314.

Rapee, R.M. and Heimberg, R.G. (1997) A cognitive-behavioral model of anxiety in social phobia. Behav. Res. Ther., 35(8): 741–756.

Roozendaal, B. (2000) 1999 Curt P. Richter award: glucocorticoids and the regulation of memory consolidation. Psychoneuroendocrinology, 25(3): 213–238.

Roozendaal, B., Griffith, Q.K., Buranday, J., de Quervain, D.J. and McGaugh, J.L. (2003) The hippocampus mediates glucocorticoid-induced impairment of spatial memory retrieval: dependence on the basolateral amygdala. Proc. Natl. Acad. Sci. U.S.A., 100(3): 1328–1333.

Roozendaal, B., Hahn, E.L., Nathan, S.V., de Quervain, D.J. and McGaugh, J.L. (2004a) Glucocorticoid effects on memory retrieval require concurrent noradrenergic activity in the hippocampus and basolateral amygdala. J. Neurosci., 24(37): 8161–8169.

Roozendaal, B., de Quervain, D.J., Schelling, G. and McGaugh, J.L. (2004b) A systemically administered beta-adrenoceptor antagonist blocks corticosterone-induced impairment of contextual memory retrieval in rats. Neurobiol. Learn. Mem., 81(2): 150–154.

Sajadi, A.A., Samaei, S.A. and Rashidy-Pour, A. (2007) Blocking effects of intra-hippocampal naltrexone microinjections on glucocorticoid-induced impairment of spatial memory retrieval in rats. Neuropharmacology, 52(2): 347–354.

Sakamoto, H., Fukuda, R., Okuaki, T., Rogers, M., Kasai, K., Machida, T., Shirouzu, I., Yamasue, H., Akiyama, T. and Kato, N. (2005) Parahippocampal activation evoked by masked traumatic images in posttraumatic stress disorder: a functional MRI study. NeuroImage, 26(3): 813–821.

Schelling, G., Briegel, J., Roozendaal, B., Stoll, C., Rothenhausler, H.B. and Kapfhammer, H.P. (2001) The effect of stress doses of hydrocortisone during septic shock on posttraumatic stress disorder in survivors. Biol. Psychiatry, 50(12): 978–985.

Schelling, G., Kilger, E., Roozendaal, B., de Quervain, D.J., Briegel, J., Dagge, A., Rothenhausler, H.B., Krauseneck, T., Nollert, G. and Kapfhammer, H.P. (2004) Stress doses of hydrocortisone, traumatic memories, and symptoms of posttraumatic stress disorder in patients after cardiac surgery: a randomized study. Biol. Psychiatry, 55(6): 627–633.

Soravia, L.M., Heinrichs, M., Aerni, A., Maroni, C., Schelling, G., Ehlert, U., Roozendaal, B. and de Quervain, D.J. (2006) Glucocorticoids reduce phobic fear in humans. Proc. Natl. Acad. Sci. U.S.A., 103(14): 5585–5590.

Squire, L.R. (1992) Memory and the hippocampus: a synthesis from findings with rats, monkeys, and humans. Psychol. Rev., 99(2): 195–231.

Wolf, O.T., Convit, A., McHugh, P.F., Kandil, E., Thorn, E.L., De Santi, S., McEwen, B.S. and de Leon, M.J. (2001) Cortisol differentially affects memory in young and elderly men. Behav. Neurosci., 115(5): 1002–1011.

Yehuda, R. (2002a) Current status of cortisol findings in posttraumatic stress disorder. Psychiatr. Clin. N. Am., 25(2): 341–368 vii.

Yehuda, R. (2002b) Post-traumatic stress disorder. N. Engl. J. Med., 346(2): 108–114.

Yehuda, R., Kahana, B., Binder-Brynes, K., Southwick, S.M., Mason, J.W. and Giller, E.L. (1995) Low urinary cortisol excretion in Holocaust survivors with posttraumatic stress disorder. Am. J. Psychiatry, 152(7): 982–986.

Yehuda, R., McFarlane, A.C. and Shalev, A.Y. (1998) Predicting the development of posttraumatic stress disorder from the acute response to a traumatic event. Biol. Psychiatry, 44(12): 1305–1313.

Young, E.A. and Breslau, N. (2004) Cortisol and catecholamines in posttraumatic stress disorder: an epidemiologic community study. Arch. Gen. Psychiatry, 61(4): 394–401.

Discussion: Chapter 17

QUESTION: Do you know anything about HPA-axis reactivity of those patients?

DE QUERVAIN: No, we did not assess HPA-axis reactivity in this study but we now plan a larger study in PTSD patients where we will perform a corticotropin stimulation test and dexamethasone suppression test before and after cortisol treatment.

PITMAN: I wonder how memory would be involved in this in spite of the situation without seeing a picture of the stimulus at the time of the testing. So they might have to remember the focal stimulus right in front of them.

DE QUERVAIN: There might be a fear network that is stored in the brain, perhaps based on fearful experiences earlier in life, which is activated by fearful cues. Such a process involves memory retrieval.

JOËLS: I was wondering about the test where you expose these social phobia patients to a social stress test. I do have a hard time seeing that as a memory situation. Also, assuming in this case cortisol levels are high during retrieval, what could happen is that by giving extra cortisol you are just blunting the patient's own HPA-axis.

DE QUERVAIN: What we found is less fear both after cortisol administration and after endogenous HPA-axis activation. Memory comes into play after the written introduction to the stress test, when the fear network is activated by this cue and when patients also may start to think about past failures in such situations, which would additionally trigger the fear response.

SANDI: My question is related to the test applied in the experiment in which you find a correlation between cortisol and subsequent fear. To what is this fear? Is it fear to the Trier test?

DE QUERVAIN: It is subjective fear related to the Trier test measured at different time points.

GUNNAR: It is interesting, because in the Trier test usually you don't find any association with fear and, thus, whatever is going on perhaps is specific to those phobic patients. Also, Pruessner found that when he subjected individuals to the Trier test repeatedly most individuals habituated. Those who didn't habituate were the ones who had more fearful, anxious personalities. It isn't clear how your results fit with his. His seem to show that fear is associated with continued glucocorticoid respondings, yours that elevated glucocorticoids reduce fear. Perhaps there is something specific to phobic patients.

DE QUERVAIN: Yes, fear memory — and we assume that glucocorticoids reduce phobic fear by reducing fear memory retrieval in these patients.

LIBERZON: From our data, we have actually used traumatic memory in Roger Pitman's paradigm of autobiographic scripts in PTSD and normal controls. We did get significant response in PTSD both in cortisol and in ACTH but it was very small. Interestingly the ACTH response was present in combat controls as well. We also have activation in the brain associated with these responses. But it is very difficult to show a robust response in PTSD to any psychological stimuli other than TSST.

BREMNER: You can show that in women with abuse-related PTSD. One of the slides that I presented was from Bernet Elzinga, who is first author on this, was reading a script of childhood sexual abuse the patients had a threefold higher increase of cortisol during the reading of the traumatic script. In that study you can measure intrusive memory at the time of the slides or sounds or traumatic scripts and patients report an increase in traumatic memory, so we sort of — the point that I wanted to make — for saying that for saying that increase relates in glucocorticoids is a good thing and I am not necessarily going to argue that specifically but then you are — increase glucocorticoids release enhances the encoding of the, or the consolidation of emotional memory, you are saying that the patients were administered glucocorticoids right after the trauma, how could that be good to enhance the consolidation of the traumatic memory? And another thing I wanted to ask if you can tell me what exactly is the citation for the humans that glucocorticoids enhance consolidation. From our data it actually impairs consolidation of emotional memory based on a negative correlation on the emotional words remembered a couple of hours.

DE QUERVAIN: There are a couple of studies I can give you the references of, especially for emotional memory consolidation. To go back to your first point, we have looked at the effects of glucocorticoids in established, chronic PTSD. Here, glucocorticoids reduce the retrieval of traumatic memories.

E.R. de Kloet, M.S. Oitzl & E. Vermetten (Eds.)
Progress in Brain Research, Vol. 167
ISSN 0079-6123

CHAPTER 18

Commentary: synthesis and perspectives

R.K. Pitman*

Massachusetts General Hospital-East, Room 2616 Building 149, 13th Street, Charleston, MA 02129, USA

Keywords: stress disorders; post-traumatic; hormones; memory; conditioning

We are more likely to remember significant life events than trivial ones. Evolution appears to have accomplished this feat of adaptation by the natural selection of effects exerted by stress hormones on the consolidation of memory traces, or alternately stated, on the acquisition of conditioned responses. The modulatory effect of stress hormones on memory was discovered approximately forty years ago, independently by David de Wied (1997) in the Netherlands and McGaugh and Roozendaal (2002) in the USA. The Dutch effort focused on the hypothalamic-pituitary-adrenal (HPA) cortical axis, including the roles of pituitary peptides and glucocorticoids, whereas the American effort focused on catecholamines, especially epinephrine and norepinephrine.

Based upon this groundbreaking research, we proposed a translational model of the pathogenesis of post-traumatic stress disorder (PTSD) (Pitman et al., 2000; Pitman and Delahanty, 2005). According to this model, a psychologically traumatic event (unconditioned stimulus, UCS) overstimulates endogenous stress hormones as part of a physiological unconditioned response (UCR). These in turn overly strengthen consolidation of the memory of the event, leading to a powerful and persistent memory (or conditioned response, CR) that is too easily activated, with consequent anxiety and dysfunction. The model further posits that after the memory of the traumatic event has been consolidated, free or cue-induced recall of the event (conditioned stimulus, CS) causes re-releases of stress hormones that further enhance the strength of the memory trace (or CR), increasing the likelihood that it will be re-recalled with yet further stress hormone releases, thereby resulting in a positive feedback loop, or vicious cycle, that promotes the development of PTSD.

The purpose of this commentary is neither to critique the above model, nor to review the status of hormonal research in PTSD. Rather, its purpose is to use this model to illustrate at the theoretical level the numerous, and sometimes opposing, potential influences of the HPA axis on the pathogenesis of PTSD (Fig. 1). For example, most HPA axis peptides, including corticotropin releasing hormone (CRH), adrenocorticotropin (ACTH), and arginine vasopressin (AVP), potentiate the acquisition of a conditioned fear response in animals (de Wied, 1997; Croiset et al., 2000). This effect is not solely mediated through the release of corticosterone. Rather, these peptides, and other memory-enhancing stress hormones, appear to interact with noradrenaline in the amygdala in generating this effect (Roozendaal et al., 2002). According to the model, CRH, ACTH, and AVP (as well as epinephrine and norepinephrine) would be expected to facilitate the formation of a traumatic memory after an intensely frightening event in humans, and thereby be potentially involved in

*Corresponding author. Tel.: +1-617-726-5333;
Fax: +1-617-726-4078;
E-mail: roger_pitman@hms.harvard.edu

DOI: 10.1016/S0079-6123(07)67018-6

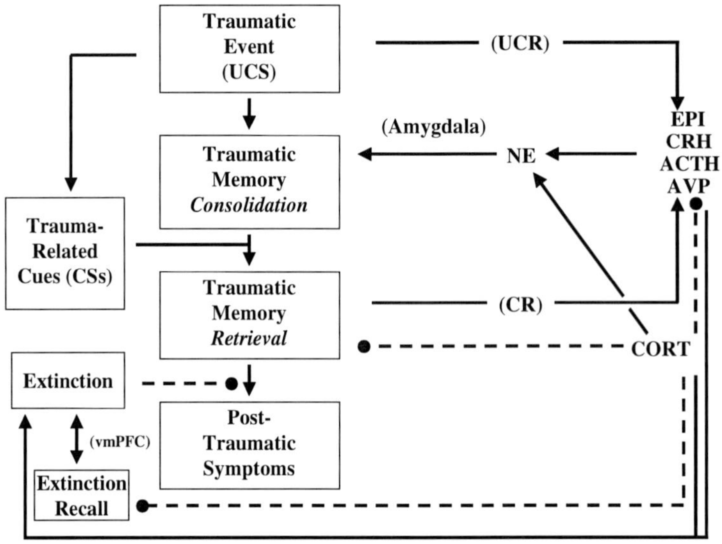

Fig. 1. Theoretical model of the pathogenesis of post-traumatic stress disorder based upon conditioning, with potential points of hypothalamic-pituitary-adrenal (HPA) axis influence. ACTH: adrenocorticotropin, AVP: arginine vasopressin, CORT: cortisol, CRH: corticotropin-releasing hormone, EPI: epinephrine, NE: norepinephrine; UCR: unconditioned response, CR: conditioned response, CS: conditioned stimulus; vmPFC: ventromedial prefrontal cortex; solid line with arrowhead: excitatory, dashed lined with ball head: inhibitory.

the formation of PTSD. Administering them in the immediate aftermath of a traumatic event would be expected to be counter-therapeutic. However, a notable exception is the peptide oxytocin, which reduces acquisition of a conditioned fear response in animals (de Wied, 1980) has amnestic effects, and might be expected to have preventive value for PTSD if given immediately after the traumatic event.

With regard to the glucocorticoids, the situation is more complex. On the one hand, corticosterone also potentiates the acquisition of a conditioned fear response in animals by interacting with noradrenaline in the amygdala (Roozendaal et al., 2006), as does its analog cortisol in humans (van Stegeren et al., 2007). For this reason, cortisol would be expected to increase traumatic memory formation in humans and be counter-therapeutic for PTSD immediately after the traumatic event. However, the glucocorticoids also decrease pituitary peptides through negative feedback. Glucocorticoids have also been found to attenuate the memory-enhancing effects of catecholamines (Borrell et al., 1984). Thus through these indirect mechanisms, cortisol could have the opposite effect, i.e., to decrease traumatic memory formation, and according to this line of reasoning it could potentially be therapeutic for PTSD when given immediately after the traumatic event. These indirect effects of cortisol are also of relevance to the question of HPA axis negative feedback sensitivity, which has been reported as being altered in PTSD. On the one hand, if this sensitivity in increased, pituitary peptides may be more easily shut off, and the risk of developing PTSD thereby diminished. On the other hand, if the most salient result of increased HPA axis sensitivity is to reduce the level of circulating cortisol itself, its relative absence could reduce the braking effect on catecholamines and thereby increase PTSD risk. In considering the potential effect of cortisol over the longer term, one must also take into account the finding that cortisol appears to decrease memory retrieval (as opposed to consolidation) (Het et al., 2005). By reducing the likelihood that the traumatic event will be recalled, cortisol has the potential to interrupt the vicious cycle described above and thereby be therapeutic for PTSD (de Quervain, 2006).

Finally, the acquisition of a conditioned fear response, or the formation of a traumatic memory, may not be the last word in PTSD's pathogenesis. It has been suggested that most persons experience PTSD symptoms in the acute aftermath of a traumatic event, and that PTSD results from the individual's inability to recover from these symptoms. Crucial to such recovery may be the phenomenon of extinction, i.e., learning not to be afraid after having once learned to be afraid. Extinction is thought to play a role in the action of cognitive behavior therapy (CBT). Like acquisition learning, extinction learning too must be consolidated if it is to have any lasting effect. The consolidation of fear acquisition is thought to involve the amygdala. Although extinction learning is also thought to involve the amygdala, the consolidation of extinction learning is thought to involve the ventromedial prefrontal cortex (vmPFC) (Milad et al., 2006). HPA axis influences that affect the pathogenesis of PTSD in one direction or the other through their effect on the consolidation of acquisition may turn out to have opposite effects when they affect the consolidation of extinction. For example, HPA axis peptides, cortisol, and even catecholamines, if released or exogenously administered during or immediately after a CBT session, might enhance extinction retention and thereby aid in reducing PTSD symptoms. But what if cortisol were to act to reduce the subsequent retrieval of the extinction memory in the vmPFC? In this case, it would be expected to prevent, rather than aid, recovery.

This commentary serves to illustrate the multiple and complex potential influences of the HPA axis on pathogenic mechanisms involved in PTSD. It will be seen that the effects of a hormone such as cortisol on this disorder (not to mention the effects of PTSD on cortisol), which is currently a matter of disagreement, will never be resolved by theoretical considerations alone. Rather, such questions can only be resolved by the collection of data in empirical investigations.

References

Borrell, J., de Kloet, E.R. and Bohus, B. (1984) Corticosterone decreases the efficacy of adrenaline to affect passive avoidance retention of adrenalectomized rats. Life Sci., 34: 99–104.

Croiset, G., Nijsen, M.J. and Kamphuis, P.J. (2000) Role of corticotropin-releasing factor, vasopressin and the autonomic nervous system in learning and memory. Eur. J. Pharmacol., 405: 225–234.

Het, S., Ramlow, G. and Wolf, O.T. (2005) A meta-analytic review of the effects of acute cortisol administration on human memory. Psychoneuroendocrinology, 30: 771–784.

McGaugh, J.L. and Roozendaal, B. (2002) Role of adrenal stress hormones in forming lasting memories in the brain. Curr. Opin. Neurobiol., 12: 205–210.

Milad, M.R., Rauch, S.L., Pitman, R.K. and Quirk, G.J. (2006) Fear extinction in rats: implications for human brain imaging and anxiety disorders. Biol. Psychol., 73: 61–71.

Pitman, R.K. and Delahanty, D.L. (2005) Conceptually driven pharmacologic approaches to acute trauma. CNS Spectr., 10: 99–106.

Pitman, R.K., Shalev, A.Y. and Orr, S.P. (2000) Post-traumatic stress disorder: emotion, conditioning, and memory. In: Gazzaniga M.S. (Ed.), The Cognitive Neurosciences. MIT Press, Cambridge, MA, pp. 1133–1147.

de Quervain, D.J. (2006) Glucocorticoid-induced inhibition of memory retrieval: implications for posttraumatic stress disorder. Ann. N.Y. Acad. Sci., 1071: 216–220.

Roozendaal, B., Brunson, K.L., Holloway, B.L., McGaugh, J.L. and Baram, T.Z. (2002) Involvement of stress-released corticotropin-releasing hormone in the basolateral amygdala in regulating memory consolidation. Proc. Natl. Acad. Sci. U.S.A., 99: 13908–13913.

Roozendaal, B., Hui, G.K., Hui, I.R., Berlau, D.J., McGaugh, J.L. and Weinberger, N.M. (2006) Basolateral amygdala noradrenergic activity mediates corticosterone-induced enhancement of auditory fear conditioning. Neurobiol. Learn. Mem., 86: 249–255.

van Stegeren, A.H., Wolf, O.T., Everaerd, W., Scheltens, P., Barkhof, F. and Rombouts, S.A. (2007) Endogenous cortisol level interacts with noradrenergic activation in the human amygdala. Neurobiol. Learn. Mem., 87: 57–66.

de Wied, D. (1980) Behavioural actions of neurohypophysial peptides. Proc. R. Soc. Lond. B Biol. Sci., 210: 183–195.

de Wied, D. (1997) Neuropeptides in learning and memory processes. Behav. Brain Res., 83: 83–90.

General Discussion: Section IV

LIBERZON: Roger, you gave a nice illustration of the fact that we cannot talk simply about low and high cortisol, we cannot talk simply about memory, because these are much too general.

PITMAN: I never said we couldn't use the concept of memory. Also, although low or high cortisol or enhanced decreased negative feedback, lead to ambiguous predictions, it may be possible through hypothesis-driven investigation to see which of the predictions are confirmed and which are refuted.

LIBERZON: In order to do that, I'm asking for better definitions of what it exactly is that we are testing. There's a need for us to use the same terminology. When we talk about recall, this can be measured in animals. Nightmares don't exactly equate with recall, although they are intrusion phenomena. It would indeed be hard to study nightmares in animals. My challenge to the panel is that we define our terms so that clinical investigators and basic scientists are sure they're studying the same phenomena.

DE KLOET: I felt very comfortable with the last analysis, even though you may argue about the different end points that are measured in the PTSD patient. I felt comfortable because the disease was broken down into manageable and testable ingredients. The paradigms you use are not complicated. They have been around for some time and can be used in animal studies as well as in the clinic. Besides the pituitary-adrenal axis, the many neuropeptides released in the brain also affect aspects of memory consolidation and retrieval. These hormones and neuropeptides probably operate in a balanced fashion in the initial stress reaction and in the later adaptive phases. I think that many of the apparent contradictions you observed can thereby be explained.

PITMAN: An example of a question we have is the following. If you have a hypersensitive receptor and a low hormone concentration, you may have either a functionally increased or a functionally decreased state depending on which is primary.

DE KLOET: I understand the chicken or egg question. Maybe one way to get a step further is to look at the temporal patterns of the various hormones involved, and the different phases of the stress response these hormones operate in to reestablish homeostasis.

LIBERZON: Larry Squire told me 15 years ago that memory researchers would rather use each other's toothbrush than each other's terminology. It is much easier to describe things in your own way. However, when you have to fit your findings into somebody else's findings, into somebody else's paradigm, you may be forced to use their terminology. That is much more difficult, but it is essential to create a common language.

JOËLS: How do we know that cortisol given for many weeks specifically affects the retrieval of traumatic memories? It could do so many things.

DE QUERVAIN: I did not show the data, but we observed that with cortisol traumatic memory retrieval falls to a certain level and then stays stable through the whole month. It is not something that takes weeks to develop.

JOËLS: I'm sure that that's true. However, my point it that cortisol affects many things in the brain. We know from animals that when we treat them, let's say with three weeks of corticosterone, these animals have completely different brains. Maybe not at the histological level, but every cellular function you look at changes. That must be true in your patients as well.

DE QUERVAIN: Memory retrieval is what I measured. And retrieval of the traumatic memory is affected by cortisol in the first few days. Of course cortisol may induce other changes, but we did not measure them.

YEHUDA: Memory retrieval seems an awkward term to describe intrusive symptoms in patients. The issue is whether they're bothered by these thoughts. What troubles me about the memory consolidation theory is that traumatic events produce a memory in everybody who experiences them, not only in those who develop PTSD. We want a traumatic memory. People who are dissociated or can't remember their trauma are in worse shape, because they don't know what it is that they need to be afraid of and hence have more intense general fear.

PITMAN: On the one hand, there is a call for clinicians to speak a common language with basic behavioral scientists. However, when we try to develop a common language, terms such as memory retrieval may not exactly fit clinicians' ideas is about the complex phenomena they deal with, so they become unacceptable.

LIBERZON: Reduction in symptoms does not necessarily mean decreased memory retrieval.

DE QUERVAIN: If a traumatic memory is in your mind, it has to have been retrieved.

ROELOFS: Can it be the case that because the behavior changes after cortisol, memory recall also changes?

SHALEV: Might there be psychological manifestations of administering cortisol that could allow the dissociation of memory from hyperarousal, which are not the same thing?

RICHTER-LEVIN: Clinically the issue is not the retrieval of the memory; the issue is suppressing the memory. More specifically, the adaptive thing is to be able to suppress the response to the memory. It becomes maladaptive when you lose the ability to regulate it. I can remember what happened in World War II, but the affect is gone, thank God!

GUNNAR: It might be fascinating in your paradigm to put cardiac monitors on patients on the first day. If they have an intrusive memory but you don't get a heart rate response, it may be because it is not as scary and frightening.

DE KLOET: In my mind the physiological function of cortisol is to eliminate a response that no longer has relevance. The fact that memory retrieval is suppressed does not imply that memory is impaired. Rather it may simply mean that new information has been stored at the expense of something that is no longer relevant. I would explain the beneficial effects of cortisol in suppressing unwanted memories in this way.

RICHTER-LEVIN: It may be that the goal is not actually to erase memories but to detach the emotion from them. We may not need to search for memory mechanisms but rather for detachment mechanisms, perhaps along the line of the two systems that Chris Brewin has been talking about.

SECTION V

Short Communications

E.R. de Kloet, M.S. Oitzl & E. Vermetten (Eds.)
Progress in Brain Research, Vol. 167
ISSN 0079-6123

CHAPTER 19

Strain specific fear behaviour and glucocorticoid response to aversive events: modelling PTSD in mice

V. Brinks*, E.R. de Kloet and M.S. Oitzl

Division of Medical Pharmacology, LACDR/LUMC, Gorlaeus Laboratories, P.O. Box 9502, Leiden University, 2300 RA Leiden, The Netherlands

Abstract: "Pavlovian" fear conditioning in rodents allows studying the formation and extinction of fear memories. Male C57BL/6J but not BALB/c mice showed differential fear memory performance expressed as freezing and scanning behaviour for context and cue. Glucocorticoid stress hormones modulate the processing of fear-related stimuli. The augmented corticosterone response of BALB/c mice to conditioning and testing, therefore, might have contributed to the strain-dependent formation of fear memories. We propose that modulation of extinction processes by glucocorticoids can be relevant in modelling anxiety disorders.

Keywords: PTSD; fear conditioning; C57BL/6J and BALB/c mice; stress system

Introduction

Increasing evidence suggests that susceptibility to post-traumatic stress disorder (PTSD) has a genetic basis, but its onset is clearly stress-related. The behavioural characteristics of PTSD are intrusive persistent memories of the trauma, avoidance behaviour and hyperarousal. Besides behavioural symptomatology, changes in the hypothalamus-pituitary-adrenal axis (HPA-axis) activity also occur. Individuals suffering from PTSD have low basal cortisol levels, increased sensitivity to stress and enhanced glucocorticoid negative feedback (Yehuda, 2006). Furthermore, the volume of the hippocampus, a brain target of glucocorticoid stress hormones, is smaller in PTSD patients as compared to controls having similar experiences but who did not develop the disorder (Bremner et al., 1995; Tischler et al., 2006), although other studies point to such features as part of a susceptibility trait (Pitman et al., 2006). These findings substantiate the notion that by definition the stress system/HPA-axis activity should be a key player in PTSD research.

The enhanced memory for the adverse event or stimuli is a central feature that might be modelled in rodent fear-conditioning paradigms. Corticosterone is known to facilitate the formation of context-related fear memories (Cordero et al., 2003). In fear conditioning of mice, an unexpected aversive stimulus such as an electric shock (unconditioned stimulus, UCS), is given once or several times in association with a non-aversive stimulus (cue; conditioned stimulus, CS) such as a light and/or tone, in a distinct environment (context). This is the well-known Pavlovian-conditioning paradigm. The animal will form and remember the association between the announcing cue and aversive stimulus and also the environment in

*Corresponding author. Tel.: +31 71 5274428;
Fax: +31 71 5274715; E-mail: v.brinks@lacdr.leidenuniv.nl

DOI: 10.1016/S0079-6123(07)67019-8

which the aversive stimulus was experienced. Fear behaviour in the mouse generally consists of scanning and freezing.

The present experiment is designed to develop an animal model for PTSD that allows: (i) to follow the formation of memory (acquisition and retrieval) for the negative event by measuring, scanning and freezing behaviour; (ii) to test memory retrieval processes for both context and cue after a delay of 24 h; (iii) to measure corticosterone before and in response to conditioning and (iv) to show the influence of the genetic background on acquisition and retrieval of the negative event. By using two mouse strains, we expect our results to be more representative for the individual differences in vulnerability to stress-related disorders.

Methods

Male BALB/c and C57BL/6J mice ($n = 8$ per group; 3-month-old) were subjected to a specific fear-conditioning paradigm that allowed to differentiate context and context/cue-related responses in the same setting. This included 10 light/tone + shock pairings with a 1 min interval on day 1. Pairings were as follows; light (260 lux) and tone (70 dB) were given simultaneously for 20 s of which an additional shock (0.4 mA) was administrated at the last 2 s. Scanning and freezing was measured when the animals were placed in the setting (Fig. 1, point 1) and during the 1 min intervals after light/tone + shock pairings (Fig. 1, points 2–11). Scanning is defined as immobility of the body, while the head is moving horizontally from side to side. The animal is still actively interacting with its environment. Freezing is defined as immobility of the body and head and is devoid of interaction with the environment. Although scanning and freezing are interdependent, they express a different quality of fear. With automatic scoring, both scanning and freezing are measured as immobility behaviour.

Fear memories, expressed as scanning and freezing were estimated 24 h later on day 2. Mice were returned to the same box: 3 min exposure to the setting (context only) was followed by 2 min of light/tone exposure (context/cue) and ended with 2 min exposure to the setting (context only). Plasma corticosterone was estimated at several time points: on the day before conditioning, after conditioning and after retention testing (see Table 1). ANOVA with repeated measures was used to test for significant progression of scanning and freezing over conditioning intervals on day 1. Student's t-test was used to compare percentage scanning and freezing for context and context/cue between strains on day 2 and corticosterone concentrations of the two strains at each time point and to basal corticosterone values before the experiment started.

Results

Figure 1A and B present the percentages of freezing and scanning in both strains on days 1 and 2.

Acquisition of fear behaviour

On day 1, when the light/tone + shock pairings took place (conditioning), the percentage of freezing progressively increased for both C57BL/6J and BALB/c mice ($F(10,140)$ 25.710, $p = 0.000$), albeit to a different degree ($F(10,140)$ 4.860, $p = 0.000$). BALB/c mice displayed a faster increase in freezing resulting in a plateau at ~70%, while freezing in C57BL/6J mice reached ~30–40%. Also scanning behaviour increased in both mouse strains ($F(10,140)$ 12.279, $p = 0.000$) to different degrees ($F(10,140)$6.662 $p = 0.000$). Group differences for scanning and freezing appear at separate time points: distinct scanning behaviour starts at interval 5, while freezing percentages differ from interval 2 onwards. There is also a strain-dependent main effect for freezing and scanning: C57BL/6J mice display high scanning and lower freezing behaviour, while BALB/c mice have high freezing and low scanning behaviour. Interestingly, total immobility measured by scanning and freezing together is the same for C57BL/6J and BALB/c mice.

Fear memories

On day 2, BALB/c mice displayed more freezing compared to C57BL/6J mice when first exposed to

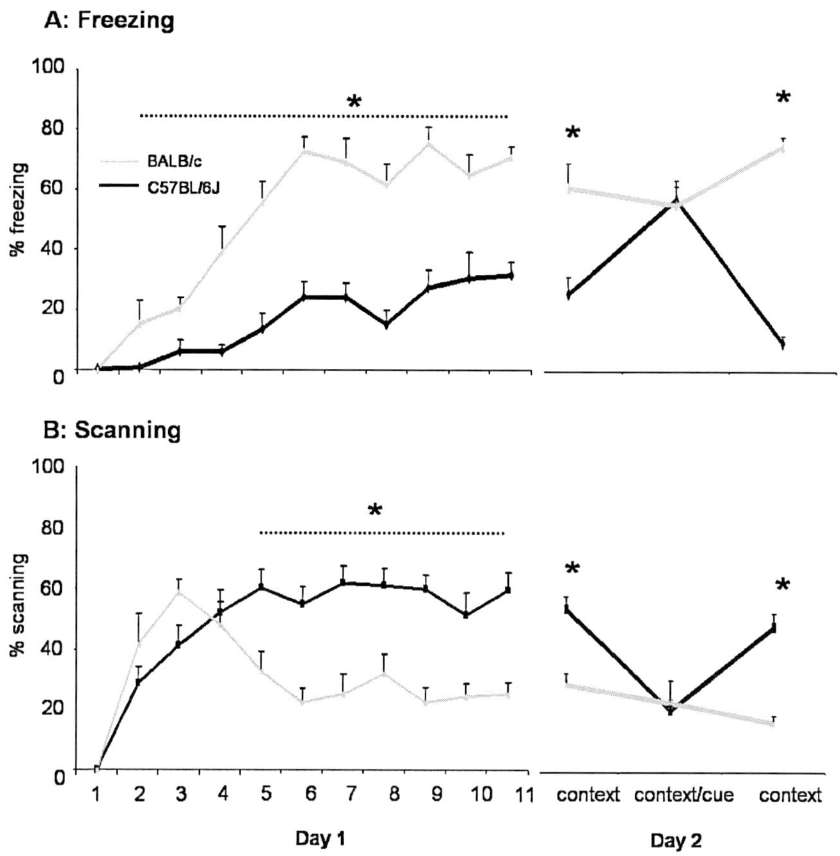

Fig. 1. (A) Percentage freezing of C57BL/6 (black line) and BALB/c mice (grey line). (B) Percentage scanning of C57BL/6J (black line) and BALB/c mice (grey line). Day 1: acquisition; time point 1 represents scanning and freezing during the first minute in the setting; time points 2–11 represent scanning and freezing in the 1 min intervals between the 10 light/tone + shock pairings. Day 2: memory/retrieval; scanning and freezing during context (3 min), context/cue (2 min) and context (2 min) exposure is presented. Data are presented as mean (±SEM) percentage of behaviour. Horizontal lines and asterisks indicate significant differences between groups. Significance was accepted at $p<0.05$.

Table 1. Plasma concentrations of corticosterone in ng/ml (mean ± SEM) of C57BL/6J and BALB/c mice measured on the day before conditioning (basal morning values), day 1 (30 and 60 min after the start of conditioning) and day 2 (30 and 60 min after start of retention test)

	Day 1: Acquisition			Day 2: Retrieval	
	Basal	30 min	60 min	30 min	60 min
C57BL/6J	11.2 ± 2.4	172.5 ± 10.0*	94.9 ± 15.5*	84.0 ± 9.6*	50.4 ± 12.5
BALB/c	9.0 ± 1.3	370.4 ± 12.6*	189.0 ± 10.0*	137.4 ± 15.7*	54.3 ± 8.8

Notes: Corticosterone assay was performed with the use of a commercially available radio immune assay kits (MP Biomedicals Inc., CA, USA). Data are represented as mean ± SEM. Significant differences: all 30 and 60 min samples compared to basal concentrations.
*C57BL/6J versus BALB/c mice. Significance was accepted at $p<0.05$.

the context ($F(1,14)$ 10.551, $p = 0.001$). For both strains, this percentage of freezing is comparable to the last freezing response on day 1. Next, the light/tone cue was presented. This resulted in an increased amount of freezing in the BALB/c mice resulting in comparable amount of freezing in C57BL/6J and BALB/c mice ($F(1,14)$ 12.921, $p = 0.857$). Subsequently switching off the cue (and thus only exposure to the context) again separated the strains ($F(1,14)$ 12.988, $p = 0.000$). C57BL/6J mice reduced their freezing while BALB/c mice increased their freezing response to the context.

Scanning of BALB/c mice was lower when first exposed to the setting compared to C57BL/6J ($F(1,14)$ 9.873, $p = 0.008$). However both strains were comparable to the last scanning data on day 1. When presenting the light/tone cue, scanning behaviour in C57BL/6J significantly decreased and displayed a similar low percentage of scanning as BALB/c mice ($F(1,14)$ 13.688, $p = 0.689$). The difference in scanning behaviour between strains occurred again when the cue was turned off ($F(1,14)$ 9.930, $p = 0.000$).

Plasma corticosterone concentrations in response to acquisition and retrieval of fear memories differed significantly between C57BL/6J and BALB/c mice (see Table 1).

At 30 and 60 min after onset of conditioning on day 1, corticosterone concentrations were twofold higher in BALB/c compared to C57BL/6J mice (30 min: $F(1,6)$ 5.761, $p =$ 0.000, 60 min: $F(1,6)$ 5.111, $p =$ 0.002). On day 2, corticosterone concentrations of BALB/c mice were increased compared to C57BL/6J mice at 30 min ($F(1,6)$ 4.972, $p = 0.027$), but returned to comparable low levels at 60 min. Basal morning resting corticosterone concentrations were comparable between strains ($F(1,14)$ 10.589, $p = 0.483$) and significantly lower than all 30 and 60 min data.

Discussion

The two inbred mouse strains show distinctly different fear responses during conditioning and retrieval of fear memories. While scanning is the main fear response to conditioning in C57BL/6J mice, BALB/c mice display a fast increase and higher level of freezing. During retrieval, both strains have the same degree of freezing and scanning to the cue, but interestingly, differ in their response to context.

BALB/c mice show a comparable high amount of freezing to context and cue. Apparently, they lack the discriminative ability between context and cue, indicating a generalised and even potentiated fear response. C57BL/6J mice clearly differentiate between context and cue. They freeze less and scan more when the cue is switched off. Previous studies that did not differentiate between freezing and scanning and tested the cue-related fear memory in another environment, reported generalised immobility in C57BL/6J, but not BALB/c mice (Balogh and Wehner, 2003) and an impairment of cue-related fear memories in BALB/c mice (Schimanski and Nguyen, 2005). Apparently, methodological differences like the severity of the conditioning protocol (number of shocks) and the way of memory testing (separate or combined context and cue retrieval) strongly affect the expression of fear. We used low shock levels during conditioning and measured scanning and freezing as fear behaviours. This approach revealed an up-to-now unknown strain difference in coping with fear-related stimuli. If this holds true for extinction, it will be assessed in further studies.

Our paradigm allows a differentiation between the more active fear behaviour expressed by scanning and the rather passive fear behaviour indicated by freezing. Recognising the light/tone stimulus (cue) as threat and freezing in anticipation of the electric shock can be considered as an adaptive response. Likewise, increased scanning in the context indicates a more active coping strategy that might prepare for escaping the expected aversive event (King, 1999). C57BL/6J may be considered as mice that cope with fear in a more active way than BALB/c mice, which indiscriminately respond with freezing. Most literature refers to BALB/c as mice with augmented unconditioned fear (Belzung and Griebel, 2001). It is likely that a fearful trait potentiates conditioned fear.

The increased corticosterone secretion of BALB/c mice compared to C57BL/6J mice to conditioning and retention testing might be related to the known memory-strengthening effects of corticosterone (Cordero et al., 2003; Cai et al., 2006). Since different brain areas are involved in cue (amygdala) and context-related (hippocampus) fear memories (Maren, 2001), it is conceivable that corticosterone differentially affects the formation of cue and context-related fear memory in these two mouse strains.

In conclusion, BALB/c mice acquire fear and remember fear-related conditions differently from C57BL/6J. It is of interest that patients suffering from PTSD or simple phobias often display symptoms in the presence of particular stimuli, whereas in generalised anxiety disorder, behavioural and emotional reactions often emerge in the absence of a particular stimulus, or are interconnected with a more complex environment. With respect to "PTSD-like" symptomatology, C57Bl/6J mice seem to be more vulnerable to cue-specific "flashbacks", while BALB/c mice present more the phenotype of generalised fear memory.

Abbreviations

CS	conditioned stimulus
HPA-axis	hypothalamus-pituitary-adrenal axis
PTSD	post-traumatic stress disorder
UCS	unconditioned stimulus

Acknowledgements

This study was supported by the Netherlands Organization of Scientific Research NWO-Cognition 051.02.010 and NWO-Aspasia 015.01.076 grant. We thank Dennis Tax for technical assistance and Harm Krugers and Olof Wiegert, University of Amsterdam, for scientific discussions.

References

Balogh, S.A. and Wehner, J.M. (2003) Inbred mouse strain differences in the establishment of long-term fear memory. Behav. Brain Res., 140(1–2): 97–106.

Belzung, C. and Griebel, G. (2001) Measuring normal and pathological anxiety-like behaviour in mice: a review. Behav. Brain Res., 125(1–2): 141–149.

Bremner, J.D., Randall, P., Scott, T.M., Bronen, R.A., Seibyl, J.P., Southwick, S.M., Delaney, R.C., McCarthy, G., Charney, D.S. and Innis, R.B. (1995) MRI-based measurement of hippocampal volume in patients with combat-related posttraumatic stress disorder. Am. J. Psychiatry, 152(7): 973–981.

Cai, W.H., Blundell, J., Han, J., Greene, R.W. and Powell, C.M. (2006) Postreactivation glucocorticoids impair recall of established fear memory. J. Neurosci., 26(37): 9560–9566.

Cordero, M.I., Venero, C., Kruyt, N.D. and Sandi, C. (2003) Prior exposure to a single stress session facilitates subsequent contextual fear conditioning in rats: evidence for a role of corticosterone. Horm. Behav., 44(4): 338–345.

King, S.M. (1999) Escape-related behaviours in an unstable, elevated and exposed environment. II. Long-term sensitization after repetitive electrical stimulation of the rodent midbrain defence system. Behav. Brain Res., 98(1): 127–142.

Maren, S. (2001) Neurobiology of Pavlovian fear conditioning. Annu. Rev. Neurosci., 24: 897–931.

Pitman, R.K., Gilbertson, M.W., Gurvits, T.V., May, F.S., Lasko, N.B., Metzger, L.J., Shenton, M.E., Yehuda, R. and Orr, S.P. (2006) Clarifying the origin of biological abnormalities in PTSD through the study of identical twins discordant for combat exposure. Ann. N.Y. Acad. Sci., 1071: 242–254.

Schimanski, L.A. and Nguyen, P.V. (2005) Mouse models of impaired fear memory exhibit deficits in amygdalar LTP. Hippocampus, 15(4): 502–517.

Tischler, L., Brand, S.R., Stavitsky, K., Labinsky, E., Newmark, R., Grossman, R., Buchsbaum, M.S. and Yehuda, R. (2006) The relationship between hippocampal volume and declarative memory in a population of combat veterans with and without PTSD. Ann. N.Y. Acad. Sci., 1071: 405–409.

Yehuda, R. (2006) Advances in understanding neuroendocrine alterations in PTSD and their therapeutic implications. Ann. N.Y. Acad. Sci., 1071: 137–166.

E.R. de Kloet, M.S. Oitzl & E. Vermetten (Eds.)
Progress in Brain Research, Vol. 167
ISSN 0079-6123

CHAPTER 20

Interaction of endogenous cortisol and noradrenaline in the human amygdala

Anda H. van Stegeren[1,*], Oliver T. Wolf[2], Walter Everaerd[1] and Serge A.R.B. Rombouts[3]

[1]*Department of Clinical Psychology and Cognitive Science Center, University of Amsterdam, Roetersstraat 15, 1018 WB Amsterdam, The Netherlands*
[2]*Universität Bielefeld, Abteilung für Psychologie,T5-221, Universitätsstrasse. 25, 33615, Bielefeld, Germany*
[3]*Department of Radiology, Leiden University Medical Center; Department of Psychology, Leiden University and Leiden Institute for Brain and Cognition (LIBC), Leiden, The Netherlands*

Abstract: Animal studies show that glucocorticoid effects on memory depend on noradrenergic activation within an intact amygdala. Testing this model in humans is the subject of the present fMRI study. Healthy subjects watched emotional and neutral stimuli after having received a betablocker or placebo. Cortisol levels of all subjects were determined and served as a marker of the subject's (endogenous) cortisol level during the experiment. Viewing emotional pictures resulted in increased amygdala activation compared to neutral pictures and this effect was enhanced in subjects with a high versus low cortisol level under placebo condition. Betablockade with propranolol, lowering the noradrenergic level in the amygdala, disrupted this effect and apparently the interaction with cortisol. These data support the hypothesis that high endogenous cortisol levels at the time of encoding interact with noradrenergic activation in the amygdala in man.

Keywords: amygdala; fMRI; noradrenaline; cortisol; human

Introduction

Memories of emotional or traumatic events tend to be better remembered than are daily returning neutral or non-emotional events. Why are incidents like the dramatic school shooting on the Virginia Tech campus in Blacksburg or the terrorist suicide attacks on September 11, 2001 so well remembered? One of the explanations for this memory difference is the extent to which the body is physiologically aroused. Studying the role of stress hormones on emotional memory is the focus of our recent research.

What was already known in human emotional information processing can be summarized as follows: first, several imaging studies show that confronting humans with emotionally valenced information (visual, auditory, or aversive tastes) activates the amygdala (Phan et al., 2002). The amygdala appears to be a core structure in all forms of emotional information and memory processing.

Second, the search for neurotransmitters and hormones involved in memory for stressful events provide clear evidence for a role of noradrenaline (NA). Studies with rats show that noradrenergic

*Corresponding author. Tel.: +31205256799;
Fax: +31206391369; E-mail: A.H.vanStegeren@uva.nl

DOI: 10.1016/S0079-6123(07)67020-4

agonists when injected systemically or locally in the basolateral nucleus of the amygdala (BLA) lead to better memory performance on stress tasks. Doing the opposite, so blocking the NA receptors with antagonists like betablockers (BBs), leads to decreased memory performance. This last aspect is also shown in humans (Cahill et al., 1994; van Stegeren et al., 1998).

Finally, it is shown in fMRI studies that NA is playing a role in this process by its effect on the noradrenergic receptors within the amygdala. Blocking the NA receptors with the BB propranolol when watching emotional and neutral pictures, not only blocks long term emotional memory performance, but also reduces amygdala activation when watching emotional pictures (Hurlemann et al., 2005; van Stegeren et al., 2005). Herewith NA is regarded as an essential neurotransmitter within the human amygdala.

The role of cortisol (CORT), as a second(ary) stress hormone involved in traumatic events or situations, has also been extensively studied. The effects of CORT or corticosteroids on memory can best be described as 'confusing': facilitating effects of CORT as well as impairing effects on cognitive performance have been described (Buchanan and Lovallo, 2001; Abercrombie et al., 2003; Lupien et al., 2005). Studying the effect of stress on memory, individual differences such as personal history, gender and age are important whereas context and memory phase during which stress is experienced, are also playing a role (de Kloet et al., 1999; Joels et al., 2006).

The relation between NA and corticosterone (in animals) can be described by a model (Roozendaal, 2000) that implies that the role of the amygdala, with its noradrenergic receptors in the BLA, is essential for any glucocorticoid-mediated effect on memory. This is determined by selective lesions of the BLA or infusions of beta-adrenoreceptor antagonists into the BLA that blocks the memory-modulating effects of systemic injections of glucocorticoids. So, glucocorticoid effects on memory depend on noradrenergic activation within an intact BLA that can interact with other brain regions.

Testing this model in humans is the subject of the present study. We present a new and additional analysis of a recent imaging study (van Stegeren et al., 2005) to investigate endogenous CORT effects on amygdala activation in humans. Healthy subjects watched emotional and neutral stimuli after having received a BB or placebo (PL). CORT levels of all subjects are measured before and immediately after the scanning procedure providing us with a marker of the subject's (endogenous) CORT level during the experiment. If CORT interacts with NA also in the human amygdala, then viewing emotional pictures should result in increased amygdala activation in subjects with a High_CORT versus Low_CORT level under PL condition. Betablockade, lowering the noradrenergic level in the amygdala, should then disrupt this effect. This hypothesis is tested here.

Methods and procedure

The first two sessions took place on two consecutive days where 14 male (M) and 14 female (F) subjects (age 20.93 ± 2.38) without medical or psychiatric history came to the scanning department of the Free University Medical Centre (VUMC). Informed consent was obtained from all subjects. Double blind, they received either a BB (propranolol, 80 mg) or PL in random order over the two days. Salivary samples for CORT measurements were obtained using Salivettes sampling devices (Sarstedt, Rommelsdorf, Germany) at baseline (t0), immediately before entering the scanner (t1) and 30 min later, immediately after the scanning procedure (t2). Trying to obtain a good estimation of the CORT level during the scanning procedure, salivary CORT levels just before and immediately after the scanning were averaged with $[\text{CORT(t1)} + \text{CORT(t2)}]/2$. Two groups (High_CORT versus Low_CORT) were then obtained by a median split.

In the scanner subjects were presented with a set of 92 pictures containing random assortments across 4 emotional categories. After each picture (presentation time 3 s) subjects were asked on screen for an affective rating of the previous picture with 1 being 'not emotional at all' (CAT1) (e.g. domestic subjects) to 4 being 'extremely emotional' (CAT4) (e.g. mutilation). These individual emotional ratings were used to classify the pictures for further event-related fMRI analysis. Two weeks later recognition memory was tested by presenting subjects with the original stimulus sets combined with filler sets that had comparable emotional valence. Memory was expressed as the percentage correctly recognized pictures.

fMRI acquisition and analysis

Imaging was carried out on a 1.5 T Sonata MR scanner (Siemens, Erlangen, Germany). A T1-weighted structural MRI-scan was acquired before functional imaging began [see (van Stegeren et al., 2005, 2007) for more extensively described methods and fMRI specifications]. All fMRI analyses were carried out using FEAT 5.4, part of FSL 3.2; www.fmrib.ox.ac.uk/fsl.

Amygdala activation was analyzed with contrasts comparing increasing emotional categories (CAT2, 3 and 4) with the neutral CAT1 (CAT2 > CAT1; CAT3 > CAT1; CAT4 > CAT1). At higher level the effect of CORT level on all contrasts was calculated, as well as the difference between the groups (CORT_High > CORT_Low; and the inverse contrast CORT_Low > CORT_High). Statistic images were first cluster corrected ($Z > 2.3$; $p = 0.05$). Given the small sizes of the amygdalae as regions of interest (ROI), these statistic images were thresholded using $Z > 2.3$, uncorrected.

Results

Cortisol

No significant difference in CORT level is found between t_1 and t_2 ($F(1, 27) = 1.13$; $p = 0.30$), meaning that CORT levels are not affected by the scanning procedure itself. As can be expected based on a median split, CORT level during scanning in the High_CORT versus Low_CORT group is significantly different (mean ± SD: High_CORT = 8.78 ± 1.97 nmol/l; Low_CORT = 5.40 ± 1.11 nmol/l, $p < 0.001$).

fMRI data: placebo only

Starting with the analysis of only the PL condition, a significant effect for the High_CORT > Low_CORT groups on amygdala activation is found, when subsequent emotional categories are compared (CAT2 > 1: trend $p = 0.07$; CAT3 > 1: $p < 0.05$; CAT4 > 1: $p < 0.05$). So increased amygdala activation for the emotional pictures, compared to the neutral CAT1 pictures, is significantly higher in the High_CORT compared to the Low_CORT group. This points at an interaction between NA levels, which we hypothesize to be related to emotional intensity, and endogenous CORT levels.

PL > BB interaction with CORT_High > CORT_Low

Support for the model would be even stronger if this interaction effect between NA and endogenous CORT would be disturbed under BB condition, in which NA levels are substantially lowered. This was exactly what we found: amygdala activation did not pass the threshold under BB condition for the High_CORT > Low_CORT groups with emotional pictures. Note that interpreting the results should be done with caution, since the High_CORT and Low_CORT groups refer to different absolute levels within each drug group. However, when calculating the difference in activation of PL > BB, interacting with CORT_High > CORT_Low concentrations, a cluster was found in the right amygdala for CAT3 > 1 ($Z = 2.35$; max. voxel at x, y, z = 20, −8, −24) and for the CAT4 > 1 contrast ($Z = 3.01$; $p = 0.05$; max. voxel at x, y, z = 14, 0, −12) (see Figs. 1 and 2).

Memory

Memory performance was assessed 2 weeks after scanning. As has been reported earlier (van Stegeren et al., 2005) memory performance for the emotional CAT3 and CAT4 stimuli was better than that of the neutral CAT1 pictures under PL condition. Betablockade with propranolol

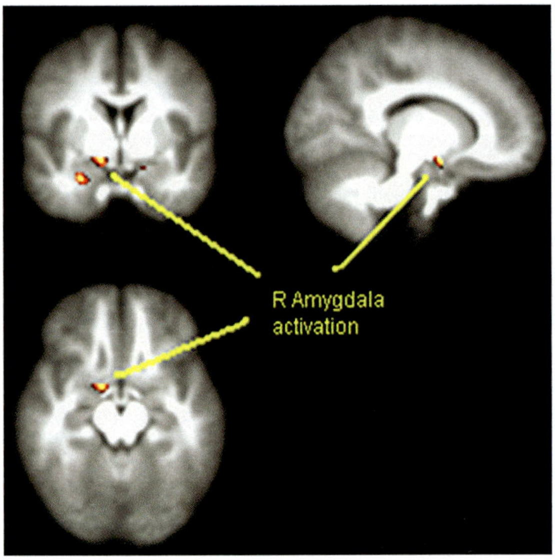

Fig. 1. PL > BB interacting with CORT_High > CRT_Low: a significant cluster in the R amygdala remains when comparing the most emotional CAT4 with CAT1 pictures ($Z = 3.01$; x, y, $z = 14, 0, -12$). R in picture = L in brain.

significantly obliterated this effect for the CAT3 > 1 difference. But no main or interaction effect of the endogenous CORT level during scanning on memory performance 2 weeks later was found.

Discussion

These data support our hypothesis that high endogenous CORT levels at the time of encoding interact with noradrenergic activation in the amygdala in man. They also agree with the model obtained in rats (Roozendaal, 2000, 2002), recently underlined in a study that focused on the synergistic actions of glucocorticoids and emotional arousal-induced noradrenergic activation of the BLA. This interaction constitutes a neural mechanism by which glucocorticoids may selectively enhance memory consolidation for emotionally arousing experiences (Roozendaal et al., 2006).

Although we were able to show that NA levels rose during the scanning, by measuring salivary alpha amylase levels as an indicator

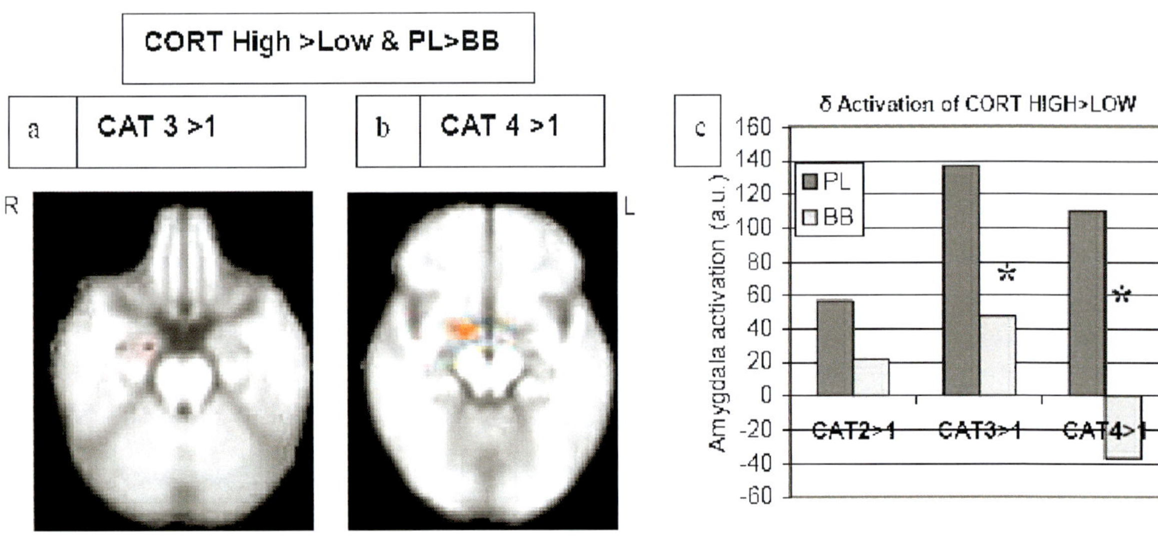

Fig. 2. Interaction of placebo (PL) > betablocker (BB) condition with High_CORT > Low_CORT groups. Significant activation for the PL > BB comparison remains in the R amygdala when High_CORT–Low_CORT group activation is contrasted. A small cluster in the amygdala is visible comparing CAT3 > 1 activation (a) and a bigger cluster for CAT4 > 1 activation (b). This means that amygdala activation that is present in subjects with High_CORT levels under PL condition during emotional CAT4 pictures is decreased when under BB condition. Delta scores (differences) in activation for High_CORT versus Low_CORT groups are plotted for emotional picture categories contrasted with neutral (CAT1) picture activation by drug groups (c). Reprinted from van Stegeren et al. (2007) Neurobiol. Learn. Mem. Copyright (2007), with permission from Elsevier.

for sympathetic-adreno-medullar system (SAM-system) activity (van Stegeren et al., 2006), it might seem surprising that CORT levels were not affected by the scanning procedure itself. A large meta-analysis reviewing laboratory studies to see which procedures do and which ones do not lead to CORT responses (Dickerson and Kemeny, 2004) shows that tasks containing both uncontrollable and social-evaluative elements are associated with the largest CORT and adrenocorticotropin hormone changes and the longest times to recovery, but that emotion induction tasks per se do not elicit a significant CORT response. High_CORT and Low_ CORT levels in this study should then be more regarded as a difference in endogenous 'state' than as a difference in a task-related CORT response.

The BB-induced decrease in amygdala activation cannot be explained by changes in brain hemodynamics in general, since no main effect of propranolol on brain activation is found during all four emotional categories. It is the selective reduction of increased amygdala activation during the emotional CAT3 pictures (hypothesized by us to be related to increased noradrenergic activation) that is shown here. Decreasing NA levels with this dosage of propranolol wipes out the differential effect of the High_CORT > Low_CORT concentrations that leads to increased amygdala activation in the PL group.

Focussing solely on the interaction of NA and CORT is of course a simplification of what happens during stress in reality. First, the stress system has two modes of operation, a fast and a slower mode, with a variety of stress hormones and agents [such as corticotropin releasing hormone (CRH), vasopressin and urocortins] playing a role in CORT release. Second, corticosteroids operate in both modes by means of mineralocorticoid receptor (MR) and glucocorticoid receptor (GR) in different ratios and with distinctive as well as interacting functions (de Kloet et al., 2005). Finally, in a recent overview the facilitating and impairing effects of stress on memory are unified in a theory that explains both functions of corticosteroids (Joels et al., 2006). The authors propose that if convergence in time and space takes place, stress hormones help to store the information attached to the event for future use. They make a distinction in rapid, non-genomic effects of stress hormones such as NA, CRH and CORT that can facilitate the encoding of information when (a) they act in the same areas that are involved in processing of the information to be remembered and (b) do so around the time that synaptic strengthening in these areas takes place. Additionally, CORT or corticosterone initiates a slower genomic signal that will suppress unrelated information reaching these circuits some time after the stressful event. This dual effect of corticosterone serves to enhance the signal-to-noise ratio of important information (Joels et al., 2006).

It is clear that the presented results on the interaction between NA and CORT in humans should be investigated in a new experimental study where drug application and control groups are designed in a way to rule out some of the drawbacks of this study — such as the post-hoc nature of the CORT grouping. Hence, although this study is supportive for this solidly tested model in animals, the ultimate support for a comparable model in humans should be furnished by well-designed glucocorticoid challenge studies.

Abbreviations

BB	betablocker
BLA	basolateral nucleus of amygdala
CORT	cortisol
CRH	corticotropin releasing hormone
GR	glucocorticoid receptor
NA	noradrenaline
MR	mineralocorticoid receptor
PL	placebo
ROI	regions of interest
SAM-system	sympathetic-adreno-medullar system

References

Abercrombie, H.C., Kalin, N.H., Thurow, M.E., Rosenkranz, M.A. and Davidson, R.J. (2003) Cortisol variation in humans affects memory for emotionally laden and neutral information. Behav. Neurosci., 117: 505–516.

Buchanan, T.W. and Lovallo, W.R. (2001) Enhanced memory for emotional material following stress-level cortisol treatment in humans. Psychoneuroendocrinology, 26: 307–317.

Cahill, L., Prins, B., Weber, M. and McGaugh, J.L. (1994) Beta-adrenergic activation and memory for emotional events. Nature, 371: 702–704.

Dickerson, S.S. and Kemeny, M.E. (2004) Acute stressors and cortisol responses: a theoretical integration and synthesis of laboratory research. Psychol. Bull., 130: 355–391.

Hurlemann, R., Hawellek, B., Matusch, A., Kolsch, H., Wollersen, H., Madea, B., Vogeley, K., Maier, W. and Dolan, R.J. (2005) Noradrenergic modulation of emotion-induced forgetting and remembering. J. Neurosci., 25: 6343–6349.

Joels, M., Pu, Z., Wiegert, O., Oitzl, M.S. and Krugers, H.J. (2006) Learning under stress: how does it work? Trends Cogn. Sci., 10: 152–158.

de Kloet, E.R., Joels, M. and Holsboer, F. (2005) Stress and the brain: from adaptation to disease. Nat. Rev. Neurosci., 6: 463–475.

de Kloet, E.R., Oitzl, M.S. and Joels, M. (1999) Stress and cognition: are corticosteroids good or bad guys? Trends Neurosci., 22: 422–426.

Lupien, S.J., Fiocco, A., Wan, N., Maheu, F., Lord, C., Schramek, T. and Tu, M.T. (2005) Stress hormones and human memory function across the lifespan. Psychoneuroendocrinology, 30: 225–242.

Phan, K.L., Wager, T., Taylor, S.F. and Liberzon, I. (2002) Functional neuroanatomy of emotion: a meta-analysis of emotion activation studies in PET and fMRI. Neuroimage, 16: 331–348.

Roozendaal, B. (2000) 1999 Curt P. Richter award. Glucocorticoids and the regulation of memory consolidation. Psychoneuroendocrinology, 25: 213–238.

Roozendaal, B. (2002) Stress and memory: opposing effects of glucocorticoids on memory consolidation and memory retrieval. Neurobiol. Learn. Mem., 78: 578–595.

Roozendaal, B., Okuda, S., Van der Zee, E.A. and McGaugh, J.L. (2006) Glucocorticoid enhancement of memory requires arousal-induced noradrenergic activation in the basolateral amygdala. Proc. Natl. Acad. Sci. U.S.A., 103: 6741–6746.

van Stegeren, A.H., Everaerd, W., Cahill, L., McGaugh, J.L. and Gooren, L.J. (1998) Memory for emotional events: differential effects of centrally versus peripherally acting beta-blocking agents. Psychopharmacology (Berl.), 138: 305–310.

van Stegeren, A.H., Goekoop, R., Everaerd, W., Scheltens, P., Barkhof, F., Kuijer, J.P. and Rombouts, S.A. (2005) Noradrenaline mediates amygdala activation in men and women during encoding of emotional material. Neuroimage, 24: 898–909.

van Stegeren, A., Rohleder, N., Everaerd, W. and Wolf, O.T. (2006) Salivary alpha amylase as marker for adrenergic activity during stress: effect of betablockade. Psychoneuroendocrinology, 31: 137–141.

van Stegeren, A.H., Wolf, O.T., Everaerd, W., Scheltens, P., Barkhof, F. and Rombouts, S.A. (2007) Endogenous cortisol level interacts with noradrenergic activation in the human amygdala. Neurobiol. Learn. Mem., 87: 57–66.

E.R. de Kloet, M.S. Oitzl & E. Vermetten (Eds.)
Progress in Brain Research, Vol. 167
ISSN 0079-6123

CHAPTER 21

Corticosteroid hormones, synaptic strength and emotional memories: corticosteroid modulation of memory — a cellular and molecular perspective

Olof Wiegert*, Marian Joëls and Harm J. Krugers

Swammerdam Institute for Life Sciences, SILS-CNS, Universiteit van Amsterdam, Kruislaan 320, 1098 SM, Amsterdam, The Netherlands

Abstract: Emotionally loaded and stressful events modulate cognitive performance. This modulation of cognitive performance is at least partially dependent on corticosteroid hormones that are released in high amounts during emotional or stressful events. Corticosterone both strengthens and suppresses cognitive performance and synaptic plasticity. These effects may critically depend on the timing of the stressful event and corticosteroid exposure with respect to the learning situation. Based on recent findings we propose a model in which corticosterone can rapidly enhance synaptic plasticity. Later, corticosterone may stabilize synaptic efficacy, possibly at the expense of reduced synaptic plasticity.

Emotionally arousing and stressful memories are among the ones that are very well remembered in general. There is profound evidence that corticosteroid hormones, which are released in large amounts during and after the stressful situation, modulate the memories for these events. In the present view, stress and glucocorticoids, briefly released within the context of a task, promote long-term consolidation of information (for a review, see Joels et al., 2006). More specifically, in various species, including man, acute post-training glucocorticoid receptor (GR) activation enhances memory consolidation whereas administration of selective GR antagonists impairs consolidation (Joels et al., 2006). A point mutation in the mouse GR, which selectively prevents dimerization and DNA-binding of the GR, impairs spatial memory performance (Oitzl et al., 2001). By contrast, GR activation or additional stressors that are applied before acquisition training or retention testing and which are not in context with the learning situation, impair acquisition and retrieval of spatial information (Joels et al., 2006). A major question that remains to be addressed is why and, more importantly, how memories for emotional and stressful events are remembered so well, and what exactly the cellular and molecular mechanisms are that are targeted by corticosteroid hormones and GR activation.

Glutamatergic synapses are key-role players in synaptic plasticity as well as in learning and memory processes. α-Amino-3-hydroxy-5-methyl-4-isoxazolepropionic acid (AMPA) type glutamate receptors mediate fast excitatory amino acid mediated synaptic transmission. Together with *N*-methyl-D-aspartate (NMDA) receptors they play an essential role in long-term potentiation (LTP) a derivative of synaptic plasticity, and probably the best-studied cellular mechanism for

*Corresponding author. Tel.: +31 20 525 7638; Fax: +31 20 525 7709; E-mail: owiegert@science.uva.nl

DOI: 10.1016/S0079-6123(07)67021-6

learning and memory. In addition, recent evidence supports the view that AMPA receptors play a crucial role in maintaining synaptic strength after induction of LTP; after induction of LTP a long-term increase in AMPA receptor mediated synaptic transmission is evident. This effect can be accomplished either via phosphorylation of existing synaptic AMPA receptors, and or by increasing the number of (new) synaptic AMPA receptors. In particular, there is evidence that GluR1 subunits become inserted into synapses after LTP induction (Malinow et al., 2000). In addition, the subsequent ability to change synaptic plasticity is dependent on AMPA receptor subunit composition (Shi et al., 2001). Importantly, AMPA receptors have been reported to play a crucial role in working memory (Reisel et al., 2005) and emotional memories (Rumpel et al., 2005). Taken together, there is considerable evidence that AMPA receptors are not only essential for both the induction and maintenance of synaptic strength, but they also determine whether and how synaptic plasticity is regulated later on (Stein et al., 2003).

Examining how corticosteroid hormones modulate these essential mechanisms underlying synaptic plasticity and efficiency may increase our understanding of how these hormones modulate learning and memory processes. Looking in detail at the synaptic level, it has been found that corticosteroid hormones rapidly enhance AMPA receptor mediated synaptic transmission via the mineralocorticoid receptor (MR), by enhancing the frequency of hippocampal synaptic events (Karst et al., 2005). Most likely this effect reflects an increase in neurotransmitter release probability. At the same time, corticosterone, when applied at the moment of LTP induction also promotes synaptic plasticity, i.e. enhances synaptic strength (Wiegert et al., 2006b). In addition, elevated corticosterone levels have been reported to slowly enhance AMPA receptor mediated synaptic transmission in ventral tegmental area and the hippocampus (Karst and Joels, 2005). These effects could be blocked by RU 38486 indicating that they are mediated by activation of the GR. These experiments indicate an enhancement of functional AMPA receptors in hippocampal CA1 synapses, due to either via insertion of novel receptors or by changes in AMPA receptor subunit composition (Wiegert et al., 2006a). Whether and how this insertion of AMPA receptors is related to the ability to maintain stored information and/or encode novel information remains to be addressed.

In summary, we hypothesize that corticosteroid hormones dynamically modulate synaptic efficacy. This effect may depend critically on duration of exposure as well as the time after corticosterone exposure and involves different cellular and molecular targets for synaptic plasticity. First, during the early phase of the stress response, when corticosteroid levels rise, these hormones excite neurons and promote LTP, possibly enhancing the encoding of information. Second, later, presumably via a genomic action, corticosteroid hormones increase AMPA receptor mediated synaptic efficacy, possibly reflecting a stabilization of a memory trace. Third, at the same time, (i.e. after corticosteroid exposure) NMDA-receptor dependent LTP, via a GR-mediated mechanism, is suppressed (Wiegert et al., 2005). This effect may be the result of GR-mediated changes in endogenous AMPA receptor function, which could preclude subsequent exogenously applied LTP (Stein et al., 2003), but could also be elicited by changes in AMPA receptor subunit composition (Wiegert et al., 2006a) as well as attenuated NMDA-receptor function.

We speculate that these slow effects of corticosterone on synaptic efficacy and synaptic plasticity may promote consolidation of emotional and declarative memories. Thus, during exposure to a stressful event synaptic efficacy is enhanced, while later on the ability to overwrite existing information is impaired. It will be a major challenge to examine whether indeed these changes in AMPA receptor mediated synaptic efficiency also underlies the strong memories for stressful and emotional memories.

References

Joels, M., Pu, Z., Wiegert, O., Oitzl, M.S. and Krugers, H.J. (2006) Learning under stress: how does it work? Trends Cogn. Sci., 10: 152–158.

Karst, H., Berger, S., Turiault, M., Tronche, F., Schutz, G. and Joels, M. (2005) Mineralocorticoid receptors are indispensable for nongenomic modulation of hippocampal glutamate transmission by corticosterone. Proc. Natl. Acad. Sci. U.S.A., 102: 19204–19207.

Karst, H. and Joels, M. (2005) Corticosterone slowly enhances miniature excitatory postsynaptic current amplitude in mice CA1 hippocampal cells. J. Neurophysiol., 94: 3479–3486.

Malinow, R., Mainen, Z.F. and Hayashi, Y. (2000) LTP mechanisms: from silence to four-lane traffic. Curr. Opin. Neurobiol., 10: 352–357.

Oitzl, M.S., Reichardt, H.M., Joels, M. and de Kloet, E.R. (2001) Point mutation in the mouse glucocorticoid receptor preventing DNA binding impairs spatial memory. Proc. Natl. Acad. Sci. U.S.A., 98: 12790–12795.

Reisel, D., Bannerman, D.M., Deacon, R.M., Sprengel, R., Seeburg, P.H. and Rawlins, J.N. (2005) GluR-A-dependent synaptic plasticity is required for the temporal encoding of nonspatial information. Behav. Neurosci., 119: 1298–1306.

Rumpel, S., LeDoux, J., Zador, A. and Malinow, R. (2005) Postsynaptic receptor trafficking underlying a form of associative learning. Science, 308: 83–88.

Shi, S., Hayashi, Y., Esteban, J.A. and Malinow, R. (2001) Subunit-specific rules governing AMPA receptor trafficking to synapses in hippocampal pyramidal neurons. Cell, 105: 331–343.

Stein, V., House, D.R., Bredt, D.S. and Nicoll, R.A. (2003) Postsynaptic density-95 mimics and occludes hippocampal long-term potentiation and enhances long-term depression. J. Neurosci., 23: 5503–5506.

Wiegert, O., Joels, M., Holman, D., Henley, J. and Krugers, H. (2006a) Corticosteroid modulation of hippocampal strength: Glutamate receptors on the move, 2006 Meeting of the Society for Neuroscience, 232.18 Atlanta, GA, USA and The 16th Neuropharmacology Conference, Long Term Potentiation, P50, Atlanta, GA, U.S.A.

Wiegert, O., Joels, M. and Krugers, H. (2006b) Timing is essential for rapid effects of corticosterone on synaptic potentiation in the mouse hippocampus. Learn. Mem., 13: 110–113 10.1101/lm.87706.

Wiegert, O., Pu, Z., Shor, S., Joels, M. and Krugers, H. (2005) Glucocorticoid receptor activation selectively hampers *N*-methyl-D-aspartate receptor dependent hippocampal synaptic plasticity in vitro. Neuroscience, 135: 403–411.

E.R. de Kloet, M.S. Oitzl & E. Vermetten (Eds.)
Progress in Brain Research, Vol. 167
ISSN 0079-6123

CHAPTER 22

Does trauma cause lasting changes in HPA-axis functioning in healthy individuals?

Ellen R. Klaassens*, Tineke van Veen and Frans G. Zitman

Department of Psychiatry, Leiden University Medical Center (LUMC), Albinusdreef 2, 2333 ZA Leiden, The Netherlands

Abstract: Although the majority of people who are exposed to traumatic events do not develop psychopathology, trauma has often been associated with increased vulnerability to psychiatric disorders. In addition, alterations in the HPA-axis have been demonstrated in patients with trauma-related psychiatric disorders. We hypothesize that trauma causes dysregulation of the HPA-axis. Therefore, we will compare HPA-axis functioning of traumatized and non-traumatized healthy individuals from the same gender and age from two categories: military and railroad personnel. In addition, a group of women with a history of childhood trauma was included. We will investigate for the putative role of attachment style and psychological resilience factors such as coping. In this article, we present the rationale for this study.

Keywords: HPA-axis; trauma; childhood trauma; long term; healthy adults; veterans; railroad personnel; coping

Introduction

In 1986, Mason et al. were the first to describe low urinary free-cortisol levels in posttraumatic stress disorder (PTSD) patients (Mason et al., 1986). In a subsequent study the research group replicated the results (Yehuda et al., 1990). They concluded that the findings suggest a physiological adaptation of the hypothalamic-pituitary-adrenal (HPA) axis to chronic stress. Later, comparable results were found with respect to day curves of cortisol in blood and saliva, by the same group as well as others (Boscarino, 1996; Yehuda et al., 1996; Neylan et al., 2005).

However, the story does not end here. The relationship between trauma, psychopathology and HPA-axis functioning is more complicated. First, PTSD is not always accompanied by low cortisol levels: some studies found high levels (Pitman and Orr, 1990; Hawk et al., 2000), whereas other studies found no evidence for cortisol changes in PTSD (Young et al., 2004). Second, trauma or stressful events not always induce PTSD. Other psychiatric disorders like mood and somatoform disorders may develop as well (Kendler et al., 1999; Heim et al., 2006). Besides, after a trauma most people continue to live their lives seemingly unaffected, without signs of psychopathology (Breslau et al., 1998). Third, traumatized people who remain psychologically healthy nevertheless may have HPA-axis disturbances. Heim et al. (2000) showed that women with childhood sexual abuse had elevated cortisol levels in response to a cognitive stress challenge, even when they were without psychopathology. Likewise, in a study by Nicolson

*Corresponding author. Tel.: +31-(0)71-5263785;
Fax: +31-(0)71-5248156; E-mail: E.R.Klaassens@LUMC.nl

DOI: 10.1016/S0079-6123(07)67022-8

(2004), parental loss before age 17 was associated with higher cortisol throughout the day in healthy male adults. This suggests that parental death during childhood may have lasting effects on the HPA-axis, even in the absence of psychopathology. In a recent study on neurobiological alterations in peacekeeping veterans, de Kloet et al. (2007) found no difference in cortisol suppression after dexamethasone between veterans with PTSD and traumatized psychologically healthy veterans, but both groups had significantly lower cortisol levels compared to healthy, non-traumatized civilian controls.

These studies suggest that HPA-axis dysfunction is not specifically linked to psychopathology but to trauma; elevated cortisol levels to childhood trauma (Heim et al., 2000; Nicolson, 2004) and decreased cortisol to adult trauma (de Kloet et al., 2007). Trauma may cause HPA-axis dysfunctions, but also an underlying factor may exist that causes HPA-axis dysfunction as well as a propensity to engage in situations with a high risk at traumatization. This is especially true for veterans. de Kloet et al. (2007) suggested that the enhanced cortisol suppression to dexamethasone in the traumatized veterans is related to trauma exposure rather than to PTSD. However, the use of a non-military healthy control group makes it impossible to distinguish between effects of trauma exposure and other military-related factors. Individuals that have voluntarily chosen a military career may well differ from non-military personnel with respect not only to psychological profile but also HPA-axis responsivity.

Hypothesis

We developed a series of studies in which we tried to eliminate the influence of such factors as much as possible. First, we designed a study to compare HPA-axis function in traumatized and non-traumatized psychologically healthy peacekeeping veterans. Second, we did the same in psychologically healthy traumatized and non-traumatized train personnel and third, in psychologically healthy readers of a women's magazine with and without a history of childhood trauma. In all cases we hypothesized that the traumatized people showed altered HPA-axis functioning compared to non-traumatized people.

Study groups

The veterans for this study will come from a randomly selected group of peacekeepers, most of which have been deployed in Lebanon (1979–1985) and former Yugoslavia (1993–1995). Even though peacekeepers are not active in an actual combat zone, reports of being shot at, seeing others are being killed, and witnessing human suffering are numerous. Military personnel have chosen a stressful and sometimes dangerous profession, possibly driven by risk-taking behavior. Frequent exposure to psychological and physical stress is part of the military training during which the recruits are prepared to withstand high levels of alertness. In this group, we will investigate the effect of adult trauma on HPA-axis functioning in healthy individuals (men) with risk-taking behavior and stress-resistance.

The second group of individuals consists of train personnel. Unlike military veterans, most railroad personnel did not choose their job driven by risk-taking behavior. In the last two decades, however, there has been an increase in serious adverse events experienced by train personnel. Every year, hundreds of train personnel are subjected to traumatic events, involving severe aggression, life-threatening situations, and 'person-under-train' accidents. Therefore, in this group, we address the effect of adult trauma on HPA-axis functioning in healthy individuals who are at risk of encountering unpredictable and infrequent traumatic events as part of their every day job, without being especially selected or trained for these encounters.

The last group in our study consists of women with a history of childhood trauma. Early in life, the HPA-axis is in development, and trauma may have different effects than during adulthood. Several studies show that HPA-axis activity in early human development is under strong psychosocial regulation and that attachment style is an important protective factor from developing alterations in HPA-axis functioning. Young children with a history of insensitive, unresponsive care, are more susceptible to cortisol elevations after stressful events than children who are securely attached to a caregiver (in Tarullo and Gunnar, 2006). In our sample of women, we assess the effect of childhood trauma on HPA-axis functioning in healthy individuals by comparing women who were traumatized during childhood with women without a history of childhood trauma. In addition, attachment style will be assessed, since it plays an important role in the development of altered HPA-axis functioning. We will also look at psychological resilience factors, including coping styles, in order to check whether the two groups of women are indeed comparable in this respect. We will do the same for the two groups of veterans and the two groups of train personnel. By further comparing the veterans with the train personnel and the traumatized women we will check whether indeed these groups differ in coping styles.

The common denominator in our study groups is that we looked at the long-term effects of trauma on HPA-axis functioning. In the peacekeeper veterans, the deployment-related trauma happened 10–25 years before inclusion in the study. For train personnel, this period of time was at least 5 years. In the group of traumatized women, the trauma took place 11–45 years before inclusion.

Each of the comparisons in this study will provide information about the effects of trauma on HPA-axis functioning in healthy individuals. This will elucidate the relationship between trauma and HPA-axis activity in individuals in different settings.

Abbreviations

HPA-axis	hypothalamic-pituitary-adrenal axis
PTSD	posttraumatic stress disorder

References

Boscarino, J.A. (1996) Posttraumatic stress disorder, exposure to combat, and lower plasma cortisol among Vietnam veterans: findings and clinical implications. J. Consult. Clin. Psychol., 64: 191–201.

Breslau, N., Kessler, R.C., Chilcoat, H.D., Schultz, L.R., Davis, G.C. and Andreski, P. (1998) Trauma and posttraumatic stress disorder in the community: the 1996 Detroit area survey of trauma. Arch. Gen. Psychiatry, 55: 626–632.

Hawk, L.W., Dougall, A.L., Ursano, R.J. and Baum, A. (2000) Urinary catecholamines and cortisol in recent-onset posttraumatic stress disorder after motor vehicle accidents. Psychosom. Med., 62: 423–434.

Heim, C., Newport, D.J., Heit, S., Graham, Y.P., Wilcox, M.M., Bonsall, R., Miller, A.H. and Nemeroff, C.B. (2000) Pituitary-adrenal and autonomic responses to stress in women after sexual and physical abuse in childhood. JAMA, 284: 592–597.

Heim, C., Wagner, D., Maloney, E., Papanicolaou, D.A., Solomon, L., Jones, J.F., Unger, E.R. and Reeves, W.C. (2006) Early adverse experience and risk for chronic fatigue syndrome: results from a population-based study. Arch. Gen. Psychiatry, 63: 1258–1266.

Kendler, K.S., Karkowski, L.M. and Prescott, C.A. (1999) Causal relationship between stressful life events and the onset of major depression. Am. J. Psychiatry, 156: 837–841.

de Kloet, C.S., Vermetten, E., Heijnen, C.J., Geuze, E., Lentjes, E.G.W.M. and Westenberg, H.G.M. (2007) Enhanced cortisol suppression in response to dexamethasone administration in traumatized veterans with and without posttraumatic stress disorder. Psychoneuroendocrinology, 32: 215–226.

Mason, J.W., Giller, E.L., Kosten, T.R., Ostroff, R.B. and Podd, L. (1986) Urinary free-cortisol levels in posttraumatic stress disorder patients. J. Nerv. Ment. Dis., 174: 145–149.

Neylan, T.C., Brunet, A., Pole, N., Best, S.R., Metzler, T.J., Yehuda, R. and Marmar, C.R. (2005) PTSD symptoms predict waking salivary cortisol levels in police officers. Psychoneuroendocrinology, 30: 373–381.

Nicolson, N.A. (2004) Childhood parental loss and cortisol levels in adult men. Psychoneuroendocrinology, 29: 1012–1018.

Pitman, R.K. and Orr, S.P. (1990) Twenty-four hour urinary cortisol and catecholamine excretion in combat-related posttraumatic stress disorder. Biol. Psychiatry, 27: 245–247.

Tarullo, A.R. and Gunnar, M.R. (2006) Child maltreatment and the developing HPA axis. Horm. Behav., 50: 632–639.

Yehuda, R., Southwick, S.M., Nussbaum, G., Wahby, V., Giller Jr., E.L. and Mason, J.W. (1990) Low urinary cortisol excretion in patients with posttraumatic stress disorder. J. Nerv. Ment. Dis., 178: 366–369.

Yehuda, R., Teicher, M.H., Trestman, R.L., Levengood, R.A. and Siever, L.J. (1996) Cortisol regulation in posttraumatic stress disorder and major depression: a chronobiological analysis. Biol. Psychiatry, 40: 79–88.

Young, E.A., Tolman, R., Witkowski, K. and Kaplan, G. (2004) Salivary cortisol and posttraumatic stress disorder in a low-income community sample of women. Biol. Psychiatry, 55(6): 621–626.

E.R. de Kloet, M.S. Oitzl & E. Vermetten (Eds.)
Progress in Brain Research, Vol. 167
ISSN 0079-6123

CHAPTER 23

Need for alternative ways of phenotyping of mood, anxiety, and somatoform disorders in biological research

G. Veen[1,*], I.M. van Vliet[1], R.H. de Rijk[2] and F.G. Zitman[1]

[1]*Department of Psychiatry, Leiden University Medical Center, P.O. Box 9600, 2300 RC Leiden, The Netherlands*
[2]*Department of Medical Pharmacology, Leiden/Amsterdam Center for Drug Research and Leiden University Medical Center, P.O. Box 9502, 2300 RA Leiden, The Netherlands*

Abstract: Variation in psychiatric symptomatology is continuous and does not coalesce into fairly well-defined categorical DSM-IV clusters. As a consequence, DSM-IV fails to meaningfully integrate information generated by neuroendocrine research. Continuous psychological dimensions selected for their predictiveness with respect to endophenotypes, as biological intermediate factors, are proposed to be the best ways in reaching an understanding of the causations in mood, anxiety, and somatoform disorders.

Keywords: hypothalamic-pituitary-adrenal-axis; mood disorders; anxiety disorders; somatoform disorders; phenotype; endophenotype; psychological dimensions

Introduction

Nowadays, psychopathology is mostly described in terms of diagnostic categories according to the Diagnostic and Statistical Manual of Mental Disorders-IV (DSM-IV). An important advantage of this system is that it yields reliable diagnoses, especially with respect to classical psychiatric disorders like depression and panic disorder, which are subsumed under axis I in the DSM-IV. However, the validity is open to debate. Firstly, the majority of patients shows a complex presentation of a wide range of psychiatric symptoms, often leading to more than one axis I diagnoses, simultaneously. Therefore, the face validity of the categorical approach of the DSM-IV has been questioned. Secondly, in general, each DSM-IV diagnosis requires the presence of a minimum number of symptoms out of a list of symptoms characterizing the disorder. However, the threshold level is mostly chosen arbitrarily, but above the mean number of symptoms found in the general population. As a consequence, the DSM-IV excludes a large group of persons with below-threshold psychopathology. Thirdly, as a diagnosis does not require the presence of all symptoms listed for the diagnosis, patients with the same DSM-IV disorder may differ greatly with respect to their symptoms. For example, two depressive patients may suffer from opposite symptoms, e.g. hyposomnia versus hypersomnia. By using DSM-IV classification, this clinical heterogeneity is not specified or adequately described. Fourthly, no close relationship between the DSM-IV axis I diagnoses and biological markers has been found.

*Corresponding author. Tel.: +31-71-5263785;
Fax: +31-71-5248156; E-mail: G.Veen@lumc.nl

DOI: 10.1016/S0079-6123(07)67023-X

For instance, notwithstanding the indications that stress plays an important role in the development of mood, anxiety, and somatoform (MAS) disorders, only in about half of the patients hypothalamic-pituitary-adrenal (HPA)-axis dysregulations are found. Furthermore, often opposite findings are found within one diagnostic entity, e.g. hyper- and hypocortisolism in respectively melancholic and atypical depression (First, 2005; Widiger and Samuel, 2005).

Does this imply that we look at the wrong biological markers or do we make the wrong groupings of the phenotype? In this article we explore the latter possibility and propose the need for alternative ways of phenotyping of MAS disorders in biological research.

Phenotype: diagnosing MAS disorders

In 1990, Van Praag proposed a new diagnostic approach, named functionalization and verticalization. Functionalization comprises converting categorical diagnoses into the psychic dysfunctions underlying the psychopathological symptoms. This enables the verticalization, by which is meant connecting the psychic dysfunction with the underlying neurobiological substratum. To do so, a sequential analysis is required, i.e. determination of the sequence of appearance of symptoms, because it is hypothesized that the first symptoms, called front runners by Van Praag, carry a primary character with respect to neuroendocrine dysfunctions. Examples are the associations between serotonergic dysfunctions and disturbances in anxiety, aggression regulation and impulse control, and between dopaminergic dysfunctions and disturbances in motoricity (Van Praag, 1990; Van Praag et al., 1990; Coccaro, 1992).

Unfortunately, for many types of psychic dysfunctions the front runners are unknown or difficult to determine. Instead of the front runners, the dimensions underlying the psychic dysfunctions may also be an appropriate link between psychopathology and neuroendocrine dysfunctions. Dimensional models, in contrast to functionalization and verticalization, do not require a sequentional analysis of psychic dysfunctions, because it is hypothesized that for each patient assessment on all dimensions that cover the psychopathology is sufficient for meaningful integration with the information generated by neuroendocrine research. Several dimensional models have been proposed for assessing mood and anxiety disorders, such as the tripartite model, approach-withdrawal model, and valence-arousal model. All these models posit that mood and anxiety disorders share a common distress dimension, but they also can be distinguished from each other by particular characteristics (Shankman and Klein, 2003). A shortcoming of these models is that they still use the DSM-IV classification as frame of reference by proposing dimensions with assumed predictiveness for DSM-IV diagnoses instead of looking for dimensions with a high concordance with biological markers, the so-called endophenotypes. The development of a new dimensional model, independent of DSM-IV diagnoses, and external validated with endophenotypes, is needed.

Endophenotype: the crucial link in between

An endophenotype is a biological marker of a phenotype closer to relevant gene action than the phenotype itself. Endophenotypes should be continuously quantifiable and predict disorders probabilistically. In the case of psychopathology, endophenotypes may be neurophysiological, biochemical, endocrinological, neuroanatomical, cognitive, or neuropsychological in nature. As MAS disorders are linked to stress, it is hypothesized that dysfunction of one of the important stress systems, the HPA-axis, is an endophenotype of these disorders.

Indeed some indications have been found that HPA-axis dysfunction is an endophenotype of MAS disorders diagnosed according to the DSM-IV. About half of the patients with a major depressive disorder show a hyperactivity of the HPA-axis. Studies of anxiety disorders revealed less robust HPA-axis dysregulations. Some, but not all patients with posttraumatic stress disorders, show hypocortisolism. Hypocortisolims has been reported in 20–25% of patients with somatoform disorders (Almasy and Blangero, 2001; Gottesman and Gould, 2003; Flint and Munafo, 2007). Given the questionable validity of diagnoses based on the DSM-IV, no large correlations between diagnoses of this type and biological markers are to be expected.

A few studies have examined HPA-axis activity in relation to psychic dysfunctions, instead of DSM-IV classification. Hyperactivity of HPA-axis is considered to play an important role for individual symptoms, such as enhanced anxiety, decreased responsiveness to the environment, decreased diurnal variation, disturbed sleep, altered psychomotor functions, decreased appetite and libido, and impaired cognition. Reduced HPA-axis activity, mediated by an enhanced negative feedback, is associated with symptoms, such as hypersomnia, hyperphagia, lethargy, and fatigue (Ehlert et al., 2005; McLean et al. 2005; Gur et al., 2004).

The relationship between dimensional models and HPA-axis activity has, so far known, never been studied.

A model to study dimensions of mood, anxiety, and somatisation and HPA-axis functioning

We propose that the development of a dimensional model that covers the symptomatology of all three MAS disorders is needed to reach more insight in its biological substrate. By using psychological questionnaires that assess a broad spectrum of symptoms, one can look for underlying dimensions that adequately and precisely describe MAS psychopathology. Dimensions do not need to have predictive value for separate DSM-IV diagnoses, but should be externally validated with biological markers, such as HPA-axis function. Basal HPA-axis activity can be measured by assessment of the cortisol diurnal pattern. HPA-axis reactivity can be examined with challenge tests like the combined dexamethasone/corticotrophin-releasing hormone (CRH) challenge test, which proved to be a sensitive measure (above 80%) in differentiating depressive patients from healthy controls (Heuser et al., 1994). It is used to examine HPA reactivity under the condition of suppressed glucocorticoid feedback as a reflection of the sensibility and responsivity of the pituitary. We hypothesize that combining these phenotypic and endophenotypic data will lead to more clarity about psychopathological processes in MAS disorders.

Abbreviations

CRH	corticotrophin-releasing hormone
DSM-IV	Diagnostic and Statistical Manual of Mental Disorders-IV
HPA-axis	hypothalamic-pituitary-adrenal-axis
MAS disorders	mood, anxiety, and somatoform disorders

References

Almasy, L. and Blangero, J. (2001) Endophenotypes as quantitative risk factors for psychiatric disease: rationale and study design. Am. J. Med. Genet., 105: 42–44.

Coccaro, E.F. (1992) Impulsive aggression and central serotonergic system function in humans: an example of a dimensional brain—behavior relationship. Int. Clin. Psychopharmacol., 7: 3–12.

Ehlert, U., Nater, U.S. and Böhmelt, A. (2005) High and low unstimulated salivary cortisol levels correspond to different symptoms of functional gastrointestinal disorders. J. Psychosom. Res., 59: 7–10.

First, M.B. (2005) Mutually exclusive versus co-occurring diagnostic categories: the challenge of diagnostic comorbidity. Psychopathology, 38: 206–210.

Flint, J. and Munafo, M.R. (2007) The endophenotype concept in psychiatric genetics. Psychol. Med., 37(2): 163–180.

Gur, A., Cevik, R., Colpan, L. and Em, S. (2004) Hypothalamic-pituitary-gonadal axis and cortisol in young women with primary fibromyalgia: the potential roles of depression, fatigue, and sleep disturbances in the occurrence of hypocortisolism. Ann. Rheum. Dis., 63: 1504–1506.

Gottesman, I.I. and Gould, T.D. (2003) The endophenotype concept in psychiatry: etymology and strategic interventions. Am. J. Psychiatry, 160(4): 636–645.

Heuser, I., Yassouridis, A. and Holsboer, F. (1994) The combined dexamethasone/CRH test: a refined laboratory test for psychiatric disorders. J. Psychiatr. Res., 28(4): 341–356.

McLean, S.A., Williams, D.A., Harris, R.E., Kop, W.J., Groner, K.H., Ambrose, K., Lyden, A.K., Gracely, R.H., Crofford, L.J., Geisser, M.E., Sen, A., Biswas, P. and Clauw, D.J. (2005) Momentary relationships between cortisol secretion and symptoms in patients with fibromyalgia. Arthritis Rheum., 52(11): 3660–3669.

Shankman, S.A. and Klein, D.N. (2003) The relation between depression and anxiety: an evaluation of the tripartite, approach-withdrawal and valence-arousal models. Clin. Psychol. Rev., 23: 605–637.

Van Praag, H.M. (1990) Two-tier diagnosing in psychiatry. Psychiatry Res., 34: 1–11.

Van Praag, H.M., Asnis, G.M., Kahn, R.S., et al. (1990) Monoamines and abnormal behavior. A multi-aminergic perspective. Br. J. Psychiatry, 157: 723–734.

Widiger, T.A. and Samuel, D.B. (2005) Diagnostic categories or dimensions? A question for diagnostic and statistical manual of mental disorders: fifth edition. J. Abnorm. Psychol., 114(4): 494–504.

E.R. de Kloet, M.S. Oitzl & E. Vermetten (Eds.)
Progress in Brain Research, Vol. 167
ISSN 0079-6123

CHAPTER 24

The HPA-axis and immune function in burnout

Paula M.C. Mommersteeg[1,2,3,*], Cobi J. Heijnen[1], Annemieke Kavelaars[1] and Lorenz J.P. van Doornen[2]

[1]*Laboratory for Psychoneuroimmunology, Division of Perinatology and Gynaecology, University Medical Center Utrecht, Homebox KC 03.068.0, P.O. Box 85090, 3508 AB Utrecht, The Netherlands*
[2]*Department of Clinical and Health Psychology, Utrecht University, P.O. Box 80.140, 3508 TC Utrecht, The Netherlands*
[3]*Department of Medical Psychology, Tilburg University, Faculty of Social and Behavioral Sciences, Warandelaan 2, P.O. Box 90153, 5000 LE Tilburg, The Netherlands*

Abstract: Burnout results from chronic work stress. Its complaints may be related to HPA-axis disturbances or changes in immune function. In our studies the salivary cortisol awakening response, day-curve, and the suppressed level after dexamethasone intake were not different in a burned-out group compared to a control group. Nor was there a change in cortisol after a treatment period. Higher levels of DHEAS and the monocyte released anti-inflammatory cytokine IL-10 were observed, however T-cell stimulated and dexamethasone inhibited cytokine release were not affected. The increased IL-10 level may be related to an increased sensitivity for infections.

Keywords: burnout; chronic stress; cortisol; cytokines; dexamethasone suppression test; DHEAS; follow-up

Introduction

Burnout is the ultimate outcome of a chronic process in which work stress is supposed to play a decisive role. People with burnout feel extremely fatigued, have become alienated from their work, experience reduced professional competence, and report a whole range of complaints such as depressed mood, increased irritability, inability to relax, disrupted sleep, somatic complaints such as aching muscles, headaches, gastro-intestinal problems, and concentration and memory problems (Maslach et al., 2001). When we assume that burnout is a stress-related syndrome, one may expect to find a disturbance in hypothalamus pituitary adrenal (HPA)-axis functioning. Inadequate glucocorticoid signaling has been suggested for other stress-related syndromes like post-traumatic stress disorder (PTSD), chronic fatigue syndrome (CFS), and major depression disorder (MDD). Reviewing the literature on burnout and related stress-syndromes has led to the hypothesis that the fatigue symptoms in burnout are related to a state of hypocortisolism, and increased feedback sensitivity of the HPA-axis (Heim et al., 2000). On the other hand, the depressive symptoms would suggest a hypercortisolemic state, and a relative non-suppression in response to dexamethasone (DEX) (Raison and Miller, 2003). Assuming a disturbance of the HPA-axis in burnout, we expected a reduction in burnout complaints to be related to a recovery of this disturbance. A longitudinal study was set up to correlate changes in complaints with changes in salivary cortisol parameters.

*Corresponding author. Tel.: +31 (0) 13 466 2175;
Fax: +31 (0) 13 466 2067;
E-mail: P.M.C.Mommersteeg@uvt.nl

DOI: 10.1016/S0079-6123(07)67024-1

Glucocorticoids play a decisive role in immune functioning. Cortisol inhibits pro-inflammatory cytokine release, e.g., TNF-α, IFN-γ, interleukin (IL)-6 and IL-1, and stimulates anti-inflammatory IL-10 and IL-4 release (Elenkov and Chrousos, 2002). Chronic psychosocial stress has been related to impaired immune functioning leading to physical illness. This process may be mediated by glucocorticoids through affecting the balance between pro- and anti-inflammatory cytokines (Kiecolt-Glaser et al., 2002).

Results

The major finding of our study was the absence of a disturbance in salivary cortisol parameters in burnout. A burnout group ($n = 74$) was compared to a healthy control group ($n = 38$). The burnout persons were on sick leave, and had received a clinical diagnosis for work-related neurasthenia according to International Statistical Classification of Diseases and Related Health Problems (ICD-10) criteria. Primary Diagnostic and Statistical Manual of Mental Disorders Edition IV (DSM-IV) disorders such as MDD or anxiety disorder were excluded. The cortisol awakening response (CAR) was measured on 2 days at 0, 15, and 30 min after awakening, and at noon, 18:00 h and 22:30 h to assess the diurnal cortisol course. A low-dose (0.5 mg) DEX was taken to test the feedback sensitivity of the HPA-axis. The suppressed cortisol level after DEX intake was measured at 0, 15, and 30 min after awakening. The cortisol CAR, day-curve and suppressed DEX level were not different between the burnout and control group (Mommersteeg et al., 2006a, b) (Fig. 1). Cortisol was not related to fatigue or depression complaints within the burnout group, thus showing no indication of an opposing hypo- or hyperfunction of the HPA-axis, potentially masking the effect in burnout (Mommersteeg et al., 2006a).

Because there is considerable variation in cortisol levels between and within persons, it is quite well possible that within a group burnout persons the reduction of the burnout complaints will covary with the cortisol parameters after a treatment and a follow-up period. This possibility was studied in the longitudinal part of the previous study (Mommersteeg et al., 2006). Burnout complaints were significantly reduced after a treatment period, without a further reduction at follow-up. Complaints remained substantially higher than norm

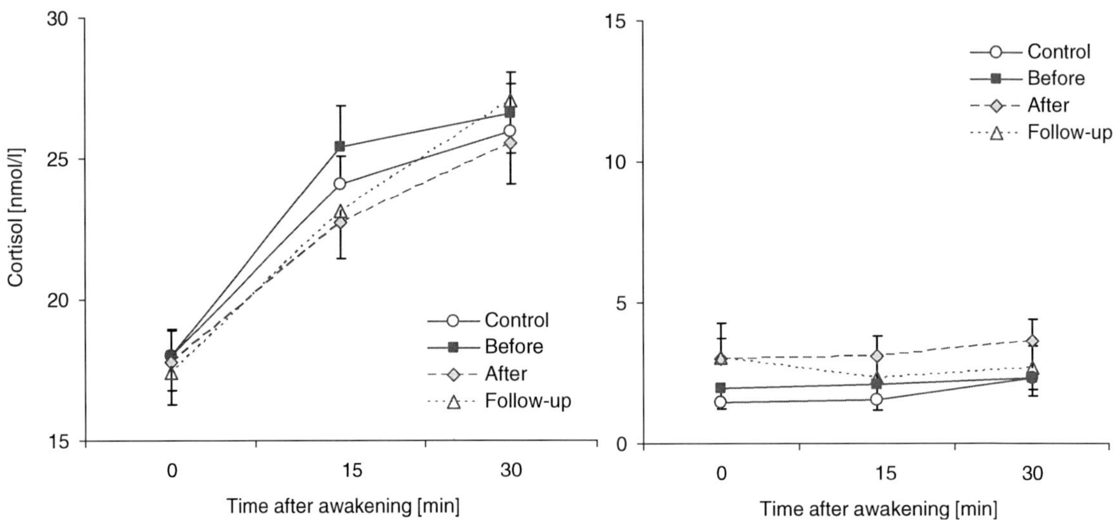

Fig. 1. Cortisol awakening response (CAR, left) and the suppressed CAR after dexamethasone intake (right) in the burnout group before treatment, after treatment and at follow-up, and in the control group. There are no differences between the groups or within the burnout group at consecutive measurements. Means and SEM are shown.

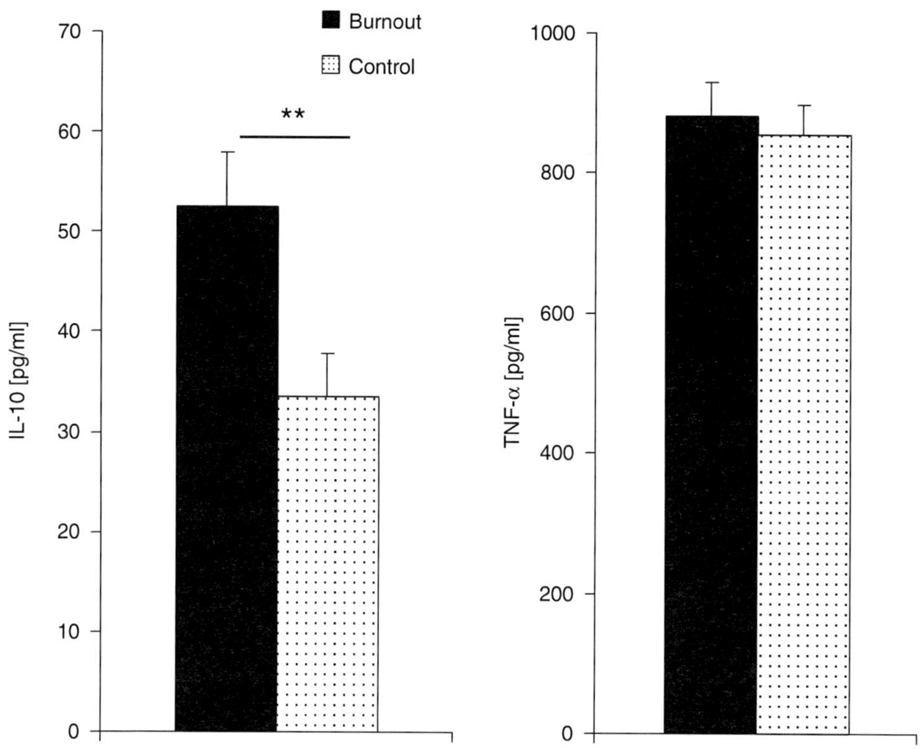

Fig. 2. Anti-inflammatory IL-10 (left) and pro-inflammatory TNF-α release (right) of LPS stimulated monocytes in the burnout and control group. The burnout group had significantly higher levels of stimulated IL-10 ($F_{(1,83)} = 9.01$, $p = 0.004$). Means and SEM are shown.

scores for a healthy population. Cortisol after awakening and after DEX intake (Fig. 1) showed, however, no parallel changes with complaint reduction. Some isolated associations emerged; the CAR (averaged over the three measurements) was significantly correlated with initial exhaustion level. A decrease in depressive symptoms correlated with an *increased* CAR, whereas the decrease in fatigue in time correlated with a *decrease* of the CAR over the three measurements (Mommersteeg et al., 2006b). The latter findings are in contradiction to the supposed hyper- and hypoactive state of the HPA-axis in MDD and CFS, respectively, and moreover explained only a minor part of the variance in complaints within (3%) and between (4%) the burnout individuals.

Immune and endocrine variables were studied in another burnout group ($n = 56$) and compared to 38 controls (Mommersteeg et al., 2006c). Again no deviations in the cortisol CAR, or in the DEX suppression test (DST) were observed. The dehydroepiandrosterone-sulphate (DHEAS) level (but not the cortisol/DHEAS-ratio) was significantly elevated in the burnout group. The burnout group had significantly higher levels of the anti-inflammatory cytokine IL-10 produced by LPS stimulated monocytes (Fig. 2). The IL-10 production of stimulated T-cells, however, was not different from the control group, and neither were there differences in the pro-inflammatory cytokine release of monocyte TNF-α (Fig. 2) or T-cell IFN-γ. The capacity of DEX to modulate pro- and anti-inflammatory cytokine release in vitro did not differ between the burnout and the control group, nor was there a change in number of whole blood counts of T-cells, B-cells, and NK-cells.

Discussion

The results show that there is no discernable disturbance of salivary cortisol in burnout. There is, however, an increased production of IL-10 and salivary DHEAS. These findings in a rather large sample of clinical burnout persons raise doubts about the existence of a relevant neuroendocrine dysregulation in burnout as suggested by some earlier studies. Still a variety of (neuroendocrine) factors may show modest disturbances, altogether leading to a state of 'allostatic load' in burnout patients. Though studies in burnout and CFS that included allostatic load parameters do not point in that direction (Cleare, 2003; Grossi et al., 2003; Schnorpfeil et al., 2003), this type of approach may be a viable option for further research.

Another option is that central mechanisms are dysregulated in burnout. To test this possibility the combined DEX/corticotrophin releasing hormone (CRH) test, or CRH or adrenocorticotropic hormone (ACTH) infusion are useful techniques. One may doubt however whether these invasive techniques are acceptable as a research tool for this (relatively) mild syndrome. Our results point toward an increased stimulated monocyte IL-10 release and increased DHEAS levels in burnout. DHEAS has immunostimulatory effects, and at the same time its non-sulphatized form DHEA has been found to reduce susceptibility to viral, bacterial, and protozoan infections (Chen and Parker, 2004). Thus the relevance of the increased DHEAS level in burnout for immune function remains to be determined. Macrophage IL-10 release inhibits T-cell proliferation and suppresses the release of pro-inflammatory cytokines like the anti-viral IFN-γ. People with burnout report more common cold and flu-like infections (Mohren et al., 2003). Moreover, vital exhaustion is related to an increased pathogen burden, with higher IL-10 serum levels (van der Ven et al., 2003). Therefore an increased IL-10 response in burnout may be related to an increased sensitivity for viral infections. Future studies should reveal the relevance of these findings.

When we started this research project we hypothesized that the HPA-axis should show disturbances in burnout. The results showed the absence of any obvious peripheral deviation in salivary cortisol, nor feedback by DEX in burnout. The correlational effects observed in the longitudinal study are too modest to represent any clinical or diagnostic value. Overall we conclude that in this study no obvious disturbance of the HPA-axis in burnout was demonstrated. The possibility of some disturbance in immune function and the hormone DHEAS in burnout deserves further attention, especially in relation to the sensitivity for infections.

Abbreviations

ACTH	adrenocorticotropic hormone
CAR	cortisol awakening response
CFS	chronic fatigue syndrome
CRH	corticotrophin releasing hormone
DEX	dexamethasone
DHEAS	dehydroepiandrosterone-sulphate
DSM-IV	Diagnostic and Statistical Manual of Mental Disorders Edition IV
DST	dexamethasone suppression test
HPA-axis	hypothalamus pituitary adrenal axis
ICD-10	International Statistical Classification of Diseases and Related Health Problems
IL	interleukin
MDD	major depression disorder
PTSD	post-traumatic stress disorder

Acknowledgments

This research was granted by the Netherlands Organisation for scientific Research (NWO) as part of the Netherlands research program 'fatigue at work' (NWO grant 580-02.108).

References

Chen, C.C. and Parker, C.R.J. (2004) Adrenal androgens and the immune system. Semin. Reprod. Med., 22(4): 369–377.

Cleare, A.J. (2003) The neuroendocrinology of chronic fatigue syndrome. Endocr. Rev., 24(2): 236–252.

Elenkov, I.J. and Chrousos, G.P. (2002) Stress hormones, proinflammatory and antiinflammatory cytokines, and autoimmunity. Ann. N.Y. Acad. Sci., 966(1): 290–303.

Grossi, G., Perski, A., Evengard, B., Blomkvist, V. and Orth-Gomer, K. (2003) Physiological correlates of burnout among women. J. Psychosom. Res., 55(4): 309–316.

Heim, C., Ehlert, U. and Hellhammer, D.H. (2000) The potential role of hypocortisolism in the pathophysiology of stress-related bodily disorders. Psychoneuroendocrinology, 25(1): 1–35.

Kiecolt-Glaser, J.K., McGuire, L., Robles, T.F. and Glaser, R. (2002) Psychoneuroimmunology and psychosomatic medicine: back to the future. Psychosom. Med., 64(1): 15–28.

Maslach, C., Schaufeli, W.B. and Leiter, M.P. (2001) Job burnout. Annu. Rev. Psychol., 52(1): 397–422.

Mohren, D.C.L., Swaen, G.M.H., Kant, I., van Amelsvoort, L.G.P.M., Borm, P.J.A. and Galama, J.M.D. (2003) Common infections and the role of burnout in a Dutch working population. J. Psychosom. Res., 55(3): 201–208.

Mommersteeg, P.M.C., Heijnen, C.J., Verbraak, M.J.P.M. and van Doornen, L.J.P. (2006a) Clinical burnout is not reflected in the cortisol awakening response, the day-curve or the response to a low-dose dexamethasone suppression test. Psychoneuroendocrinology, 31(2): 216–225.

Mommersteeg, P.M.C., Heijnen, C.J., Verbraak, M.J.P.M. and van Doornen, L.J.P. (2006b) A longitudinal study on cortisol and complaint reduction in burnout. Psychoneuroendocrinology, 31(7): 793–804.

Mommersteeg, P.M.C., Heijnen, C.J., Kavelaars, A. and Doornen, L.J.P.V. (2006c) Immune and endocrine function in burnout syndrome. Psychosom. Med., 68(6): 879–886.

Raison, C.L. and Miller, A.H. (2003) When not enough is too much: the role of insufficient glucocorticoid signaling in the pathophysiology of stress-related disorders. Am. J. Psychiatry, 160(9): 1554–1565.

Schnorpfeil, P., Noll, A., Schulze, R., Ehlert, U., Frey, K. and Fischer, J.E. (2003) Allostatic load and work conditions. Soc. Sci. Med., 57(4): 647–656.

van der Ven, A., van Diest, R., Hamulyak, K., Maes, M., Bruggeman, C. and Appels, A. (2003) Herpes viruses, cytokines, and altered hemostasis in vital exhaustion. Psychosom. Med., 65(2): 194–200.

E.R. de Kloet, M.S. Oitzl & E. Vermetten (Eds.)
Progress in Brain Research, Vol. 167
ISSN 0079-6123

CHAPTER 25

Elevated plasma corticotrophin-releasing hormone levels in veterans with posttraumatic stress disorder

C.S. de Kloet[3,*], E. Vermetten[1], E. Geuze[1], E.G.W.M. Lentjes[5], C.J. Heijnen[4], G.K. Stalla[6] and H.G.M. Westenberg[2]

[1]*Military Mental Health Research Centre, Central Military Hospital, Utrecht, The Netherlands*
[2]*Rudolf Magnus Institute of Neurosciences, Department of Psychiatry, University Medical Center, Utrecht, The Netherlands*
[3]*Altrecht Institute for Mental Health Care, Zeist, The Netherlands*
[4]*Laboratory of Psychoneuroimmunology, University Medical Center Utrecht, Utrecht, The Netherlands*
[5]*Laboratory of Endocrinology, University Medical Center Utrecht, Utrecht, The Netherlands*
[6]*Max Planck Institute of Psychiatry, Munich, Germany*

Abstract: Posttraumatic stress disorder (PTSD) is associated with alterations in corticotrophin-releasing hormone (CRH) secretion. Plasma CRH levels, which are easily acquired, might serve as a predictor of hypothalamic CRH levels. Assessment of plasma CRH, adrenocorticotrophin hormone (ACTH), and cortisol levels in 31 veterans with PTSD, 30 traumatized veterans without PTSD matched on age, year, and region of deployment (traumacontrols), and 28 age-matched healthy controls (HCs) was carried out. Plasma CRH levels were higher in PTSD patients compared to both HCs ($p = 0.005$) and traumacontrols ($p = 0.007$). This led to our conclusion, that elevated plasma CRH levels are specifically related to PTSD and not to exposure to traumatic stress during deployment.

Keywords: posttraumatic stress disorder; corticotrophin-releasing factor; adrenocorticotropic hormone; cortisol; veterans; stress

Introduction

During the last decades, clinical studies have provided evidence for dysregulation of the HPA-axis in posttraumatic stress disorder (PTSD) (Yehuda, 2005). Corticotrophin-releasing hormone (CRH) neurons originating in the paraventricular nucleus of the hypothalamus initiate the neuroendocrine responses to stress by activating the hypothalamic-pituitary-adrenal axis (HPA-axis) (Nemeroff, 1992). CRH expressing neurons are also present in extra-hypothalamic regions of the brain, where they are involved in behavioral and autonomic responses to stress (McNally and Akil, 2002). In addition, CRH is produced peripherally.

Thus far, CRH levels in PTSD have only been assessed in cerebrospinal fluid (CSF). Three out of four studies reported elevated CSF CRH levels in patients with PTSD (for review, see de Kloet et al., 2006) and one study reported elevated CSF CRH levels in PTSD with psychotic symptoms only (Sautter et al., 2003).

*Corresponding author. Tel.: +31655895963/+31612273361; Fax: +31306965371; E-mail: c.de.kloet@altrecht.nl

DOI: 10.1016/S0079-6123(07)67025-3

Previous clinical and preclinical studies showed no significant correlations between CSF CRH levels and plasma adrenocorticotrophin hormone (ACTH) and cortisol levels in clinical and preclinical studies (Kalin et al., 1989; Geracioti et al., 1997; Baker et al., 2005). Therefore, CSF CRH levels are supposed to reflect the overall central nervous system release and not specifically hypothalamic secretion. A disadvantage of measurements in CSF is that lumbal puncture is a stressfull and invasive procedure. It is therefore of interest to evaluate whether plasma CRH levels, which are easily acquired can provide additional information on central CRH secretion and especially hypothalamic CRH release.

The aim of the present study was to assess plasma CRH levels in a homogeneous sample of patients with PTSD. Based on the behavioral effects of CRH, we hypothesize enhanced levels of plasma CRH in patients with PTSD compared to healthy controls (HCs). To control for environmental factors such as military training, deployment, and trauma on outcome measures we also compared veterans with PTSD to matched traumatized veterans without PTSD. To get an indication whether plasma CRH levels are related to PTSD symptoms, the correlations between plasma CRH and PTSD symptom scores as assessed by the Clinician Administered PTSD Scale (CAPS) were measured.

Methods

PTSD patients were recruited from the Department of Military Psychiatry at the Central Military Hospital, The Netherlands. Trauma controls (TCs) were selected from a group of registered male veterans. They were matched with the PTSD group for age, year, and region of deployment. Healthy male controls were matched for age. Only HCs with no report of traumatic experiences and with a low score on the Hopkins Symptom Checklist (SCL-90) and absence of PTSD symptoms, measured with the Dutch Self Inventory for PTSD (ZIL), were included in this study. All veterans were screened for psychiatric illness using the structured clinical interview for Diagnostic and Statistical Manual for Mental Disorders IV (DSM IV) axis I disorders (SCID-I). The diagnosis of PTSD was confirmed by the CAPS. Only patients who had a CAPS score above 50 were included. TCs were included if they met A1 criteria for PTSD, but had a CAPS score below 25 and did not meet DSM IV criteria for PTSD or any other current axis I disorder. Subjects with a serious somatic illness or a psychiatric illness other than mood and anxiety disorders were excluded.

This study was approved by the Institutional Review Board of the University Medical Center Utrecht, The Netherlands.

Biochemical assessments

Blood was drawn by venepuncture between 8.30 and 9.30 am, and immediately put on ice and centrifuged at 4°C. Plasma was stored at −80°C. Plasma concentrations of CRH were measured using a radioimmunoassay after an extraction procedure previously described (Stalla et al., 1986). The lower limit of detection was 10 pg/ml and the intra- and inter-assay coefficients of variation (CV) were below 9%. Plasma ACTH concentrations were analyzed using a sandwich assay with luminescent detection (Nichols Advantage) (Vogeser et al., 2000); reference values: 10–50 ng/l, intra-assay CV below 5%, inter-assay CV 7–8.5%. Plasma cortisol concentrations were analyzed using a competitive lumino-immuno-assay (Nichols Advantage); reference values: 0.10–0.4 μmol/l, intra-assay CV below 3.5–9%, inter-assay CV 6.5–12%. All samples were analyzed in one batch.

Statistical analysis

Demographics, psychometric scores, and endocrine parameters were compared between the three included groups. Parametric tests (analysis of variance, ANOVA) and post-hoc tests (Scheffe) were used when appropriate. Where necessary a logarithmic transformation was used to normalize the distribution. In addition endocrine parameters were compared between PTSD patients with and without co-morbid major depressive disorder (MDD). Correlation analyses (Pearson correlation) were performed between CAPS total score and plasma CRH levels in all subjects and within

Table 1. Demographic characteristics and test variables of PTSD patients, trauma controls (TCs), and healthy controls (HCs)

Demographic characteristics	PTSD ($n = 31$)			Trauma controls ($n = 30$)			Healthy controls ($n = 28$)			p
	Mean	SD	Range	Mean	SD	Range	Mean	SD	Range	
Age (years)	33.4	5.6	25–44	33.5	5	25–40	32.7	5.4	25–45	0.75
Year deployment	1992	5.8	1980–2002	1992	5.7	1980–2002				0.64
CAPS	75.7	14.8	54–102	7.6	7	0–25				<0.001
Hamilton D	15.8	5.6	3–29	0.9	1.4	0–5				<0.001
Hamilton A	19	7.2	4–29	1.5	1.5	0–5				<0.001
Co-morbid disorders (lifetime)	MDD ($n = 15$)			MDD ($n = 3$)						
	Bipolar disorder ($n = 2$)									
	Alcohol dependence ($n = 3$)									
	Alcohol abuse ($n = 3$)									
	Substance abuse ($n = 2$)									
	Substance dependence ($n = 1$)									
	Panic disorder ($n = 3$)									
	Somatoform disorder ($n = 5$)									
Country deployment	Bosnia ($n = 18$)	Bosnia ($n = 18$)								
	Lebanon ($n = 6$)	Lebanon ($n = 6$)								
	Cambodia ($n = 4$)	Cambodia ($n = 5$)								
	Afghanistan ($n = 2$)	Afghanistan ($n = 1$)								
	Kosovo ($n = 1$)									

Notes: Outcome variables were compared between three groups using parametric (ANOVA) or non-parametric tests (Kruskal-Wallis-test) depending on the equality of variance. Variables are displayed as mean, standard deviation (SD), and range.

the PTSD group and between plasma CRH and ACTH levels in all subjects.

All statistical analyses were performed with statistical package for social sciences (SPSS) 12.0 for Windows (SPSS, Chicago, IL). The statistical threshold for significance for all measures was set at $p<0.05$.

Results

The demographic characteristics of PTSD patients ($n = 31$), TCs ($n = 30$), and HCs ($n = 28$) are displayed in Table 1. The reported traumatic events (A1 criteria) were comparable in PTSD patients and TCs. Thirteen PTSD patients were diagnosed with a current depressive episode. Twenty-five patients were naïve for psychotropic medication; all other patients were free from medication for at least 4 weeks.

We observed a significant group difference in plasma CRH levels (ANOVA: $F_{2,86} = 7.11$; $p = 0.001$). Post-hoc tests showed elevated plasma CRH levels in PTSD patients (CRH: 64.7±7.9 pg/ml) compared to both HCs (CRH: 58.4±7.2 pg/m; $p = 0.005$) and TCs (CRH: 58.7±6.8 pg/ml; $p = 0.007$). ACTH levels were log-transformed to normalize the distribution. No significant group differences were observed for plasma ACTH (PTSD: 29.7±17.8 ng/l, TC: 24.9±10.2 ng/l, HC: 25.2±8.8 ng/l; ANOVA: $F_{2,86} = 1.33$; $p = 0.269$) and plasma cortisol levels (PTSD: 0.39±0.12 μmol/l, TC: 0.41±0.07 μmol/l, HC: 0.37±0.12 μmol/l; ANOVA: $F_{2,86} = 1.48$; $p = 0.234$). Plasma CRH levels in PTSD patients with a current depressive episode (CRH: 65.5±8.1 pg/ml) did not differ from those without a current depressive episode (CRH: 64.1±7.8 pg/ml) ($t = 0.51$; df = 29; $p = 0.61$).

A weak, but statistically significant, correlation was found between plasma CRH and plasma ACTH (Pearson correlation: $r = 0.25$; $p = 0.017$). Within the PTSD group no significant correlation was observed between CAPS total score and plasma CRH, ACTH, and cortisol levels.

Discussion

We observed significantly higher plasma levels of CRH in veterans with PTSD compared to traumatized veterans without PTSD and non-military HCs. This confirmed the hypothesis of hypersecretion of CRH in patients with PTSD and supported the notion that enhanced plasma CRH levels are specifically related to PTSD and not to exposure to traumatic events during deployment or other military-related factors.

A weak, but significant, correlation between plasma CRH and ACTH levels was observed. Our results are in line with the reported correlations between diurnal plasma CRH and ACTH levels (Sasaki et al., 1987; Watabe et al., 1987), indicating that plasma CRH is, at least partly, of hypothalamic origin. The study of Galard et al. (2002) who reported a reduction of plasma CRH levels in depressed patients after dexamethasone administration and the observation of lower plasma CRH levels in patients with Cushings disease and after steroid therapy (Sasaki et al., 1987) also supported this notion. However, the observation of lower plasma CRH levels in endocrine disorders with a normal adrenocortical function and a rise of plasma CRH after glucose administration, in absence of an increase in ACTH and cortisol (Sasaki et al., 1987), as well as the insensitivity of plasma CRH to dexamethasone in healthy volunteers (Galard et al., 2002), also suggests an extra-hypothalamic origin of plasma CRH.

When we consider the fact that central administration of CRH resulted in behaviors consistent with fear and anxiety (Kalin et al., 1989; Servatius et al., 2005), the observation of both elevated plasma CRH levels and CSF CRH levels in PTSD is of great interest. The fact that Brunner et al. (2001) reported no correlation between CSF and plasma CRH levels suggest that plasma CRH levels might provide additional information on altered central CRH secretion in PTSD.

Some limiting factors need to be taken into account. First, CRH, ACTH, and cortisol measurement was confined to a single time point. Second, 13 out of 31 patients were diagnosed with a current depressive episode. However, plasma CRH levels in PTSD patients with and without co-morbid MDD were not different, suggesting that co-morbidity did not influence our results. Finally, blood was collected by venepuncture that is known to be a significant stressor.

In conclusion, this study suggests that enhanced plasma CRH levels are specifically associated with PTSD and not with trauma exposure. The observed correlation between plasma CRH and ACTH suggests that plasma CRH partly reflects PTSD-related alterations in the HPA-axis.

Studies in a prospective design are necessary to answer the question, whether the observed higher plasma CRH level in PTSD is related to psychopathology or a vulnerability factor.

Abbreviations

ACTH	adrenocorticotrophin hormone
ANOVA	analysis of variance
CAPS	Clinician Administered PTSD Scale
CRH	corticotrophin-releasing hormone
CSF	cerebrospinal fluid
CV	coefficients of variation
DSM-IV	Diagnostic and Statistical Manual for Mental Disorders IV
HCs	healthy controls
HPA-axis	hypothalamic-pituitary-adrenal axis
MDD	major depressive disorder
PTSD	posttraumatic stress disorder
SCID-I	structured clinical interview for DSM IV axis I disorders
SCL-90	Hopkins symptom checklist
SPSS	statistical package for social sciences
TCs	trauma controls
ZIL	Dutch Self Inventory for PTSD

References

Baker, D.G., Ekhator, N.N., Kasckow, J.W., Dashevsky, B., Horn, P.S., Bednarik, L., et al. (2005) Higher levels of basal serial CSF cortisol in combat veterans with posttraumatic stress disorder. Am. J. Psychiatry, 162: 992–994.

Brunner, J., Stalla, G.K., Stalla, J., Uhr, M., Grabner, A., Wetter, T.C., et al. (2001) Decreased corticotropin-releasing hormone (CRH) concentrations in the cerebrospinal fluid of eucortisolemic suicide attempters. J. Psychiatr. Res., 35: 1–9.

Galard, R., Catalan, R., Castellanos, J.M. and Gallart, J.M. (2002) Plasma corticotropin-releasing factor in depressed patients before and after the dexamethasone suppression test. Biol. Psychiatry, 51: 463–468.

Geracioti Jr., T.D., Loosen, P.T. and Orth, D.N. (1997) Low cerebrospinal fluid corticotropin-releasing hormone concentrations in eucortisolemic depression. Biol. Psychiatry, 42: 165–174.

Kalin, N.H., Shelton, S.E. and Barksdale, C.M. (1989) Behavioral and physiologic effects of CRH administered to infant primates undergoing maternal separation. Neuropsychopharmacology, 2: 97–104.

de Kloet, C.S., Vermetten, E., Geuze, E., Kavelaars, A., Heijnen, C.J. and Westenberg, H.G. (2006) Assessment of HPA-axis function in posttraumatic stress disorder: pharmacological and non-pharmacological challenge tests, a review. J. Psychiatr. Res., 40: 550–567.

McNally, G.P. and Akil, H. (2002) Role of corticotropin-releasing hormone in the amygdala and bed nucleus of the stria terminalis in the behavioral, pain modulatory, and endocrine consequences of opiate withdrawal. Neuroscience, 112: 605–617.

Nemeroff, C.B. (1992) New vistas in neuropeptide research in neuropsychiatry: focus on corticotropin-releasing factor. Neuropsychopharmacology, 6: 69–75.

Sasaki, A., Sato, S., Murakami, O., Go, M., Inoue, M., Shimizu, Y., et al. (1987) Immunoreactive corticotropin-releasing hormone present in human plasma may be derived from both hypothalamic and extrahypothalamic sources. J. Clin. Endocrinol. Metab., 65: 176–182.

Sautter, F.J., Bissette, G., Wiley, J., Manguno-Mire, G., Schoenbachler, B., Myers, L., et al. (2003) Corticotropin-releasing factor in posttraumatic stress disorder (PTSD) with secondary psychotic symptoms, nonpsychotic PTSD, and healthy control subjects. Biol. Psychiatry, 54: 1382–1388.

Servatius, R.J., Beck, K.D., Moldow, R.L., Salameh, G., Tumminello, T.P. and Short, K.R. (2005) A stress-induced anxious state in male rats: corticotropin-releasing hormone induces persistent changes in associative learning and startle reactivity. Biol. Psychiatry, 57: 865–872.

Stalla, G.K., Stalla, J., Schopohl, J., von Werder, K. and Muller, O.A. (1986) Corticotropin-releasing factor in humans: I. CRF stimulation in normals and CRF radioimmunoassay. Horm. Res., 24: 229–245.

Vogeser, M., Engelhardt, D. and Jacob, K. (2000) Comparison of two automated adrenocorticotropic hormone assays. Clin. Chem., 46: 1998–2000.

Watabe, T., Tanaka, K., Kumagae, M., Itoh, S., Hasegawa, M., Horiuchi, T., et al. (1987) Diurnal rhythm of plasma immunoreactive corticotropin-releasing factor in normal subjects. Life Sci., 40: 1651–1655.

Yehuda, R. (2005) Neuroendocrine aspects of PTSD. Handbook Exp. Pharmacol.: 371–403.

E.R. de Kloet, M.S. Oitzl & E. Vermetten (Eds.)
Progress in Brain Research, Vol. 167
ISSN 0079-6123

CHAPTER 26

Precuneal activity during encoding in veterans with posttraumatic stress disorder

Elbert Geuze[1,2,*], Eric Vermetten[1,2], Carien S. de Kloet[1] and Herman G.M. Westenberg[2]

[1]*Department of Military Psychiatry, Central Military Hospital, Ministry of Defense, Utrecht, The Netherlands*
[2]*Department of Psychiatry, Rudolf Magnus Institute of Neuroscience, Utrecht University Medical Center, Utrecht, The Netherlands*

Abstract: Impaired attention and memory are symptoms frequently associated with posttraumatic stress disorder (PTSD). Previous studies have identified fronto-temporal alterations during encoding in patients with PTSD. We examine the role of the precuneus (located in the posteromedial parietal lobe) that is known to play a role in memory, but has largely been neglected in PTSD research. Male veterans with and without PTSD ($n = 12$ per group) were subjected to fMRI during encoding of 12 neutral, non-trauma related word pairs. The precuneus was less activated in veterans with PTSD, which correlated significantly with the severity of PTSD. Like fronto-temporal regions the precuneus is differentially activated during memory formation in veterans with PTSD.

Keywords: PTSD; memory; fMRI; precuneus; encoding

Introduction

Exposure to traumatic events can lead to the development of psychopathology, such as posttraumatic stress disorder (PTSD). Patients with PTSD do not only experience recurrent intrusive thoughts and (sometimes vivid) memories of the traumatic event, but also symptoms of hyperarousal, avoidance and numbing, and difficulties of attention and memory (Thygesen et al., 1970).

Neuroimaging has become an important technique for understanding brain function and is widely used to examine both functional and morphological changes in neuropsychiatric disorders, including PTSD. Patients with PTSD frequently report memory difficulties and empirical research provides support for a memory deficit in PTSD. In a previous study, we have investigated associative memory processing in PTSD with functional magnetic resonance imaging (fMRI) using the encoding of 12 word-pair associates as a neurocognitive task in Dutch veterans with PTSD and without PTSD (Geuze et al., 2007). Although the precuneus usually receives little attention because of its hidden location in the posteromedial parietal lobe, it is known to play a role in memory (Taylor et al., 2000). In addition, the precuneus is part of a network of brain areas involved in self-consciousness, and thus may be of interest in psychiatric disorders (Cavanna and Trimble, 2006). In this study we examine the activity of the precuneus during encoding of neutral (i.e., not trauma related) material in patients with PTSD compared to controls.

*Corresponding author. Tel.: +31 30 250 2585;
Fax: +31 30 250 2586; E-mail: s.g.geuze@umcutrecht.nl

DOI: 10.1016/S0079-6123(07)67026-5

Methods

Subjects

Twelve veterans with PTSD, and 12 veterans without PTSD, were recruited. PTSD patients were recruited from the Department of Military Psychiatry at the Central Military Hospital in Utrecht. Control subjects were recruited via direct mail to veterans who were registered at the Veterans Institute in Doorn, the Netherlands. All participants were male Dutch veterans who had served in United Nations (UN) peacekeeping missions in Lebanon, Cambodia, and Bosnia. None of the included veterans were physically injured at the time of deployment. Control veterans were matched to the patient group with respect to age, handedness, year of deployment, and country of deployment.

PTSD was diagnosed using Diagnostic and Statistical Manual of Mental Disorders (version IV) (DSM-IV) criteria, and confirmed using the Clinician-Administered PTSD Scale (CAPS; Blake et al., 1995) and by consensus with three clinicians (EG, EV, and CdK). Only patients with CAPS scores >50 were included in the study. Comorbid disorders were examined utilizing the Structured Clinical Interview for DSM-IV (SCID; First et al., 1997). Control subjects were also assessed with both the SCID and the CAPS. Control subjects met the A1 criterion for PTSD (i.e., they had all experienced a traumatic event). All subjects received a physical examination by a physician. Subjects were excluded if they had any clinical significant abnormality of a clinical laboratory test, a history of psychiatric illness (controls only) or neurological dysfunction (all subjects), a history of alcohol and/or drug abuse (DSM-IV criteria) within 6 months prior to the study, or claustrophobia. None of the participants was taking psychotropic drugs at the time of the study. Written informed consent was obtained from all subjects who participated in the study after a complete written and verbal description of the study. The study was performed between August 2005 and February 2006. This study was approved by the Medical Ethical Review Boards of the University Medical Centre of Utrecht, the Netherlands, and the Central Institute of Mental Health, Mannheim, Germany.

Experimental procedure

fMRI was carried out on a 1.5 T Scanner (Magnetom Vision, Siemens, Erlangen, Germany) at the Central Institute of Mental Health in Mannheim, Germany. Functional data was assessed using a functional echoplanar series (EPI) using 25 contiguous transverse slices (thickness, 5 mm; 1 mm gap; field of view, $220 \times 220\,mm^2$; matrix, 64×64 voxels; slice acquisition time, 104 ms; volume acquisition time, 2600 ms; repetition time, 3089 ms).

The fMRI protocol consisted of the encoding of 12 word-pairs (a slightly adapted form of the paradigm used by Ino et al. (2004)) (see Fig. 1). The task consisted of 12 21 s blocks (6 stimuli per block, 3.5 s ISI) in which encoding blocks (presentations of word-pairs, e.g., rose-flower) were alternated with control blocks (presentations of successive numbers of two figures, e.g., 31–32), starting with the latter. The subjects were required to memorize word-pairs in the encoding block and to silently repeat the two figures in the control block. The number of word-pairs (12 pairs) was presented in the same order three times.

The anatomical three-dimensional (3D) magnetization prepared rapid acquisition gradient echo (MPRAGE) scan with a voxel size of $1 \times 1 \times 1\,mm^3$ and a field of view of $256 \times 256\,mm^2$ was acquired after encoding.

Data analysis

All the image data preparation and preprocessing steps as well as statistical analyses and the map volumetric projection were performed in Brain Voyager QX 1.6 (Brain Innovation, Maastricht, the Netherlands). 3D data preprocessing included slicescan time correction, linear trend removal, temporal high-pass filtering, and 3D motion correction. All functional imaging data were smoothed with a 4 mm full-width at half maximum (FWHM) Gaussian kernel. Structural 3D and functional 4D data sets were transformed into the standard space corresponding to the atlas of Talairach and Tournoux (1988).

The stimulation protocol was convoluted with a hemodynamic response function (Boynton et al., 1996) to account for the expected delay and generic

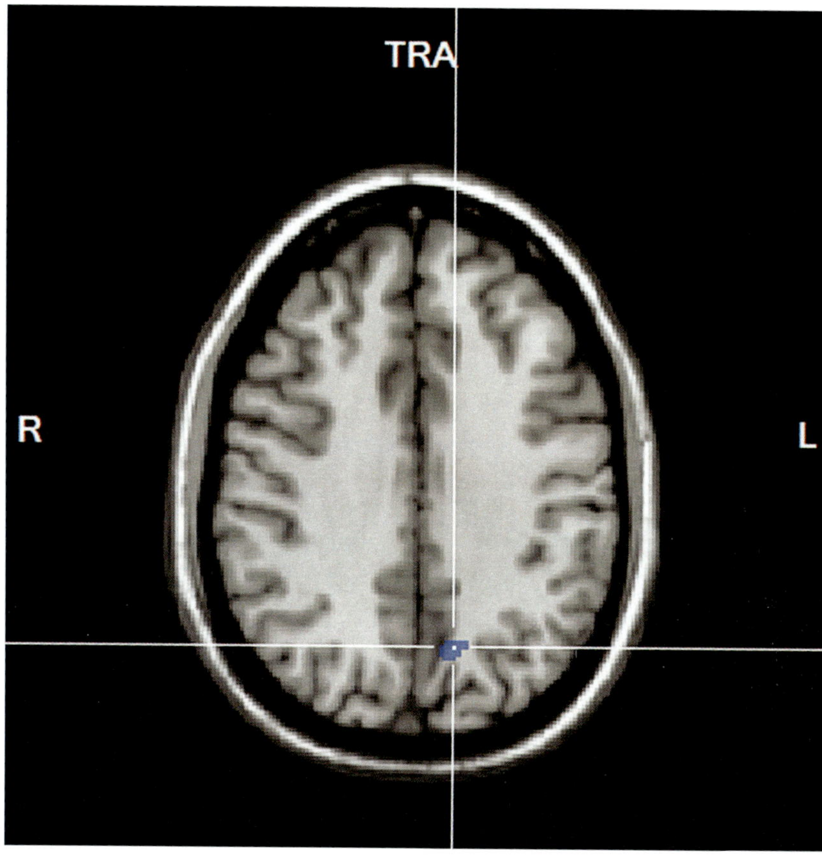

Fig. 1. This picture is a difference-map of activation displayed on a standard brain, showing that veterans with PTSD displayed reduced activation in the precuneus compared to veterans without PTSD.

shape of the blood oxygen level dependent (BOLD) signal. In order to correct for multiple comparisons, the false discovery rate (FDR) controlling procedure was applied on the resulting p values for all voxels ($q = 0.05$) (Benjamini and Hochberg, 1995; Genovese et al., 2002). Voxel level and region of interest (ROI) level inter-group linear contrasts were computed using two-tailed t-tests. 3D statistical maps were overlaid on the Talairach-transformed Montreal Neurological Institute T1-weighted brain template (http://www.bic.mni.mcgill.ca).

Results

Psychometric data

PTSD patients and control veterans were optimally matched with respect to age [34.8 (5.8) vs. 34.7 (3.7), $p > 0.05$]. According to the SCID, the PTSD group met lifetime (past) DSM-IV diagnostic criteria for major depressive disorder ($n = 2$), alcohol abuse ($n = 2$), alcohol dependence ($n = 1$), substance abuse ($n = 1$), substance dependence ($n = 1$), and panic disorder without agoraphobia ($n = 1$). None of the patients with PTSD had any current comorbid disorder. Among the control subjects, the SCID did not reveal any current or lifetime psychiatric disorders.

Main effects encoding

In all subjects solid activations were seen in the bilateral dorsolateral prefrontal cortex (DLPFC), medial prefrontal cortex (mPFC), anterior cingulate cortex (ACC), parietal lobe, lingual gyri, left temporal lobe, left parahippocampal gyrus, left hippocampus, and the bilateral globus pallidus. In the patient group, additional significant activation was seen in the bilateral orbitofrontal cortex (OFC).

fMRI group analysis encoding

During the encoding condition, veterans with PTSD showed altered activity in fronto-temporal areas. In addition, veterans with PTSD showed decreased activation in the left precuneus (t_{patients}, -0.04 ± 1.46, t_{controls}, 1.78 ± 1.45; see Fig. 1). In the PTSD group, CAPS scores were significantly correlated with activity in the left precuneus BA7 (Pearson's $r = -0.791$; $p < 0.005$). In addition, interregional correlational analyses showed that that in patients with PTSD activity in the precuneus was highly correlated with activity in the right ACC, the right superior temporal gyrus, right parahippocampal gyrus, and the right middle temporal gyrus.

Discussion

During encoding of the word pairs, both patients and controls revealed solid activations in fronto-temporal regions, consistent with previous neuroimaging studies (Taylor et al., 2000; Ino et al., 2004; Law et al., 2005). In a previous publication, we have focused on altered activity during encoding in fronto-temporal brain areas in veterans with PTSD compared to veterans without PTSD (Geuze et al., 2007). In this paper we specifically wanted to focus on the altered activity in the precuneus, which veterans with PTSD also displayed during encoding.

Precuneus activity is usually associated with source memory processing, or spatial location encoding (Lundstrom et al., 2005; Frings et al., 2006). In addition, the left precuneus is also proposed to be involved in mental imagery and buffering of working memory (Taylor et al., 2000). Ineffectual activation of the precuneus in PTSD during encoding could possibly contribute to the retrieval deficit in PTSD.

Within the veterans with PTSD, activity in the precuneus showed a significant negative correlation with severity of PTSD symptoms (as indicated by the CAPS score). The precuneus and surrounding posteromedial areas are amongst the brain structures displaying the highest resting metabolic rates and are characterized by transient decreases in the tonic activity during engagement in non-self-referential goal-directed actions (default mode of brain function). Therefore, it has recently been proposed that the precuneus is involved in the interwoven network of the neural correlates of self-consciousness, engaged in self-related mental representations during rest (Cavanna and Trimble, 2006). Perhaps patients are also less self-conscious when engaged in a mental activity. In a positron emission tomography (PET) study which investigated neural correlates of hypnosis, decreased activity was found in the precuneus, the right inferior parietal lobe, and in the posterior cingulate (Rainville et al., 1999). Perhaps dissociative symptoms of patients with PTSD during fMRI scanning may also play a role in the decreased activation of the precuneus. In a previous fMRI study, patients with a dissociative response to a traumatic script also revealed altered activity in the precuneus (Lanius et al., 2002).

Interregional correlational analyses of precuneus activity in veterans with PTSD with other brain areas which displayed altered activity, showed that activity in the precuneus was highly correlated with activity in the right ACC, the right superior temporal gyrus, right parahippocampal gyrus, and the right middle temporal gyrus. This provides support for the idea that the precuneus is part of a larger network displaying altered activity in veterans with PTSD.

Our finding of decreased activation of the precuneus in veterans with PTSD during encoding warrants further investigation of this area in PTSD. It would also be of great interest if future studies would investigate activity in posteromedial areas during dissociation or during baseline metabolism.

Abbreviations

ACC	anterior cingulate cortex
BOLD	blood oxygen level dependent

CAPS	Clinician-Administered PTSD scale
3D	three-dimensional
DLPFC	dorsolateral prefrontal cortex
DSM-IV	Diagnostic and Statistical Manual of Mental Disorders (version IV)
FDR	false discovery rate
fMRI	functional magnetic resonance imaging
FWHM	full-width at half maximum
mPFC	medial prefrontal cortex
MPRAGE	magnetization prepared rapid acquisition gradient echo
OFC	orbitofrontal cortex
PET	positron emission tomography
PTSD	posttraumatic stress disorder
ROI	region of interest
SCID	Structured Clinical Interview for DSM-IV
UN	United Nations

Acknowledgments

This work was financially supported by the Dutch Ministry of Defence. The authors would also like to thank the Division of Neuroimaging of the Central Institute of Mental Health in Mannheim for acquisition of fMRI scans, Christian Schmahl, MD for facilitating fMRI acquisition in Mannheim, Arthur Rademaker, MSc for clinical assessments, the Dutch Veterans Institute in Doorn for help with acquisition of control veterans, and Anja Jochims, MSc for help with data acquisition.

References

Benjamini, Y. and Hochberg, Y. (1995) Controlling the false discovery rate: a practical and powerful approach to multiple testing. J. R. Stat. Soc. B., 57: 289–300.

Blake, D.D., Weathers, F.W., Nagy, L.M., Kaloupek, D.G., Gusman, F.D., Charney, D.S. and Keane, T.M. (1995) The development of a Clinician-Administered PTSD Scale. J. Trauma. Stress, 8: 75–90.

Boynton, G.M., Engel, S.A., Glover, G.H. and Heeger, D.J. (1996) Linear systems analysis of functional magnetic resonance imaging in human V1. J. Neurosci., 16: 4207–4221.

Cavanna, A.E. and Trimble, M.R. (2006) The precuneus: a review of its functional anatomy and behavioural correlates. Brain, 129: 564–583.

First, M.B., Spitzer, R.L., Gibbon, M. and Williams, J.B.W. (1997) Structured Clinical Interview for DSM-IV Axis I Disorders. SCID-I/P Manual. Biometrics Research, New York.

Frings, L., Wagner, K., Quiske, A., Schwarzwald, R., Spreer, J., Halsband, U. and Schulze-Bonhage, A. (2006) Precuneus is involved in allocentric spatial location encoding and recognition. Exp. Brain Res., 173: 661–672.

Genovese, C.R., Lazar, N.A. and Nichols, T. (2002) Thresholding of statistical maps in functional neuroimaging using the false discovery rate. Neuroimage, 15: 870–880.

Geuze, E., Vermetten, E., Ruff, M., de Kloet, C.S. and Westenberg, H. (2007) Neural Correlates of Associative Memory Processing in Veterans with PTSD. J. Psychiatr. Res., Aug 13.

Ino, T., Doi, T., Kimura, T., Ito, J. and Fukuyama, H. (2004) Neural substrates of the performance of an auditory verbal memory: between-subjects analysis by fMRI. Brain Res. Bull., 64: 115–126.

Lanius, R.A., Williamson, P.C., Boksman, K., Densmore, M., Gupta, M., Neufeld, R.W., Gati, J.S. and Menon, R.S. (2002) Brain activation during script-driven imagery induced dissociative responses in PTSD: a functional magnetic resonance imaging investigation. Biol. Psychiatry, 52: 305–311.

Law, J.R., Flanery, M.A., Wirth, S., Yanike, M., Smith, A.C., Frank, L.M., Suzuki, W.A., Brown, E.N. and Stark, C.E. (2005) Functional magnetic resonance imaging activity during the gradual acquisition and expression of paired-associate memory. J. Neurosci., 25: 5720–5729.

Lundstrom, B.N., Ingvar, M. and Petersson, K.M. (2005) The role of precuneus and left inferior frontal cortex during source memory episodic retrieval. Neuroimage, 27: 824–834.

Rainville, P., Hofbauer, R.K., Paus, T., Duncan, G.H., Bushnell, M.C. and Price, D.D. (1999) Cerebral mechanisms of hypnotic induction and suggestion. J. Cogn. Neurosci., 11: 110–125.

Talairach, J. and Tournoux, P. (1988) Coplanar Stereotaxic Atlas of the Human Brain. 3-Dimensional Proportional System: An Approach to Cerebral Imaging. Thieme Medical, New York.

Taylor, J.G., Horwitz, B., Shah, N.J., Fellenz, W.A., Mueller-Gaertner, H.W. and Krause, J.B. (2000) Decomposing memory: functional assignments and brain traffic in paired word associate learning. Neural Netw., 13: 923–940.

Thygesen, P., Hermann, K. and Willanger, R. (1970) Concentration camp survivors in Denmark persecution, disease, disability, compensation. A 23-year follow-up. A survey of the long-term effects of severe environmental stress. Dan. Med. Bull., 17: 65–108.

E.R. de Kloet, M.S. Oitzl & E. Vermetten (Eds.)
Progress in Brain Research, Vol. 167
ISSN 0079-6123

CHAPTER 27

Posttraumatic stress disorder with secondary psychotic features: neurobiological findings

Mario H. Braakman[1,*], Frank A.M. Kortmann[2], Wim van den Brink[3] and Robbert J. Verkes[2]

[1]*Institute of Mental Health Care, Wolfheze, the Netherlands and Department of Psychiatry, Radboud University Nijmegen Medical Center, Wolfheze 2, 6874 BE Wolfheze, The Netherlands*
[2]*Department of Psychiatry, Radboud University Nijmegen Medical Center, PO Box 9101, NL-6500 HB Nijmegen, The Netherlands*
[3]*Department of Psychiatry, Academic Medical Center, University of Amsterdam, PO Box 75867, 1070 AW Amsterdam, The Netherlands*

Abstract: The neurobiological knowledge on the potentially new diagnostic entity "posttraumatic stress disorder with secondary psychotic features" (PTSD-SP) is reviewed. Studies published between 1980 and 2006 were traced focussing on adult patients suffering from this "syndrome". Studies on cortisol, corticotrophin releasing hormone, dopamine beta-hydroxylase, smooth pursuit eye movements and psychopharmacology are described and potential pathophysiological mechanisms briefly discussed. More research is needed to validate the nosological status of PTSD-SP in order to promote neurobiological research and adequate therapeutic interventions.

Keywords: PTSD; psychosis; PTSD-SP; neurobiology; corticotrophin releasing hormone (CRH); cortisol; dopamine beta-hydroxylase (DbH); smooth pursuit eye movement (SPEM)

Introduction

A number of publications express the emergence of a new syndrome: posttraumatic stress disorder with secondary psychotic symptoms (PTSD-SP). This syndrome meets DSM-IV TR criteria of posttraumatic stress disorder, followed by psychotic features, especially hallucinations and delusions. These features are not confined to episodes of re-experiencing. The content of these psychotic features is often trauma-related, paranoid in nature, and not bizarre. There is no formal thought disorder. PTSD-SP appears to have a chronic course. Comorbid major depressive disorders occur frequently. There is no history of psychotic episodes, prior to the traumatic event(s). No relationship has been found between the nature or the severity of the traumatic events and the subsequent manifestation of psychotic features in patients suffering from PTSD. In first degree relatives there is an increased prevalence of major depression, but not of psychotic disorders (Braakman et al., submitted).

In this paper all studies on this topic are reviewed to determine the state of neurobiological knowledge on this potential syndrome and its pathophysiological underpinnings. The main findings of

*Corresponding author. Tel.: +3126-4833145; Fax: +3126-4833173; E-mail: m.braakman@degelderseroos.nl

DOI: 10.1016/S0079-6123(07)67027-7

these studies are reported as well as the varied spectrum of hypotheses put forward.

Methodology

Studies published between 1980 and 2006 were traced using the databases MEDLINE, PsycINFO, EMBASE Psychiatry, and Published International Literature On Traumatic Stress (PILOTS), with the full-text terms: 'PTSD', 'posttraumatic stress disorder', 'stress disorders, posttraumatic', 'psychological trauma', each combined with 'psychosis', 'psychotic', 'schizophrenia', 'hallucination', 'hallucinations', 'delusion', 'delusions', and 'reality testing'. Publications were screened, using the title and the abstract for relevance to the topic, i.e. PTSD and psychotic features. Reference lists were screened to find additional studies.

Results

A total of 45 publications were detected dealing with adult patients suffering from PTSD-SP. Only nine out of these studies focused on neurobiological issues and surpassed the evidence level of case studies; five are mainly psychopharmacologically oriented. The main findings (summarized in Table 1) concern corticotrophin releasing hormone (CRH) plasma levels of cortisol, dopamine beta-hydroxylase (DbH), smooth pursuit eye movement (SPEM), and pharmacotherapy.

Corticotrophin releasing hormone

Based on the assumption that PTSD-SP could be a severe subtype of PTSD, enhanced hyperactivity of the CRH system can be expected. Therefore, Sautter et al. (2003) assessed the cerebrospinal fluid concentrations of CRH. Their results demonstrate that patients with primary PTSD and subsequent appearing psychotic symptoms have higher cerebrospinal fluid concentrations of CRH than patients suffering from PTSD without psychosis and healthy comparison subjects.

Cortisol

Abnormal cortisol levels have been found in a wide array of psychiatric disorders, including psychosis, PTSD, and depressive disorders. Manguno-Mire et al. (in prep.) demonstrate that subjects with PTSD and secondary psychotic features show significantly higher baseline cortisol levels than subjects with PTSD without psychotic features and control subjects. In contrast to the non-suppression associated with major depressive disorder with psychotic features (Nelson and Davis, 1997), PTSD-SP subjects show hypersuppression of cortisol following dexamethasone admission. This study indicates that PTSD-SP has a neuroendocrine profile different from PTSD and depressive disorder with psychotic features.

Dopamine beta-hydroxylase

DbH converts dopamine into norepinephrine. During synaptic transmitter release it enters the extracellular space and hence becomes present in the cerebrospinal fluid and in blood plasma. DbH activity in plasma is a very stable heritable trait but varies extensively across unrelated individuals (Cubells and Zabetian, 2004). Hamner and Gold

Table 1. Summary of findings

Studies	PTSD	PTSD-SP	Schizophrenia
Baseline cortisol	Equal[a]	Elevated[a]	
CRH in CSF	Equal[a]	Elevated[a]	
Plasma DbH	Equal[a]	Elevated[a]	
SPEM performance		High velocity impairment	Low velocity impairment
SPEM deficits		Deficits in continuation	Deficits in initiation
Risperidone		Positive psychotic features do not improve[b]	

[a]Compared to healthy controls.
[b]Compared to placebo.

(1998) observed that plasma DbH-activity was elevated in PTSD patients with psychotic features as compared to both PTSD patients without psychotic features and healthy control subjects.

Smooth pursuit eye movement

SPEM refers to the movement of the eye that smoothly tracks slowly moving objects in the visual field. SPEM deficits are a well established phenomenon in schizophrenia (Thaker et al., 2003). Cerbone et al. (2003) studied SPEM in patients suffering from PTSD-SP and found marked differences compared to schizophrenia. The performance of patients with PTSD and secondary psychotic symptoms differed significantly from controls and from patients with schizophrenia in terms of the percentage of time in smooth pursuit. Patients with PTSD-SP showed impaired SPEM performance at higher velocity as compared to normal controls, while schizophrenia subjects were deficient in low velocity SPEM. PTSD-SP subjects showed deficits in the continuation of smooth pursuit, while schizophrenia was associated with deficits in the initiation of smooth pursuit.

Pharmacotherapy

Only one of the available pharmacological studies on PTSD-SP meets an adequate methodological quality, i.e. a randomized, double-blind, placebo-controlled trial. In this study risperidone or placebo were added to a standard regimen of antidepressant treatment (Hamner et al., 2003). The risperidone group improved significantly more than the placebo group in terms of the total Positive and Negative Syndrome Scale (PANSS). A more detailed analysis revealed that the level of significance was reached due to the improvement of the 'general psychopathology subscale' of the PANSS. Neither the 'positive symptoms subscale' nor the 'negative symptoms subscale' of the PANSS improved significantly. Thus, while general psychopathological symptoms improved, positive symptoms like delusions and hallucinations improved to the same degree in both groups.

Pathophysiological hypotheses

A variety of pathophysiological hypotheses on PTSD-SP have been proposed in the abovementioned studies. Sautter et al. (2003) proposed several hypotheses based on increased activation of CRH-circuitry.

1. Increased activation of hypothalamic CRH would produce increased cortisol secretion from the adrenal gland, which in turn increases CNS dopamine-activity of the mesocortical dopamine system.
2. Higher levels of CRH could lead to psychotic symptoms through the mechanism of CRH at the cyclic adenosine monophosphate (cAMP) level in the frontal cortex: high levels of CRF in PTSD-SP augment dopaminergic stimulation of cAMP in the frontal cortex because both CRF receptor subtypes use G-protein stimulatory heterotrimeric receptors that increase cAMP levels when activated by CRF.
3. Activation of CRH systems located outside the HPA-axis (e.g. loecus coeruleus, amygdalae, and the hippocampus) could, due to an increased frontal circuit dopamine activity, result in increased secretion of CRH in PTSD-SP subjects.
4. Another hypothesis, not based on dopaminergic activity, but on noradrenergic hyperactivity has been proposed by Hamner and Gold (1998): higher DbH activity could be expected to facilitate increased noradrenaline synthesis and might contribute to psychosis.

Discussion

There is ample of evidence that stress and cortisol is involved in dopaminergic alterations in the brain and hypercortisolemia as in Morbus Cushing can lead to frank psychosis. Thus a dopamine-based pathophysiology is worthwhile exploring although the lack of antipsychotic activity of risperidone points toward a pathophysiological mechanism that is (at least partially) different from the one proposed in schizophrenia.

It should be noted that all published research findings await replication and the presented findings should, therefore, be met with caution and reliable pathophysiological hypotheses are still preliminary. Finally no direct comparison data exist comparing PTSD-SP with major depressive disorder with psychotic features.

Conclusion

Research focusing on the validation of PTSD-SP and the delineation of clear diagnostic criteria is of great importance for the promotion of neurobiological and pathophysiological research and the study of (pharmaco-) therapeutic interventions of this complex disorder. Neurobiological studies focusing on PTSD-SP are limited.

Nonetheless, the divergent topics, and the PTSD-SP specific findings of these studies, strengthen preliminary phenomenological evidence that PTSD-SP can be delineated from other related disorders like schizophrenia (differences in smooth pursuit and DbH), PTSD (differences in DbH and cerebrospinal fluid levels of CRH) as well as depressive disorder with psychotic features (differences in DbH activity).

Acknowledgments

This work was supported by grants from de Gelderse Roos, Institute of Mental Health Care, the Netherlands, and ZonMw (project no. 100-002-018).

References

Braakman, M.H., Kortmann, F.A.M. and Brink, W.v.d. (submitted) Validity of 'posttraumatic stress disorder with psychotic features': a review of the evidence.

Cerbone, A., Sautter, F.J., Manguno-Mire, G., Evans, W.E., Tomlin, H., Schwartz, B. and Myers, L. (2003) Differences in smooth pursuit eye movement between posttraumatic stress disorder with secondary psychotic symptoms and schizophrenia. Schizophr. Res., 63: 59–62.

Cubells, J.F. and Zabetian, C.P. (2004) Human genetics of plasma dopamine beta-hydroxylase activity: applications to research in psychiatry and neurology. Psychopharmacology, 174: 463–476.

Hamner, M.B., Faldowski, R.A., Ulmer, H.G., Frueh, B.C., Huber, M.G. and Arana, G.W. (2003) Adjunctive risperidone treatment in post-traumatic stress disorder: a preliminary controlled trial of effects on comorbid psychotic symptoms. Int. Clin. Psychopharmacol., 18: 1–8.

Hamner, M.B. and Gold, P.B. (1998) Plasma dopamine beta-hydroxylase activity in psychotic and non-psychotic post-traumatic stress disorder. Psychiatry Res., 77: 175–181.

Manguno-Mire, G.M., Sautter, F.J., Johnson, J.E., Myers, L., Bryan, F., Nenov, N., Ross, R., Bissette, G. and O'Neill, P. (in prep.) Cortisol response to dexamethasone in PTSD patients with psychotic symptoms.

Nelson, J.C. and Davis, J.M. (1997) DST studies in psychotic depression: a meta-analysis. Am. J. Psychiatry, 154: 1497–1503.

Sautter, F.J., Bissette, G., Wiley, J., Manguno-Mire, G., Schoenbachler, B., Myers, L., Johnson, J.E., Cerbone, A. and Malaspina, D. (2003) Corticotropin-releasing factor in posttraumatic stress disorder (PTSD) with secondary psychotic symptoms, nonpsychotic PTSD, and healthy control subjects. Biol. Psychiatry, 54: 1382–1388.

Thaker, G.K., Avila, M.T., Hong, E.L., Medoff, D.R., Ross, D.E. and Adami, H.M. (2003) A model of smooth pursuit eye movement deficit associated with the schizophrenia phenotype. Psychophysiology, 40: 277–284.

E.R. de Kloet, M.S. Oitzl & E. Vermetten (Eds.)
Progress in Brain Research, Vol. 167
ISSN 0079-6123

CHAPTER 28

Neuroendocrine dysregulations in sexually abused children and adolescents: a systematic review

I.A.E. Bicanic[1,*], M. Meijer[2], G. Sinnema[3], E.M. van de Putte[2] and M. Olff[4]

[1]*University Medical Center Utrecht, Center for the Treatment of Psychotrauma in Children and Adolescents, KA.00.004.0, P.O. Box 85090, 3508 AB Utrecht, The Netherlands*
[2]*University Medical Center Utrecht, Department of Pediatrics, KE.04.133.1, P.O. Box 85090, 3508 AB Utrecht, The Netherlands*
[3]*University Medical Center Utrecht, Department of Pediatric Psychology, KA.00.004.0, P.O. Box 85090, 3508 AB Utrecht, The Netherlands*
[4]*Academic Medical Center Amsterdam/De Meren, Department of Psychiatry, Head Center for Psychological Trauma, Meibergdreef 5, 1105 AZ Amsterdam, The Netherlands*

Abstract: Several studies provided evidence for neuroendocrine dysregulations in adults with a history of child sexual abuse. This review focuses on neuroendocrine studies in sexually abused children and adolescents, dating from January 1, 1990 to January 1, 2007 and obtained from a systematic Medline Indexed literature search to identify endocrine correlates of child sexual abuse. Results from studies on hypothalamic-pituitary-adrenal axis (re)activity showed to be inconclusive. Studies on the sympathetic nervous system provided evidence for a higher baseline activity of this system in sexually abused children and adolescents. Factors contributing to divergent outcomes will be discussed and suggestions for future research will be presented.

Keywords: sexual abuse; PTSD; depression; HPA axis; CRH; ACTH; cortisol

Introduction

The body's major stress systems involve the sympathetic nervous system (SNS) and the hypothalamic-pituitary-adrenal (HPA) axis. In recent years, several neurobiological investigations have provided evidence for dysregulations of these systems in adults who have experienced sexual abuse in childhood, such as *lower* basal cortisol levels and *higher* adrenocorticotropin hormone (ACTH) response to psychological stressors. However, less attention has been paid to neurobiological research in sexually abused children and adolescents. More research is considered important since improved understanding of the correlates of child sexual abuse (CSA), both psychological and biological, may lead to better treatment options for children affected by CSA. The objective of this review is to evaluate systematically selected neuroendocrine studies in sexually abused children and adolescents specifically focussing at identifying endocrine correlates of CSA.

Methodology

A Medline Indexed systematic literature search, limited to publications dating from January 1,

*Corresponding author. Tel.: +31 30 2504113;
Fax: +31 30 2505325; E-mail: i.a.e.bicanic@umcutrecht.nl

DOI: 10.1016/S0079-6123(07)67028-9

1990 to January 1, 2007, was performed for MeSH keywords "child abuse" OR rape OR "sex offenses" AND hydrocortisone OR catecholamines OR "neurosecretory systems" OR glucocorticoids OR corticotropin OR amylases OR "adrenal cortex hormones" OR "pituitary hormones" and for free text words "sexual abuse" OR "sexual trauma" OR "child abuse" OR rape OR "sex offenses" AND ACTH OR cortisol OR CRH OR "HPA axis" OR "hypothalamic-pituitary-adrenocortical axis" OR catecholamines. The search was limited to studies on children aged 0–18 years and resulted in 127 publications. A first selection of publications was performed by analysis of title and abstract by two independent reviewers. One hundred seventeen publications were excluded based on age of subjects (34), other biological parameters (26), other type of abuse or patient groups (15), and type of publications such as reviews (36) and letters or replies (6). Ten original publications were studied in detail. Finally, three studies were excluded since the number of sexually abused subjects was too small (<17%), not being representative for the objective of this review.

Results

The selected studies focused on baseline functioning of the SNS system and on baseline HPA axis activity as well as reactivity to pharmacological challenge tests. Table 1 summarizes the primary results in each of the seven studies.

After administration of ovine corticotropin-releasing hormone (CRH), De Bellis et al. (1994a) measured *lower* ACTH, but normal cortisol concentrations in 13 sexually abused girls aged 7–15, compared to 13 non-abused controls.

Table 1. Summary of the primary results in each of the seven studies

	Studies						
	De Bellis et al. (1994a)	De Bellis et al. (1994b)	Kaufman et al. (1997)	Kaufman et al. (1998)	De Bellis et al. (1999)	King et al. (2001)	Duval et al. (2004)
Number of cases	13	12	13	10	18	10	14
Females (%)	100	100	54	60	44	100	86
Mean age/age range in years	11.2/7–15	11.5/8–15	9.6/7–13	10.3/7–13	10.4/8–13	6.4/5–7	16.2/13–19
CSA (%)	100	100	77	77	83	100	100
PTSD (%)	0	8	83	50	100	10	100
Age of onset CSA (yr)	6.3±2.4	6.3±2.4	n.s.	n.s.	4.7±3.0	n.s.	10.6±3.9
Duration CSA	22.1±26.6 months	22.1±26.6 months	n.s.	n.s.	2.4±1.8 yr	<1 yr	n.s.
Time since CSA elapsed (yr)	4.7±3.5	5.2±3.5	n.s.	n.s.	n.s.	n.s.	5.6±4.0
Baseline							
Cortisol	=		=	=	↑	↓	=
ACTH	↓		=				=
Catechol		↑			↑		
Prolactin				=			
Challenge test							
Cortisol	=		=	=			=
ACTH	↓		↑				↓
Prolactin				↑			

Notes: n.s., not specified; ↑, significantly higher compared to controls; ↓, significantly lower compared to controls; and =, no difference compared to controls. Different challenge tests were used: oCRH in De Bellis et al. (1994a); CRH in Kaufman et al. (1997); L-5-hydroxytryptophan in Kaufman et al. (1998); and dexamethasone in Duval et al. (2004).

Fifty-four percent of the sexually abused had dysthymia and histories of suicidal behavior, but no current post-traumatic stress disorder (PTSD). In contrast, in a study by Kaufman et al. (1997), a challenge test with human CRH provoked a *higher* ACTH response and normal cortisol response in 13 depressed abused children aged 7–13, compared to 13 depressed non-abused and 13 healthy controls. Eight out of 13 depressed abused children also met the criteria for PTSD and 10 were sexually abused. The contradictory findings between these two studies may be clarified by a subanalysis in the Kaufman study, revealing that ACTH hyperresponding to CRH only occurred in depressed abused children, who were exposed to chronic ongoing adversity. The CRH challenge was performed on the second day of a multitest psychobiological protocol. On the third day of the protocol, after exclusion of children with nausea, an L-5-hydroxytryptophan challenge was performed, provoking a *higher* prolactin response and normal cortisol response in 10 depressed abused children, compared to 10 depressed non-abused and 10 healthy controls (Kaufman et al., 1998). In the last challenge study that was reviewed, 14 sexually abused adolescents with PTSD exhibited a *lower* ACTH response to 1 mg dexamethasone suppression test (DST) compared to 14 hospitalized controls (Duval et al., 2004). Cortisol response to DST was also lower in the sexually abused adolescents, but this difference did not reach statistical significance. Of the sexually abused subjects, 2 had comorbid major depressive disorder and 10 had depressive symptoms.

Unfortunately, findings from studies on baseline HPA activity are also inconsistent and actually do not allow comparison across studies because of methodological differences. Some studies found *higher* basal cortisol in sexually abused children and adolescents (De Bellis et al., 1999), others found *lower* basal cortisol (King et al., 2001), and again others found *normal* cortisol values (Kaufman et al., 1997, 1998; Duval et al., 2004). Additionally, for ACTH levels, some found *lower* basal evening levels (De Bellis et al., 1994a) and others found *normal* values (Kaufman et al., 1997; Duval et al., 2004).

Results from SNS studies in sexually abused children and adolescents are more consistent. De Bellis et al. (1994b) found a group of 12 dysthymic sexually abused girls aged 8–15 to secrete significantly *higher* concentrations of urinary catecholamines compared to nine non-abused controls. Similar results were found in 18 prepubertal abused children with PTSD and a high degree of comorbid psychiatric disorders (De Bellis et al., 1999). This group, of whom 15 had experienced sexual abuse, secreted significantly *higher* concentrations of urinary dopamine and norepinephrine compared to non-abused patients with overanxious disorder (OAD) and to healthy controls. Urinary epinephrine was significantly *higher* in the PTSD group compared to the OAD group, but not to healthy controls.

Discussion and conclusion

Our goal to identify endocrine correlates of CSA from the existing literature is attained with limited success due to several reasons. First, it is difficult to identify subjects with exclusively *sexual* abuse experiences, since various types of abuse tend to coexist and chronic sexual abuse usually occurs in the context of affective neglect. With respect to this coexistence, it should be recognized that the HPA axis is under strong psychosocial regulation in early life (Gunnar and Donzella, 2002), a period when parental sensitivity and responsivity are being challenged and variations in parent–child relationships occur. Second, most studies did not specify sexual abuse variables that may affect outcome, such as severity, duration, and frequency of the sexual abuse experiences, nature of the relation to the perpetrator, use of physical force, age at sexual abuse, and years since the sexual abuse elapsed. Third, all studies were characterized by a small number of subjects and therefore had limited statistical power, which also limited the potential to examine age and gender effects. Fourth, not all studies have adequately considered confounding factors, such as exposure to chronic ongoing adversity, daylight, menstrual cycle, pubertal status, medication, awakening time, smoking, and food intake. Furthermore, from the results it is not possible to state with certainty whether the biological findings reflect the effects of sexual abuse or of PTSD or even of comorbid psychopathology. In future research, samples of sexually abused children with and without PTSD — and preferably without comorbidity — should be included in order to better understand the relation between stress hormones and PTSD.

Finally, inconsistencies in the studies may be associated with the sampling method, such as type of assay procedure, time point of sampling, number of sampling days, number of samples used for determining endocrine outcomes, sampling restrictions, sampling in plasma, urine, or saliva, and type of challenge test. Future studies can learn from previous research by realizing uniformity in sampling and analytical procedures, so that results may be compared across studies.

In conclusion, this review showed that findings from studies to date on HPA axis (re)activity in sexually abused children and adolescents have been both varied and contradictory, possibly due to methodological shortcomings and the developmental status of the neuroendocrine system. Without the presence of a specific neuroendocrine profile in this group, it is not possible to draw conclusions to what extent the HPA axis is affected in relation to CSA. Interestingly, from the limited data we can conclude that CSA appears to be related to higher catecholamine secretion, suggesting the presence of a higher baseline activity of the SNS.

To successfully identify abnormal endocrine profiles of CSA in future research, we suggest that larger studies should be performed in both sexually abused boys and girls, characterized by specification of sexual abuse variables, structural evaluation of PTSD and comorbid psychopathology, uniform sampling procedures, and standardized methods for neuroendocrine assessments.

Abbreviations

ACTH	adrenocorticotropin hormone
CRH	corticotropin-releasing hormone
CSA	child sexual abuse
DST	dexamethasone suppression test
HPA	hypothalamic-pituitary-adrenal
OAD	overanxious disorder
PTSD	post-traumatic stress disorder
SNS	sympathetic nervous system

References

De Bellis, M.D., Baum, A.S., Birmaher, B., Keshavan, M.S., Eccard, C.H., Boring, A.M., Jenkins, F.J. and Ryan, N.D. (1999) A.E. Bennett research award. Developmental traumatology. Part I: Biological stress systems. Biol. Psychiatry, 45(10): 1259–1270.

De Bellis, M.D., Chrousos, G.P., Dorn, L.D., Burke, L., Helmers, K., Kling, M.A., Trickett, P.K. and Putnam, F.W. (1994a) Hypothalamic-pituitary-adrenal axis dysregulation in sexually abused girls. J. Clin. Endocrinol. Metab., 78(2): 249–255.

De Bellis, M.D., Lefter, L., Trickett, P.K. and Putnam, F.W. (1994b) Urinary catecholamine excretion in sexually abused girls. J. Am. Acad. Child Adolesc. Psychiatry, 33(3): 320–327.

Duval, F., Crocq, M.A., Guillon, M.S., Mokrani, M.C., Monreal, J., Bailey, P. and Macher, J.P. (2004) Increased adrenocorticotropin suppression following dexamethasone administration in sexually abused adolescents with posttraumatic stress disorder. Psychoneuroendocrinology, 29(10): 1281–1289.

Gunnar, M.R. and Donzella, B. (2002) Social regulation of the cortisol levels in early human development. Psychoneuroendocrinology, 27(1–2): 199–220.

Kaufman, J., Birmaher, B., Perel, J., Dahl, R.E., Moreci, P., Nelson, B., Wells, W. and Ryan, N.D. (1997) The corticotropin-releasing hormone challenge in depressed abused, depressed nonabused, and normal control children. Biol. Psychiatry, 42(8): 669–679.

Kaufman, J., Birmaher, B., Perel, J., Dahl, R.E., Stull, S., Brent, D., Trubnick, L., al-Shabbout, M. and Ryan, N.D. (1998) Serotonergic functioning in depressed abused children: clinical and familial correlates. Biol. Psychiatry, 44(10): 973–981.

King, J.A., Mandansky, D., King, S., Fletcher, K.E. and Brewer, J. (2001) Early sexual abuse and low cortisol. Psychiatry Clin. Neurosci., 55(1): 71–74.

E.R. de Kloet, M.S. Oitzl & E. Vermetten (Eds.)
Progress in Brain Research, Vol. 167
ISSN 0079-6123

CHAPTER 29

Volume of discrete brain structures in complex dissociative disorders: preliminary findings

T. Ehling[1,*], E.R.S. Nijenhuis[2] and A.P. Krikke[3]

[1]*Neuroimaging Center (NIC/BCN), University of Groningen, Groningen, and Mental Health Care Drenthe, Department of Outpatient Services, Altingerweg 1, 9411 PA Beilen, The Netherlands*
[2]*Mental Health Care Drenthe, Top Referent Trauma Center, Beilerstraat 197, 9401 PJ Assen, The Netherlands*
[3]*Wilhelmina Hospital Assen, Department of Radiology, Europaweg-Zuid 1, 9401 RK Assen, The Netherlands*

Abstract: Based on findings in traumatized animals and patients with posttraumatic stress disorder, and on traumatogenic models of complex dissociative disorders, it was hypothesized that (1) patients with complex dissociative disorders have smaller volumes of hippocampus, parahippocampal gyrus, and amygdala than normal controls, (2) these volumes are associated with severity of psychoform and somatoform dissociative symptoms, and (3) patients who recovered from dissociative identity disorder (DID) have more hippocampal volume that patients with florid DID. The preliminary findings of the study are supportive of these hypotheses. Psychotherapy for dissociative disorders may affect hippocampal volume, but longitudinal studies are required to document this potential causal relationship.

Keywords: dissociative disorders; hippocampus; parahippocampal gyrus; amygdala; volumetry

Introduction

According to the theory of structural dissociation, traumatization involves a division of personality structure into two or more different, but more or less intensely interacting prototypical subsystems, each with its own distinct psychobiological underpinnings (Nijenhuis et al., 2002). This lack of integration is primarily due to exposure to highly stressful events, limitations of integrative capacity, and lack of social support. In its basic form, the personality becomes divided into a metaphorical "apparently normal" part (ANP) that is focused on fulfilling functions in daily living, and an "emotional" part (EP) that is largely fixated in physical defence from major threats, in particular, threat to the integrity of the body (Reinders et al., 2006).

Mounting evidence indeed suggests that recurrent exposure to major threat and neglect interferes with integrative mental actions and with the normative development and function of integrative brain structures. Thus, adult patients with chronic PTSD have smaller hippocampal volume than mentally healthy controls and individuals who have been exposed to potentially traumatizing events but who did not develop a trauma-related mental disorder (Kitayama et al., 2005). However, hippocampal volume is also smaller in a range of other mental and somatic diseases (Geuze et al., 2005).

The hippocampus (HC) has a function in integration, in short-term memory, and in long-term

*Corresponding author. Tel.: +31-593-535350;
Fax: +31-593-535349; E-mail: thomas.ehling@ggzdrenthe.nl

DOI: 10.1016/S0079-6123(07)67029-0

memory up to 1 year. HC is also involved in spatial and contextual learning, as well as in inhibition of emotional responses. Research in mammals has documented serious structural and functional damage of HC on chronic stress exposure. Structures of HC, especially dendrites, appear to be sensitive to chronically elevated levels of glucocorticoids as released by the hypothalamic-pituitary-adrenal axis (HPA-axis) (Zhao et al., 2007).

Little is known about HC volume loss in dissociative identity disorder (DID) and dissociative disorder not otherwise specified (DDNOS), type 1. As mentioned, HC is sensitive to extreme stress and hyperarousal, which individuals with DID and DDNOS have often encountered in early life. Tsai et al. (1999) found that HC volume was significantly smaller in a female patient with DID, compared to HC volume in normal female adults. Vermetten et al. (2006) documented that hippocampal volume was 19.2% smaller and amygdalar volume was 31.6% smaller in 15 female patients with DID, compared to healthy subjects. There is some preliminary evidence of recovery of HC volume after successful treatment. Vermetten et al. (2003) reported a HC volume recovery of 4.6% bilaterally after successful treatment with paroxetine. However, Lindauer et al. (2005) found no effect of short-term psychotherapy on hippocampal volume in PTSD.

Because of the low integrative capacity and persistent dissociative symptoms of chronically traumatized individuals, including dissociative amnesia, we hypothesized that compared to healthy controls HC is smaller in complex dissociative disorders, but smaller in DID than in DDNOS, and that HC volume is larger in patients who, due to long-term psychotherapy, recovered from DID compared to HC volume in florid DID. We also hypothesized that HC volume is more strongly correlated with dissociative symptoms than with general psychopathological symptoms. Finally, assuming that chronic stress exposure seriously affects the maturation of the brain more generally, we hypothesized that the parahippocampal gyrus (PHG), which serves as a interface between HC and neocortex, and the amygdala (AM) which is involved in perceiving and reacting to emotional stimuli, are also smaller in patients with dissociative disorders compared to controls.

Methods

A sample was taken of 10 DID-patients, 13 DDNOS-patients, 10 DID-patients who completely recovered from DID after phase-oriented psychotherapy with an average duration of 4.5 years and 20 healthy controls. All were female and all were matched post hoc for age and educational level. The Diagnostic and Statistical Manual of Mental Disorders 4th Edition (DSM-IV) criteria for DID and DDNOS present serious problems in diagnosis. Experts thus urge clinicians and researchers to select cases using the Structured Clinical Interview for DSM-IV Dissociative Disorders (SCID-D; Steinberg, 1994), which is a psychometrically sound diagnostic instrument. Participants were thus diagnosed using the SCID-D. They underwent structural magnetic resonance imaging (MRI) at the Wilhelmina Hospital Assen, The Netherlands. Images were acquired with a 1.0 T scanner (Philips Gyroscan Intera System) obtaining 100–112 1.5 mm gapless contiguous coronal slices (echo time, 4.7 ms; repetition time, 30 ms; flip angle 30°; field of view (FOV), 100%; matrix size, 256 × 192 mm) of the whole brain (fast field echo scans, T1-weighted, 3D). All participants completed the Dissociation Questionnaire (DIS-Q), the Somatoform Dissociation Questionnaire (SDQ-20), the Traumatic Experiences Checklist (TEC), the PTSD self-reporting list (PTSD-sr), and the Hopkins Symptom Checklist (HSCL-90). Manual tracing of region of interest (ROI) was performed according to the protocols of Pantel et al. (2000) and Pruessner et al. (2000, 2002) regarding anatomical hallmarks of HC, PHG, and AM. Statistical analysis, e.g. repeated analysis of variance (ANOVA), was conducted using software package SPSS 11.0.

Results

There were no statistically significant differences among (ex-)patients and controls for age and educational level. Moreover, there was no main effect for side (right/left) and no interaction between side and diagnosis (patients versus controls). Compared to controls, women with DID had less HC (right 25% and 26% left) (see Fig. 1) and PHG

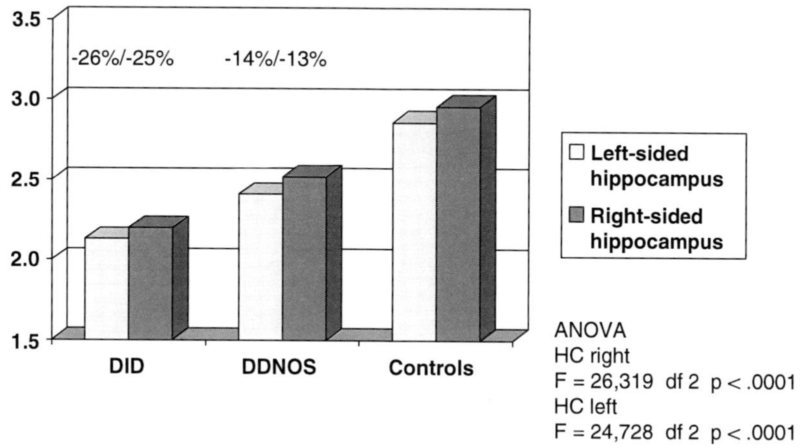

Fig. 1. Hippocampal volumes in patients with dissociative disorders and controls. Adopted from Ehling, Nijenhuis and Krikke (2005).

volume (20% bilaterally). Women who recovered from DID had more HC (right 18% and left 9%), but not more PHG volume, compared to women with florid DID. Bilateral HC volume of women with DDNOS was smaller (right 14% and left 13%), and PHG volume was bilaterally 19% smaller than in healthy controls. In both DID and DDNOS, AM volume was 10–12% smaller than in healthy controls.

HC volumes (corrected for estimated brain volume) were strongly and negatively correlated with cumulative reporting of potentially traumatizing events (TEC; Spearman $\rho = -.73, p = .001$), psychoform dissociation (DIS-Q; Spearman $\rho = -.63$, $p = .004$), somatoform dissociation (SDQ-20; Spearman $\rho = -.78$, $p < .0001$), posttraumatic stress symptoms (PTSD-sr; Spearman $\rho = -.56, p = .028$), and general psychopathology (HSCL-90; Spearman $\rho = -.53$, $p = .015$). PHG volumes were also negatively correlated with psychoform dissociation (DIS-Q; Spearman $\rho = -.62$, $p < .005$), somatoform dissociation (SDQ-20; Spearman $\rho = -.45$, $p = < .005$), and posttraumatic stress symptoms (PTSD-sr; Spearman $\rho = -.46, p < .005$).

Discussion

HC, PHG, and AM volumes were smaller in patients with complex dissociative disorders, and HC was smaller in DID than in DDNOS. The findings are in line with the findings of Vermetten et al. (2006) regarding HC and AM volumes in DID. Volume of HC and PHG were strongly correlated with psychoform and somatoform dissociative symptoms in the current study, and more strongly than with a measure of general psychopathology. HC and PHG volume were also strongly correlated with reported exposure to potentially traumatizing events.

Limitations of the study are that statistical control for alcohol abuse, independent measurements by a second blinded rater, and more precise correction for total brain volume are still to be done. No checks were done regarding the veracity of the reported potentially traumatizing events. Pending these limitations, all hypotheses of the study were supported by the findings that are consistent with the traumatogenetic models of complex dissociative disorders. The validity can be doubted of comparing patients with DID or DDNOS who report massive exposure to potentially traumatizing events and healthy controls who commonly do not report such exposure to potentially traumatizing events. However, mentally healthy controls who report exposure to childhood emotional neglect and emotional, physical, and sexual abuse that is as intense and extensive as reports of such events by patients with complex dissociative disorders are exceptional if not nonexistent. Controlling for reported exposure to potentially traumatizing events in a study in which dissociative patients and controls are compared would therefore not be feasible. One way to tease out the influence of reported potential traumatization on the relationship of dissociative disorders and smaller HC, PHG, and AM, would be to compare patients with dissociative disorders and patients with different mental disorders, who, as can be expected from prior findings, will probably report considerable to severe exposure to potentially traumatizing childhood events.

Abbreviations

AM	amygdala
ANOVA	analysis of variance
ANP	apparently normal part
DDNOS	dissociative disorder not otherwise specified
DID	dissociative identity disorder
DIS-Q	Dissociation Questionnaire
DSM-IV	Diagnostic and Statistical Manual of Mental Disorders 4th Edition
EP	emotional part
FOV	field of view
HC	hippocampus
HPA-axis	hypothalamic-pituitary-adrenal axis
HSCL-90	Hopkins Symptom Checklist
MRI	magnetic resonance imaging
PHG	parahippocampal gyrus
PTSD-sr	PTSD self reporting list
ROI	region of interest
SCID-D	Structured Clinical Interview for DSM-IV Dissociative Disorders
SDQ-20	Somatoform Dissociation Questionnaire
TEC	Traumatic Experiences Checklist

References

Geuze, E., Vermettten, E. and Bremner, J.D. (2005) MR-based in vivo hippocampal volumetrics: 2. Findings in neuropsychiatric disorders. Mol. Psychiatry, 10(2): 160–184.

Kitayama, N., Vaccarino, V., Kutner, M., Weiss, P. and Bremner, J.D. (2005) Magnetic resonance imaging (MRI) measurement of hippocampal volume in posttraumatic stress disorder: a meta-analysis. J. Affect. Disord., 88(1): 79–86.

Lindauer, R.J., Vlieger, E.J., Jalink, M., Olff, M., Carlier, I.V., Majoie, C.B., Den Heeten, G.J. and Gersons, B.P. (2005) Effects of psychotherapy on hippocampal volume in outpatients with posttraumatic stress disorder: a MRI investigation. Psychol. Med., 35(10): 1421–1431.

Nijenhuis, E.R.S., Van der Hart, O. and Steele, K. (2002) The emerging psychobiology of trauma-related dissociation and dissociative disorders. In: D'Haenen H., Den Boer J.A. and Willner P. (Eds.), Biological Psychiatry, Vol. 2, Wiley, London, pp. 1079–1098.

Pantel, J., O'Leary, D.S., Cretsinger, K., Bockholt, H.J., Keefe, H., Magnotta, V.A. and Andreasen, N.C. (2000) A new method for the in vivo volumetric measurement of the human hippocampus with high neuroanatomical accuracy. Hippocampus, 10(6): 752–758.

Pruessner, J.C., Kohler, S., Crane, J., Pruessner, M., Lord, C., Byrne, A., Kabani, N., Collins, D.L. and Evans, A.C. (2002) Volumetry of temporopolar, perirhinal, entorhinal and parahippocampal cortex from high-resolution MR images: considering the variability of the collateral sulcus. Cereb. Cortex, 12(12): 1342–1353.

Pruessner, J.C., Li, L.M., Serles, W., Pruessner, M., Collins, D.L., Kabani, M., Lupien, S. and Evans, A.C. (2000) Volumetry of hippocampus and amygdala with high-resolution MRI and three-dimensional analysis software: minimizing the discrepancies between laboratories. Cereb. Cortex, 10(4): 433–442.

Reinders, A.A.T.S., Nijenhuis, E.R.S., Quak, J., Korf, J., Paans, A.M.J., Haaksma, J., Willemsen, A.T.M. and Den Boer, J.A. (2006) Psychobiological characteristics of dissociative identity disorder: a symptom provocation study. Biol. Psychiatry, 60(7): 730–740.

Tsai, G.E., Condie, D., Wu, M.T. and Chang, I.W. (1999) Functional magnetic resonance imaging of personality switches in a woman with dissociative identity disorder. Harv. Rev. Psychiatry, 7(2): 119–122.

Vermetten, E., Schmahl, C., Lindner, S., Loewenstein, R.J. and Bremner, J.D. (2006) Hippocampal and amygdalar volumes in dissociative identity disorder. Am. J. Psychiatry, 163(4): 630–636.

Vermetten, E., Vythilingam, M., Southwick, S.M., Charney, D.S. and Bremner, J.D. (2003) Long-term treatment with paroxetine increases verbal declarative memory and hippocampal volume in posttraumatic stress disorder. Biol. Psychiatry, 54(7): 693–702.

Zhao, H., Xu, H., Xu, X. and Young, D. (2007) Predatory stress induces hippocampal cell death by apoptosis in rats. Neurosci. Lett., 421: 115–120.

E.R. de Kloet, M.S. Oitzl & E. Vermetten (Eds.)
Progress in Brain Research, Vol. 167
ISSN 0079-6123

CHAPTER 30

Epilogue

Eric Vermetten

The field of neuroendocrinology of PTSD is fueled with intriguing findings and paradoxes. As pointed out in the preface, the relevant question for this Colloquium and book obviously has been why some individuals become ill, but others actually gain strength from identical stressful situations. Are vulnerability and resilience opposite sides of a coin? The endocrine fingerprint of PTSD has been described as a downregulated cortisol (CORT) secretion occurring in the aftermath of severe childhood trauma, but infants strongly attached to an overprotective mother usually also display attenuated CORT reactivity later. Animal models of maternal deprivation show enhanced CORT reactivity in later life, but in some models an initial hyporeactivity of CORT secretion at younger ages is observed. There are questions that remain unanswered: Is attenuated CORT reactivity actually a predisposing factor for PTSD? Or is inadequate CORT release unable to contain an exaggerated initial stress reaction? As the chapters in this book indicated: the neuroendocrinology of traumatic stress is more alive than ever! The field has developed much since the first notions of Hans Selye in 1936 when he discussed the theory of the General Adaptation Syndrome.

It is 60 years after Grinker and Spiegel published their volume *Men Under Stress*, in which they described adaptation and breakdown patterns on the battlefield during WOII. Much has been learned from Vietnam War and later wars. Natural and man-made disasters, but also child abuse have confronted clinicians with the psychiatric sequelae of traumatic stress. DSMIII has embraced and incorporated PTSD in 1980. The numbers are known, the psychobiology is well described, and the long-lasting consequences clear. Intrusive recollections of a traumatic event, hyperarousal, and avoidance of clues associated with the trauma are hallmark symptoms of PTSD. Already a decade ago initial hypotheses were formulated that the functional neuroanatomy of traumatic stress comprises a circuit of brain areas involved in both stress and memory function, including the hippocampus, amygdala, cingulate cortex, medial and dorsolateral prefrontal cortex. As chapters in this book showed, an abundance of studies demonstrated the way in which these brain areas are affected in PTSD. Studies with positron emission tomography (PET) and functional magnetic resonance imaging (fMRI) demonstrated a hyperreactive amygdala to trauma-related stimuli. Exaggerated startle responses and flashbacks have been related to a failure of frontal regions to dampen the symptoms of arousal and distress that were mediated as reminders of the traumatic event. The model in PTSD that attributed the intrusive re-experiencing in PTSD to a failure of inhibitory processes over a fear-motivated hyperresponsive limbic system has seen supported by many studies. This research has initiated the use of psychopharmacological agents that are targeted to affect limbic activation as well as decreased frontal lobe functions. Also, the use of extinction-based exposure therapies that are so critical in PTSD treatment have been supported by findings of prefrontal cortical function.

Clinicians as well as researchers need a framework for understanding the disease concepts. In 1993 Charney et al. wrote a landmark paper on the

DOI: 10.1016/S0079-6123(07)67030-7

psychobiology of PTSD in which they described long-lasting alterations in three psychobiological domains: (1) stress sensitization, (2) fear conditioning, and (3) failure in extinction (Charney et al., Arch. Gen. Psychiatry, 50: 295–305, 1993). Their conceptual notions called for clinical empirical support which has been provided now by excellent studies, as several chapters in this book illustrate. Another fruitful model that has provided a framework for understanding PTSD has been proposed by McEwen. He coined the 'allostasis' concept, i.e., maintaining homeostasis through change (McEwen, Ann. N.Y. Acad. Sci., 840: 33–44, 1998). In this model the cost of allostasis or allostatic load represented the outcome of wear and tear of the organism. From this point of view PTSD is manifested in, e.g., noradrenergic hyperreactivity, dysregulation of the HPA axis (basal hypocortisolaemia, reactive hypercortisolaemia, altered glucocorticoid receptors, increased responsiveness, or sensitivity to stressors), dysregulated responsiveness of amygdala, orbitofrontal cortex dysfunction resulting in failure of extinction, and in reduced hippocampal volume. The brain is the key organ of allostatic regulation because its cognitive operations determine what is threatening and therefore stressful and also how the physiological and behavioral responses occur. These conceptual notions, frameworks, as well as other related research questions have been addressed by a panel of international scientists.

Unique element of this book is the combination of fundamental research of glucocorticoids from gene to behavior with clinical research of PTSD. As the various contributions demonstrated, major advances have occurred in understanding the role of CORT. We have currently more systematic measures of the HPA-axis, with more knowledge of comorbid risk factors, such as disorganized or insecure attachment in childhood, early physical and sexual abuse, and depression. We have a better understanding of the activation of stress response systems, including sympathetic nervous system arousal, impaired parasympathetic soothing, and HPA dysregulation including initial hyperactivation followed by low CORT levels, loss of normal diurnal variation, and hyperresponsiveness to subsequent even minor stressors. This HPA abnormality has been associated with a dysregulatory cascade either associated with or causing hippocampal damage. Smaller hippocampal volume appeared associated with PTSD and dissociative symptoms. While debate continues regarding the possibility that the smaller volume is a genetically based vulnerability factor for PTSD and dissociation or a result of glucocorticoid abnormalities, the association has become clear.

Imaginative use of fMRI and other new brain imaging technologies have provided fascinating glimpses into the activity of brain regions involved in dissociative symptoms. The normal integration of systems that regulate emotion (dorsolateral prefrontal cortex, anterior cingulate, subgenual cingulate, and amygdala), autonomic nervous system activity (insula, thalamus), sensation (parietal, temporal, occipital cortex, and cerebellum), attention (thalamus, anterior cingulate, frontal cortex), and memory (hippocampus) can be disrupted in specific ways that compromise normal adaptation.

Zooming in on the situation of exposure to traumatic stress we may translate findings from preclinical studies as if it is often better for an individual not to get the 'big picture' at a time of extreme danger, but rather to focus attention on control of responses that might elicit more harm, or to find safe means of exiting the threatening situation. There seems to be a post-stress time period of days to weeks in which most individuals engage in processing of traumatic experiences, allowing them to acknowledge, bear, and put into perspective traumatic experiences and their implications. This may allow for a modulation of the emotional and physiological impact of the trauma, providing a gradual desensitization of traumatic stimuli. However, the clinical perspective of persistent stress responses seems to set some individuals on a different path. While traumatic stress may promote memory for the arousing impact of life-threatening stress, it can have the paradoxical effect of non-consolidation in the absence of sufficient CORT. This may transform physical responses at the time of the trauma into psychological effects over the periodic incursions of traumatic memories and associations into consciousness. Individuals may often feel they have memorized all events at the time of the trauma, yet experience intrusive thoughts, flashbacks, nightmares, numbing, amnesia, and hyperarousal as a kind of retraumatization. In many cases these symptoms seem to sensitize rather than produce habituation to traumatic experiences, perpetuating further PTSD and other stress-related symptoms.

The field of neuroendocrinology and PTSD has come to a phase of which we may not yet realize the potential impact. We would like to make two last notions at the end of this volume. First, the notion of early life adversities as laboratory model for neglect and abuse. By modeling gene–environment interactions animal models that mimic cognitive and emotional symptoms of vulnerable phenotypes for PTSD are currently generated. Such animal models have demonstrated to provide powerful approaches for translation purposes. However, in translation to human life, it poses moral imperatives specifically on prevention of early life adversity. What strategies are devised to prevent abuse? Is history of abuse a screening element for stressful jobs? We may need more armamentarium to fully translate the preclinical work, but we are moving to a point where this is a valid and necessary effort. In other domains of clinical research (e.g., parasitology, ophthalmology) the concepts are more trivial, but translation is valid and full. Second, since the clinical phenotype of PTSD is now well known and carefully described in terms of psychoneurobiology, the challenge that lies ahead is to better understand the developmental trajectory to enable the design of rational and timely implemented combined psychotherapeutical and pharmacological strategies that prevent or minimize the occurrence of full-blown PTSD after exposure. In other words, while research on tertiary preventive efforts to reduce the burden of traumatic stress is ongoing, a true challenge for the next decade of clinical research in the field of PTSD lies in primary and secondary prevention. It is unlikely that traumatic stress will be eliminated from our society.

This book summarizes the efforts of 3 days, which my colleagues Professors Ron de Kloet and Melly Oitzl and I feel have been very exciting. It has been an excellent conference. We hope that reading this book will inspire new sets of experiments, hypotheses, and clinical research studies. We also hope that the concepts spelled out in this book will form a basis for the next years to move ahead in this field. Researchers from many disciplines are interested in contributing to the interface of neuroendocrinology and PTSD. The field has now reached a critical mass that enables interdisciplinary meetings. We hope that the participants will have a pleasant recollection of the meeting and that the readers of this book are inspired by the richness of the theme.

Subject Index

Edwards Brothers Malloy
Thorofare, NJ USA
May 13, 2014